Graduate Texts in Physics

Series Editors

Kurt H. Becker, NYU Polytechnic School of Engineering, Brooklyn, USA

Jean-Marc Di Meglio, Matière et Systèmes Complexes, Bâtiment Condorcet, Université Paris Diderot, Paris, France

Sadri Hassani, Department of Physics, Illinois State University, Normal, USA

Morten Hjorth-Jensen, Department of Physics, Blindern, University of Oslo, Oslo, Norway

Bill Munro, Okinawa Institute of Science and Technology Graduate University, Onna-son, Japan

Richard Needs, Cavendish Laboratory, University of Cambridge, Cambridge, UK

William T. Rhodes, Department of Electrical Engineering and Computer Science, Florida Atlantic University, Boca Raton, USA

Susan Scott, Australian National University, Acton, Australia

H. Eugene Stanley, Department of Physics, Center for Polymer Studies, Boston University, Boston, USA

Martin Stutzmann, Walter Schottky Institute, Technical University of Munich, Garching, Germany

Andreas Wipf, Institute of Theoretical Physics, Friedrich-Schiller-University Jena, Jena, Germany

Bassano Vacchini

Open Quantum Systems

Foundations and Theory

 Springer

Bassano Vacchini

Dipartimento di Fisica "Aldo Pontremoli"
Università degli Studi di Milano
Milan, Italy

Istituto Nazionale di Fisica Nucleare
Milan, Italy

ISSN 1868-4513   ISSN 1868-4521   (electronic)
Graduate Texts in Physics
ISBN 978-3-031-58220-2   ISBN 978-3-031-58218-9   (eBook)
https://doi.org/10.1007/978-3-031-58218-9

La mia conversïone, omè!, fu tarda;
ma, come fatto fui roman pastore,
così scopersi la vita bugiarda.

Vidi che lì non s'acquetava il core,
né più salir potiesi in quella vita;
per che di questa in me s'accese amore.

Dante (Comedìa, Purgatorio, Canto XIX)

To my wife Laura and my sons
Federico, Martino and Matilde
to whom my affection will remain unaltered
for deep reasons hidden in these tercets

Preface

This book has grown out of my lectures on the theory of open quantum systems delivered to master students of the University of Milan over the years, almost a decade by now. It also benefits from lectures given at the "19th Chris Engelbrecht Summer School in Theoretical Physics" in Salt Rock (South Africa) in 2007, at the "Abdus Salam International Centre for Theoretical Physics" of Trieste (Italy) in 2011, at the "51 Winter School of Theoretical Physics" in Lądek Zdrój (Poland) in 2015, at the "International Centre for Theoretical Sciences" of Bangalore (India) in 2016, at the "QUSTEC Summer School 2022" in Freiburg (Germany) in 2022, and at the "Faculty of Science and Technology" of Al Hoceima (Morocco) in 2023.

Having the opportunity to teach to classes of highly motivated and genuinely engaged students is such a significant experience that I found it worth to give it a legacy and a future, transforming it from the *white and black* of the blackboard to the *black and white* of the paper. The guiding idea was to prepare lectures for students attracted by theoretical physics, open to consider the mathematical aspects of a topic.

The theory of open quantum systems was born with the theory of measurement in quantum mechanics, trying to better understand the situation in which a quantum system of interest interacts with other degrees of freedom in such a way as to obtain predictions on some observable quantity of the system itself. Setting the proper mathematical framework for this description has led to important advancements in applications, opening the way to many achievements in quantum information theory, and in many respects laying the ground for the development of quantum technologies. The aim of this presentation of the theory of open quantum systems is therefore to put into evidence and connect concepts coming from the foundations of quantum theory and from quantum information. The book insists on the endeavor to more deeply and realistically describe the notion of state and observable in quantum mechanics, quantum transformations with a measurement character, and the effect of an environment on the dynamics of a system. I leave it to the readers to decide to what extent this attempt has been successful.

Open quantum systems and the formalism of their description have attracted a lot of attention in the last decades. This fact provides another motivation for putting together these notes for the sake of teaching, despite the existence of outstanding books on the subject, such as the fundamental monograph *The Theory of Open Quantum Systems* appeared in 2002 [1], that gave an important twist to the field. Many excellent monographs and books have appeared focusing on different aspects of open quantum systems and foundations of quantum theory. We here mention those that have more influenced the writing of these notes [2–12]. It appears that within the years an open system approach has started to be of interest for many areas of physics beyond the more traditional molecular and atomic optics, chemical physics, quantum information, or condensed matter theory, e.g., nuclear and high-energy physics. Indeed, the formalism of open quantum systems is steadily extended and is now finding new applications and developments in dealing with many-body systems and in the blooming field of quantum thermodynamics. In this framework, researchers are concerned with fundamental questions, such as the very definition of heat, work, and internal energy at a microscopic level and in the presence of strong coupling. At the same time, application-oriented questions, such as the ultimate efficiency in heat to work conversion in cyclic operations, are thoroughly investigated. This approach reflects the two main souls of studies in open quantum systems, that address both technologically motivated investigations and fundamentally oriented questions. Indeed, tools and concepts from open quantum system theory play a role in the investigation of theories alternative to quantum mechanics, such as dynamical reduction models. In this perspective, these notes devote a consistent fraction of the presentation to the very introduction of the modern formalism of quantum mechanics, nicely presented in [13–17].

In the study of open quantum systems, many fundamental notions have been introduced, that are slowly becoming part of the standard presentation of quantum mechanics, and that we try to emphasize in the exposition. These include the framework of decoherence to describe the insurgence of a classical description in experiments devised to put into evidence typical quantum behaviors, the notion of positive operator-valued measure, the structure of a master equation in GKSL form, often called for simplicity Lindblad master equation, and the condition of complete positivity. Indeed, the use of the tensor product in the description of composite systems in quantum mechanics leads to the property of entanglement, in turn strictly related to the notion of complete positivity, as stressed in the presentation. On the other hand, the relevance of complete positivity is strictly related to non-commutativity of the algebra of observables in quantum mechanics, thus showing a nice interplay between a local and a non-local unveiling of a typical quantum feature.

The book is divided into two parts and written in decreasing universality order in that, roughly speaking, the breadth of the exposed topics narrows along the way, at the same time moving towards more recently established results. Part I provides a construction of quantum mechanics as a probability theory, to be compared with classical probability. The construction leads to states as statistical operators and observables as positive operator-valued measures. It is then natural to investigate the physically relevant transformation of these spaces in themselves, namely quantum maps. Part II more directly dwells with open quantum system theory, starting from the consequences of having a fixed bipartition within a set of

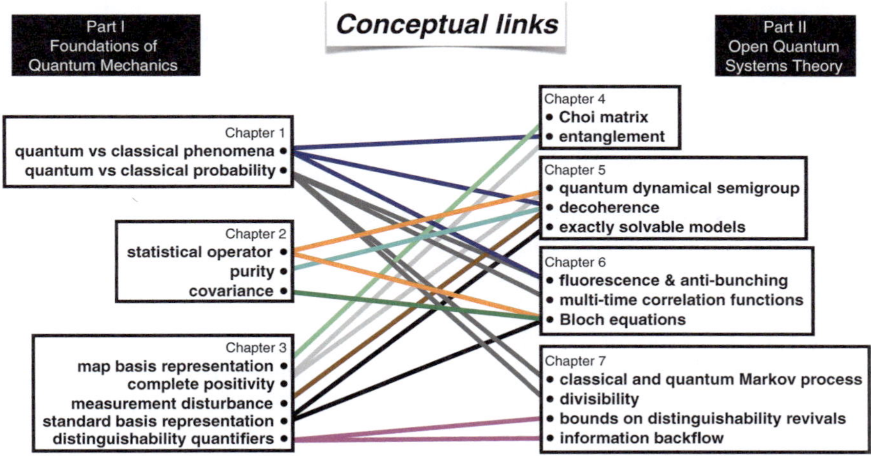

Fig. 1 Schematic visualization of key concepts and phenomena considered in the book and of some of their conceptual links, for the sake of thinking about them and of a non-sequential reading of the book

interacting quantum degrees of freedom, usually called system and environment. The very notion and possible description of a reduced dynamics are then considered, first by means of simple but paradigmatic examples, that allow to a large extent for an analytic treatment. An approximate but general treatment is then considered, relying on projection operator techniques, worked out in detail for the case of weak coupling between system and reservoir. In the final part of the book, a possible characterization of reduced quantum dynamics featuring memory effects is considered, analyzing in this perspective the previously considered models and highlighting the different structures of master equations needed in order to consider situations in which strong-coupling or low-temperature regimes become important. A number of conceptual links between the two parts of the book are shown in Fig. 1, for the sake of a non-sequential reading of the book.

Apart from the last chapter on memory effects, the manuscript does not claim to provide the reader with material that cannot be found in other books. It suggests however a path to introduce, in lectures or in personal study, a physicist or a mathematician interested in quantum theory and its foundations to the study of open quantum systems.

In many respects, writing a book is an important occasion to learn about your own ignorance, with the appealing perspective to reduce it a little bit, which compensates for the efforts needed to bring the project to a conclusion.

Needless to say, *I have to* and *I want to* thank a number of friends and colleagues for what they taught me about quantum mechanics and especially open quantum systems. Too many of them deserve special mentions, so that I prefer to resort to an alphabetical order: Albarelli Francesco, Alicki Robert, Barchielli Alberto, Bassi Angelo, Belgiorno Francesco, Benedetti Claudia, Buchleitner Andreas, Campbell

Steve, Carmeli Claudio, Cassinelli Gianni, Chruściński Dariusz, Cialdi Simone, Ciccarello Francesco, Diósi Lajos, Fagnola Franco, Ferialdi Luca, Genoni Marco, Gregoratti Matteo, Guarnieri Giacomo, Hellwig Karl-Eberhard, Holevo Alexander, Hornberger Klaus, Huelga Susana, Lorenzo Salvatore, Maniscalco Sabrina, Martinazzo Rocco, Megier Nina, Melsheimer Olaf, Olivares Stefano, Palma Massimo, Paris Matteo, Paternostro Mauro, Petruccione Francesco, Piilo Jyrki, Strunz Walter, Tamascelli Dario, Toigo Alessandro, Uchiyama Chikako, and Zanghí Pierantonio. I have to make an exception for my Ph.D. supervisor Ludovico Lanz, who introduced me to research in physics, and Heinz-Peter Breuer without whose advice and collaboration through the years my research and understanding of the subject wouldn't have been by far the same. A special thank also goes to Andrea Smirne, for the careful reading of a preliminary version of the manuscript and the many enriching exchanges. Last but not least, my warm gratitude goes to all the students that attended the lectures through the years, providing important stimuli to shape the notes that led to the present book.

Milan, Italy   Bassano Vacchini
February 2024

References

1. H.-P. Breuer, F. Petruccione, *The Theory of Open Quantum Systems* (Oxford University Press, Oxford, 2002)
2. E.B. Davies, *Quantum Theory of Open Systems* (Academic Press, London, 1976)
3. K. Kraus, *States, Effects, and Operations*, Lect. Not. Phys., **190** (Springer, Berlin, 1983). https://doi.org/10.1007/3-540-12732-1
4. R. Alicki, K. Lendi, *Quantum Dynamical Semigroups and Applications*, Lect. Not. Phys., **286** (Springer, Berlin, 1987)
5. M. Weissbluth, *Photon-atom Interactions* (Academic Press, Burlington, MA, London, 1989)
6. U. Weiss, *Quantum Dissipative Systems* (World Scientific, Singapore, 1993)
7. M.O. Scully, M.S. Zubairy, *Quantum Optics* (Cambridge University Press, Cambridge, 1997)
8. C.W. Gardiner, P. Zoller, *Quantum Noise* (Springer, New York, 2000)
9. R. Alicki, M. Fannes, *Quantum Dynamical Systems* (Oxford University Press, Oxford, 2001)
10. E. Joos, H.D. Zeh, C. Kiefer, D. Giulini, J. Kupsch, I.O. Stamatescu, *Decoherence and the Appearance of a Classical World in Quantum Theory*, 2nd edn. (Springer, Berlin, 2003). https://doi.org/10.1007/978-3-662-05328-7
11. B. Schumacher, M. Westmoreland, *Quantum Processes Systems, and Information* (Cambridge University Press, Cambridge, 2010). https://doi.org/10.1017/CBO9780511814006
12. A. Rivas, S.F. Huelga, *Open Quantum Systems: An Introduction* (Springer, Berlin, 2012). https://doi.org/10.1007/978-3-642-23354-8
13. G. Ludwig, *Einführung in die Grundlagen der Theoretischen Physik* (Vieweg, Braunschweig, 1976)
14. A.S. Holevo, *Probabilistic and Statistical Aspects of Quantum Theory* (North-Holland, Amsterdam, 1982)
15. P. Busch, M. Grabowski, P. Lahti, *Operational Quantum Physics*, Lect. Not. Phys., **31** (Springer-Verlag, 1995). https://doi.org/10.1007/978-3-540-49239-9
16. A.S. Holevo, *Statistical Structure of Quantum Theory*, Lect. Not. Phys., **67** (Springer, Berlin, 2001). https://doi.org/10.1007/3-540-44998-1
17. T. Heinosaari, M. Ziman, *The Mathematical Language of Quantum Theory* (Cambridge University Press, Cambridge, 2011)

Contents

Acronyms and Symbols

c.o.n.s.	Complete orthonormal system
GKSL	Gorini, Kossakowski, Sudarshan and Lindblad
POVM	Positive operator-valued measure
PVM	Projection-valued measure
$\lvert A\rvert$	Modulus of a bounded operator
$\lVert A\rVert$	Uniform norm of an operator
$\lVert A\rVert_1$	Trace norm of an operator
$\lVert A\rVert_2$	Hilbert-Schmidt norm of an operator
$\langle A, B\rangle_{\mathrm{HS}}$	Scalar product on the space of Hibert-Schmidt operators
$\log$	Natural logarithm
$\mathbb{1}_A$	Identity operator on $\mathcal{H}_A$
$\mathbb{1}_{\mathbb{C}^n}$	Identity matrix in dimension n
$\mathbb{1}_{\mathcal{H}}$	Identity operator on $\mathcal{H}$
$\mathbb{1}_{\mathcal{M}_n(\mathbb{C})}$	Identity map on $\mathcal{M}_n(\mathbb{C})$
$\mathcal{B}(\mathcal{H})$	Banach space of bounded operators
$\mathcal{B}'(\mathcal{H})$	Dual of the Banach space of bounded operators
$C(\mathcal{H})$	Set of compact operators
$\mathcal{E}(\mathcal{H})$	Convex set of effect operators
$\mathcal{M}_n(\mathbb{C})$	Linear space of $n \times n$ matrices with complex entries
$\mathcal{P}(\mathcal{H})$	Lattice of projection operators
$\mathcal{P}_1(\mathcal{H})$	Set of rank one projection operators
$\mathcal{S}(\mathcal{H})$	Convex set of quantum states
$\widetilde{\mathcal{S}}(\mathcal{H})$	Convex subset of subcollections
$\mathcal{T}(\mathcal{H})$	Banach space of trace class operators
$\mathcal{T}'(\mathcal{H})$	Dual of the Banach space of trace class operators
$\mathcal{T}_+(\mathcal{H})$	Subset of positive trace class operators
$\mathcal{T}_s(\mathcal{H})$	Subset of self-adjoint trace class operators
$\mathfrak{F}_t$	Semigroup of transformations
$\mathfrak{L}(\mathcal{B}(\mathbb{C}^n), \mathcal{B}(\mathbb{C}^n))$	Space of linear maps of $\mathcal{B}(\mathbb{C}^n)$ into itself
$\mathfrak{L}$	Generator of classical Markov semigroup
$\mathfrak{S}$	Distinguishability quantifier between pairs of quantum states
$\mathfrak{T}_t$	Conditional transition probability

$\mathfrak{U}$	Stochastic matrix
$HS(\mathcal{H})$	Banach space of Hibert-Schmidt operators
$\mathrm{Prob}(\Omega)$	Convex set of probability distributions on the measure space Ω
$\mathrm{Tr}_E[\cdot]$	Partial trace with respect to the Hilbert space $\mathcal{H}_E$
$\mathrm{Tr}_E\{\cdot\}$	Trace with respect to the Hilbert space $\mathcal{H}_E$

List of Figures

List of Remarks

Part I
Foundations of Quantum Mechanics

These are my principles.
If you don't like them, well, I have others.
—Attributed to Groucho Marx

The aim of the Part I is to introduce quantum mechanics as a probability theory, building on the duality between the space of states and observables, to be identified in quantum theory as statistical operators and positive operator-valued measures respectively. We consider in particular the relevant transformations on these spaces, stressing the emergence of the typical quantum property of complete positivity, in its relation to the non-commutativity of the considered spaces. These transformations are further compared in view of their measurement character, relying on concepts from quantum information theory.

Quantum Mechanics as Quantum Probability

1

Abstract

Quantum mechanics is naturally to be seen as a probability theory for the description of statistical experiments, also valid at a microscopic level. To highlight this fact, we first address the general formulation of a statistical theory, which is relevant for the description of physical systems in which the outcomes of an experiment have to be described by a probability distribution. We briefly outline the frameworks corresponding to a classical and a quantum formulation, starting to analyze the distinguishing features of the latter.

1.1 Introduction

In the description of physical phenomena we are often faced with the evidence that a statistical description is necessary, namely we cannot provide exact predictions for single realizations of an experiment, but rather for suitable averages obtained in many repetitions. According to classical physics the basic underlying description of phenomena is deterministic, so that the statistical features simply express our incapability to quantitatively describe in detail the behavior of all microscopic degrees of freedom involved in a specific experiment, as well as to assign them a deterministic initial condition. A statistical description then imposes itself both because of convenience and limited knowledge. In quantum mechanics two new aspects appear. In the first instance a statistical description becomes mandatory. On top of this, the classical probabilistic formulation does no more apply and a new mathematical framework has to be introduced to properly describe statistical experiments involving microscopic physical systems.

The general structure of a statistical theory can be outlined considering two basic sets, the set of states S and the set of observables $\mathcal{B}$. For any $p \in S$ and any $X \in \mathcal{B}$ we must be able to construct a probability measure μ_p^X defined on the

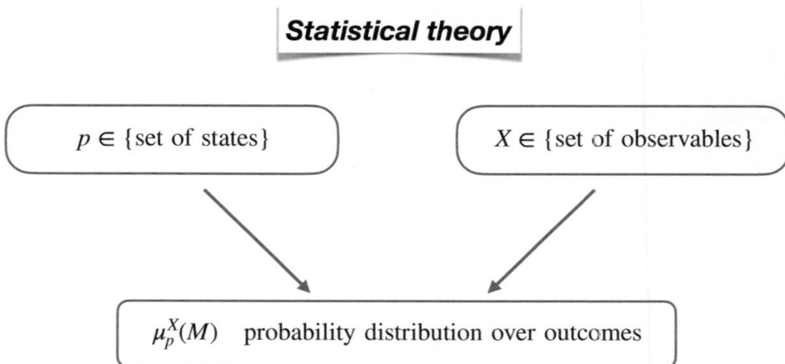

Fig. 1.1 General framework for the theoretical description of statistical experiments. Each element of the set of states can be combined with any observable providing the probability distribution for the possible outcomes. The theory should define the nature of these two sets and how their elements can be combined via a probability formula to determine the frequency of the different outcomes

σ-algebra of the Borel sets over $\mathbb{R}$, let us denote it by $\mathcal{B}(\mathbb{R})$, such that for any $M \in \mathcal{B}(\mathbb{R})$ the number $\mu_p^X(M)$ provides the probability that X takes values in M if the state is p, as schematically depicted in Fig. 1.1. The whole analysis of course relies on the premise that the statistical experiments are reproducible, and the frequencies with which the different outcomes appear are well defined. The structure based on these three basic elements—the set of states and observables together with a probability formula to combine each state with each observable to provide a probability distribution—is realized in two different mathematical frameworks in the classical and quantum case. It is interesting to notice that the formalization of the two theories, by Kolmogorov and von Neumann respectively, took place around the same time [1,2]. As noted in [3], while classical probability is based on measure theory, assigning probabilities to volumes, quantum probability is based on Pythagoras' theorem, assigning probabilities to orthogonal directions.

1.2   Classical and Quantum Probability

In the classical case, as formalized by Kolmogorov, probability theory is naturally embedded within abstract measure theory. We briefly recall the relevant framework. The basic elements are a set Ω and a σ-algebra $\mathcal{F}$ over Ω, see Remark 1.1, together with a measure μ defined on $\mathcal{F}$ and taking values in $\mathbb{R}_+ \cup \{\infty\}$ such that σ-additivity holds, namely $\mu(\cup_i A_i) = \sum_i \mu(A_i)$ provided the collection of sets $\{A_i\} \in \mathcal{F}$ is such that $A_i \cap A_j = \emptyset$ for $\forall i \neq j$.

> **1.1 Definition of σ-Algebra**
>
> Given a set Ω we define as σ-algebra over Ω a family $\mathcal{F}$ of subsets of Ω such that
>
> - the set Ω belongs to $\mathcal{F}$
> - if $M \in \mathcal{F}$ then $M^C \in \mathcal{F}$
> - if $\{M_n\}_{n\in\mathbb{N}} \subseteq \mathcal{F}$ then $\cup_{n=1}^{\infty} M_n \in \mathcal{F}$
>
> If $\mathcal{F}$ is a σ-algebra over Ω, then Ω is said to be a measurable space and the elements of $\mathcal{F}$ are the measurable sets in Ω.

The triple $(\Omega, \mathcal{F}, \mu)$ is called a measure space, and given a real measurable function f, that is a function such that $f^{-1}(M)$ is a measurable set in $\mathcal{F}$ for all open sets $M \in \mathcal{B}(\mathbb{R})$, we can consider integrals of the form $\int_A f \, d\mu$ for $A \in \mathcal{F}$. If we further ask the measure to be finite and normalized, in the sense that $\mu(\Omega) = 1$, we call it a probability measure. Denoting such measure with p we say that the triple $(\Omega, \mathcal{F}, p)$ is a probability space. As suggested by Kolmogorov, this is the natural framework to describe a probability theory in the classical setting, so that a probability model is fixed by specifying a measure space which consists of the space of elementary events, a σ-algebra on this space characterizing the events to which a probability can be assigned, and a probability measure on it.

In the Kolmogorov model of classical probability theory we start from a measurable space and consider the following construction of states and observables to give a statistical theory, as exemplified in Remark 1.2. The set of states $\mathcal{S}$ is identified with the convex set of probability measures on $\mathcal{F}$, and the set of observables $\mathcal{B}$ is identified with the set of measurable functions, also called random variables. The probability formula which provides the distribution for the possible values of the random variable X given the state p is expressed as

$$\mu_p^X(M) = \int_{X^{-1}(M)} dp \tag{1.1}$$

$$= p(X^{-1}(M)). \tag{1.2}$$

The mean values can then be obtained as follows

$$\langle X \rangle_p = \int_{\mathbb{R}} x \, d\mu_p^X(x) \tag{1.3}$$

$$= \int_{\mathbb{R}} x \, dp(X^{-1}(x)) \tag{1.4}$$

$$= \int_{\Omega} X(y) dp(y). \tag{1.5}$$

Note that while the equality (1.3) is always meaningful in a statistical theory, corresponding to the very definition of mean value as first moment of the probability distribution, the last expression (1.5) relies on the existence of an underlying event space, which is given only in the classical case, and whenever available can be the most convenient for calculations.

We can thus compactly write

$$\langle X \rangle_p = \int_\Omega X \mathrm{d}p, \tag{1.6}$$

and similarly for higher moments, so that in general for a measurable function f of the observable we have

$$\langle f(X) \rangle_p = \int_\Omega f(X) \mathrm{d}p. \tag{1.7}$$

1.2 Massive Point Particle

For the case of a massive point particle the measure space is the usual phase-space $\mathbb{R}^3 \times \mathbb{R}^3$. The probability measure can be expressed via a probability density $f(\boldsymbol{x}, \boldsymbol{p})$, i.e. a positive and normalized element of $L^1(\mathbb{R}^3 \times \mathbb{R}^3)$, the set of absolutely integrable functions on the phase-space, namely such that

$$\int_{\mathbb{R}^3 \times \mathbb{R}^3} \mathrm{d}\boldsymbol{x} \mathrm{d}\boldsymbol{p} |f(\boldsymbol{x}, \boldsymbol{p})| < +\infty. \tag{1.8}$$

The observables are described as random variables and can be taken to be elements of the dual space $L^\infty(\mathbb{R}^3 \times \mathbb{R}^3)$, that is the set of functions X such that $\sup_{\boldsymbol{x} \in \mathbb{R}^3, \boldsymbol{p} \in \mathbb{R}^3} |X(\boldsymbol{x}, \boldsymbol{p})| < +\infty$. The mean value of any such observable is given by the duality formula between the spaces given by

$$\langle X \rangle_f = \int_{\mathbb{R}^3 \times \mathbb{R}^3} \mathrm{d}\boldsymbol{x} \mathrm{d}\boldsymbol{p} \, X(\boldsymbol{x}, \boldsymbol{p}) f(\boldsymbol{x}, \boldsymbol{p}). \tag{1.9}$$

Any observable X taking values in $\mathbb{R}$ defines a probability measure on this space according to the formula

$$\mu_f^X(M) = \int_{X^{-1}(M)} \mathrm{d}\boldsymbol{x} \mathrm{d}\boldsymbol{p} \, f(\boldsymbol{x}, \boldsymbol{p}), \tag{1.10}$$

where M is a Borel set in the outcome space $\mathbb{R}$ of the observable. In particular, the expectation value of any observable, i.e. random variable, is obtained from the very same probability density $f(x, p)$. Considering the case of the position observable $X(x, p) = x$ we can check that the probability measure μ_f^x can be expressed through the density corresponding to the marginal

$$f^x(x) = \int_{\mathbb{R}^3} d\boldsymbol{p}\, f(x, p). \tag{1.11}$$

1.2.1 Classical and Quantum Description of a Finite system

To give a first highlight of the relationship between classical and quantum probability we briefly compare following [4] the classical and quantum description of a finite system. We consider first a classical setting taking as measure space $\Omega = \{\omega_1, \dots, \omega_N\}$, together with the σ-algebra $\mathcal{F}$ given by the power set of Ω. The generic state can be identified with a probability vector, so that

$$\mathcal{S} = \left\{ \boldsymbol{p} = (p_1, \dots, p_N) | p_i \geqslant 0, \sum_{i=1}^{N} p_i = 1 \right\}. \tag{1.12}$$

Note that introducing the extreme points of this convex set as $(\mathbf{e}_i)_j = \delta_{ij}$, so that e.g. $\mathbf{e}_1 = (1, 0, \dots, 0)$, for each state we have the convex decomposition

$$\boldsymbol{p} = \sum_{i=1}^{N} p_i \mathbf{e}_i, \tag{1.13}$$

whose uniqueness follows from the fact that the extreme points also provide a linear basis. Real random variables corresponding to observables can be identified with N-tuples of real numbers

$$\mathcal{B} = \{ \boldsymbol{x} = (x_1, \dots, x_N) | x_i \in \mathbb{R} \}, \tag{1.14}$$

and their commutativity under point-wise multiplication is immediately apparent. The probability formula, predicting the outcome x_i for the random variable $\boldsymbol{x}$ provided the state is given by $\boldsymbol{p}$, then simply reads

$$\mu_{\boldsymbol{p}}^x(\{x_i\}) = p_i. \tag{1.15}$$

The mean values according to their very definition (1.3) are given by

$$\langle x \rangle_p = \sum_{i=1}^{N} x_i \mu_p^x(\{x_i\}) \tag{1.16}$$

$$= \sum_{i=1}^{N} x_i \, p_i, \tag{1.17}$$

with the integral over the sample space here replaced by the sum over the discrete index.

Let us now move to the quantum description, considering as Hilbert space $\mathbb{C}^N$. The set of states is then identified with Hermitian positive definite matrices with trace one

$$\mathcal{S} = \{S \in \mathcal{M}_N(\mathbb{C}) | S = S^\dagger, S \geqslant 0, \mathrm{Tr}\, S = 1\}, \tag{1.18}$$

while the observables according to the standard formulation of quantum mechanics are given by all Hermitian matrices

$$\mathcal{B} = \{X \in \mathcal{M}_N(\mathbb{C}) | X = X^\dagger\}. \tag{1.19}$$

Exploiting the following expression for a Hermitian matrix X

$$X = \sum_{j=1}^{N} x_j Q_j, \tag{1.20}$$

where $\{x_j\}$ are the real eigenvalues, and $\{Q_j\}$ are the corresponding projections, we can express the probability formula as

$$\mu_S^X(\{x_i\}) = \mathrm{Tr}\{S Q_i\}, \tag{1.21}$$

which provides a probability distribution over the spectrum of the possible values attained by the observable X. In quantum mechanics this is assumed to be the probability distribution of X provided the system is in the state S. The mean values can again be obtained from (1.3)

$$\langle X \rangle_S = \sum_{j=1}^{N} x_j \mu_S^X(\{x_j\}) \tag{1.22}$$

$$= \sum_{j=1}^{N} x_j \, \mathrm{Tr}\{S Q_j\} \tag{1.23}$$

$$= \mathrm{Tr}\{S X\}, \tag{1.24}$$

thus also recovering in the last line a simple formula in which the random variable X directly appears, in analogy to the classical expression (1.17), which however calls for an underlying measure space, not available in the quantum case.

Within this quantum probabilistic model we can recover the corresponding classical probabilistic model in dimension N considering suitable subsets of states and observables. This shows that a classical probabilistic model can be embedded in a quantum one with the same dimensionality, though not vice versa. To this aim consider the subset of $\mathcal{S}$ given by the states diagonal in a given basis

$$
\mathcal{S}_{\text{diag}} = \left\{ \begin{pmatrix} p_1 & \cdots & 0 \\ \vdots & \ddots & \vdots \\ 0 & \cdots & p_N \end{pmatrix} \middle| \; p_i \geqslant 0, \sum_{i=1}^{N} p_i = 1 \right\}, \tag{1.25}
$$

and note that this is a convex subset of $\mathcal{S}$. On the same footing consider the subset of $\mathcal{B}$

$$
\mathcal{B}_{\text{diag}} = \left\{ \begin{pmatrix} x_1 & \cdots & 0 \\ \vdots & \ddots & \vdots \\ 0 & \cdots & x_N \end{pmatrix} \middle| \; x_i \in \mathbb{R} \right\}, \tag{1.26}
$$

which is closed under matrix multiplication. In this case the quantum probability formula reduces to the classical one

$$
\begin{aligned}
\mu_S^X(\{x_i\}) &= \text{Tr}\{S Q_i\} \tag{1.27} \\
&= p_i \tag{1.28} \\
&= \mu_p^x(\{x_i\}), \tag{1.29}
\end{aligned}
$$

and similarly for the mean values

$$
\begin{aligned}
\langle X \rangle_S &= \sum_{j=1}^{N} x_j \mu_S^X(\{x_j\}) \tag{1.30} \\
&= \sum_{j=1}^{N} x_j p_j \tag{1.31} \\
&= \sum_{j=1}^{N} x_j \mu_p^x(\{x_i\}). \tag{1.32}
\end{aligned}
$$

1.2.2 Classical and Quantum Logic

Let us make a few remarks on the logic underlying the statistical models formulated in this way, whose analysis started with the seminal paper [5]. We call events the elements of the σ-algebra $\mathcal{F}$ over which the measure μ is defined, and note that the structure of the σ-algebra naturally provides a Boolean algebra, so that we have the basic elements of classical logic. Indeed, if the events A and B are in $\mathcal{F}$, we can also consider the events $A \cup B$, $A \cap B$ and A^C, while $\emptyset$ and Ω can be identified with the impossible and certain event respectively. The notions of inclusion, union and intersection then allow to consider on the σ-algebra a natural structure of lattice whose definition is recalled in Remark 1.3. In particular, this lattice turns out to be orthocomplemented, in the sense that to each set A we can associate its complement A^C, and distributive since for any choice of sets $A, B, C \in \mathcal{F}$ we have

$$A \cap (B \cup C) = (A \cap B) \cup (A \cap C). \tag{1.33}$$

A σ-algebra of subsets of a set Ω therefore naturally has the structure of Boolean algebra. Conversely, it can be shown that any Boolean algebra is isomorphic to a lattice of subsets of a set. The outcomes of trials or statistical experiments naturally have a lattice structure. In this framework distributivity is a crucial property, obeyed by classical but not by quantum events.

1.3 Definition of Lattice

A lattice is a set L with

1. A partial order relation $\leqslant$, i.e. a binary relation satisfying $\forall a, b, c \in L$: $a \leqslant a$ (reflexivity); $a \leqslant b, b \leqslant a$ implies $a = b$ (antisymmetry); $a \leqslant b$, $b \leqslant c$ implies $a \leqslant c$ (transitivity).
2. Two binary operations $\vee$ and $\wedge$, called join and meet, satisfying

$$a \leqslant a \vee b, \quad b \leqslant a \vee b,$$

   as well as

$$a \wedge b \leqslant a, \quad a \wedge b \leqslant b.$$

   The join and meet operations are commutative, associative and idempotent ($a \vee a = a$, $b \wedge b = b$). The existence of these two binary operations warrants that each pair of elements has a least upper bound and a upper lower bound.
3. A lattice is complete if every subset of L has a least upper bound and a upper lower bound.
4. A lattice has unit and zero element if $0 \leqslant a \leqslant 1 \; \forall a \in L$.
5. A lattice with zero and unit element is orthocomplemented if $\forall a \in L$ there exists an element a', called the complement of a, such that $a \vee a' = 1$, $a \wedge a' = 0$, $(a')' = a$ and $a \leqslant b \Rightarrow b' \leqslant a'$.

6. A lattice is distributive if

$$a \wedge (b \vee c) = (a \wedge b) \vee (a \wedge c)$$

as well as

$$a \vee (b \wedge c) = (a \vee b) \wedge (a \vee c).$$

7. A complete orthocomplemented distributive lattice with zero and unit element is called a Boolean algebra.

In the classical case the logic of events identified with the subsets belonging to the σ-algebra provides as shown a Boolean algebra, which warrants in particular the distributivity of the meet and join operations. This fact allows to consider the decomposition of an event in terms of elementary ones. This is no more true in quantum mechanics, as we can grasp from the following suggestive example [6]. We consider incompatible experimental outcomes such as perfect path determination and full visibility of interference fringes in an interferometric setup. Focusing on a two-path Mach-Zender configuration we identify the events: $a \leftrightarrow$ interference pattern; $b \leftrightarrow$ passage through path one; as well as the complementary $b' \leftrightarrow$ passage through path two. One then has

$$a \wedge (b \vee b') \neq (a \wedge b) \vee (a \wedge b'), \tag{1.34}$$

where the l.h.s. is equal to a since $b \vee b'$ is the identity or certain event, while since a is incompatible with both b and b' the r.h.s. is equal to the empty set identifying the impossible event. This fact can be rephrased stating that in quantum mechanics not every situation can be analyzed in terms of an arbitrary number of complementary elementary events. From the mathematical point of view, in the quantum case a natural lattice of events is given by the set of orthogonal projections on a given Hilbert space $\mathcal{P}(\mathcal{H})$, discussed in Remark 1.4, which despite being complete and orthocomplemented, is not distributive. Violation of distributivity can be directly checked by a suitable choice of projections.

1.4 The lattice $\mathcal{P}(\mathcal{H})$ of Orthogonal Projections

We recall a few facts about the set of orthogonal projections on a Hilbert space, which we denote by $\mathcal{P}(\mathcal{H})$. To each orthogonal projection $P \in \mathcal{P}(\mathcal{H})$ it is uniquely associated its eigenspace M_P, so that it can be identified with the projection on a subspace of the Hilbert space, namely $P \leftrightarrow \Pi_{\mathcal{M}_P}$. Defining the join and meet operations according to

$$P \vee Q = \Pi_{\overline{\mathcal{M}_P \cup \mathcal{M}_Q}} \tag{1.35}$$

$$P \wedge Q = \Pi_{\mathcal{M}_P \cap \mathcal{M}_Q}, \tag{1.36}$$

where in the first line we have considered the closure of the union of the subspaces, as well as taking the usual notion of complementary projection

$$P' = \mathbb{1} - P \tag{1.37}$$

$$= P^{\perp}, \tag{1.38}$$

this set provides an orthocomplemented complete lattice, which is however not distributive. Moreover $P \leqslant Q$ (in the sense of operator ordering, that is $\langle \psi | P \psi \rangle \leqslant \langle \psi | Q \psi \rangle$ for all $\psi \in \mathcal{H}$) iff $\mathcal{M}_P \subseteq \mathcal{M}_Q$, or equivalently $PQ = QP = P$, thus providing a partial order relation. The lattice has zero and identity, while distributivity does not hold unless the projections commute.

In the classical case the convex set of states is in particular a simplex, that is each state can be uniquely demixed in terms of extreme points of the set. We recall that the extreme points of a convex set are those elements which cannot be expressed as non-trivial convex combination of other elements in the set. E.g., in the phase-space example introduced above extreme elements are given by measures concentrated in a single point of $\mathbb{R}^3 \times \mathbb{R}^3$. This is no more true in quantum mechanics, as we will show in Sect. 2.2.5.

Another feature of the classical case, no long recovered in the quantum setting, is given by the fact that the set of random variables over a measure space, given by the measurable functions, is a commutative algebra under point-wise multiplication, while as well-known quantum observables do generally not commute.

We have thus put into evidence different aspects which are typical of classical probabilistic models: the existence of a sample space, the distributivity of the associated lattice of events, the simplex structure of the state space, the commutativity of the space of observables. All these features do not find their counterpart in the quantum description, thus providing useful perspectives in order to figure out and understand the difference between classical and quantum description. Indeed, there are many direction in which one can cross the quantum-classical border, and the better we know and understand quantum mechanics, the deeper we realize connections and differences with the classical theory. This standpoint is schematically visualized in Fig. 1.2. We support the viewpoint that a presentation of quantum mechanics as a probability theory, and the study of open quantum systems as a natural quantum counterpart of classical stochastic descriptions, can be very fruitful in this respect.

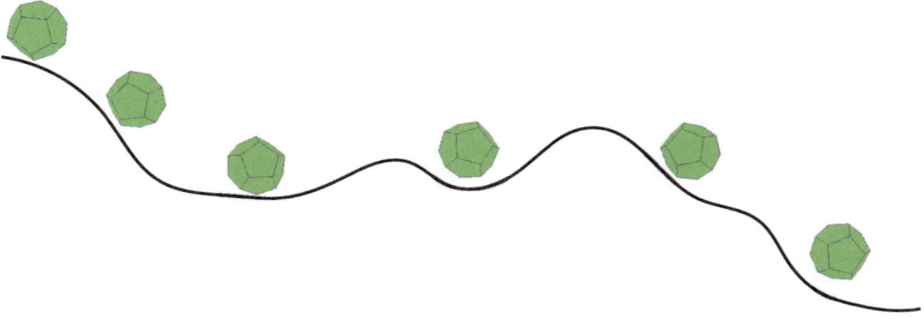

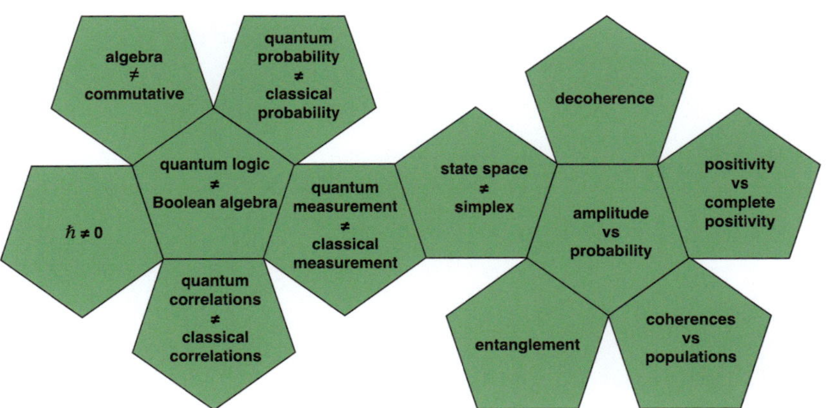

Fig. 1.2 Pictorial view of how to investigate the relationship between classical and quantum mechanics. The difference between the two theories can be understood and tested along different directions, here depicted as the many faces of a polyhedron. By exploring the description and dynamics of quantum system we let the polyhedron roll, thus bringing to the view different aspects. Some of them are mentioned in the unfolded faces

References

1. J. von Neumann, *Mathematische Grundlagen der Quantenmechanik* (Springer, Berlin, 1932)
2. A.N. Kolmogorov, *Grundbegriffe der Wahrscheinlichkeitsrechnung* (Springer, Berlin, 1933)
3. C.J. Isham, *Lectures on Quantum Theory: Mathematical and Structural Foundations* (Imperial College Press, London, 1995)
4. A.S. Holevo, *Statistical Structure of Quantum Theory, Lecture Notes in Physics*, vol. m 67 (Springer, Berlin, 2001). https://doi.org/10.1007/3-540-44998-1
5. G. Birkhoff, J.V. Neumann, Ann. Math. **37**, 823 (1936). https://doi.org/10.2307/1968621
6. F. Strocchi, *An Introduction to the Mathematical Structure of Quantum Mechanics* (World Scientific, 2005). https://doi.org/10.1142/7038

Statistical Structure of Quantum Mechanics

2

Abstract

We explore the statistical structure of quantum theory providing a construction of the general notion of state, and by duality of observable, in the quantum setting. This will lead us to consider states as statistical operators, and observables as positive operator-valued measures. The connection of these notions with state vectors and self-adjoint operators introduced in the standard formulation of quantum mechanics is considered, pointing to the novel aspects inherent to this more realistic description of the outcomes of statistical experiments.

2.1 Introduction

As a starting point we stress the fact that quantum mechanics arises in order to predict the outcomes of statistical experiments. Despite the original evidence to look for quantum mechanics was in well-known but more elaborated settings, such as the study of black body radiation and atomic spectra, having in mind to identify the basic building blocks of the theory we are led to consider the most simple statistical experiments, in terms of which all others can be formulated. We consider single particle statistical experiments, such that a single microsystem is prepared by a preparation apparatus which feeds, without any backreaction, a registration apparatus according to the simple scheme depicted in Fig. 2.1. The boxes in this graphic representation are sometimes called *Ludwig's Kisten* in honor of the German physicist Günther Ludwig that gave important contributions to the foundations of quantum mechanics. In particular, he was one of the pioneers in the introduction of effects in the formalism of quantum mechanics [1].

The considered experiments are of statistical nature in that only the frequencies of the outcomes and not the single outcomes can be predicted. Note that, in this spirit,

B. Vacchini, *Open Quantum Systems*, Graduate Texts in Physics,
https://doi.org/10.1007/978-3-031-58218-9_2

15

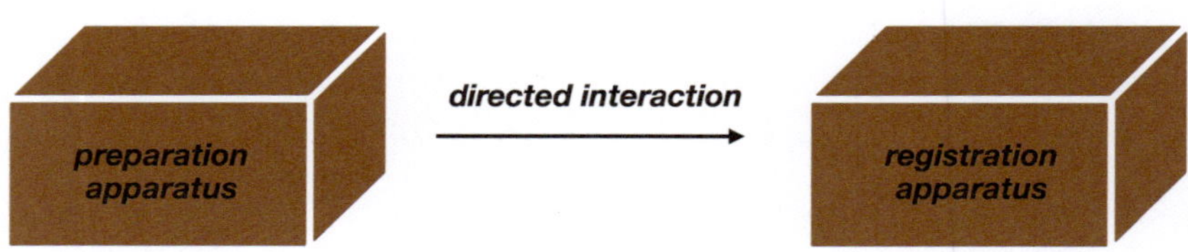

Fig. 2.1 A scheme of the most simple statistical experiment to be taken as reference for the construction of quantum mechanics. A preparation apparatus directly feeds a registration apparatus. Both apparata are depicted as boxes or Kisten in German language

the state of the system is the mathematical representative for the concrete preparation through which the considered system has been obtained in some laboratory. The basic question to be answered is therefore how to correctly obtain the statistics of the possible outcomes in a given experiment. That is to say, in analogy with the classical case, which mathematical objects describe the preparation procedure, the so-called states, and the registration procedure, the so-called observables, together with how we can combine them to obtain the relevant probability measure on the possible outcomes of the experiment, namely the probability formula. Note that these mathematical objects will characterize whole equivalence classes. States are identified with equivalence classes of preparation procedures. The elements of the equivalence class correspond to different prescriptions in the build-up of the preparation apparatus, still leading to the same preparation of the system, so that they all provide the same statistics for any subsequent measurement. Similarly, observables are identified with equivalence classes of registration procedures, where again different elements of the equivalence class correspond to different experimental apparata providing the measurement of the same physical quantity, so that they all lead to the same statistics for each state.

We will introduce the relevant mathematical objects in a bottom up approach, starting from the standard formalism of quantum mechanics.

2.2 States and Effects

In the standard presentation of quantum mechanics we introduce the Hilbert space of the system $\mathcal{H}$, and states are described by vectors $\psi \in \mathcal{H}$ with unit norm. Given the Hilbert space, as discussed in Remark 1.4 we can also consider the associated lattice of projection operators, which as argued below can be taken as events. Indeed, given any projection P with associated subspace $\mathcal{M}_P$, we can consider the probabilities

$$\mu_\psi^P = ||P\psi||^2 \tag{2.1}$$

corresponding to the norm of the projected vector, to be interpreted as the probability that the event is verified if the state is ψ. The event is certain if $\psi \in \mathcal{M}_P$, and impossible if $\psi \in \mathcal{M}_P^\perp$, namely the orthogonal complement of $\mathcal{M}_P$. For the special case in which P is a one-dimensional projection we recover the so-called Born's rule [2]

$$\mu_\psi^{P_\varphi} = |\langle\varphi|\psi\rangle|^2. \tag{2.2}$$

We now consider an observable in the standard sense of a bounded self-adjoint operator $B \in \mathcal{B}(\mathcal{H})$, where $\mathcal{B}(\mathcal{H})$ denotes the Banach space of bounded linear operators on $\mathcal{H}$. We assume for simplicity that the observable takes values in $\mathbb{R}$ and denote by $M \in \mathcal{B}(\mathbb{R})$ a generic set in the Borel σ-algebra on the real line. We thus have the following equivalent expressions for the probability formula

$$\mu_\psi^B(M) = \langle\psi|E^B(M)\psi\rangle \tag{2.3}$$
$$= ||E^B(M)\psi||^2, \tag{2.4}$$

giving the probability to obtain outcomes for the measurement of B in the set M. Here $E^B(\cdot)$ denotes a projection depending on the outcome set M. Assuming for simplicity that B has pure point spectrum, so that

$$B = \sum_i b_i P_{b_i}, \tag{2.5}$$

with P_{b_i} projection on the eigenspace corresponding to the real eigenvalue b_i, we define

$$E^B(M) = \sum_{\{i\,|\,b_i \in M\}} P_{b_i}, \tag{2.6}$$

so that in particular $E^B(\{b_i\}) = P_{b_i}$.

Note that the identity between (2.3) and (2.4) has been obtained relying on idempotency of $E^B(M)$. Note further that starting from this collection of projections upon integration we obtain all the relevant information for a statistical description, such as mean values and other moments of the probability distribution. Indeed, mean values are defined according to

$$\langle B \rangle_\psi = \int_\mathbb{R} x\,\mathrm{d}\mu_\psi^B(x) \tag{2.7}$$
$$= \sum_i b_i \mu_\psi^B(\{b_i\}) \tag{2.8}$$

$$= \sum_i b_i \langle \psi | E^B(\{b_i\}) \psi \rangle \qquad (2.9)$$

$$= \langle \psi | \sum_i b_i E^B(\{b_i\}) \psi \rangle \qquad (2.10)$$

$$= \langle \psi | B \psi \rangle, \qquad (2.11)$$

so that they can be most directly expressed as expectation value of the self-adjoint operator itself. We stress that to obtain this result we have not exploited idempotency of the projections $E^B(\cdot)$. Considering however the second moment we have

$$\langle B^2 \rangle_\psi = \int_{\mathbb{R}} x^2 \mathrm{d}\mu_\psi^B(x) \qquad (2.12)$$

$$= \sum_i b_i^2 \mu_\psi^B(\{b_i\}) \qquad (2.13)$$

$$= \sum_i \langle \psi | b_i^2 E^B(\{b_i\}) \psi \rangle \qquad (2.14)$$

$$= \langle \psi | \left(\sum_j b_j E^B(\{b_j\}) \right) \left(\sum_i b_i E^B(\{b_i\}) \right) \psi \rangle \qquad (2.15)$$

$$= \langle \psi | B^2 \psi \rangle, \qquad (2.16)$$

so that idempotency is now a crucial property in order to express the second moment of the probability distribution of the observable B as expectation value of the square of the operator. On the same premises we have

$$\langle f(B) \rangle_\psi = \int_{\mathbb{R}} f(x) \mathrm{d}\mu_\psi^B(x) \qquad (2.17)$$

$$= \langle \psi | f(B) \psi \rangle, \qquad (2.18)$$

and in particular

$$\mathrm{Var}_\psi(B) = \langle B^2 \rangle_\psi - \langle B \rangle_\psi^2. \qquad (2.19)$$

2.2.1 Pure States and Mixtures

Having in mind that a state should describe a concrete preparation procedure, let us consider how to possibly obtain the norm one vector ψ as state, and insofar such a vector corresponds to a general preparation. A vector ψ can be singled out in the very special situation in which we measure a complete set of commuting observables, removing all possible degeneracy and thus uniquely fixing a common eigenvector, so that we can denote $\psi = u_a$, where a is a vector fixing the eigenvalue of each of the commuting observables. This is however not the most general preparation, since this set of states is not closed under convex mixtures. Indeed, let us consider two such preparation procedures leading to two orthogonal eigenvectors u_a and u_b, corresponding to distinct eigenvalues fixed by the vectors $\boldsymbol{a}$ and $\boldsymbol{b}$. We can prepare a new state mixing these apparata with frequencies N_1/N and N_2/N respectively, with $N_1 + N_2 = N$, in the sense that N_1 out of N times we use the first apparatus, while N_2 out of N times the second one. The overall statistics of the measured outcomes can be obtained relying on conditioning and using the relation

$$P(x) = \sum_y P(x, y) \tag{2.20}$$

$$= \sum_y P(x|y)P(y), \tag{2.21}$$

where $P(\cdot, \cdot)$ denotes a joint probability and $P(\cdot|\cdot)$ a conditional probability. We thus have, according to (2.3) for the probability that the quantity B takes values in the set M

$$P(B \in M) = \mu^B_{u_{a_1}}(M)\frac{N_1}{N} + \mu^B_{u_{a_2}}(M)\frac{N_2}{N} \tag{2.22}$$

$$= \langle u_{a_1}|E^B(M)u_{a_1}\rangle \frac{N_1}{N} + \langle u_{a_2}|E^B(M)u_{a_2}\rangle \frac{N_2}{N}. \tag{2.23}$$

Importantly, the probability to obtain an outcome in M given this preparation procedure cannot be written as before in the form $\|E^B(M)\psi\|^2$ with ψ a state vector. This can be seen from the fact that the state of the system should account for an arbitrary final measurement, and therefore $E^B(M)$ can be taken to be an arbitrary projection.

Similarly for the mean values using (2.21) we have

$$\langle X \rangle = \sum_{x,y} x P(x, y) \tag{2.24}$$

$$= \sum_{x,y} x P(x|y)P(y) \tag{2.25}$$

$$= \sum_y \left[\sum_x x P(x|y)\right] P(y), \tag{2.26}$$

so that

$$\langle B \rangle = \langle u_{a_1}|Bu_{a_1}\rangle \frac{N_1}{N} + \langle u_{a_2}|Bu_{a_2}\rangle \frac{N_2}{N}. \tag{2.27}$$

We are thus led to consider an operator of the form

$$\rho = \lambda P_{u_{a_1}} + (1 - \lambda)P_{u_{a_2}}, \tag{2.28}$$

where P_φ denotes the one-dimensional projection on the subspace generated by the vector φ, so that

$$P_\varphi \psi = \varphi \langle \varphi | \psi \rangle, \tag{2.29}$$

or equivalently in Dirac notation

$$P_\varphi = |\varphi\rangle\langle\varphi|, \tag{2.30}$$

and we have introduced the mixing parameter

$$\lambda = N_1/N \tag{2.31}$$

such that $0 \leqslant \lambda \leqslant 1$ and $N_2/N = 1 - \lambda$. Exploiting (2.28) we can write

$$P(B \in M) = \mathrm{Tr}\{\rho E^B(M)\} \tag{2.32}$$

as well as

$$\langle B \rangle = \mathrm{Tr}\{\rho B\}, \tag{2.33}$$

and more generally

$$\langle f(B) \rangle = \int f(x)\mathrm{d}\mu_\rho^B(x), \tag{2.34}$$

with

$$\mu_\rho^B(M) = \mathrm{Tr}\{\rho E^B(M)\}. \tag{2.35}$$

Let us recall what we mean by trace of an operator, whose condition of existence will be clarified in Sect. 2.2.2. Denoting with $\{u_n\}$ a basis in $\mathcal{H}$, that is a complete orthonormal system (c.o.n.s.) of vectors, we have by definition

$$\mathrm{Tr}\{A\} = \sum_n \langle u_n | A u_n \rangle, \tag{2.36}$$

so that the trace is a linear map, further satisfying invariance under a cyclic permutation, namely

$$\mathrm{Tr}\{AB\} = \mathrm{Tr}\{BA\}. \tag{2.37}$$

An operator ρ like the one introduced in (2.28) is called a statistical operator.

We have been led to consider statistical operators from a simple analysis of states as representatives of preparation procedures and noticing that the set of preparation procedures is stable under mixing. Other paths also lead to consider statistical operators, such as the treatment of problems in statistical mechanics, investigation of the classical limit in the description of macroscopic systems, analysis of bipartite systems, namely systems described on a Hilbert space with a tensor product structure. In this respect, we stress that in our approach we have made reference to the Hilbert space of the considered system only, providing an intrinsic motivation for the introduction of statistical operators. The very notion of mixing of preparation procedures belongs to any statistical theory. We now want to explore the special features of the obtained class of states in the framework of quantum probability.

2.2.2 Trace Class, Hilbert-Schmidt and Compact Operators

We here better specify the functional space in which a statistical operator can be described, thus characterizing its properties. Indeed, instead of mixtures of two one-dimensional orthogonal projections, as in the previous example, we can obviously consider a higher number of terms as well as an arbitrary convex combination, even a continuous one, of nonorthogonal states. The latter kind of mixture cannot be read as classical mixture of mutually exclusive events.

We first give a precise meaning to the trace operation. Let $A \in \mathcal{B}(\mathcal{H})$ be a positive operator, namely $\langle \psi | A \psi \rangle \geqslant 0$ for all $\psi \in \mathcal{H}$, and consider the series of positive numbers

$$\mathrm{Tr}\{A\} = \sum_n \langle u_n | A u_n \rangle, \tag{2.38}$$

for a given basis $\{u_n\}$. If the series converges the result is independent of the choice of basis, indeed considering another basis $\{v_k\}$ we have

$$\sum_n \langle u_n | A u_n \rangle = \sum_n \| A^{1/2} u_n \|^2 \tag{2.39}$$

$$= \sum_n \left(\sum_k |\langle v_k | A^{1/2} u_n \rangle|^2 \right) \tag{2.40}$$

$$= \sum_k \left(\sum_n |\langle u_n | A^{1/2} v_k \rangle|^2 \right) \tag{2.41}$$

$$= \sum_k \| A^{1/2} v_k \|^2 \tag{2.42}$$

$$= \sum_k \langle v_k | A v_k \rangle. \tag{2.43}$$

The independence from the choice of basis justifies the definition. For a generic operator $A \in \mathcal{B}(\mathcal{H})$ not necessarily positive, we can consider the positive operator

$$|A| = \sqrt{A^\dagger A}, \tag{2.44}$$

and if $\mathrm{Tr}\,|A| < \infty$ we say that A is a trace class operator. We recall that an operator A is said to be positive if

$$\langle \varphi | A\varphi \rangle \geqslant 0 \tag{2.45}$$

for any $\varphi \in \mathcal{H}$. If the inequality in (2.45) holds with the greater sign the operator will be called strictly positive. In the present book the term positive has always to be understood as non-negative. The space $\mathcal{T}(\mathcal{H})$ of trace class operators on a given Hilbert space turns out to be a Banach space with respect to the trace norm defined for $A \in \mathcal{T}(\mathcal{H})$ as

$$\|A\|_1 = \mathrm{Tr}\,|A|. \tag{2.46}$$

For a trace class operator the expression (2.38) is thus a finite and basis independent quantity. The space of trace class operators can also be obtained as closure with respect to the trace norm of the set of degenerate operators, that is operators of the form

$$A = \sum_{i=1}^{n} |u_i\rangle\langle v_i|, \tag{2.47}$$

where the sum runs over a finite index set and $\{u_i\}$, $\{v_i\}$ are linearly independent families of vectors.

The space $\mathcal{T}(\mathcal{H})$ is also a bilateral ideal of $\mathcal{B}(\mathcal{H})$, that is to say a subset of $\mathcal{B}(\mathcal{H})$ invariant under left and right multiplication, namely $\forall A \in \mathcal{B}(\mathcal{H})$, $\forall T \in \mathcal{T}(\mathcal{H})$ we have $AT, TA \in \mathcal{T}(\mathcal{H})$ and in particular

$$\|AT\|_1 \leqslant \|A\|\|T\|_1, \tag{2.48}$$

where $\|\cdot\|$ denotes the uniform norm on the space of operators

$$\|A\| = \sup_{\psi \in \mathcal{H}} \frac{\|A\psi\|}{\|\psi\|}. \tag{2.49}$$

Another important bilateral ideal of $\mathcal{B}(\mathcal{H})$ is given by the space of Hilbert-Schmidt operators. Given $A \in \mathcal{B}(\mathcal{H})$ we can consider the quantity $\sum_n \langle u_n | A^\dagger A u_n \rangle$, whose value as shown above does not depend on the basis. If this quantity is finite the operator is said to be in the Hilbert-Schmidt class, and the expression

$$\|A\|_2 = \sqrt{\mathrm{Tr}\,A^\dagger A} \tag{2.50}$$

actually defines a norm which makes the space of Hilbert-Schmidt operators $\mathrm{HS}(\mathcal{H})$ a Banach space. In particular, it is the closure with respect to the Hilbert-Schmidt norm of the set of degenerate operators. Due to the relation

$$\sqrt{\operatorname{Tr}|A|^2} \leqslant \operatorname{Tr}|A| \tag{2.51}$$

we have in particular that $\mathrm{HS}(\mathcal{H}) \supseteq \mathcal{T}(\mathcal{H})$. This set is also a bilateral ideal of $\mathcal{B}(\mathcal{H})$, indeed $\forall A \in \mathcal{B}(\mathcal{H})$, $\forall T \in \mathrm{HS}(\mathcal{H})$ we have $AT, TA \in \mathrm{HS}(\mathcal{H})$ and in particular

$$\|AT\|_2 \leqslant \|A\|\|T\|_2. \tag{2.52}$$

Most importantly, the space of Hilbert-Schmidt operators $\mathrm{HS}(\mathcal{H})$ is also a Hilbert space with respect to the following inner product

$$\langle A, B \rangle_{\mathrm{HS}} = \operatorname{Tr}\{A^\dagger B\}, \tag{2.53}$$

well-defined for any pair of operators since the product of two Hilbert-Schmidt operators is trace class. In this setting the Cauchy-Schwarz inequality becomes

$$|\operatorname{Tr}\{A^\dagger B\}| \leqslant \|A\|_2\|B\|_2 \tag{2.54}$$

with $A, B \in \mathrm{HS}(\mathcal{H})$, to be compared with the inequality

$$|\operatorname{Tr}\{A^\dagger B\}| \leqslant \|A\|\|B\|_1 \tag{2.55}$$

with $A \in \mathcal{B}(\mathcal{H})$ and $B \in \mathcal{T}(\mathcal{H})$, which follows from (2.48) and plays an important role: it warrants that the expression of mean values as considered in (2.33) is always well-defined. The latter relationship can be better understood noting that $\mathcal{B}(\mathcal{H}) = \mathcal{T}'(\mathcal{H})$, namely the space of bounded operators is the dual space with respect to the space of trace class operators, that is to say the space of continuous linear functionals on $\mathcal{T}(\mathcal{H})$. Each bounded operator can be seen to act on $\mathcal{T}(\mathcal{H})$ via the trace operation, so that the map $B \to \operatorname{Tr}\{B\cdot\}$ is an isometric isomorphism of $\mathcal{B}(\mathcal{H})$ onto $\mathcal{T}'(\mathcal{H})$. To better clarify this fact let us consider the simple case in which $\dim \mathcal{H} = n$. We can consider in $\mathcal{T}(\mathcal{H})$ the canonical basis of matrices $\{E^{ij}\}$ with matrix elements $(E^{ij})_{kl} = \delta_{ik}\delta_{jl}$, and write any $T \in \mathcal{T}(\mathcal{H})$ as $T = \sum_{i,j=1}^{n} T_{ij} E^{ij}$. Any continuous linear functional φ defined on $\mathcal{T}(\mathcal{H})$ can then be identified with a matrix $B \in \mathcal{B}(\mathcal{H})$. Indeed, upon defining

$$\varphi(E^{ij}) = B_{ji}, \tag{2.56}$$

we have $\forall T \in \mathcal{T}(\mathcal{H})$

$$\varphi(T) = \varphi\left(\sum_{i,j=1}^{n} T_{ij} E^{ij}\right) \tag{2.57}$$

$$= \sum_{i,j=1}^{n} T_{ij}\varphi(E^{ij}) \tag{2.58}$$

$$= \sum_{i,j=1}^{n} T_{ij} B_{ji} \tag{2.59}$$

$$= \mathrm{Tr}\{TB\}, \tag{2.60}$$

so that the linear functional φ can be identified with the matrix B, and the duality relation with the trace operation, which in the finite-dimensional setting coincides with the standard notion of trace of a matrix. In the case of a generic Hilbert space given a fixed $B \in \mathcal{B}(\mathcal{H})$ we have recalled that $\mathrm{Tr}\{BT\}$ is a well-defined complex number $\forall T \in \mathcal{T}(\mathcal{H})$, since the subset of trace class operators is a bilateral ideal of $\mathcal{B}(\mathcal{H})$. Furthermore thanks to (2.55) we have

$$|\mathrm{Tr}\{BT\}| \leqslant \|B\|\|T\|_1. \tag{2.61}$$

Each bounded operator can therefore be identified with a linear continuous functional on the space $\mathcal{T}(\mathcal{H})$ by means of the trace operation. Also the opposite can be shown to hold, so that we have indeed that $\mathcal{B}(\mathcal{H}) = \mathcal{T}'(\mathcal{H})$. Along the same lines, it can be shown that $\mathcal{T}(\mathcal{H}) \subseteq \mathcal{B}'(\mathcal{H})$, and in the infinite-dimensional case the inclusion is actually strict. The space $\mathrm{HS}(\mathcal{H})$ instead is a Hilbert space, and therefore coincides with its dual, as follows from (2.53) and (2.54).

It is an important fact that both trace class and Hilbert-Schmidt operators are not only bilateral ideals of the set of bounded operators, but are also contained within the set of compact operators, which we denote as $C(\mathcal{H})$, so that we have the following chain of inclusions

$$\mathcal{B}(\mathcal{H}) \supseteq C(\mathcal{H}) \supseteq \mathrm{HS}(\mathcal{H}) \supseteq \mathcal{T}(\mathcal{H}), \tag{2.62}$$

which all become strict in the case of an infinite-dimensional Hilbert space, while the equal sign is recovered for the finite-dimensional case. Notice that in accordance to (2.62) we also have

$$\| \cdot \| \leqslant \| \cdot \|_2 \leqslant \| \cdot \|_1. \tag{2.63}$$

The space of compact operators $C(\mathcal{H})$ is the closure of the set of degenerate operators defined in (2.47) with respect to the uniform norm. An operator is compact if and only if it can be written as

$$C = \sum_{n=1}^{\infty} \lambda_i |f_i\rangle\langle g_i|, \tag{2.64}$$

where the positive constants λ_i are called the singular values. A compact operator transforms weakly converging series of vectors in norm convergent series, that is $\langle \varphi_n, \psi \rangle \xrightarrow{n \to \infty} \langle \varphi, \psi \rangle$ for all $\psi \in \mathcal{H}$ implies $\|C\varphi_n - C\varphi\| \xrightarrow{n \to \infty} 0$. Also the set of compact operators is a bilateral ideal of $\mathcal{B}(\mathcal{H})$.

The set of compact operators is of special importance because of the following fundamental theorem [3].

Theorem 2.1 *Let $A \in \mathcal{B}(\mathcal{H})$ be self-adjoint and compact, then there exists a c.o.n.s. $\{u_n\}$ of eigenvectors of A with corresponding eigenvalues $\{a_n\}$, such that A admits the following spectral representation*

$$A = \sum_n a_n |u_n\rangle\langle u_n|, \tag{2.65}$$

where the eigenspaces of the non zero eigenvalues are finite-dimensional and $\lim_{n \to \infty} a_n = 0$. If in particular $A \in \mathrm{HS}(\mathcal{H})$ then

$$\sqrt{\sum_n |a_n|^2} = \|A\|_2, \tag{2.66}$$

and for $A \in \mathcal{T}(\mathcal{H})$ we also have

$$\sum_n |a_n| = \|A\|_1. \tag{2.67}$$

We are now in the position to properly introduce the set of states in the quantum probabilistic setting for a system described in a Hilbert space $\mathcal{H}$. This set will be denoted by $\mathcal{S}(\mathcal{H})$ and is given by the convex set of statistical operator

$$\mathcal{S}(\mathcal{H}) = \{\rho \in \mathcal{T}(\mathcal{H}) | \rho = \rho^\dagger, \rho \geqslant 0, \mathrm{Tr}\{\rho\} = 1\}. \tag{2.68}$$

The convexity of the space $\mathcal{S}(\mathcal{H})$ implies that any convex combination of elements of $\mathcal{S}(\mathcal{H})$ is still in the set, i.e.

$$\{\rho_i\} \in \mathcal{S}(\mathcal{H}), \;\; \mu_i \geqslant 0, \;\; \sum_{i=1}^n \mu_i = 1 \Rightarrow \sum_{i=1}^n \mu_i \rho_i \in \mathcal{S}(\mathcal{H}). \tag{2.69}$$

It is enough to consider mixtures of two elements only, so that convexity is implied by

$$\rho_1, \rho_2 \in \mathcal{S}(\mathcal{H}), \;\; 0 \leqslant \mu \leqslant 1 \Rightarrow \mu \rho_1 + (1 - \mu)\rho_2 \in \mathcal{S}(\mathcal{H}), \tag{2.70}$$

as follows by induction from the identity

$$\sum_{i=1}^{n} \mu_i \rho_i = \mu_1 \rho_1 + (1 - \mu_1) \left(\frac{\mu_2}{1 - \mu_1} \rho_2 + \ldots + \frac{\mu_n}{1 - \mu_1} \rho_n \right). \tag{2.71}$$

Since trace class implies compact, and positive implies self-adjoint, according to Theorem 2.1 we have the following fundamental representation of a statistical operator

$$\rho = \sum_i p_i |\varphi_i\rangle\langle\varphi_i|, \tag{2.72}$$

which provides an orthogonal decomposition since the rank one projections $\{|\varphi_i\rangle\langle\varphi_i|\}$ are orthogonal and thus identify an orthogonal resolution of the identity

$$\mathbb{1} = \sum_i |\varphi_i\rangle\langle\varphi_i|. \tag{2.73}$$

Note that since the projections are one-dimensional the p_i are generally not distinct. Here, the $\{\varphi_i\}$ do form a c.o.n.s. and the $\{p_i\}$ provide a probability distribution: $p_i \geqslant 0$, $\sum_i p_i = 1$. The decomposition of an operator in terms of orthogonal rank-one projections, that is in terms of extreme states, is called Schatten decomposition [4]. We stress the important fact that the distribution of the eigenvalues of a statistical operator, given by the probability distribution $\{p_i\}$, is uniquely fixed, so that features of this distribution can be used to characterize the operator. We further stress that while the representation (2.72) always exists, it is in general not unique, and many others can be considered, as we shall see in Sect. 2.2.5.

2.2.3   Purity and Entropy

We now want to provide some further insight on the structure of the set $\mathcal{S}(\mathcal{H})$. To this aim we will put into evidence two features, namely purity and von Neumann entropy, related to the mixedness and information content of a state, which allow to introduce two distinct ordering on the set of states. These features are naturally expressed also in terms of the probability distribution uniquely determined by the eigenvalues of a statistical operator. In this respect, they correspond to special instances of one-parameter families of functions defined on this probability distribution, so-called Rényi and Tsallis entropies [5], extracting different insights in information theory and statistical physics respectively from this probability distribution.

2.2.3.1   Purity

We define pure states to be the extreme points of the convex set $\mathcal{S}(\mathcal{H})$, i.e. those states that cannot be written as a non-trivial convex mixture, in the sense that a representation of the form

$$\rho = \mu\rho_1 + (1 - \mu)\rho_2 \tag{2.74}$$

only holds if $\mu \in \{0, 1\}$ or $\rho_1 = \rho_2 = \rho$. All other states are called mixed. According to (2.72), which expresses a statistical operator as a mixture of one-dimensional projections, we have that pure states are given by one-dimensional projections. We can formalize these facts as follows

$$\text{extreme}(\mathcal{S}(\mathcal{H})) = \mathcal{P}_1(\mathcal{H}) \subseteq \mathcal{P}(\mathcal{H}), \tag{2.75}$$

where $\mathcal{P}_1(\mathcal{H})$ is the set of one-dimensional projections on $\mathcal{H}$, and equality in the last relation holds iff $\dim \mathcal{H} = 2$. This feature can also be captured at the level of the eigenvalues by considering the square of the state. We therefore introduce the purity of a state by means of the expression

$$\mathcal{P}(\rho) = \text{Tr}\{\rho^2\} = \sum_i p_i^2. \tag{2.76}$$

From the properties of a statistical operator or equivalently of a probability distribution $\{p_i\}$ we have

$$0 < \mathcal{P}(\rho) \leqslant 1. \tag{2.77}$$

The higher the purity, the closer a state is to a pure state. We notice that $\mathcal{P}(\rho) = \|\rho\|_2^2$ so that the value zero cannot be attained, corresponding to the null operator, while the value one is only taken by pure states. If the Hilbert space has finite dimension n, the purity attains its minimum value $1/n$ on the maximally mixed state, which is proportional to the identity. We thus have the equivalence of the statements: *(i)* ρ is pure; *(ii)* ρ is a one-dimensional projection; *(iii)* $\text{Tr}\{\rho^2\} = 1$. Moreover the very definition (2.76) implies that $\mathcal{P}(U\rho U^\dagger) = \mathcal{P}(\rho)$, with U any unitary operator, thanks to invariance of the eigenvalues of an operator under a unitary transformation. An important property of the purity, being defined on a convex set, is its behavior with respect to mixing. We have that purity is a convex map, i.e. it decreases upon mixing

$$\mathcal{P}(\lambda\rho_1 + (1 - \lambda)\rho_2) \leqslant \lambda\mathcal{P}(\rho_1) + (1 - \lambda)\mathcal{P}(\rho_2) \quad \forall\rho_1, \rho_2 \in \mathcal{S}(\mathcal{H}) \quad \forall\lambda \geqslant 0 \tag{2.78}$$

and equality holds iff $\rho_1 = \rho_2$, so that purity is strictly convex. The proof of this statement can be obtained along the same lines as the proof of (2.104), to which we refer the reader.

If $\dim \mathcal{H} = 2$, the Hilbert space is isomorphic to $\mathbb{C}^2$, and a generic state can be written as

$$\rho = \frac{1}{2}(\mathbb{1} + \boldsymbol{r} \cdot \boldsymbol{\sigma}) \tag{2.79}$$

with $\boldsymbol{r} \in \mathbb{R}^3$, $\|\boldsymbol{r}\| \leqslant 1$, and $\boldsymbol{r} \cdot \boldsymbol{\sigma} = \sum_{i=1}^3 r_i\sigma_i$ with $\{\sigma_i\}_{i=1,2,3}$ the usual Pauli matrices, whose definition and properties are recalled in Remark 2.1. The vector

r is usually called Bloch vector. This representation allows a one to one mapping between points in the sphere $\mathbb{S}^2 = \{r \in \mathbb{R}^3, \|r\| \leqslant 1\}$ and statistical operators in $\mathbb{C}^2$. This sphere is called Bloch sphere, or also Poincaré sphere in optics.

2.1 Pauli Matrices

We here recall the very definition of the Pauli matrices and their basic properties. The Pauli matrices are given by

$$\sigma_x = \begin{pmatrix} 0 & 1 \\ 1 & 0 \end{pmatrix} \quad \sigma_y = \begin{pmatrix} 0 & -i \\ i & 0 \end{pmatrix} \quad \sigma_z = \begin{pmatrix} 1 & 0 \\ 0 & -1 \end{pmatrix}, \tag{2.80}$$

and their commutation relations follow from the fundamental commutation relations for the triple of angular momentum operators

$$S_i = \frac{\hbar}{2}\sigma_i, \tag{2.81}$$

namely

$$[S_i, S_j] = i\hbar \sum_{k=1}^{3} \varepsilon_{ijk} S_k, \tag{2.82}$$

where ε_{ijk} denotes the completely antisymmetric tensor in three dimension, so that we have

$$[\sigma_i, \sigma_j] = 2i \sum_{k=1}^{3} \varepsilon_{ijk} \sigma_k. \tag{2.83}$$

We further have

$$\sigma_i \sigma_j = \delta_{ij} + i \sum_{k=1}^{3} \varepsilon_{ijk} \sigma_k. \tag{2.84}$$

These identities imply the useful relations

$$e^{i\theta n \cdot \sigma} = \cos\theta + i n \cdot \sigma \sin\theta \tag{2.85}$$

and

$$(a \cdot \sigma)(b \cdot \sigma) = a \cdot b + i(a \times b) \cdot \sigma, \tag{2.86}$$

with a, b real vectors and n an arbitrary versor.

In analogy to the raising and lowering operators in the theory of angular momentum, we are led to introduce the operators

$$\sigma_+ = \frac{1}{2}(\sigma_x + i\sigma_y) \tag{2.87}$$

$$\sigma_- = \frac{1}{2}(\sigma_x - i\sigma_y) \tag{2.88}$$

such that

$$[\sigma_z, \sigma_\pm] = \pm\hbar\sigma_\pm. \tag{2.89}$$

Denoting with

$$|+\rangle = \begin{pmatrix} 1 \\ 0 \end{pmatrix} \tag{2.90}$$

and

$$|-\rangle = \begin{pmatrix} 0 \\ 1 \end{pmatrix} \tag{2.91}$$

the eigenvectors of the σ_z operator relative to the eigenvalues $+1$ and -1 respectively, we have

$$\sigma_+|-\rangle = |+\rangle \quad \sigma_-|+\rangle = |-\rangle \tag{2.92}$$

as well as

$$\sigma_+|+\rangle = 0 \quad \sigma_-|-\rangle = 0, \tag{2.93}$$

corresponding to the representation

$$\sigma_+ = |+\rangle\langle-| = \begin{pmatrix} 0 & 1 \\ 0 & 0 \end{pmatrix} \quad \sigma_- = |-\rangle\langle+| = \begin{pmatrix} 0 & 0 \\ 1 & 0 \end{pmatrix}. \tag{2.94}$$

For the sake of later convenience we also recall the explicit expression of the operator

$$\boldsymbol{n} \cdot \boldsymbol{\sigma} = \begin{pmatrix} \cos\theta & \sin\theta e^{-i\phi} \\ \sin\theta e^{+i\phi} & -\cos\theta \end{pmatrix}, \tag{2.95}$$

with $\boldsymbol{n} = (\sin\theta\cos\phi, \sin\theta\sin\phi, \cos\theta)$ an arbitrary versor, and of its corresponding normalized eigenvectors

$$|\boldsymbol{n}_+(\theta,\phi)\rangle = \begin{pmatrix} \cos\frac{\theta}{2}e^{-i\frac{\phi}{2}} \\ \sin\frac{\theta}{2}e^{+i\frac{\phi}{2}} \end{pmatrix} \tag{2.96}$$

and

$$|\boldsymbol{n}_-(\theta,\phi)\rangle = \begin{pmatrix} -\sin\frac{\theta}{2}e^{-i\frac{\phi}{2}} \\ \cos\frac{\theta}{2}e^{+i\frac{\phi}{2}} \end{pmatrix}. \tag{2.97}$$

The geometry of the quantum states reflects in this case the geometry of the sphere, in the sense that the set of extreme states coincides with its boundary, in particular each mixed state can be demixed in terms of two extreme states. This is not generally true. Indeed, for $\dim\mathcal{H} \geqslant 3$, the extreme states, namely one-dimensional projections, do not coincide anymore with the boundary of the set of states $\mathcal{S}(\mathcal{H})$. To show this we rely on an intrinsic characterization of the boundary of a convex set. We introduce the interior $\widetilde{\mathcal{S}}(\mathcal{H})$ as the subset of the elements ρ such that $\forall\sigma \in \mathcal{S}(\mathcal{H})$ there exists a $\lambda > 1$ such that

$$\lambda\rho + (1-\lambda)\sigma \in \mathcal{S}(\mathcal{H}), \tag{2.98}$$

as shown pictorially in Fig. 2.2. We define then the boundary as $\partial\mathcal{S}(\mathcal{H}) = \mathcal{S}(\mathcal{H})\backslash\widetilde{\mathcal{S}}(\mathcal{H})$, that is $\rho \in \partial\mathcal{S}(\mathcal{H})$ iff there exists $\sigma \in \mathcal{S}(\mathcal{H})$ such that $\forall\lambda > 1$

$$\lambda\rho + (1-\lambda)\sigma \notin \mathcal{S}(\mathcal{H}). \tag{2.99}$$

As a consequence, if ρ has a zero eigenvalue, it belongs to the boundary. Indeed, let $\varphi \in \mathcal{H}$ be an eigenvector of ρ corresponding to the zero eigenvalue, then taking $\sigma = |\varphi\rangle\langle\varphi|$ for $\lambda > 1$ we have $\langle\varphi|((1-\lambda)\sigma + \lambda\rho)\varphi\rangle = (1-\lambda) < 0$, so that $(1-\lambda)\sigma + \lambda\rho$ is not a positive operator. We can now easily show that the boundary does not generally coincide with the set of pure states. If $\dim\mathcal{H} \geqslant 3$ we can consider in $\mathcal{H}$ an orthogonal triple $\{\varphi_1, \varphi_2, \varphi_3\}$ as well as the statistical operator

$$\rho = \lambda P_{\varphi_1} + (1-\lambda)P_{\varphi_2}, \tag{2.100}$$

where $0 < \lambda < 1$, so that it is not pure. Nevertheless $\rho\varphi_3 = 0$, so that it has a zero eigenvalue and therefore it belongs to the boundary. The distinction between boundary and subset of extreme states becomes clear thinking about different convex subsets of the Euclidean space, such as a circle or a triangle. Actually, in finite

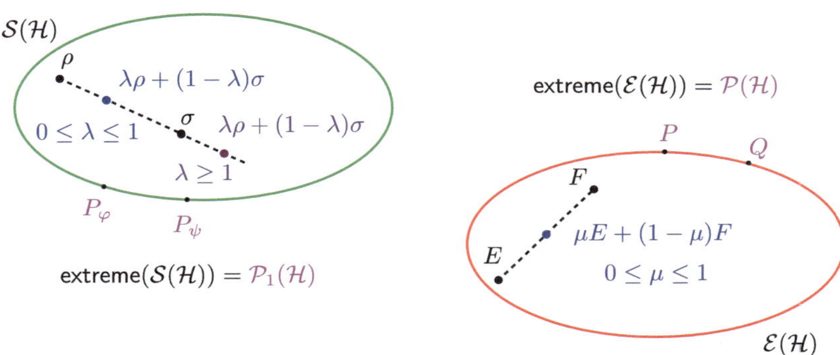

Fig. 2.2 Schematic representation recalling the convex nature of both sets of states and effects. For the set of states the extreme elements are the one-dimensional projections. For the set of effects all orthogonal projections are extreme elements

dimensions it can be shown that the boundary of $\mathcal{S}(\mathcal{H})$ coincides with the statistical operators admitting at least a zero eigenvalue [6].

2.2.3.2 Entropy

We now concentrate on the information content of a state. As we already stressed, quantum mechanics is a quantum probability, concerned in providing predictions for suitable statistical experiments. For a stochastic experiment, the information content related to a certain event is a measure of the uncertainty related to its appearance before we know it, or equivalently of the information gained upon learning its value. In information theory, the information content of a certain event is related to the number of binary questions necessary to ascertain it. Suppose the sample space Ω is a set with $N = 2^n$ elements $\omega_1, \ldots, \omega_N$, which identify the elementary events. An element of Ω can then be characterized by a sequence of n zeros or one. A realization of the stochastic event has an information content I equal to $n = \log_2 N$ bits. More generally, we set $I = \log_2 N$, even if N is not a power of two. Suppose now the elementary events have equal probability $p_i = 1/N$, then the information of the realization of the event ω_i has information content $I = \log_2(N) = -\log_2(p_i)$, and we adopt this quantity as information content of an event taking place with probability p_i for a distribution which is not necessarily uniform. Indeed, the requirement of equal probability of the events in the heuristic argument can be easily overcome. The average information gained by a stochastic experiment measuring events with probability p_i is thus quantified by the expression

$$H(\{p_i\}) = -\sum_{i=1}^{N} p_i \log_2(p_i), \tag{2.101}$$

which is known as Shannon entropy, and plays a central role in classical information theory [7]. In this simple-minded presentation, it specifies the amount of bits needed to store the mean information associated to a stochastic experiment.

In the quantum setting this quantity is naturally generalized to the von Neumann entropy

$$S(\rho) = - \operatorname{Tr}\{\rho \log \rho\} = - \sum_i p_i \log p_i, \qquad (2.102)$$

where $\{p_i\}$ denotes the probability distribution uniquely determined by ρ via its orthogonal decomposition (2.72). Note that here and in the following we will consider the logarithm in the natural basis, in view of its connection with the thermodynamic entropy. In other words, the von Neumann entropy of ρ is equal to the Shannon entropy of its eigenvalues. According to its role as state, the statistical operator ρ allows to introduce a probability distribution for the outcomes of any observable B. In the notation of (2.5) we denote this probability distribution as $\{p_{b_j}\}$, with $p_{b_j} = \operatorname{Tr}\{\rho P_{b_j}\}$. We can thus also consider the associated Shannon entropy which satisfies

$$H(\{p_{b_j}\}) \geqslant S(\rho). \qquad (2.103)$$

This result is often called majorization theorem and tells us that the most informative measurement on a state is obtained for observables commuting with ρ and therefore diagonal in the same basis [5]. From its definition in terms of the eigenvalues (2.102), we have that the von Neumann entropy is invariant under unitary transformations $S(U\rho U^\dagger) = S(\rho)$. Moreover, as it can be easily shown, we have the equivalence of the statements: *(i)* ρ is pure; *(ii)* ρ is a one-dimensional projection; *(iii)* $S(\rho) = 0$.

Importantly, we can investigate the behavior of the von Neumann entropy with respect to mixing and show that it is a concave map, i.e. it increases upon mixing

$$S(\lambda\rho_1 + (1-\lambda)\rho_2) \geqslant \lambda S(\rho_1) + (1-\lambda)S(\rho_2) \quad \forall \rho_1, \rho_2 \in \mathcal{S}(\mathcal{H}) \quad \forall \lambda \geqslant 0 \quad (2.104)$$

where equality holds iff $\rho_1 = \rho_2$, so that entropy is strictly concave.

Proof Let us consider the function $\eta(x) = -x \log x$, put equal to zero in $x = 0$ and strictly concave for $x \geqslant 0$. According to (2.71) any mixture of states can always be expressed as a convex combination of two of them, so that it is enough to consider concavity with respect to

$$\rho = \lambda\rho_1 + (1-\lambda)\rho_2. \qquad (2.105)$$

We consider the usual decomposition (2.72) in terms of rank-one orthogonal projections, where in particular the eigenvalues can be obtained as the following expectation values

$$p_i = \langle \varphi_i | \rho \varphi_i \rangle. \qquad (2.106)$$

We thus have from (2.102) for the expression of the von Neumann entropy

$$S(\rho) = \sum_i \eta(p_i) \tag{2.107}$$

$$= \sum_i \eta(\langle \varphi_i | \rho \varphi_i \rangle) \tag{2.108}$$

$$\geq \sum_i [\lambda \eta(\langle \varphi_i | \rho_1 \varphi_i \rangle) + (1 - \lambda) \eta(\langle \varphi_i | \rho_2 \varphi_i \rangle)], \tag{2.109}$$

where in the last line we have exploited (2.105) and used once the concavity of η. We now exploit it once more observing that for a concave function

$$\eta(\langle B \rangle) \geq \langle \eta(B) \rangle, \tag{2.110}$$

where $B \in \mathcal{B}(\mathcal{H})$ is a self-adjoint operator and $\langle B \rangle$ its expectation value with respect to an arbitrary state, say σ. To show this, introducing according to (2.5) the representation

$$B = \sum_i b_i P_{b_i}, \tag{2.111}$$

and denoting with

$$p_{b_i} = \mathrm{Tr}\{\sigma P_{b_i}\} \tag{2.112}$$

the probability of the outcome b_i for the quantity associated to B if the state of the system is σ, we obtain, exploiting the fact that the mean value of an observable is nothing but a convex mixture of its possible outcomes

$$\eta(\langle B \rangle) = \eta \left(\sum_i \mathrm{Tr}\{\sigma P_i\} b_i \right) \tag{2.113}$$

$$= \eta \left(\sum_i p_{b_i} b_i \right), \tag{2.114}$$

so that using the concavity of the function η as visualized in Fig. 2.3 and the spectral theorem we obtain

$$\eta(\langle B \rangle) \geq \sum_i p_{b_i} \eta(b_i) \tag{2.115}$$

$$= \sum_i \mathrm{Tr}\{\sigma P_i\} \eta(b_i) \tag{2.116}$$

$$= \mathrm{Tr}\{\sigma \eta(B)\} \tag{2.117}$$

$$= \langle \eta(B) \rangle. \tag{2.118}$$

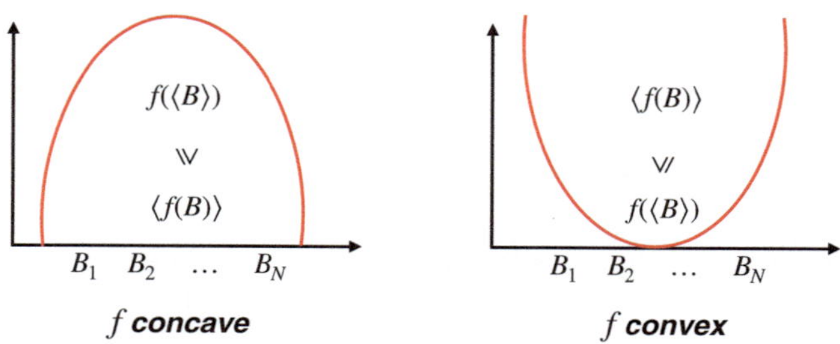

Fig. 2.3 Pictorial visualization of concave and convex functions as used for the derivation of important inequalities

This result is a special instance of the Jensen's inequality applied to concave functions. In probability theory Jensen's inequality can be generally formulated as follows: if X is a random variable and φ is a convex function

$$\varphi(\mathbb{E}[X]) \leqslant \mathbb{E}[\varphi(X)], \tag{2.119}$$

where the symbol $\mathbb{E}$ denotes the average over the underlying measure space. Inserting this result in the previous expression we finally come to

$$S(\rho) \geqslant \sum_i [\lambda \langle \varphi_i | \eta(\rho_1) \varphi_i \rangle + (1 - \lambda) \langle \varphi_i | \eta(\rho_2) \varphi_i \rangle] \tag{2.120}$$

$$= \lambda S(\rho_1) + (1 - \lambda) S(\rho_2), \tag{2.121}$$

and due to strict concavity of the function $\eta(x)$ equality is attained iff the states ρ_1 and ρ_2 coincide. $\qquad \Box$

Another useful property of the von Neumann entropy is obtained considering two statistical operators $\rho, w \in \mathcal{S}(\mathcal{H})$, such that $\mathrm{supp}(\rho) \subseteq \mathrm{supp}(w)$, namely

$$S(\rho) \leqslant - \mathrm{Tr}\{\rho \log w\}, \tag{2.122}$$

where the equality sign only holds iff $\rho = w$, and the relation is known as Klein's inequality.

Proof To prove the statement we notice that it is equivalent to

$$\mathrm{Tr}\{\rho \log w\} - \mathrm{Tr}\{\rho \log \rho\} \leqslant 0. \tag{2.123}$$

According to (2.72) we consider the representations

$$\rho = \sum_j \rho_j |\varphi_j\rangle\langle\varphi_j| \tag{2.124}$$

$$w = \sum_k w_k |\psi_k\rangle\langle\psi_k|, \tag{2.125}$$

leading to

$$\mathrm{Tr}\{\rho \log w\} - \mathrm{Tr}\{\rho \log \rho\} = \sum_{j,k} \rho_j |\langle\varphi_j|\psi_k\rangle|^2 (\log w_k - \log \rho_j) \tag{2.126}$$

where the expression is well-defined thanks to $\mathrm{supp}(\rho) \subseteq \mathrm{supp}(w)$. We now restrict the sum to the ρ_j different from zero, and exploit the inequality $\log y \leqslant y - 1$, valid for $y > 0$ and becoming an equality iff $y = 1$, so that we obtain

$$\mathrm{Tr}\{\rho \log w\} - \mathrm{Tr}\{\rho \log \rho\} = \sum_{j,k} \rho_j |\langle\varphi_j|\psi_k\rangle|^2 \log \frac{w_k}{\rho_j} \tag{2.127}$$

$$\leqslant \sum_{j,k} \rho_j |\langle\varphi_j|\psi_k\rangle|^2 \left(\frac{w_k}{\rho_j} - 1\right) \tag{2.128}$$

$$= \sum_{j,k} |\langle\varphi_j|\psi_k\rangle|^2 (w_k - \rho_j) \tag{2.129}$$

$$= \sum_k w_k - \sum_j \rho_j \tag{2.130}$$

equal to zero thanks to normalization. $\qquad\qquad\qquad\qquad\qquad\qquad\square$

A crucial consequence of Klein's inequality is positivity of the quantity

$$S(\rho, w) = \mathrm{Tr}\{\rho \log \rho\} - \mathrm{Tr}\{\rho \log w\}, \tag{2.131}$$

set equal to $+\infty$ when the condition $\mathrm{supp}(\rho) \subseteq \mathrm{supp}(w)$ is not satisfied. This quantity, that we shall discuss in more detail in Sect. 3.2.5.2, is known as quantum relative entropy and plays a crucial role in quantum information and quantum statistical mechanics.

Purity and entropy provide two different partial orders on the set of states $\mathcal{S}(\mathcal{H})$, as can also be seen considering their range for the case of a finite-dimensional Hilbert space with $\dim \mathcal{H} = n$

$$0 \leqslant S(\rho) \leqslant \log n \tag{2.132}$$

$$\frac{1}{n} \leqslant \mathcal{P}(\rho) \leqslant 1. \tag{2.133}$$

The purity of a state makes reference to the notion of extremality, while the von Neumann entropy is related to its information content. Pure states saturate the maximum and minimum value of purity and entropy respectively, while the maximally mixed statistical operator corresponding to $\rho = \frac{1}{2}\mathbb{1}$ saturates the maximum value of entropy and the minimum value of purity.

As stressed below (2.102) the von Neumann entropy of a state coincides with the Shannon entropy of the distribution of its eigenvalues, and therefore with the Rényi and Tsallis entropies of order 1, while Rényi and Tsallis entropies of order 2, which provide a lower bound to the von Neumann entropy, can be expressed in terms of the purity according to $\mathcal{R}_2(\rho) = -\log\mathcal{P}(\rho)$ and $\mathcal{T}_2(\rho) = 1 - \mathcal{P}(\rho)$, where the latter is also known as linear entropy [5].

2.2.4 Maximum-Entropy State

We have stressed that a statistical operator has to describe an equivalence class of preparation procedures, that is of actual strategies for the preparation of a quantum system in the laboratory. We already hinted to a few possible strategies. In Sect. 2.2.1 we considered how to prepare pure states, corresponding to the most refined preparation procedures, while in Sect. 2.2.5 taking a somehow complementary perspective we will suggest how to prepare a maximally mixed state, corresponding to maximum lack of information. We consider now a more realistic situation, of interest in particular for large systems, in which the information on a state does not correspond to the fact that it has been obtained by performing a projective measurement on an impinging state, possibly selecting according to the outcomes. The available information rather has the form of knowledge about mean values of a collection of observables, which we shall call relevant observables. Note that variances can also be included, considering the mean values of squares of operators. This collection will generally be a restricted set, since knowledge of all possible mean values would suffice to fix the state uniquely. Relevant observables are typically given by conserved quantities or the integral of densities of conserved quantities over suitable macroscopic space regions. We denote with

$$M = \{R_i\}_i \tag{2.134}$$

the set of relevant observables, so that each R_i is a self-adjoint operator, and let

$$m = \{r_i\}_i \tag{2.135}$$

be the set of real numbers corresponding to the assigned mean values. Let us further introduce the convex set of statistical operators compatible with these mean values

$$\mathcal{M}(\mathcal{H}) = \{\rho \in \mathcal{S}(\mathcal{H})\,|\, \mathrm{Tr}\{\rho R_i\} = r_i \quad \forall R_i \in M\}\,. \tag{2.136}$$

Among all the states in $\mathcal{M}(\mathcal{H})$ we want to characterize the state which takes into account the available information on mean values but includes no further bias, namely as discussed above the state which satisfies these constraints and has maximum von Neumann entropy. This strategy is known as principle of maximum entropy [8–10]. We therefore face a maximization problem with constraints, given by the assigned mean values as well as normalization. We will see that the solution to this problem indeed is a statistical operator, in particular a strictly positive operator. Given the expression (2.102) for the von Neumann entropy we look for stationary points of the functional

$$\tilde{S}(\rho) = S(\rho) + (\Omega + 1)\langle \mathbb{1} \rangle_\rho - \sum_i \beta_i \langle R_i \rangle_\rho \tag{2.137}$$

$$= -\operatorname{Tr}\left\{ \rho \left[\log \rho - (\Omega + 1)\mathbb{1} + \sum_i \beta_i R_i \right] \right\}, \tag{2.138}$$

where the Lagrange parameters $\{\beta_i\}$ are real. Given the general expression of a statistical operator (2.72), we have to consider variations both with respect to the basis of its eigenvectors, which can be implemented via unitary transformations, and with respect to the distribution of the eigenvalues. Changes of basis are induced by unitary transformations, which we write in the form $\exp(iW\theta)$ with W a self-adjoint operator, so that we obtain for an infinitesimal transformation the expression

$$\delta\rho = i[W, \rho]\delta\theta. \tag{2.139}$$

We therefore come to

$$\delta\tilde{S}(\rho) = -\operatorname{Tr}\left\{ \delta\rho \left[\log \rho - (\Omega + 1)\mathbb{1} + \sum_i \beta_i R_i \right] + \delta\rho \right\} \tag{2.140}$$

$$= -\operatorname{Tr}\left\{ \delta\rho \left[\log \rho + \sum_i \beta_i R_i \right] \right\}, \tag{2.141}$$

where we have used the fact that $\delta\rho$ is traceless. Further exploiting the explicit expression of $\delta\rho$ and the useful relation

$$\operatorname{Tr}\{[A, B]C\} = \operatorname{Tr}\{A[B, C]\}, \tag{2.142}$$

coming from cyclic invariance of the trace operation, we have from (2.139)

$$\delta\tilde{S}(\rho) = -\operatorname{Tr}\{i[W, \rho]\log \rho\}\delta\theta - \operatorname{Tr}\left\{ i[W, \rho]\sum_i \beta_i R_i \right\}\delta\theta \tag{2.143}$$

$$= -i\operatorname{Tr}\left\{ W \left[\rho, \sum_i \beta_i R_i \right] \right\}\delta\theta. \tag{2.144}$$

Asking the variation to be zero for arbitrary self-adjoint W we can make in particular the choice

$$W = i \left[\rho, \sum_i \beta_i R_i \right], \tag{2.145}$$

so that a necessary condition for the functional $\tilde{S}(\rho)$ to be stationary is given by

$$\left[\rho, \sum_i \beta_i R_i \right] = 0, \tag{2.146}$$

so that there exists a common spectral decomposition of the commuting self-adjoint operators ρ and $\sum_i \beta_i R_i$. Expressing the spectral representation of the latter operator in the form

$$\sum_i \beta_i R_i = \sum_\alpha R_\alpha |\alpha\rangle \langle \alpha|, \tag{2.147}$$

with $\{|\alpha\rangle\}$ basis in the Hilbert space of the system, this implies that the ρ we are looking for can be written as

$$\rho = \sum_\alpha \rho_\alpha |\alpha\rangle \langle \alpha|, \tag{2.148}$$

for suitable weights ρ_α. The commutativity of $\sum_i \beta_i R_i$ and ρ is actually also a sufficient condition for stationarity of the functional. Indeed, suppose

$$\rho = \sum_\gamma \rho_\gamma |\gamma\rangle \langle \gamma|, \tag{2.149}$$

with an arbitrary basis $\{|\gamma\rangle\}$, we would then have

$$\tilde{S}(\rho) = -\sum_\gamma \rho_\gamma \left\{ \log \rho_\gamma - (\Omega + 1) + \langle \gamma| \sum_i \beta_i R_i |\gamma\rangle \right\} \tag{2.150}$$

and the latter expression due to the variational theorem in quantum mechanics is stationary provided the $|\gamma\rangle$ are eigenvectors of $\sum_i \beta_i R_i$. Indeed, a functional of the form $\Gamma[\varphi] = \langle \varphi | H \varphi \rangle$, with H a selfadjoint operator, is stationary if φ is an eigenvector of H. Let us consider variations of the functional obtained considering variations of the state vector which have to preserve the norm, namely

$$\delta(\Gamma[\varphi] - \lambda \|\varphi\|^2) = \langle \delta\varphi | H\varphi \rangle + \langle \varphi | H\delta\varphi \rangle - \lambda \langle \delta\varphi | \varphi \rangle - \lambda \langle \varphi | \delta\varphi \rangle \tag{2.151}$$
$$= 2\Re \langle \delta\varphi | (H - \lambda)\varphi \rangle, \tag{2.152}$$

with λ a real parameter. The variation is equal to zero $\forall \delta\varphi$ if $(H - \lambda)|\varphi\rangle = 0$, so that φ is an eigenvector of H. As a result the commutativity condition (2.146) is both necessary and sufficient in order to have stationarity with respect to variations of the basis. We can thus conclude that

$$\tilde{S}(\rho) = -\sum_\alpha \rho_\alpha (\log \rho_\alpha - (\Omega + 1) + R_\alpha), \qquad (2.153)$$

and varying with respect to the eigenvalues we obtain

$$\delta\tilde{S}(\rho) = -(\log \rho_\beta - \Omega + R_\beta)\delta\rho_\beta, \qquad (2.154)$$

equal to zero for all β if

$$\rho_\alpha = e^{\Omega - R_\alpha}, \qquad (2.155)$$

so that the eigenvalues of the statistical operator are indeed non-negative, and furthermore strictly positive. The statistical operator solution of the stationarity problem is therefore given by

$$\rho = \sum_\alpha e^{\Omega - R_\alpha} |\alpha\rangle\langle\alpha| \qquad (2.156)$$

$$= e^{\Omega} e^{-\sum_i \beta_i R_i}, \qquad (2.157)$$

where normalization implies

$$e^{-\Omega} = \mathrm{Tr}\left\{ e^{-\sum_i \beta_i R_i} \right\}. \qquad (2.158)$$

According to standard notation in statistical mechanics we introduce the quantity

$$Z = \mathrm{Tr}\left\{ e^{-\sum_i \beta_i R_i} \right\}, \qquad (2.159)$$

so that the state which makes the functional $\tilde{S}$ stationary is identified to be

$$\rho = \frac{e^{-\sum_i \beta_i R_i}}{Z}. \qquad (2.160)$$

We now show that among the states in $\mathcal{M}(\mathcal{H})$, that is with the assigned mean values, it has indeed maximum entropy. Suppose $\sigma \in \mathcal{M}$, exploiting Klein's inequality (2.122) we have

$$\mathrm{Tr}\{\sigma \log \sigma\} \geqslant \mathrm{Tr}\{\sigma \log \rho\} \qquad (2.161)$$

$$= \mathrm{Tr}\left\{ \sigma \left[-\sum_i \beta_i R_i - \log Z \right] \right\} \qquad (2.162)$$

$$= \mathrm{Tr}\left\{\rho\left[-\sum_i \beta_i R_i - \log Z\right]\right\} \tag{2.163}$$

$$= \mathrm{Tr}\{\rho \log \rho\} \tag{2.164}$$

since both these operators have the same mean values for the relevant observables. As a consequence

$$S(\rho) \geqslant S(\sigma) \tag{2.165}$$

for all $\sigma \in \mathcal{M}(\mathcal{H})$. We have in particular the expression

$$S(\rho) = \sum_i \beta_i r_i + \log Z. \tag{2.166}$$

We have thus identified the state which maximizes the entropy for fixed mean values. This state is in Gibbs form and its expression has now been obtained adopting the maximum information principle to identify the state. Given its structure in terms of the relevant variables and the Lagrange parameters, let us consider the equations which implicitly fix the parameters in terms of the mean values. Starting from

$$\Omega = -\log \mathrm{Tr}\left\{\mathrm{e}^{-\sum_i \beta_i R_i}\right\} \tag{2.167}$$

we have

$$\frac{\partial \Omega}{\partial \beta_i} = -\frac{\partial \log Z}{\partial \beta_i} \tag{2.168}$$

$$= \frac{\mathrm{Tr}\left\{R_i \mathrm{e}^{-\sum_j \beta_j R_j}\right\}}{\mathrm{Tr}\left\{\mathrm{e}^{-\sum_j \beta_j R_j}\right\}} \tag{2.169}$$

$$= \langle R_i \rangle \tag{2.170}$$

$$= r_i, \tag{2.171}$$

where in order to evaluate the derivatives we have actually used the formula

$$\frac{\partial}{\partial \lambda} \mathrm{Tr}\{\mathrm{e}^{A(\lambda)}\} = \mathrm{Tr}\left\{\frac{\partial A(\lambda)}{\partial \lambda}\mathrm{e}^{A(\lambda)}\right\}, \tag{2.172}$$

which follows from the general formula (2.176) obtained in Remark 2.2 thanks to invariance of the trace under cyclic permutations.

2.2 Derivative of Exponential Operators

We consider the derivative of the exponential of an operator depending on a parameter, say $A(\lambda)$. The operator and its derivative do generally not commute, so that the standard formulae do not apply. Building on the Leibniz rule we have

$$\frac{\partial e^{A(\lambda)}}{\partial \lambda} = \sum_{n=0}^{\infty} \frac{1}{(n+1)!} \sum_{k=0}^{n} A(\lambda)^k \frac{\partial A(\lambda)}{\partial \lambda} A(\lambda)^{n-k}, \tag{2.173}$$

and exploiting the integral relation

$$\int_0^1 \mathrm{d}x\, x^r (1-x)^s = \frac{r!\,s!}{(r+s+1)!} \tag{2.174}$$

valid for any pair of integers r and s we obtain

$$\frac{\partial e^{A(\lambda)}}{\partial \lambda} = \sum_{n=0}^{\infty} \sum_{k=0}^{n} \frac{1}{k!(n-k)!} \int_0^1 \mathrm{d}x\, [x A(\lambda)]^k \frac{\partial A(\lambda)}{\partial \lambda} [(1-x) A(\lambda)]^{n-k},$$

$$\tag{2.175}$$

which can be recognized to be the Cauchy product of two exponential series, thus coming to

$$\frac{\partial}{\partial \lambda} e^{A(\lambda)} = \int_0^1 \mathrm{d}u\, e^{u A(\lambda)} \frac{\partial A(\lambda)}{\partial \lambda} e^{(1-u)A(\lambda)}. \tag{2.176}$$

The actual parameters appearing in the expression of the statistical operator are still to be determined and are implicitly fixed by the equations

$$\begin{cases} \Omega = -\log \mathrm{Tr}\left\{ e^{-\sum_i \beta_i R_i} \right\} \\ \frac{\partial \Omega}{\partial \beta_i} = r_i \end{cases}, \tag{2.177}$$

which can have solutions provided the set $m = \{r_i\}$ is a consistent set of expectation values compatible with quantum mechanical requirements.

2.2.5   Decompositions

We have shown that each statistical operator thanks to the spectral theorem has at least one orthogonal decomposition in terms of pure states, which is also called Schatten decomposition. It is unique iff all eigenvalues are distinct, while in general there are many, known as canonical convex decompositions. The freedom is here encoded in an arbitrary change of basis in the eigenspaces. All refer to the same statistical operator, to be understood as mathematical representative of an equivalence class of preparation procedures, and lead to the very same predictions for any statistical experiment, according to the probability formula (2.35). Each of them however, can be seen as a possible alternative way to prepare such a state by mixing other preparations.

2.3 Maximally Mixed State

Let us take the finite-dimensional Hilbert space $\mathbb{C}^2$, describing e.g. polarization degrees of freedom, and consider the maximally mixed state $\rho = \frac{1}{2}\mathbb{1}$, proportional to the identity operator. We fix a basis $\{|+\rangle, |-\rangle\}$, whose elements can be understood as eigenvectors of the polarization along a fixed direction, which we can take as z-axis. Exploiting this basis, according to (2.96) and (2.97) of Remark 2.1, for any $\theta \in [0, \pi]$ and $\phi \in [0, 2\pi]$ fixing a versor $\boldsymbol{n}$ we obtain a new basis

$$|\boldsymbol{n}_+(\theta, \phi)\rangle = +\cos\frac{\theta}{2}\mathrm{e}^{-i\frac{\phi}{2}}|+\rangle + \sin\frac{\theta}{2}\mathrm{e}^{+i\frac{\phi}{2}}|-\rangle \tag{2.178}$$

$$|\boldsymbol{n}_-(\theta, \phi)\rangle = -\sin\frac{\theta}{2}\mathrm{e}^{-i\frac{\phi}{2}}|+\rangle + \cos\frac{\theta}{2}\mathrm{e}^{+i\frac{\phi}{2}}|-\rangle, \tag{2.179}$$

which describes polarization eigenvectors in an arbitrary direction, and recovers the reference one for $\theta, \phi = 0$. For the maximally mixed state we can then consider the following orthogonal decompositions

$$\rho = \frac{1}{2}\mathbb{1} \tag{2.180}$$

$$= \frac{1}{2}(P_+ + P_-) \tag{2.181}$$

$$= \frac{1}{2}(P_{\boldsymbol{n}_+(\theta,\phi)} + P_{\boldsymbol{n}_-(\theta,\phi)}), \tag{2.182}$$

as well as the nonorthogonal decomposition

$$\rho = \frac{1}{4\pi}\int_0^{2\pi} \mathrm{d}\phi \int_0^{\pi} \mathrm{d}\theta \, \sin\theta \, P_{\boldsymbol{n}_+(\theta,\phi)}, \tag{2.183}$$

in which the one-dimensional projections

$$P_{n_+(\theta,\phi)} = \begin{pmatrix} \cos^2\frac{\theta}{2} & \sin\frac{\theta}{2}\cos\frac{\theta}{2}e^{-i\phi} \\ \sin\frac{\theta}{2}\cos\frac{\theta}{2}e^{+i\phi} & \sin^2\frac{\theta}{2} \end{pmatrix} \qquad (2.184)$$

are generally not orthogonal for distinct values of θ and ϕ. The validity of the latter representation can be checked directly or obtained observing that the only state invariant under rotations must be proportional to the identity, with a proportionality factor fixed by the trace of the operator.

Indeed, we generally have an uncountable number of distinct convex decompositions in terms of pure states, not necessarily orthogonal between them. In dimension n, at most n orthogonal pure states are required, but more are allowed. To simply realize this fact consider an arbitrary statistical operator $\rho \in \mathcal{S}(\mathcal{H})$, together with any c.o.n.s. $\{\chi_i\}$, allowing to write the completeness relation

$$\mathbb{1} = \sum_i |\chi_i\rangle\langle\chi_i|. \qquad (2.185)$$

We can now consider the following chain of identities

$$\rho = \rho^{1/2}\mathbb{1}\rho^{1/2} \qquad (2.186)$$

$$= \rho^{1/2}\sum_i |\chi_i\rangle\langle\chi_i|\rho^{1/2} \qquad (2.187)$$

$$= \sum_i \|\rho^{1/2}\chi_i\|^2 \frac{\rho^{1/2}|\chi_i\rangle}{\|\rho^{1/2}\chi_i\|} \frac{\langle\chi_i|\rho^{1/2}}{\|\rho^{1/2}\chi_i\|} \qquad (2.188)$$

$$= \sum_i \lambda_i |\psi_i\rangle\langle\psi_i|, \qquad (2.189)$$

where we have introduced the weights

$$\lambda_i = \|\rho^{1/2}\chi_i\|^2 = \langle\chi_i|\rho\chi_i\rangle, \qquad (2.190)$$

which are positive and sum up to one, and for each non zero λ_i we have defined the vectors

$$\psi_i = \frac{\rho^{1/2}\chi_i}{\sqrt{\lambda_i}}, \qquad (2.191)$$

which are of norm one but generally nonorthogonal. This simple construction shows that the state space is not a simplex, that is, there is no unique convex decomposition of a state in terms of extreme states, at variance with the classical state considered in Sect. 1.2.

2.4 Schrödinger-Gisin-HJW Theorem

We here introduce a result originally due to Schrödinger [11], and later further considered and enriched in its meaning in [12, 13] (see [14] for an history of this result), which connects the possible orthogonal and nonorthogonal finite decompositions of a state with finite-dimensional range. This result will be further exploited in Theorem 3.6 to prove non uniqueness of the Kraus representation. In Theorem 4.1 of Sect. 4.3.1 we will further show how these decompositions can be related to preparation schemes obtained performing different measurements on other correlated degrees of freedom, providing an alternative formulation of the result. For the complete characterization of all the decompositions of a given state we refer the reader to [15].

Theorem 2.2 (Schrödinger, 1936) *Let us consider a state $\rho \in S(\mathcal{H})$ admitting the spectral decomposition*

$$\rho = \sum_{i=1}^{d} p_i |\varphi_i\rangle\langle\varphi_i|, \tag{2.192}$$

with $\{\varphi_i\}$ a set of normalized vectors orthogonal to each other in $\mathcal{H}$, and the weights $\{p_i\}$ providing a probability distribution, namely $p_i \geqslant 0$ and $\sum_i^d p_i = 1$. Then we can write

$$\rho = \sum_{k=1}^{M} \mu_k |\psi_k\rangle\langle\psi_k|, \tag{2.193}$$

with $\mu_k \geqslant 0$, $\sum_k^M \mu_k = 1$ and $\{\psi_k\}$ a collection of $M \geqslant d$ vectors of norm one, not necessarily orthogonal to each other, iff there exists a $M \times M$ unitary matrix U such that

$$\psi_k = \frac{1}{\sqrt{\mu_k}} \sum_{i=1}^{d} U_{ki} \sqrt{p_i} \varphi_i. \tag{2.194}$$

Proof The *if* statement is proven by direct inspection. Let us assume (2.194). Then according to (2.193) we have the following representation of the considered state

$$\rho = \sum_{k=1}^{M} \sum_{i,j=1}^{d} U_{ki} \sqrt{p_i} |\varphi_i\rangle \langle \varphi_j| \sqrt{p_j} U_{kj}^{*}, \tag{2.195}$$

and thanks to unitarity

$$\sum_{k=1}^{M} U_{kj}^{*} U_{ki} = (U^{\dagger} U)_{ji} = \delta_{ij} \tag{2.196}$$

we obtain (2.192).

In order to prove the *only if* statement let us define the matrix elements

$$U_{ki} = \sqrt{\frac{\mu_k}{p_i}} \langle \varphi_i | \psi_k \rangle, \tag{2.197}$$

which provide the first d columns of an $M \times M$ matrix and warrant (2.194) as immediately follows from direct inspection, thanks to the fact that $\{\varphi_i\}$ is a c.o.n.s. and thus obeys the completeness relation (2.185). To verify that these columns of length M obtained from (2.197) for $i = 1, \ldots, d$ can be completed to form an $M \times M$ unitary matrix we check their orthogonality. We have

$$\sum_{k=1}^{M} U_{kj}^{*} U_{ki} = \frac{1}{\sqrt{p_i p_j}} \langle \varphi_i | \left(\sum_{k=1}^{M} \mu_k |\psi_k\rangle \langle \psi_k| \right) \varphi_j \rangle, \tag{2.198}$$

so that thanks to (2.193) and the fact that the representation (2.192) is orthogonal we indeed verify (2.196). $\square$

The proven relations fix the freedom available in expressing a statistical operator as convex linear combination of pure states. The vectors appearing in any such decomposition must be linearly dependent on the eigenvectors of the considered statistical operator.

The existence of infinitely many decompositions, orthogonal or nonorthogonal, for the same statistical operator shows that these decompositions do not have a direct physical meaning. Indeed, they cannot be unveiled by any subsequent measurement performed on the system. The state is described but not characterized by an ensemble $\{\lambda_i, \psi_i\}$ obtained as in (2.189). We cannot describe a quantum state as arising by classical ignorance, quantified by the probability distribution of the eigenvalues, of a known set of alternative events described by extreme elements. Indeed, the decomposition is highly non-unique and moreover the different projections are generally nonorthogonal, so that they cannot be identified with exclusive alternatives. For a characterization of the different decompositions and their relationships see also Remark 2.4 and Theorem 4.1. The different decompositions describe or suggest different possible preparation procedures leading via mixing to the same state. Note, however, that these preparation procedures might even be incompatible, in the sense that they cannot be performed together with a single apparatus. To visualize this crucial feature it is enough to come back to the example of the maximally mixed polarization state in $\mathbb{C}^2$, describing an unpolarized beam. This beam can be obtained by equal mixture of beams polarized in opposite directions along the same axis. The specific choice of axis is irrelevant, and there is no way, by means of measurements performed on the final beam, to find along which axis the polarization was performed. We find here substantiated and exemplified the fact discussed at the beginning of this chapter that states have to be identified with equivalence classes of preparation procedures. In this case the considered preparations are in particular incompatible, in the sense that they cannot be obtained by a single apparatus, because this would amount to prepare common eigenvectors of non-commuting observables, such as the polarization in different directions, with a single apparatus.

2.2.6 Effects

We have thus identified statistical operators as the mathematical representatives of equivalence classes of preparation procedures. As discussed in Sect. 1.2, to complete the statistical picture we need to identify the space of observables and the relevant probability formula. We therefore move on to characterize the mathematical representatives of equivalence classes of registration procedures, essentially following [16]. In the spirit of Sect. 2.1 we first consider registrations or measurements corresponding to the simplest events, corresponding to dichotomic alternatives. Such apparata provide yes-no answers to fixed questions, e.g. whether a certain measured quantity takes values in a given interval. Such elementary events or basic building blocks of an observable can be naturally identified with maps associating to any state the probability of a positive answer

$$\mathsf{E} : \mathcal{S}(\mathcal{H}) \rightarrow [0, 1]. \tag{2.199}$$

As we already stressed the space state is convex, so that a mixture of states is again a state, and in particular a state may be represented and prepared according to many different convex decompositions. To ensure coherence of the predictions we therefore have to ask that such maps preserve convex mixtures, i.e. they are affine, namely

$$\mathsf{E}(\lambda\rho_1 + (1 - \lambda)\rho_2) = \lambda\mathsf{E}(\rho_1) + (1 - \lambda)\mathsf{E}(\rho_2) \tag{2.200}$$

for any $0 \leqslant \lambda \leqslant 1$ and any pair of states ρ_1 and ρ_2. It is important to stress that convex combinations of preparation procedures is all that can be obtained in a laboratory. A statistical experiment consists of a large number N of repetitions under the very same experimental conditions. If a preparation procedure identifies an apparatus for the preparation of a certain state of the system, considering a convex combination amounts to use different apparata for fractions of the number of repetitions, e.g. λN times the apparatus preparing the system in the state ρ_1 and $(1 - \lambda)N$ times the apparatus preparing the system in the state ρ_2. We call effects such affine maps from $\mathcal{S}(\mathcal{H})$ to the interval $[0, 1]$, and we now show that they can be identified with the positive bounded operators in the interval between the zero and the identity operator. Let us consider a self-adjoint operator $B \in \mathcal{B}(\mathcal{H})$. As we have seen in Sect. 2.2.2, thanks to $\mathcal{B}(\mathcal{H}) = \mathcal{T}'(\mathcal{H})$ each bounded operator identifies a linear functional on the space of trace class operators, so that the assignment

$$\rho \mapsto \mathrm{Tr}\{\rho B\} \tag{2.201}$$

provides an affine functional on the convex set of states. Noting that for $\rho = P_\psi$ the identity $\mathrm{Tr}\{\rho B\} = \langle \psi | B \psi \rangle$ holds, we have that $\mathrm{Tr}\{\rho B\} \geqslant 0$ for all ρ iff $B \geqslant 0$. In the same way we have $\mathrm{Tr}\{\rho B\} \leqslant 1$ for all ρ iff $1 - \mathrm{Tr}\{\rho B\} = \mathrm{Tr}\{\rho(\mathbb{1} - B)\} \geqslant 0$, so that $B \leqslant \mathbb{1}$. Therefore, a map complying with (2.199) and (2.200) is obtained via the assignment

$$\rho \mapsto \mathrm{Tr}\{\rho E\} \tag{2.202}$$

for $E \in \mathcal{B}(\mathcal{H}), 0 \leqslant E \leqslant \mathbb{1}$. Actually, any map complying with (2.199) and (2.200) can be written in this form, as we now show.

Theorem 2.3 *Given an affine map*

$$\mathsf{E} : \mathcal{S}(\mathcal{H}) \to [0, 1] \tag{2.203}$$
$$\rho \mapsto \mathsf{E}(\rho)$$

there exists a bounded self-adjoint operator E such that the representation

$$\mathsf{E}(\rho) = \mathrm{Tr}\{\rho E\} \tag{2.204}$$

holds with $0 \leqslant E \leqslant \mathbb{1}$.

Proof Let us first give the idea of the proof. We extend the affine functional E to a linear bounded functional on the whole linear space $\mathcal{T}(\mathcal{H})$ to exploit the duality relation $\mathcal{B}(\mathcal{H}) = \mathcal{T}'(\mathcal{H})$ and therefore identify the map with a bounded operator. This uniquely identified operator then has to lie between zero and one, to comply with the range of the functional. To extend E to a uniquely defined linear map on $\mathcal{T}(\mathcal{H})$ we proceed in three steps. We first extend the map from the convex set of states $\mathcal{S}(\mathcal{H})$ to the cone of positive operators $\mathcal{T}_+(\mathcal{H})$, then to the set of self-adjoint operators $\mathcal{T}_s(\mathcal{H})$, and finally to the whole linear space $\mathcal{T}(\mathcal{H})$. To avoid confusion we use a different label for each of the extended maps as follows

$$\mathsf{E}_{|\mathcal{S}(\mathcal{H})} \rightsquigarrow \mathsf{E}_{+|\mathcal{T}_+(\mathcal{H})} \rightsquigarrow \mathsf{E}_{r|\mathcal{T}_s(\mathcal{H})} \rightsquigarrow \mathsf{E}_{c|\mathcal{T}(\mathcal{H})}. \tag{2.205}$$

We start with the affine functional E defined on the convex set $\mathcal{S}(\mathcal{H})$, and extend it to a functional E_+ defined on the positive cone $\mathcal{T}_+(\mathcal{H})$ by setting $\mathsf{E}_+(0) = 0$ together with

$$\mathsf{E}_+(T) = \mathrm{Tr}\{T\}\,\mathsf{E}\left(\frac{T}{\mathrm{Tr}\{T\}}\right), \tag{2.206}$$

for $T > 0$. This functional is homogeneous with respect to multiplication by a positive scalar s due to

$$\mathsf{E}_+(sT) = \mathrm{Tr}\{sT\}\,\mathsf{E}\left(\frac{sT}{\mathrm{Tr}\{sT\}}\right) \tag{2.207}$$

$$= s\,\mathsf{E}_+(T). \tag{2.208}$$

It is moreover additive since exploiting the fact that E is affine we have for $S, T \in \mathcal{T}_+(\mathcal{H})$

$$\mathsf{E}_+(S+T) = \mathrm{Tr}\{S+T\}\left(\frac{S+T}{\mathrm{Tr}\{S+T\}}\right) \tag{2.209}$$

$$= \mathrm{Tr}\{S+T\}\,\mathsf{E}\left(\frac{\mathrm{Tr}\{S\}}{\mathrm{Tr}\{S+T\}}\frac{S}{\mathrm{Tr}\{S\}} + \frac{\mathrm{Tr}\{T\}}{\mathrm{Tr}\{S+T\}}\frac{T}{\mathrm{Tr}\{T\}}\right) \tag{2.210}$$

$$= \mathrm{Tr}\{S\}\,\mathsf{E}\left(\frac{S}{\mathrm{Tr}\{S\}}\right) + \mathrm{Tr}\{T\}\,\mathsf{E}\left(\frac{T}{\mathrm{Tr}\{T\}}\right) \tag{2.211}$$

$$= \mathsf{E}_+(S) + \mathsf{E}_+(T). \tag{2.212}$$

To extend E_+ to $\mathcal{T}_s(\mathcal{H})$ we note that each bounded self-adjoint operator can be expressed as the sum of positive and negative parts, according to

$$T = T^+ - T^- \tag{2.213}$$

with

$$T^+ = \frac{1}{2}(|T| + T) \tag{2.214}$$

$$T^- = \frac{1}{2}(|T| - T), \tag{2.215}$$

implying in particular

$$|T| = T^+ + T^-. \tag{2.216}$$

The expression (2.213) is called Hahn-Jordan decomposition, and we have in particular that the supports of T^+ and T^- are orthogonal. We thus define for $T \in \mathcal{T}_s(\mathcal{H})$

$$E_r(T) = E_+(T^+) - E_+(T^-). \tag{2.217}$$

With this definition the functional is homogeneous with respect to multiplication by a real scalar and additive within $\mathcal{T}_s(\mathcal{H})$. As a last step we extend to the whole $\mathcal{T}(\mathcal{H})$ noting that each $T \in \mathcal{T}(\mathcal{H})$ can be written as linear combination of two self-adjoint operators according to $T = T_R + iT_I$ with

$$T_R = \frac{1}{2}(T + T^\dagger) \tag{2.218}$$

$$T_I = \frac{1}{2i}(T - T^\dagger). \tag{2.219}$$

We can thus set for $T \in \mathcal{T}(\mathcal{H})$

$$E_c(T) = E_r(T_R) + iE_r(T_I). \tag{2.220}$$

Again we could check that the definition makes the functional homogeneous with respect to multiplication by a complex scalar and additive within $\mathcal{T}(\mathcal{H})$. This linear functional on $\mathcal{T}(\mathcal{H})$ is further bounded. Indeed, according to (2.214) and (2.215), as well as (2.218) and (2.219), we can express any operator $T \in \mathcal{T}(\mathcal{H})$ as linear combination of positive operators

$$T = T_R^+ - T_R^- + iT_I^+ - iT_I^-. \tag{2.221}$$

Due to linearity we then have

$$|E_c(T)| \leqslant |E_c(T_R^+)| + |E_c(T_R^-)| + |E_c(T_I^+)| + |E_c(T_I^-)|, \tag{2.222}$$

and further using the action (2.206) of the map on positive operators together with $E(T/\operatorname{Tr}\{T\}) \leqslant 1$, i.e. the boundedness of the original functional E,

$$|E_c(T)| \leqslant \operatorname{Tr}\{T_R^+\} + \operatorname{Tr}\{T_R^-\} + \operatorname{Tr}\{T_I^+\} + \operatorname{Tr}\{T_I^-\}, \tag{2.223}$$

which thanks to (2.216), together with (2.218) and (2.219), leads to

$$|E_c(T)| \leqslant \frac{1}{2}\operatorname{Tr}|T + T^\dagger| + \frac{1}{2}\operatorname{Tr}|T - T^\dagger|. \tag{2.224}$$

Finally recalling the expression of the trace norm (2.46) and using the triangle inequality we come to

$$|E_c(T)| \leqslant 2\|T\|_1, \tag{2.225}$$

expressing continuity of the introduced functional. According to the duality relation $\mathcal{B}(\mathcal{H}) = \mathcal{T}'(\mathcal{H})$ this bounded functional, which we now denote again with E for simplicity, can be identified with an operator in $\mathcal{B}(\mathcal{H})$, let us call it E, via

$$\mathsf{E}(T) = \ \mathrm{Tr}\{TE\} \quad \forall T \in \mathcal{T}(\mathcal{H}), \tag{2.226}$$

and since the functional restricted to $\mathcal{S}(\mathcal{H})$ has to take values in $[0, 1]$, as shown above we must have $0 \leqslant E \leqslant \mathbb{1}$. We stress in particular that different operators do correspond to different effects. $\square$

We are now in the position to identify the set which provides the mathematical representatives of equivalence classes of elementary registration procedures, corresponding to yes-no answers in the measurement, which we call set of effects

$$\mathcal{E}(\mathcal{H}) = \ \{E \in \mathcal{B}(\mathcal{H}), 0 \leqslant E \leqslant \mathbb{1}\}. \tag{2.227}$$

We will use the term effect to denote such operators, with which the originally introduced affine maps can be identified. Even in the finite-dimensional case we have the following strict inclusions

$$\mathcal{P}(\mathcal{H}) \subset \mathcal{E}(\mathcal{H}) \subset \mathcal{B}(\mathcal{H}), \tag{2.228}$$

clarifying that orthogonal projections are only a subset of the set of effects. The name effect arises as the translation of the name *Effekt* originally introduced by Ludwig to indicate these operators, which entered the stage of the description of measurements in quantum mechanics starting from the 60' [1, 17]. In correspondence to the fact that registration procedures can be mixed in the laboratory, the set of effects is convex, namely

$$E_1, E_2 \in \mathcal{E}(\mathcal{H}), \ \ 0 \leqslant \mu \leqslant 1 \Rightarrow \ \mu E_1 + (1 - \mu)E_2 \in \mathcal{E}(\mathcal{H}), \tag{2.229}$$

as visualized in Fig. 2.2. We further have that $E \in \mathcal{E}(\mathcal{H})$ implies $E^{\perp} = (\mathbb{1} - E) \in \mathcal{E}(\mathcal{H})$, so that an effect uniquely identifies the complementary one. Importantly, the usual sum of linear operators defines a partial operation in $\mathcal{E}(\mathcal{H})$, since the sum of two effects E_1, E_2 is defined provided $E_1 + E_2 \leqslant \mathbb{1}$. Effects which can be summed do not necessarily commute, as exemplified in Remark 2.5.

2.5 Effects and Commutativity

To mark the difference between orthogonal projections and effects in the description of yes-no measurements let us consider two orthogonal states $\varphi_{1,2}$ together with the projections $P_1 = P_{\varphi_1}$ and $P_2 = P_{(\varphi_1 + \varphi_2)/\sqrt{2}}$, defined according to (2.29). We have in particular $[P_1, P_2] \neq 0$, so that $P_1 + P_2$ is no more a projection. On the other hand, setting $E_1 = \frac{1}{2}P_1$ and $E_2 = \frac{1}{2}P_2$, we still have $[E_1, E_2] \neq 0$, but $\|E_1 + E_2\| \leqslant 1$, so that $E_1 + E_2$ still is an effect identifying a valid measurement.

Within the set $\mathcal{E}(\mathcal{H})$ projections do have a special role, in that they coincide with the extreme points, namely

$$\text{extreme}(\mathcal{E}(\mathcal{H})) = \mathcal{P}(\mathcal{H}) \supseteq \mathcal{P}_1(\mathcal{H}) \tag{2.230}$$

in the notation of (2.75).

Proof Due to the spectral theorem each effect can be represented as a convex mixture of projections, so that only projections can be extreme points. On the other hand, suppose an orthogonal projection is not extreme, so that it can be demixed in a nontrivial way in terms of effects, that is $P \in \mathcal{E}(\mathcal{H}) \cap \mathcal{P}(\mathcal{H})$, so that $P = P^2$, and

$$P = \lambda E_1 + (1 - \lambda)E_2, \tag{2.231}$$

with $E_1, E_2 \in \mathcal{E}(\mathcal{H})$ and $0 < \lambda < 1$. Taking $\varphi \in \mathcal{M}_P^{\perp}$, that is, in the orthogonal complement of the eigenspace of P, so that $P\varphi = 0$, we have

$$0 = \lambda\langle\varphi|E_1\varphi\rangle + (1 - \lambda)\langle\varphi|E_2\varphi\rangle \geqslant \lambda\langle\varphi|E_1\varphi\rangle \geqslant 0, \tag{2.232}$$

so that $E_1\varphi = 0 \ \forall\varphi \in \mathcal{M}_P^{\perp}$. Similarly $\forall\psi \in \mathcal{M}_P$ we have, exploiting

$$\mathbb{1} - P = \lambda(\mathbb{1} - E_1) + (1 - \lambda)(\mathbb{1} - E_2), \tag{2.233}$$

the chain of inequalities

$$0 = \lambda(1 - \langle\psi|E_1\psi\rangle) + (1 - \lambda)(1 - \langle\psi|E_2\psi\rangle) \geqslant \lambda(1 - \langle\psi|E_1\psi\rangle) \geqslant 0, \tag{2.234}$$

implying $E_1\psi = \psi$, so that E_1 and P coincide on a complete set of vectors and therefore $E_1 = P$. $\qquad\square$

Note that positivity and constraint on the range of E are enough for (2.204) to provide a well-defined probability formula, giving the frequency with which a preparation corresponding to the state ρ triggers the registration corresponding to the effect E. We have as required for the probabilistic interpretation $0 \leqslant \text{Tr}\{\rho E\} \leqslant 1$, while idempotency of E is not necessary, and generally does not apply. Idempotency is only recovered for the extreme points of the convex set of effects, that as we have shown are to be identified with orthogonal projections. Projections thus represent the most refined measurement apparata, that allow by mixing to recover any other effect, while the opposite is not true. A projective measurement cannot be obtained by mixing measurement apparata described by effects. The set of effects in $\mathbb{C}^2$ is described in Remark 2.6.

2.6 Effects in $\mathcal{H} = \mathbb{C}^2$

For the sake of example, we can characterize effects in the simplest finite-dimensional Hilbert space, namely $\mathcal{H} = \mathbb{C}^2$, relevant for the description of qubits. A generic operator on this space can be written in terms of the Pauli operators as

$$E = \frac{1}{2}(\varepsilon \mathbb{1} + \boldsymbol{e} \cdot \boldsymbol{\sigma}), \tag{2.235}$$

with $\boldsymbol{e} \cdot \boldsymbol{\sigma} = \sum_{i=x,y,z} e_i \sigma_i$. For $\varepsilon \in \mathbb{R}$ and $\boldsymbol{e} \in \mathbb{R}^3$ the operator is self-adjoint, with eigenvalues $\frac{1}{2}(\varepsilon \pm \|\boldsymbol{e}\|)$. Due to the constraint $0 \leqslant E \leqslant \mathbb{1}$ the eigenvalues have to lie between 0 and 1 leading to $\|\boldsymbol{e}\| \leqslant 1, 0 \leqslant \varepsilon \leqslant 2$, together with the constraint

$$\|\boldsymbol{e}\| \leqslant \min\{\varepsilon, 2 - \varepsilon\}. \tag{2.236}$$

A visualization of the convex set of effects in this space is given in Fig. 2.4.

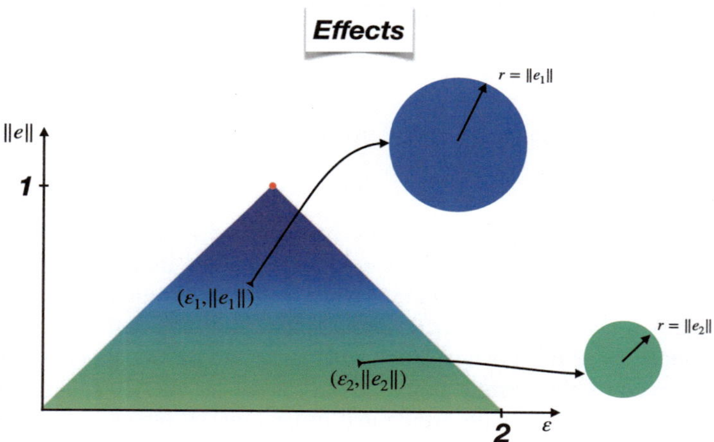

Fig. 2.4 Pictorial representation of the convex space of effects for $\mathcal{H} = \mathbb{C}^2$. The triangle visualizes the constraint (2.236). According to (2.235) to each point $(\varepsilon, \|e\|)$ of this triangle is associated the surface of a sphere with radius $r = \|e\|$. In this notation the mixture of two effects is obtained by mixing the parameters appearing in their definition. The set of extreme effects is given by the surface of the sphere of radius one associated to the red point on the top of the triangle

2.2.7 Generalized Probability Measures

According to Sects. 2.2.1 and 2.2.6 the most general preparation procedure is given by a statistical operator $\rho \in \mathcal{S}(\mathcal{H})$, and we have consequently identified the set of effects $E \in \mathcal{E}(\mathcal{H})$ as describing the most general yes-no elementary measurement. The key fact to come to this conclusion was the duality relation between the linear spaces in which these two convex sets are contained, namely $\mathcal{T}(\mathcal{H})$ and $\mathcal{B}(\mathcal{H})$ respectively. A natural question is whether we can take the reverse route. Taking effects as the space of basic events to be observed in a measurement, with its partial sum operation, we can ask how to obtain a probability measure defined on this set. Indeed, a state can be naturally identified with a probability measure on the space of observables, in particular it has to assign to each elementary event the probability of its occurrence. Let us formalize this statement introducing the notion of generalized probability measure. We say that

$$\begin{aligned} \nu : \mathcal{E}(\mathcal{H}) &\to [0, 1] \\ E &\mapsto \nu(E) \end{aligned} \tag{2.237}$$

is a generalized probability measure on the set of effects provided

1. $0 \leqslant \nu(E) \leqslant 1$
2. $\nu(\mathbb{1}) = 1$
3. $\nu\left(\sum_{i=1}^{n} E_i\right) = \sum_{i=1}^{n} \nu(E_i)$ whenever $\sum_{i=1}^{n} E_i \leqslant \mathbb{1}$, so that $\sum_{i=1}^{n} E_i$ is also an effect.

It is an important result that this generalized probability measure can be identified with a statistical operator [18, 19].

Theorem 2.4 (Busch, 2003) *Let ν be a generalized probability measure on the set of effects $\mathcal{E}(\mathcal{H})$, then there exists a statistical operator $\rho \in \mathcal{S}(\mathcal{H})$ such that $\nu(E) = \nu_\rho(E) = \mathrm{Tr}\{\rho E\}$ for all $E \in \mathcal{E}(\mathcal{H})$.*

The key point is again a duality relation. While $\mathcal{T}'(\mathcal{H}) = \mathcal{B}(\mathcal{H})$, in the infinite-dimensional case we have the strict inclusion $\mathcal{B}'(\mathcal{H}) \supset \mathcal{T}(\mathcal{H})$. The set $\mathcal{T}(\mathcal{H})$ can however be identified with the linear functionals on $\mathcal{B}(\mathcal{H})$ which besides being continuous are also normal, which is yet another regularity condition [20], that we shall discuss in Remark 3.5. The idea is therefore to show that ν can be extended as a linear functional to the whole linear space $\mathcal{B}(\mathcal{H}) \supset \mathcal{E}(\mathcal{H})$, using the same strategy as in the proof of Theorem 2.3. This functional can be further shown to be continuous and normal, so that it can be identified with an element of $\mathcal{T}(\mathcal{H})$, and the restriction of its range to the interval $[0, 1]$ implies that this element is indeed a statistical operator.

We might further wonder whether restricting the generalized probability measure to act on the set of projections, which is only a subset of the set of effects, as in the more standard formulation of quantum mechanics, leads to consider more general states than statistical operators. The answer is negative, and is due to a famous theorem by Gleason [21], whose original proof is notoriously difficult. Also in this case let us formalize this statement introducing the notion of measure on the non-commutative lattice of projections. We say that

$$\mu : \mathcal{P}(\mathcal{H}) \to [0, 1] \tag{2.238}$$
$$P \mapsto \mu(P)$$

is a generalized probability measure on the space of projections provided

1. $0 \leqslant \mu(P) \leqslant 1$
2. $\mu(\mathbb{1}) = 1$
3. $\mu\left(\sum_{i=1}^n P_i\right) = \sum_{i=1}^n \mu(P_i)$ whenever $P_i P_j = 0$ for $i \neq j$, that is the projections are orthogonal, so that $\sum_{i=1}^n P_i$ is also a projection.

It turns out that if $\dim \mathcal{H} \geqslant 3$ such a generalized probability measure can still be identified with a statistical operator.

Theorem 2.5 (Gleason, 1957) *Let μ be a generalized probability measure on the set of projections $\mathcal{P}(\mathcal{H})$ and $\dim \mathcal{H} \geqslant 3$, then there exists a statistical operator $\rho \in \mathcal{S}(\mathcal{H})$ such that $\mu(P) = \mu_\rho(P) = \mathrm{Tr}\{\rho P\}$ for all $P \in \mathcal{P}(\mathcal{H})$.*

These results show that statistical operators indeed provide the most general expression for the state of a quantum system. We recall that a state has to be understood as the mathematical representative of an equivalence class of preparation procedures, that can be realized in a laboratory and lead to the very same statistics for any subsequent observation. This statistics has to be obtained via the probability formula (2.35).

2.2.8 Dispersion-Free States

To better grasp the difference between classical and quantum probability, as well as between projections and effects, let us discuss the issue of dispersion-free states. We say that a state in a statistical theory is dispersion-free provided the variance of any observable calculated with this state is zero, so that each outcome is predicted with certainty.

Consider first the classical framework discussed in Sect. 1.2. Taking $\mathbb{R}$ as measure space the extreme elements of the space of probability measures on $\mathbb{R}$ are those with support reduced to a point, be it x_0, and they can be formally expressed through a density given by a Dirac delta function with support in x_0. As we can immediately check, this measure has zero variance, so that this state assigns to each random variable a deterministic value, the statistical aspect of the description is thus completely washed out. For this class of extreme states no observable exhibits a nontrivial statistics. On the other hand, still within the classical description, consider a finite sample space $\Omega = \{1, \ldots, N\}$, so that as considered in Sect. 1.2.1 states are of the form $\boldsymbol{p} = (p_1, \ldots, p_N)$, with $p_i \geqslant 0$ and $\sum_{i=1}^{N} p_i = 1$, with extreme states $\{\mathbf{e}_j\}_{j=1,\ldots,N}$ determined as $(\mathbf{e}_i)_j = \delta_{ij}$. In this setting, we easily identify the basic elementary observables corresponding to projections and effects. Projections are N-dimensional vectors $\boldsymbol{\pi}$ whose entries π_j are either 0 or 1, while effects are given by N-dimensional vectors $\boldsymbol{\epsilon}$ whose entries ϵ_j satisfy $0 \leqslant \epsilon_j \leqslant 1$. As a result, if we consider generalized observables in the sense of effects, also in the classical case there are no dispersion-free states. For every state $\boldsymbol{\pi}$ we can find an effect $\boldsymbol{\epsilon}$ such that the probability μ of a positive outcome for the yes-no question corresponding to $\boldsymbol{\epsilon}$, given by $\mu = \sum_{j=1}^{N} \epsilon_j \pi_j$, gives a value strictly different from either 0 or 1. However, restricting to observables given by projections we do have dispersion-free states, coinciding with the extreme states $\{\mathbf{e}_j\}_{j=1,\ldots,N}$, indeed in this case the probability μ of a positive outcome only takes the values 0 or 1.

Let us now consider the quantum case. It immediately appears that taking effects for the description of elementary observations, there are no dispersion-free states. Indeed, given any state $\rho \in \mathcal{S}(\mathcal{H})$ we can consider the effect $E = \frac{1}{2}\rho$, and the probability formula gives the result $\mathrm{Tr}\{E\rho\} = \frac{1}{2}\mathrm{Tr}\{\rho^2\} < 1$, strictly less than one as follows from (2.77). One might still ask if there exist dispersion-free states in quantum mechanics when restricting to the set of projections. To show that this is not the case suppose there exist ρ such that

$$\langle A^2 \rangle = \langle A \rangle^2 \tag{2.239}$$

for any self-adjoint operator $A \in \mathcal{B}(\mathcal{H})$, or equivalently

$$\mathrm{Tr}\{\rho A^2\} = \mathrm{Tr}\{\rho A\}^2. \tag{2.240}$$

Of course, for an observable A with a non-empty point spectrum we can certainly find a state such that the dispersion of A is equal to zero, just by taking a pure state fixed by an eigenstate of A. The point is however whether the statistical description becomes irrelevant for all observables. To show that this is not the case consider $A = P_\phi$ for a certain one-dimensional projection P_ϕ. If ρ were dispersion-free then $\forall \phi \in \mathcal{H}$ we would have

$$\mathrm{Tr}\{\rho P_\phi\}^2 = \mathrm{Tr}\{\rho P_\phi^2\} = \mathrm{Tr}\{\rho P_\phi\}, \tag{2.241}$$

so that $\mathrm{Tr}\{\rho P_\phi\}$ is equal to either 0 or 1. Since $\mathrm{Tr}\{\rho P_\phi\} = \langle \phi | \rho \phi \rangle$, it can neither be equal to 0 for any ϕ, otherwise ρ should be the zero operator, nor equal to 1 for any ϕ, otherwise ρ should be the identity operator. Accordingly, there exists at least a vector $\psi \in \mathcal{H}$ such that $\langle \psi | \rho \psi \rangle = 1$. It follows that $\mathrm{Tr}\{\rho(1 - P_\psi)\} = 0$, and in particular $\mathrm{Tr}\{\rho P_\chi\} = 0 \; \forall \chi \perp \psi$. We can now consider the vector $\Phi = \cos\alpha \chi + \sin\alpha \psi$ such that

$$\mathrm{Tr}\{\rho P_\Phi\} = \sin^2\alpha + \sin\alpha\cos\alpha[\langle \chi | \rho \psi \rangle + \langle \psi | \rho \chi \rangle]. \tag{2.242}$$

This function is continuous in α and takes the value 0 for $\alpha = 0$, and 1 for $\alpha = \pi/2$, it therefore passes through intermediate points, leading to Hilbert space vectors for which the expectation value is neither 0 nor 1. We have thus shown that there are no dispersion-free states in quantum mechanics, so that the statistical aspect of the description is an intrinsic feature of quantum mechanics.

2.3　Observables

So far we have characterized the mathematical representative of equivalence classes of basic experiments with yes-no answer. In terms of these basic building blocks, we can now construct the general notion of observable with arbitrary outcome space. Let us start from a discrete setting and suppose that each distinct outcome is represented by an effect E_j corresponding to a yes-no signature in the measurement apparatus, e.g. a detection event in a specific area of a detector. Given a state ρ each outcome takes place with probability $\mathrm{Tr}\{\rho E_j\}$, and since we are considering distinct events we have

$$\sum_j \mathrm{Tr}\{\rho E_j\} \leqslant 1, \tag{2.243}$$

for any statistical operator ρ. If the inequality is not saturated we can always add a further effect $\mathbb{1} - \sum_j E_j$, corresponding to no event in the measurement apparatus, given that the preparation apparatus has fired. Indeed, we have seen that the linear sum of effects which still lies below the identity is also an effect, so that they can be observed together, as different outcomes of the same measurement apparatus. We thus assume without loss of generality that $\sum_j \mathrm{Tr}\{\rho E_j\} = 1$. Such an observable is therefore described by a collection of operators $\{E_j\}$ such that $0 \leqslant \mathrm{Tr}\{\rho E_j\} \leqslant 1$ and $\sum_j \mathrm{Tr}\{\rho E_j\} = 1$. Note that since these relations must hold for any state ρ, they uniquely identify the collection of effects summing up to the identity.

2.3.1　Positive Operator-Valued Measures

We now put this result in a general framework, allowing to consider any possible measurement outcome. Consider the outcome space of a given physical quantity, be it $(\Omega, \mathcal{F})$, where $\mathcal{F}$ is the σ-algebra of measurable outcomes in the sample space Ω. In the case of discrete or numerable outcomes, Ω will consist of a finite or denumerable set of real numbers. In the general case $\Omega = \mathbb{R}^n$ for a certain n, and $\mathcal{F}$ is given by the Borel σ-algebra of open sets $\mathcal{B}(\mathbb{R}^n)$. We define as positive operator-valued measure or probability operator-valued measure, abbreviated as POVM, a map

$$\begin{aligned} F : \mathcal{F} &\to \ \mathcal{E}(\mathcal{H}) \subset \mathcal{B}(\mathcal{H}) \\ M &\mapsto \ F(M) \end{aligned} \tag{2.244}$$

such that

1. $0 \leqslant F(M) \leqslant \mathbb{1}$ for any $M \in \mathcal{F}$
2. $F(\emptyset) = 0$, $F(\Omega) = \mathbb{1}$
3. $F\left(\bigcup_i M_i\right) = \sum_i F(M_i)$ in the weak topology for any sequence of disjoint sets in $\mathcal{F}$, i.e. $M_i \cap M_j = \emptyset$ for $i \neq j$.

These properties express the fact that a POVM takes values in the set of positive operators, between zero and unit so as to extract probabilities, and is σ-additive with respect to the dependence on the considered interval of outcomes. The latter property justifies the use of the word measure. The weak topology, corresponding to validity of the identities at the level of matrix elements of the operator, allows for a natural connection with classical scalar measures. Indeed, given an arbitrary $\rho \in \mathcal{S}(\mathcal{H})$ and an arbitrary POVM F, we immediately obtain a classical probability measure according to the formula

$$\mu_\rho^F(M) = \ \mathrm{Tr}\{\rho F(M)\}, \tag{2.245}$$

as granted by linearity and continuity of the trace operation. The classical measure μ_ρ^F provides the probability distribution of the observable associated to F given a preparation ρ, that is the distribution of the possible outcomes once the state is ρ. Otherwise stated, the map $F(\cdot)$ is a POVM provided the map $\mathcal{F} \ni M \to \mathrm{Tr}\{\rho F(M)\}$ is a probability measure for any $\rho \in \mathcal{S}(\mathcal{H})$. A simple example of POVM is given in Remark 2.7. To substantiate the fact that positive operator-valued measures do indeed provide the general notion of observable, let us introduce the convex set $\mathrm{Prob}(\Omega)$ of all probability measures on the outcome space $(\Omega, \mathcal{F})$. We now show that each map associating to any element ρ of the convex set of quantum states $\mathcal{S}(\mathcal{H})$ an element of the convex set of classical states $\mathrm{Prob}(\Omega)$, namely an observable compatible with the statistical interpretation of quantum mechanics, can be identified with a POVM. We formulate this statement as a theorem.

Theorem 2.6 *Given a POVM $F(\cdot)$ on the outcome space $(\Omega, \mathcal{F})$ the map*

$$\Phi_F : \mathcal{S}(\mathcal{H}) \to \ \mathrm{Prob}(\Omega) \tag{2.246}$$
$$\rho \mapsto \ \mathrm{Tr}\{\rho F(\cdot)\}$$

provides an affine map from the convex set of quantum states to the convex set of probability distributions over $(\Omega, \mathcal{F})$. Vice versa, any such affine map can be realized as $\Phi = \Phi_F$ for some POVM $F(\cdot)$.

Proof Given a positive operator-valued measure $F(\cdot)$ we immediately check that for any ρ the measure $\mu_\rho^F(M) = \mathrm{Tr}\{\rho F(M)\}$ belongs to $\mathrm{Prob}(\Omega)$, and thanks to linearity of the trace operation the map is affine in its dependence on ρ. To show that the converse holds, suppose we are given an affine map

$$\Phi : \mathcal{S}(\mathcal{H}) \;\to\; \mathrm{Prob}(\Omega) \tag{2.247}$$
$$\rho \;\mapsto\; \Phi[\rho](\cdot)$$

where each $\Phi[\rho](\cdot)$ is a probability distribution over Ω. That is to say $0 \leqslant \Phi[\rho](M) \leqslant 1$, $\Phi[\rho](\emptyset) = 0$, $\Phi[\rho](\Omega) = 1$, together with the proper additive behavior $\Phi[\rho]\left(\bigcup_i M_i\right) = \sum_i \Phi[\rho](M_i)$ for disjoint sets. For fixed $M \in \mathcal{F}$ we can consider the map

$$\Phi[\cdot](M) : \mathcal{S}(\mathcal{H}) \;\to\; [0, 1] \tag{2.248}$$
$$\rho \;\mapsto\; \Phi[\rho](M)$$

which due to the properties of Φ is also affine in its dependence on ρ, and takes values in the interval $[0, 1]$. According to Theorem 2.3 such a map can be represented as an effect, let us call it $F(M)$ since it depends on M, by expressing it as $\Phi[\rho](M) = \mathrm{Tr}\{\rho F(M)\}$. It remains to show that the dependence on M is such as to make $F(\cdot)$ a positive operator-valued measure. Indeed, since $F(M)$ is an effect we already know that $0 \leqslant F(M) \leqslant \mathbb{1} \; \forall M \in \mathcal{F}$. Moreover since $\Phi[\rho](\cdot) \in \mathrm{Prob}(\Omega)$ we immediately have $F(\emptyset) = 0$ and $F(\Omega) = \mathbb{1}$. Let us now consider a sequence of disjoint sets in $\mathcal{F}$, i.e. $M_i \cap M_j = \emptyset$ for $i \neq j$. We have, since $\Phi[\rho](\cdot)$ is σ-additive

$$\mathrm{Tr}\left\{\rho F\left(\bigcup_i M_i\right)\right\} = \Phi[\rho]\left(\bigcup_i M_i\right) \tag{2.249}$$

$$= \sum_i \Phi[\rho](M_i) \tag{2.250}$$

$$= \mathrm{Tr}\left\{\rho \sum_i F(M_i)\right\}, \tag{2.251}$$

and therefore σ-additivity in the weak operator topology. $\qquad\square$

We thus take POVMs as the most general description of observables. To connect with the standard notion of observable as a self-adjoint operator let us consider the very special case in which $F(M) = F(M)^2 \; \forall M \in \mathcal{F}$, so that the measure actually takes values in the set of projections. We therefore define as projection-valued measure, abbreviated as PVM, a map

$$E : \mathcal{F} \;\to\; \mathcal{P}(\mathcal{H}) \subset \mathcal{B}(\mathcal{H}) \tag{2.252}$$
$$M \;\mapsto\; E(M)$$

such that

1. $0 \leqslant E(M) \leqslant \mathbb{1}$, $E^2(M) = E(M)$ for any $M \in \mathcal{F}$

2. $E(\emptyset) = 0$, $E(\Omega) = \mathbb{1}$
3. $E\left(\bigcup_i M_i\right) = \sum_i E(M_i)$ in the weak topology for any sequence of disjoint sets in $\mathcal{F}$, i.e. $M_i \cap M_j = \emptyset$ for $i \neq j$.

Observables which do correspond to POVMs which are in particular PVMs are called sharp observables. To stress the fact that the basic difference between POVMs and PVMs lies in idempotency and commutativity of the operators in the range, we recall, without proof, the equivalence of the following statements for an observable E with outcome space $(\Omega, \mathcal{F})$ [3,22]:

1. E is a PVM
2. $E(M)E(N) = E(M \cap N)$ for all $M, N \in \mathcal{F}$
3. $E(M)E(M^C) = 0$ for all $M \in \mathcal{F}$

where the second statement is known as multiplicativity and implies that the range of a PVM is given by commuting projections.

The crucial difference is therefore idempotency, which is not necessary in order to obtain a statistical description of the measurement outcomes in terms of a classical probability measure as in (2.245). It has however important consequences from the mathematical point of view. For the case of projection-valued measures we know that there is a one to one correspondence with self-adjoint operators. Indeed, given a projection-valued measure E^A we can construct a self-adjoint operator A according to

$$A = \int_{\mathbb{R}} x \, dE^A(x), \tag{2.253}$$

where we have assumed that the associated observable takes values in $\mathbb{R}$. An important point is that all moments of the probability distribution obtained from E^A once fixed a state ρ can be expressed directly in terms of this operator, which is therefore of great significance. For any fixed state ρ we have, denoting by $\mathrm{Mean}_\rho(E^A)$ the first moment of the classical probability distribution obtained from the projection-valued measure E^A,

$$\mathrm{Mean}_\rho(E^A) = \int_{\mathbb{R}} x \, d\mu_\rho^{E^A}(x) \tag{2.254}$$

$$= \mathrm{Tr}\{A\rho\}, \tag{2.255}$$

where $\mu_\rho^{E^A}(M) = \mathrm{Tr}\{\rho E^A(M)\}$. Thanks to the identity

$$f(A) = \int_{\mathbb{R}} f(x) \, dE^A(x), \tag{2.256}$$

so that in particular

$$A^2 = \int_{\mathbb{R}} x^2 dE^A(x),$$ (2.257)

we can further express the variance as

$$\text{Var}_\rho(E^A) = \int_{\mathbb{R}} x^2 d\mu_\rho^{E^A}(x) - \left(\int_{\mathbb{R}} x d\mu_\rho^{E^A}(x)\right)^2$$ (2.258)

$$= \text{Tr}\{A^2\rho\} - \text{Tr}\{A\rho\}^2.$$ (2.259)

Similarly, any moment or combination of moments of the statistical distribution can be directly expressed in terms of powers of the operator A. This is no more true in the case of a POVM. The integral

$$B^{(1)} = \int_{\mathbb{R}} x dF(x)$$ (2.260)

still generally defines an operator, but the latter is not necessarily self-adjoint, and even if this is the case the operator corresponding to the second moment,

$$B^{(2)} = \int_{\mathbb{R}} x^2 dF(x),$$ (2.261)

is generally not the square of the first $B^{(2)} \neq (B^{(1)})^2$. This opens the complicated moments problem [23]. As a general fact the focus remains on the positive operator-valued measure itself and cannot be simply shifted on a single operator.

As we discussed in Sect. 2.2.6, the set of effects is convex and its extreme elements do coincide with orthogonal projections. In this respect the most informative yes-no measurements are associated to the set of projections. This is no more true when considering the notion of observable in full generality. Indeed, the set of POVMs is still a convex set, but its extreme elements do not coincide with the set of PVMs [24].

2.7 POVM in $\mathcal{H} = \mathbb{C}^2$

For the sake of example let us consider an observable in the sense of a proper POVM in $\mathcal{H} = \mathbb{C}^2$. Making reference to (2.235) we introduce the operators

$$F(\{k\}) = \frac{1}{2}(\alpha_k \mathbb{1} + \boldsymbol{m}_k \cdot \boldsymbol{\sigma})$$ (2.262)

with $k = 1, \ldots, N$ with $N \geqslant 2$. According to (2.236) each of them is an effect if $\alpha_k \in \mathbb{R}$ and $\boldsymbol{m}_k \in \mathbb{R}^3$, together with $\|\boldsymbol{m}_k\| \leqslant \min\{\alpha_k, 2 - \alpha_k\}$. In order for the collection of operators $\{F(\{k\})\}_{k=1,\ldots,N}$ to be a POVM we further ask the normalization condition

$$\sum_{k=1}^{N} F(\{k\}) = \mathbb{1}, \tag{2.263}$$

implying $\sum_{k=1}^{N} \alpha_k = 2$, and $\sum_{k=1}^{N} \boldsymbol{m}_k = 0$, so that the three-dimensional vectors $\boldsymbol{m}_k$ are linearly dependent. The effect associated to two or more outcomes is fixed by additivity. A natural choice compatible with these conditions is obtained by taking $\alpha_k = 2/N$ for all k, together with a collection of coplanar vectors $\boldsymbol{m}_k$ uniformly distributed in the plane and of the same length, so that $(\boldsymbol{m}_k/\|\boldsymbol{m}_k\|) \cdot (\boldsymbol{m}_{k+1}/\|\boldsymbol{m}_{k+1}\|) = \cos(2\pi/N)$ for $k = 1, \ldots, N-1$ and $\|\boldsymbol{m}_j\| = 2/N$, as visualized in Fig. 2.5. The probability formula, if the preparation is described by a generic state ρ expressed as in (2.79), takes the form

$$P(\{k\}) = \mathrm{Tr}\{\rho F(\{k\})\} \tag{2.264}$$

$$= \mathrm{Tr}\left\{\frac{1}{2}(\mathbb{1} + \boldsymbol{r} \cdot \boldsymbol{\sigma})\frac{1}{2}(\alpha_k \mathbb{1} + \boldsymbol{m}_k \cdot \boldsymbol{\sigma})\right\} \tag{2.265}$$

$$= \frac{1}{N}(1 + \boldsymbol{r} \cdot \hat{\boldsymbol{m}}_k) \tag{2.266}$$

where $\boldsymbol{r}$ is a vector with norm less or equal to one, and $\hat{\boldsymbol{m}}_k = \boldsymbol{m}_k/\|\boldsymbol{m}_k\|$. In particular we have $P(\{k\}) \leqslant 2/N$. For the case $N = 2$ we recover a standard PVM composed of two orthogonal projections, while for $N \geqslant 3$ we have a proper POVM. Notice in particular that for $N \geqslant 3$ the outcome is never certain, whatever the state, as it should be corresponding to a single apparatus allowing for the measurement of spin along different directions. For this simple case we can also completely spell out the map Φ_F which associates to a quantum state a classical probability distribution according to (2.246). The sample space is given by $\Omega = \{1, \ldots, N\}$, so that an element $\boldsymbol{p} \in \mathrm{Prob}(\Omega)$ simply corresponds to a probability vector, and the map Φ_F explicitly acts as $\rho \mapsto \boldsymbol{p}$, with $p_k = \mathrm{Tr}\{\rho F(\{k\})\}$. In particular we have $\|\boldsymbol{p}\| \leqslant \sqrt{2/N}$, which implies that the image of Φ_F is not the whole set $\mathrm{Prob}(\Omega)$. For this choice of POVM the effects (2.262) can be simply written as

$$F(\{k\}) = \frac{2}{N} P_{\hat{\boldsymbol{m}}_k}, \tag{2.267}$$

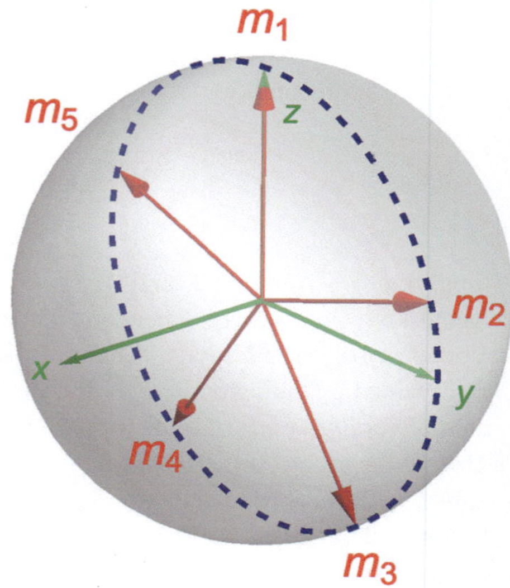

Fig. 2.5 Visualization of the directions associated to the considered POVM with N outcomes in the Hilbert space $\mathbb{C}^2$, for the case $N = 5$

where we have introduced the orthogonal projections along a direction identified by the versor $\boldsymbol{n}$ as

$$P_{\boldsymbol{n}} = \frac{1}{2}(\mathbb{1} + \boldsymbol{n} \cdot \boldsymbol{\sigma}). \tag{2.268}$$

In this case the single effects are proportional to projections along the given direction, with a multiplicative factor smaller than one. The probability formula (2.35) can therefore be written as

$$\mathrm{Tr}\{\rho F(\{k\})\} = \frac{2}{N}\frac{1}{2}(1 + \boldsymbol{r} \cdot \hat{\boldsymbol{m}}_k), \tag{2.269}$$

to be compared with the standard formula for a projective measurement

$$\mathrm{Tr}\{\rho P_{\boldsymbol{n}}\} = \frac{1}{2}(1 + \boldsymbol{r} \cdot \boldsymbol{n}) \tag{2.270}$$

which can attain the value one corresponding to a sharp statement.

2.3.1.1 Stern-Gerlach Experiment

To clarify how an unsharp spin measurement might naturally arise, let us consider the following more realistic description of the Stern-Gerlach experiment [25] depicted in Fig. 2.6. We consider as Hilbert space $\mathcal{H} = L^2(\mathbb{R}^3) \otimes \mathbb{C}^2$, where both center

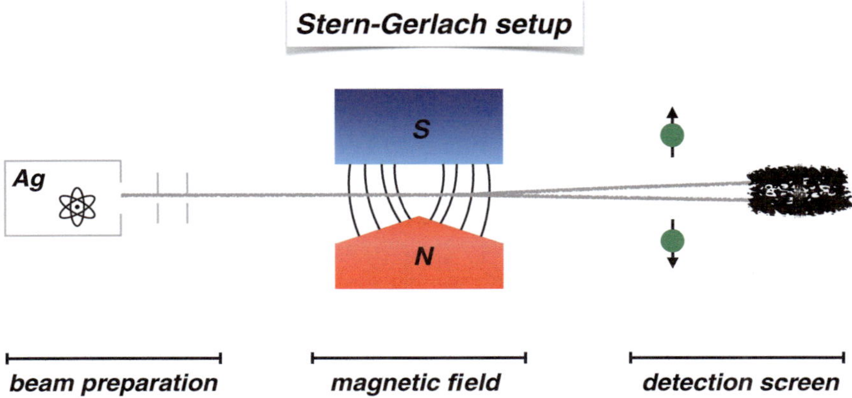

Fig. 2.6 Pictorial representation of the Stern-Gerlach experiment. The experimental setup is composed of three stages: the preparation of the beam by evaporation and collimation, the interaction with a inhomogeneous magnetic field coupling center of mass and internal degrees of freedom, the spatially resolved detection of the silver atoms leading to blurred lips

of mass and relevant internal degrees of freedom, namely magnetic moment, of the considered atom beam are described. We devise a preparation procedure made up of two steps as follows. In the first step we prepare an atom beam described by a pure state in which there is no correlation between center of mass and internal degrees of freedom, so that it can be written in the form

$$\Phi_{\text{in}} = \psi_{\text{in}} \otimes \varphi_{\text{in}}, \tag{2.271}$$

where $\psi_{\text{in}} \in L^2(\mathbb{R}^3)$ is a normalized wavepacket and φ_{in} is a state in $\mathbb{C}^2$

$$\varphi_{\text{in}} = \mu_\uparrow v_\uparrow + \mu_\downarrow v_\downarrow, \tag{2.272}$$

where the pair of vectors $\{v_\uparrow, v_\downarrow\}$ denotes a basis in $\mathbb{C}^2$, and the complex amplitudes $\{\mu_\uparrow, \mu_\downarrow\}$ obey $|\mu_\uparrow|^2 + |\mu_\downarrow|^2 = 1$, warranting also the correct normalization of Φ_{in}. In the second step the beam goes through an inhomogeneous magnetic field, impressing a direction in the spatial propagation of the atoms which depends on their magnetic moment. We describe this transformation through a unitary operation as follows

$$\Phi_{\text{out}} = U\Phi_{\text{in}} \tag{2.273}$$
$$= \mu_\uparrow \psi_\uparrow \otimes v_\uparrow + \mu_\downarrow \psi_\downarrow \otimes v_\downarrow, \tag{2.274}$$

where $\psi_\uparrow$ and $\psi_\downarrow$ denote the upward and downward deflected wavepackets, which are generally nonorthogonal $\langle \psi_\uparrow | \psi_\downarrow \rangle \neq 0$, though assumed to be normalized to one. These two steps combined together then characterize the preparation procedure leading to Φ_{out}, as depicted in Fig. 2.6, while the effect of free evolution was assumed negligible. We now suppose to observe on the overall system the following PVM, as described in (2.252), defined on the Hilbert space $L^2(\mathbb{R}^3) \otimes \mathbb{C}^2$ and with two outcomes

$$E(\{\uparrow\}) = P_\uparrow \otimes \mathbb{1}_{\mathbb{C}^2}, \tag{2.275}$$

and

$$E(\{\downarrow\}) = P_\downarrow \otimes \mathbb{1}_{\mathbb{C}^2}. \tag{2.276}$$

Here $\{P_\uparrow, P_\downarrow\} \in \mathcal{P}(L^2(\mathbb{R}^3))$ are orthogonal projections, corresponding to detection of the atom in the upper or lower half-plane of the final detection screen respectively. The probability to have a detection event in the upper half-plane thus becomes

$$P(\uparrow) = \mathrm{Tr}_{L^2(\mathbb{R}^3) \otimes \mathbb{C}^2}\{E(\{\uparrow\}) P_{\Phi_{\mathrm{out}}}\} \tag{2.277}$$

$$= \|E(\{\uparrow\})\Phi_{\mathrm{out}}\|^2_{L^2(\mathbb{R}^3) \otimes \mathbb{C}^2} \tag{2.278}$$

$$= |\mu_\uparrow|^2 \|P_\uparrow \psi_\uparrow\|^2_{L^2(\mathbb{R}^3)} + |\mu_\downarrow|^2 \|P_\uparrow \psi_\downarrow\|^2_{L^2(\mathbb{R}^3)}, \tag{2.279}$$

where we have used orthogonality of $v_\uparrow$ and $v_\downarrow$, and similarly for $P(\downarrow)$.

Crucially, due to lack of orthogonality between $\psi_\uparrow$ and $\psi_\downarrow$ the coefficients $\|P_\uparrow \psi_\uparrow\|^2_{L^2(\mathbb{R}^3)}$ and $\|P_\uparrow \psi_\downarrow\|^2_{L^2(\mathbb{R}^3)}$ are neither exactly one nor exactly zero. Their interpretation reads as follows. After going through the inhomogenous magnetic field the atoms depending on their magnetic moment will be deflected either upwards or downwards. The corresponding wavepackets $\psi_\uparrow$ and $\psi_\downarrow$, however, have by necessity support in the full space. The value of $\|P_\uparrow \psi_\uparrow\|^2_{L^2(\mathbb{R}^3)}$ will be close to one, corresponding to the fact that $\psi_\uparrow$ has its support concentrated in the upper half-plane. The tail of the associated probability distribution will however have a small but non-vanishing overlap with the lower half-plane, so that a few detection events associated to this half-plane do actually correspond to atoms whose magnetic moment is directed upwards. The broader the wavepacket, the higher the number of these detection events and therefore the factor $\|P_\uparrow \psi_\downarrow\|^2_{L^2(\mathbb{R}^3)}$. This situation corresponds to a limited confidence in associating the spatial detection event with the value of the magnetic moment. The probability of these two outcomes can be equivalently obtained from an observable with the same outcome set but defined on $\mathbb{C}^2$, that is to say relative to the internal degrees of freedom only, namely

$$F(\{\uparrow\}) = P_{v_\uparrow} \|P_\uparrow \psi_\uparrow\|^2_{L^2(\mathbb{R}^3)} + P_{v_\downarrow} \|P_\uparrow \psi_\downarrow\|^2_{L^2(\mathbb{R}^3)}, \tag{2.280}$$

and

$$F(\{\downarrow\}) = P_{v_\uparrow} \|P_\downarrow \psi_\uparrow\|^2_{L^2(\mathbb{R}^3)} + P_{v_\downarrow} \|P_\downarrow \psi_\downarrow\|^2_{L^2(\mathbb{R}^3)}, \tag{2.281}$$

which identifies a POVM according to (2.244). We have in particular

$$P(\uparrow) = \mathrm{Tr}_{\mathbb{C}^2}\{F(\{\uparrow\})P_{\varphi_{\mathrm{in}}}\}, \qquad (2.282)$$

with φ_{in} as in (2.272), and similarly for $P(\downarrow)$. We thus see that the PVM on the overall system does correspond to a POVM for the subsystem corresponding to the internal degrees of freedom only. Note that F is a proper POVM, in the sense that neither $F(\{\uparrow\})$ nor $F(\{\downarrow\})$ is a projection operator, unless $\|P_\uparrow \psi_\uparrow\|^2_{L^2(\mathbb{R}^3)}$ and $\|P_\uparrow \psi_\downarrow\|^2_{L^2(\mathbb{R}^3)}$ are exactly equal to one and zero respectively. In this case, which corresponds to a sharp localization in the upper or lower half-plane, the observable becomes a PVM.

As discussed above the unsharp situation corresponds to realistically taking into account the width of the wavepackets. A sharp observable on the other hand corresponds to a situation in which we can perfectly discriminate among the two magnetic moments, and the pattern on the screen is given by two sharp lines, with no detection of atoms in the intermediate region.

2.3.2 Naimark's Dilation

The considered treatment of the Stern-Gerlach experiment is in a sense paradigmatic, showing how a POVM measurement can naturally arise when the degrees of freedom of the system of interest are indirectly measured by coupling to other degrees of freedom, which we will call external. It provides an example of indirect measurement. In the above considered example, the degrees of freedom of the system describe the atom's magnetic moment, while the external degrees of freedom are associated to the center of mass of the atom. A projective measurement on the external degrees of freedom leads to a POVM for the statistics of an observation on the system only, keeping into account the noise induced by the interaction with the external degrees of freedom. In this framework, a PVM generally becomes a POVM. Indeed, let us consider a PVM on the external degrees of freedom, say $E_E(\cdot)$ as in (2.252) with outcome space $(\Omega, \mathcal{F})$, a factorized initial state $\rho_S \otimes \rho_E$, a coupling described by a unitary operator U_{SE}. We then have for the distribution of the outcomes

$$P(M) = \mathrm{Tr}_S \mathrm{Tr}_E\{U_{SE}(\rho_S \otimes \rho_E)U_{SE}^\dagger(\mathbb{1}_S \otimes E_E(M))\} \qquad (2.283)$$

$$= \mathrm{Tr}_S\{\rho_S \mathrm{Tr}_E\{\rho_E U_{SE}^\dagger(\mathbb{1}_S \otimes E_E(M))U_{SE}\}\} \qquad (2.284)$$

$$= \mathrm{Tr}_S\{\rho_S F_S(M)\}, \qquad (2.285)$$

with $M \in \mathcal{F}$. These predictions can be described in terms of a POVM with the same outcome space $(\Omega, \mathcal{F})$ identified by the collection of effects

$$F_S(M) = \mathrm{Tr}_E\{\rho_E U_{SE}^\dagger(\mathbb{1}_S \otimes E_E(M))U_{SE}\}. \qquad (2.286)$$

All properties required for the definition of a POVM in (2.244) can be directly verified from (2.286). With reference to the Stern-Gerlach experiment we can indeed equivalently write (2.280) in the form

$$F(\{\uparrow\}) = \ \mathrm{Tr}_{L^2(\mathbb{R}^3)}\{P_{\varphi_{\mathrm{in}}} U^\dagger E(\{\uparrow\})U\}. \tag{2.287}$$

This is a particular instance of a more general situation. As follows from the work of Naimark [26,27], any POVM can be seen to come from a PVM with the same outcome space but defined on a larger Hilbert space. This general statement will be proven in Sect. 3.2.4 as a consequence of Stinespring's theorem and relying on the notion of complete positivity. We here provide a direct constructive formulation of the result [17], restricting however the proof to the case of a denumerable set of outcomes. The notion of POVM and Naimark's result are also nicely related to the theory of frames [28,29].

Theorem 2.7 (Naimark, 1943) *Let $F(\cdot)$ be a POVM in $\mathcal{H}_S$ with outcome space $(\Omega, \mathcal{F})$. Then there exist a Hilbert space $\mathcal{H}_E$, a state in $\mathcal{H}_E$ that can be taken to be pure, and a PVM $E(\cdot)$ in $\mathcal{H}_S \otimes \mathcal{H}_E$ on the same outcome space such that*

$$\mu^E_{\rho \otimes P_\psi}(M) = \mu^F_\rho(M) \quad \forall M \in \mathcal{F} \quad \forall \rho \in \mathcal{S}(\mathcal{H}) \tag{2.288}$$

with P_ψ pure state in $\mathcal{H}_E$.

Proof In the proof we restrict to a denumerable set of outcomes, so that $\Omega \to \{\omega_1, \omega_2, \ldots\}$ and choose $\mathcal{H}_E = l_2(\mathbb{C})$, so that $\mathcal{H}_S \otimes \mathcal{H}_E = \bigoplus_i \mathcal{H}_S$. Writing vectors in $\mathcal{H}_S \otimes \mathcal{H}_E$ in the form

$$\Phi = \begin{pmatrix} \varphi_1 \\ \varphi_2 \\ \vdots \end{pmatrix} \tag{2.289}$$

with $\varphi_i \in \mathcal{H}_S$, we introduce in $\mathcal{H}_S \otimes \mathcal{H}_E$ a Hermitian sesquilinear form relying on the scalar product in $\mathcal{H}_S$

$$\langle \Phi | \Psi \rangle_{\mathcal{H}_S \otimes \mathcal{H}_E} = \sum_i \langle \varphi_i | F(\omega_i) \psi_i \rangle_{\mathcal{H}_S}, \tag{2.290}$$

which is moreover positive as follows from the positivity of the F_i, that is

$$\langle \Phi | \Phi \rangle_{\mathcal{H}_S \otimes \mathcal{H}_E} \geqslant 0 \quad \forall \Phi \in \mathcal{H}_S \otimes \mathcal{H}_E. \tag{2.291}$$

This form becomes a scalar product, so that it is non-degenerate, provided we identify vectors whose difference has zero norm. We can now introduce the map

$$V : \mathcal{H}_S \to \ \mathcal{H}_S \otimes \mathcal{H}_E \tag{2.292}$$

$$\varphi \mapsto \Phi_\varphi = \begin{pmatrix} \varphi \\ \varphi \\ \vdots \end{pmatrix},$$

which provides an isometric embedding of $\mathcal{H}_S$ in $\mathcal{H}_S \otimes \mathcal{H}_E$ according to the previously introduced scalar product, namely

$$\langle \Phi_\varphi | \Phi_\varphi \rangle_{\mathcal{H}_S \otimes \mathcal{H}_E} = \sum_i \langle \varphi | F(\omega_i) \varphi \rangle_{\mathcal{H}_S} \tag{2.293}$$

$$= \langle \varphi | \varphi \rangle_{\mathcal{H}_S}, \tag{2.294}$$

as follows from the normalization property of the POVM, that is $\sum_i F(\omega_i) = \mathbb{1}$. In particular, Φ_φ identifies a pure product state in the space $\mathcal{H}_S \otimes \mathcal{H}_E$ endowed with the previously introduced scalar product. We can now introduce a PVM in $\mathcal{H}_S \otimes \mathcal{H}_E$ considering the orthogonal projections $\{E(\omega_i)\}$ whose action is defined according to

$$(E(\omega_i)\Phi)_k = \delta_{ik}\varphi_i, \tag{2.295}$$

so that only the i-th component in (2.289) is retained and all others are set to zero. Considering finally the adjoint map

$$V^\dagger : \mathcal{H}_S \otimes \mathcal{H}_E \rightarrow \mathcal{H}_S \tag{2.296}$$

$$\Phi_\varphi \mapsto \varphi,$$

we have for any $\varphi \in \mathcal{H}_S$

$$\langle \varphi | V^\dagger E(\omega_k) V \varphi \rangle_{\mathcal{H}_S} = \langle \Phi_\varphi | E(\omega_k)\Phi_\varphi \rangle_{\mathcal{H}_S \otimes \mathcal{H}_E} \tag{2.297}$$

$$= \sum_i \langle (\Phi_\varphi)_i | F(\omega_i)(E(\omega_k)\Phi_\varphi)_i \rangle_{\mathcal{H}_S} \tag{2.298}$$

$$= \sum_i \langle \varphi | F(\omega_i)\delta_{ik}\varphi \rangle_{\mathcal{H}_S} \tag{2.299}$$

$$= \langle \varphi | F(\omega_k)\varphi \rangle_{\mathcal{H}_S}, \tag{2.300}$$

and therefore

$$V^\dagger E(\omega_k) V = F(\omega_k). \tag{2.301}$$

Taking now the vector $\psi = (1, 0, 0, \ldots)$ as state in $\mathcal{H}_E = l_2(\mathbb{C})$ we have

$$\mathrm{Tr}_E\{(\mathbb{1}_S \otimes P_\psi)E(\omega_k)\} = V^\dagger E(\omega_k) V \tag{2.302}$$

and therefore the desired result

$$\mathrm{Tr}_{S \otimes E}\{(\rho \otimes P_\psi)E(\omega_k)\} = \mathrm{Tr}_S\{\rho V^\dagger E(\omega_k) V\} \tag{2.303}$$

$$= \mathrm{Tr}_S\{\rho F(\omega_k)\}. \tag{2.304}$$

$$\square$$

2.3.3 Coexistence

We now briefly discuss the problem of considering different observables together, and see the major changes taking place in this respect when moving from PVM to POVM. Let us first introduce as range of an observable $F(\cdot)$ on the outcome space $(\Omega, \mathcal{F})$ the set

$$\mathrm{ran}(F) = \{F(M), M \in \mathcal{F}\} \subseteq \mathcal{E}(\mathcal{H}), \tag{2.305}$$

that is the collection of all operators in the range of the map. We further say that the effects $E_1, \ldots, E_n$ are coexistent if there exists an observable in the sense of POVM such that

$$\{E_1, \ldots, E_n\} \subseteq \mathrm{ran}(F), \tag{2.306}$$

that is to say they are all contained in the range of a single observable [23]. This holds obviously true if $\sum_{j=1}^{n} E_j \leqslant \mathbb{1}$, since in this case we can construct an observable adding the effect $\mathbb{1} - \sum_{j=1}^{n} E_j$. In this case the effects can be realized together within a single POVM, so that they can be measured with the same apparatus, but the requirement of coexistence is actually weaker, as we show in Remark 2.8 [16].

2.8 Coexistent Effects in $\mathcal{H} = \mathbb{C}^2$

We consider a spin direction observable with outcome space $(\Omega, \mathcal{F})$, where Ω is the surface of the sphere of radius one in three dimensions and $\mathcal{F}$ the σ-algebra naturally induced on it by the Borel σ-algebra in $\mathbb{R}^3$. As already observed in Remark 2.7, while a PVM in $\mathbb{C}^2$ is restricted to two outcomes, corresponding to the largest number of orthogonal projections, this is no more true for a POVM, allowing for unsharp measurements. We thus define the effects

$$F^{\mathrm{sphere}}(M) = \frac{1}{2\pi} \int_M \mathrm{d}^3 n\, P_n, \tag{2.307}$$

where n is a unit vector fixing a direction, P_n the associated projection operator defined in (2.268), and the integral is over the solid angle corresponding to the area M of the surface of the sphere. The defining properties of a POVM as in (2.244) can be directly verified. For any direction m we can consider the hemisphere M_+ admitting m as pole, together with M_- with $-m$ as pole, and consider a spin observable with two outcomes $\{+, -\}$ corresponding to the effects

$$F_m^{\text{hemisphere}}(\{\pm\}) = \frac{1}{2\pi} \int_{M_\pm} d^3 n\, P_n. \tag{2.308}$$

Let us notice that these effects can also be expressed as

$$F_m^{\text{hemisphere}}(\{\pm\}) = \frac{1}{2} P_{\pm m} + \frac{1}{2}\frac{1}{2} \mathbb{1}, \tag{2.309}$$

namely a convex combination of a projection and half the identity, the former corresponding to a sharp measurement and the latter corresponding to uniform noise, which reduces the accuracy of the measurement.

Starting from these expressions the defining properties of a POVM are easily verified. For any unit vector m we have different hemispheres and therefore different POVMs. All of them are obviously within the range of F^{sphere}, so that they are coexistent. However, the sum of effects corresponding to different hemispheres is in general no more an effect, failing to lie below the identity. Note moreover that, though coexistent, the effects $F_m^{\text{hemisphere}}(\{\pm\})$ and $F_n^{\text{hemisphere}}(\{\pm\})$ do not commute unless m and n are collinear.

We further mention without proof the relationship between coexistence and commutativity, which is the reference notion for joint measurability of observables in the standard framework of quantum mechanics. Let us recall that two observables are said to commute if all effects in their range commute. Given two observables we have that coexistence coincides with commutativity if at least one of the two is actually a PVM. If both are POVM, commutativity implies coexistence, but not vice versa, as already clear from the considered example. It is therefore just the release of commutativity that significantly enlarges the class of physical measurements that can be described by POVMs.

2.3.4 Covariance

We now introduce a convenient way to formulate symmetries of maps, namely the notion of map covariant under a given symmetry group G [25]. This notion is of great interest in many situations, both for the construction of POVMs, as we shall see in Sect. 2.3.5, and for the characterization of dynamical maps. The condition of covariance for maps is formalized in Remark 2.9 and its relevance will appear in Remark 5.6, as well as Sects. 5.3.3, 6.2.2 and 6.5.1. Consider a measure space X with a Borel σ-algebra of sets $\mathcal{B}(X)$. Such a space is called a G-space if there exists an action of G on X, namely a map that sends group elements $g \in G$ to transformation maps μ_g on X, realizing a group homomorphism between G and the group of automorphisms of X, namely

$$\mu : G \to \ \mathrm{Aut}(X) \tag{2.310}$$
$$g \mapsto \ \mu_g$$

so that group composition, existence of inverse and identity are preserved according to

$$\mu_g \mu_h = \mu_{gh} \quad \forall g, h \in G \tag{2.311}$$
$$\mu_e = 1_X \tag{2.312}$$
$$(\mu_g)^{-1} = \mu_{g^{-1}} \quad \forall g \in G, \tag{2.313}$$

where e denotes the identity element of G. We recall that according to standard usage we denote as automorphisms the invertible transformations of the space X into itself. If furthermore G acts transitively on X, in the sense that any two point of X can be mapped one into the other with μ_h for a suitable $h \in G$, then X is called a transitive G-space. A relevant example of interest to us is the case of $X = \mathbb{R}^3$, a typical outcome space for the measurement of position and momentum. This space is a transitive G-space with respect to the group of translations. The elements of the group are three-dimensional vectors, acting in the obvious way on the Borel sets of $\mathbb{R}^3$, i.e. $\mu_{\mathbf{a}}(M) = M + \mathbf{a}$ for all $\mathbf{a} \in \mathbb{R}^3$ and for all $M \in \mathcal{B}(\mathbb{R}^3)$. Now that we have defined the action of the symmetry group G on the measurable space X, we further consider a unitary representation of the same group G on a Hilbert space $\mathcal{H}$, that is a collection $\{U(g)\}_{g \in G}$ of unitary operators on $\mathcal{H}$ such that

$$U(g)U(h) = \ U(gh) \quad \forall g, h \in G \tag{2.314}$$
$$U(e) = \ 1_{\mathcal{H}} \tag{2.315}$$
$$U(g)^{-1} = \ U(g^{-1}) \quad \forall g \in G, \tag{2.316}$$

acting on vectors according to

$$\psi_g = U(g)\psi \quad \psi \in \mathcal{H} \quad g \in G. \tag{2.317}$$

This action induces a representation of G on a space $\mathcal{A}(\mathcal{H})$ of operators acting on $\mathcal{H}$ according to

$$A_g = U^{\dagger}(g)AU(g) \quad A \in \mathcal{A}(\mathcal{H}) \quad g \in G. \tag{2.318}$$

Given a map $\mathcal{M}$ defined on $\mathcal{B}(X)$ and taking values in $\mathcal{A}(\mathcal{H})$, namely

$$\mathcal{M} : \mathcal{B}(X) \to \ \mathcal{A}(\mathcal{H}) \tag{2.319}$$

$$X \mapsto M(X),$$

the map is said to be covariant with respect to the symmetry group G provided it obeys

$$U^\dagger(g)M(X)U(g) = M(\mu_{g^{-1}}(X)) \qquad \forall X \in \mathcal{B}(\mathcal{X}) \;\; \forall g \in G. \tag{2.320}$$

A symmetry transformation on the domain of the map is transferred to the symmetry transformation corresponding to the same group element on the range of the map. The action of G on $\mathcal{X}$ commutes with the representation of G on $\mathcal{A}(\mathcal{H})$, induced by the unitary representation of G on $\mathcal{H}$.

The notion of covariance, as described in Remark 2.9, holds for a generic map, but we will consider it now for the special case of a POVM defined on the measure space $(\Omega, \mathcal{F})$. Denoting by μ_g the action of the group G on the space Ω, the requirement of covariance can be written as

$$U^\dagger(g)F(M)U(g) = F(\mu_{g^{-1}}(M)) \qquad \forall M \in \mathcal{F} \;\; \forall g \in G, \tag{2.321}$$

and corresponds to commutativity of the diagram

$$
\begin{array}{ccc}
\mathcal{F} & \xrightarrow{\;\;F\;\;} & \mathcal{E}(\mathcal{H}) \\
\Big\downarrow{\scriptstyle \mu_{g^{-1}}} & & \Big\downarrow{\scriptstyle U^\dagger(g)\cdot U(g)}\; \\
\mathcal{F} & \xrightarrow{\;\;F\;\;} & \mathcal{E}(\mathcal{H})
\end{array}
$$

2.9 Covariant Map

We recall here that the notion of covariance introduced in Sect. 2.3.4 for maps from a measurable space to a space of operators can be extended to maps acting between operator spaces. Consider a map O defined on a linear space of operators $\mathcal{A}(\mathcal{H}_A)$ and taking values in $\mathcal{B}(\mathcal{H}_B)$

$$O : \mathcal{A}(\mathcal{H}_A) \to \mathcal{B}(\mathcal{H}_B) \tag{2.322}$$
$$X \mapsto O(X),$$

together with the unitary representations $\{U_A(g)\}_{g \in G}$ and $\{U_B(g)\}_{g \in G}$ of the group G on the Hilbert spaces $\mathcal{H}_A$ and $\mathcal{H}_B$ respectively. The map O is said to be covariant with respect to the symmetry group G provided it obeys

$$U_B^\dagger(g)O(X)U_B(g) = O(U_A^\dagger(g)XU_A(g)) \qquad \forall X \in \mathcal{A}(\mathcal{H}_A) \;\; \forall g \in G. \tag{2.323}$$

In analogy to (2.321) this corresponds to commutativity of the diagram

$$
\begin{array}{ccc}
\mathcal{A}(\mathcal{H}_A) & \xrightarrow{\ O\ } & \mathcal{B}(\mathcal{H}_B) \\
{\scriptstyle U_A^\dagger(g)\cdot U_A(g)}\big\downarrow & & \big\downarrow{\scriptstyle U_B^\dagger(g)\cdot U_B(g)} \\
\mathcal{A}(\mathcal{H}_A) & \xrightarrow{\ O\ } & \mathcal{B}(\mathcal{H}_B)
\end{array} \ .
$$

2.3.5 Position and Momentum Observables

We now consider POVM and PVM describing measurements related to position and momentum for a particle with Hilbert space $L^2(\mathbb{R}^3)$. The aim of this treatment is threefold. In the first instance, this provides an example of POVM defined on an infinite-dimensional Hilbert space. It further establishes a viewpoint allowing to introduce the very notion of position and momentum observables without reference to the standard correspondence rule, which is used to introduce them from classical mechanics. Quantum mechanics is not introduced as a new mechanics, rather as a new probability theory in which fundamental observables are obtained from symmetry principles. As we shall see, while for the case of either position or momentum the possible POVMs will be essentially given by a suitable coarse-graining with respect to the standard PVMs associated to them, if we want to provide statistical predictions for the measurement of both position and momentum together, the corresponding observable is given by necessity in terms of a POVM. Indeed, as it is well known there is no PVM observable associated to position and momentum, due to the non-commutativity of the associated self-adjoint operators. As a third important point, this example will show that some, but not all, unsharp observables arise as smeared version of a sharp observable. The unsharpness brought about by coarse-graining may or may not admit the kind of ignorance interpretation familiar from classical physics. In general, the unsharpness is the reflection of a genuine quantum indeterminacy, as it happens for the case of position and momentum. We stress that the interest of the introduced more general notion of quantum observables goes beyond foundational questions, proving relevant for quantum technologies, also in situations such as quantum estimation in which the investigated quantity cannot be identified with an observable [30].

2.3.5.1 Generalized Position Observable

Let us start looking for position observables in the framework of POVMs. As said, we do not rely on the usual correspondence principle with respect to classical mechanics, but we rather give an operational definition of position observable as describing localization measurements. We will determine them fixing the behavior with respect to the action of the relevant symmetry group, which in this case is the isochronous Galilei group, briefly discussed in Remark 2.10. In order for a measurement F^x on the outcomes space $(\mathbb{R}^3, \mathcal{B}(\mathbb{R}^3))$ to be a position measurement, we ask that the probability distribution of the outcomes of F^x performed on a state ρ and on the correspondingly transformed one $\rho_g = U(g)\rho U^\dagger(g)$ should satisfy the relation

$$\mathrm{Tr}\left\{\rho_g F^x(\mu_g(M))\right\} = \mathrm{Tr}\left\{\rho F^x(M)\right\}, \qquad (2.324)$$

for any state $\rho \in \mathcal{S}(\mathcal{H})$, any set $M \in \mathcal{B}(\mathbb{R}^3)$ and any group element $g \in \mathcal{G}$, where now $\mathcal{G}$ denotes the isochronous Galilei group. We assume that the group acts in the natural way on the measurable space $\mathcal{B}(\mathbb{R}^3)$ of the Borel sets of $\mathbb{R}^3$. That is to say, its action corresponds to either a translation in the position or momentum space, or to a rotation in both spaces. The symmetry requirement (2.324) is translated according to (2.321) in the following covariance equations

$$\begin{aligned}
U^\dagger(a)F^x(M)U(a) &= F^x(M - a) & \forall a \in \mathbb{R}^3 \;\; \forall M \in \mathcal{B}(\mathbb{R}^3) \\
U^\dagger(q)F^x(M)U(q) &= F^x(M) & \forall q \in \mathbb{R}^3 \;\; \forall M \in \mathcal{B}(\mathbb{R}^3) \\
U^\dagger(R)F^x(M)U(R) &= F^x(R^{-1}M) & \forall R \in \mathrm{SO}(3) \;\; \forall M \in \mathcal{B}(\mathbb{R}^3),
\end{aligned} \qquad (2.325)$$

with $a \in \mathbb{R}^3$ a vector with dimensions of position, $q \in \mathbb{R}^3$ a vector with dimensions of momentum and the expression of the unitary representations are detailed in Remark 2.10.

2.10 Representation of the Galilei Group

The Galilei group provides the group of the coordinate transformations between inertial frames in the Newtonian space-time $\mathbb{R}^3 \times \mathbb{R}$. An element $g \in G$ of this ten-parameter group of transformations on the space-time coordinates is of the form

$$g = (b, a, v, R), \qquad (2.326)$$

where b denotes a time, a a spatial translation, v identifies a velocity transformation, that is a so-called proper Galilei transformation, and R a rotation. The action of the group on points in space-time is defined as

$$\mu_g(x, t) = (Rx + vt + a, t + b). \qquad (2.327)$$

The inverse transformation is identified by the element

$$g^{-1} = (-b, -R^{-1}(a - bv), -R^{-1}v, R^{-1}), \qquad (2.328)$$

while the composition of two transformations is given by

$$gg' = (b + b', Ra' + b'v + a, Rv' + v, RR'). \qquad (2.329)$$

If b is set equal to zero, we recover a subgroup known as isochronous Galilei group

$$G = (\mathbb{R}^3 \times \mathbb{R}^3) \times' SO(3), \qquad (2.330)$$

expressed as the semi-direct product of the space and velocity translations with the rotation group. The semi-direct product is denoted with $\times'$ and arises because of the composition law (2.329), in which the action of rotations and translations is not independent. For the case of a non-relativistic spinless particle of mass m it can be shown that the group G acts on $L^2(\mathbb{R}^3)$ by means of an irreducible projective unitary representation U defined by the following action on wavevectors

$$(U(a, v, R)\psi)(x) = e^{imv \cdot (x - a)} \psi(R^{-1}(x - a)). \qquad (2.331)$$

The collection of these operators satisfies the composition law

$$U(g)U(g') = e^{i\omega(g, g')}U(gg'), \qquad (2.332)$$

with

$$\omega(g, g') = \frac{m}{2}(a \cdot Rv' - v \cdot R'a), \qquad (2.333)$$

so that they provide a unitary representation up to a factor, hence the name projective representation. For a proper treatment of the subject that involves a number of subtleties see [31, 32]. We use the identifications

$$U(a) = U(a, \mathbf{0}, \mathbb{1}), \qquad (2.334)$$

as well as

$$U(q) = U(\mathbf{0}, q/m, \mathbb{1}) \qquad (2.335)$$

and

$$U(R) = U(\mathbf{0}, \mathbf{0}, R) \qquad (2.336)$$

to single out the subgroups of space translations, Galilei transformations or boosts, and rotations. The unitary representation of translations is given by

$$U(\boldsymbol{a}) = \mathrm{e}^{-\frac{i}{\hbar}\boldsymbol{a}\cdot\hat{\boldsymbol{p}}}, \tag{2.337}$$

with $\boldsymbol{a} \in \mathbb{R}^3$ and $\hat{\boldsymbol{p}}$ denotes the standard triple of momentum operators, while the unitary representation of boosts is given by

$$U(\boldsymbol{q}) = \mathrm{e}^{+\frac{i}{\hbar}\boldsymbol{q}\cdot\hat{\boldsymbol{x}}}, \tag{2.338}$$

with $\boldsymbol{q} \in \mathbb{R}^3$ and $\hat{\boldsymbol{x}}$ denotes the standard triple of position operators. The unitary representation of rotations $U(R)$ in turn is obtained exponentiating the angular momentum operator $\hat{\boldsymbol{x}} \times \hat{\boldsymbol{p}}$.

An observable $F^{\boldsymbol{x}}$ can be termed a position observable if it transforms covariantly with respect to translations and rotations, and is invariant under velocity transformations. These equations can also be seen as a requirement on the possible macroscopic apparata performing the measurement associated to the observable $F^{\boldsymbol{x}}$. The apparatus used to test whether the considered system is localized in the translated region $M - \boldsymbol{a}$ should be in the equivalence class to which the translated apparatus used to test localization in the region M belongs, and similarly for rotations. Localization measurements should instead be unaffected by boost transformations, indeed the detection of a system in a given position should not depend on its velocity. We therefore define as position observable a solution to these covariance equations, that is to say a POVM complying with the covariance conditions, which is sometimes called generalized position observable.

Let us now consider the possible solutions to these equations [25]. We investigate first the special case of a PVM. In this case the solution is uniquely given by the usual spectral decomposition of the position operator, namely we can write

$$E^{\boldsymbol{x}}(M) = \int_M d^3\boldsymbol{x}\,|\boldsymbol{x}\rangle\langle\boldsymbol{x}| \tag{2.339}$$

$$= \chi_M(\hat{\boldsymbol{x}}), \tag{2.340}$$

where χ_M denotes the characteristic function of the set M. The orthogonal projections $E^{\boldsymbol{x}}(M)$ thus act in a multiplicative way in the position representation of a state $\psi \in L^2(\mathbb{R}^3)$, namely

$$\langle\boldsymbol{y}|E^{\boldsymbol{x}}(M)\psi\rangle = (E^{\boldsymbol{x}}(M)\psi)(\boldsymbol{y}) \tag{2.341}$$

$$= \chi_M(\boldsymbol{y})\psi(\boldsymbol{y}), \tag{2.342}$$

so that we recover the usual formula for the prediction of a detection event in the region M if the system is in the normalized state ψ

$$\|E^x(M)\psi\|^2 = \langle\psi|E^x(M)\psi\rangle \tag{2.343}$$

$$= \int_M d^3y|\psi(y)|^2. \tag{2.344}$$

Let us heuristically sketch the argument justifying the result (2.339). If E^x has to be a PVM, it identifies a triple of commuting self-adjoint operators. At the same time, according to Stone's theorem [3] the unitary representation $U(q)$ of the group of boosts given by (2.338) can be uniquely expressed in terms of a PVM E_U^x according to

$$U(q) = \int_{\mathbb{R}^3} e^{+\frac{i}{\hbar}q\cdot y} dE_U^x(y). \tag{2.345}$$

This PVM defines a triple of self-adjoint operators in terms of the integrals

$$\hat{x}_j = \int_{\mathbb{R}} y dE_U^{x_j}(y), \tag{2.346}$$

which are the generators of boosts. The invariance under boosts of our PVM localization observable E^x then implies that the generator of boosts $\hat{x}$ can be identified with the triple of self-adjoint operators defined by E^x, since they all commute. Covariance under rotations then warrants that this triple indeed transforms as a vector. If the measurement is identified with a PVM, all statistical predictions can be formulated in terms of the uniquely associated self-adjoint operators. In particular, for a given state ψ mean values and variances of the classical probability distribution giving the position distribution can be expressed by means of the operators $\hat{x}$

$$\text{Mean}_\psi(E^x) = \langle\hat{x}\rangle_\psi \tag{2.347}$$

$$\text{Var}_\psi(E^x) = \langle\hat{x}^2\rangle_\psi - \langle\hat{x}\rangle_\psi^2. \tag{2.348}$$

The pair (U, E), where U is the unitary representation of the symmetry group, here the isochronous Galilei group, and E a PVM covariant under the action of U, is called a system of imprimitivity.

More generally a solution of the covariance equations (2.325) can be obtained as follows. Let us consider a rotationally invariant probability density $h(x)$, that is a positive function such that $h(Rx) = h(x)$ and

$$\int d^3x\, h(x) = 1, \tag{2.349}$$

with zero mean, so that in particular the associated variance reads

$$\text{Var}(h) = \int d^3x\, x^2\, h(x). \tag{2.350}$$

we can check, as detailed in Remark 2.11, that the expression

$$F_h^x(M) = \int_M d^3y \int_{\mathbb{R}^3} d^3x\, h(\boldsymbol{x} - \boldsymbol{y})|\boldsymbol{x}\rangle\langle\boldsymbol{x}| \qquad (2.351)$$

$$= (h * \chi_M)(\hat{\boldsymbol{x}}), \qquad (2.352)$$

where the symbol $*$ denotes convolution, namely

$$(f * g)(\boldsymbol{x}) = \int_{\mathbb{R}^3} \mathrm{d}^3y\, f(\boldsymbol{x} - \boldsymbol{y})g(\boldsymbol{y}), \qquad (2.353)$$

actually is a POVM complying with the covariance equations (2.325), and in fact provides their general solution. The pair (U, F_h^x), where U is the unitary representation of the symmetry group and F_h^x a POVM covariant under its action is called system of covariance.

2.11 POVM as Smeared PVM

We now verify that for any rotationally invariant probability density $h(\boldsymbol{x})$ the expression F_h^x as defined in (2.351) is a POVM on the outcomes space $(\mathbb{R}^3, \mathcal{B}(\mathbb{R}^3))$, satisfying the covariance equations (2.325). Due to positivity of the function h the operator $F_h^x(M)$ is positive for any Borel set M. Moreover, thanks to the fact that the POVM is built in terms of an integral with an operator density, we immediately have $F_h^x(\emptyset) = 0$, $F_h^x(\mathbb{R}^3) = \mathbb{1}$, as well as σ-additivity. F_h^x is thus a POVM. Exploiting the action of the operator $F_h^x(M)$ on a generic vector in the position representation,

$$\langle\boldsymbol{y}|F_h^x(M)\psi\rangle = (F_h^x(M)\psi)(\boldsymbol{y}) \qquad (2.354)$$

$$= (h * \chi_M)(\boldsymbol{y})\psi(\boldsymbol{y}) \qquad (2.355)$$

$$= \int_M d^3x\, h(\boldsymbol{y} - \boldsymbol{x})\psi(\boldsymbol{y}), \qquad (2.356)$$

we obtain for the probability of a detection in the interval M, if the state of the system is given by the pure state ψ, the expression

$$\langle\psi|F_h^x(M)\psi\rangle = \int_M d^3x \int_{\mathbb{R}^3} d^3y\, h(\boldsymbol{x} - \boldsymbol{y})|\psi(\boldsymbol{y})|^2. \qquad (2.357)$$

It immediately appears that this result is obtained by smearing the standard prediction with the confidence function $h(\boldsymbol{x})$. The sharper the distribution $h(\boldsymbol{x})$, the closer the prediction to the standard one. The covariance properties can be directly verified starting from (2.351) and considering the action of the unitary representations of the different symmetries in $L^2(\mathbb{R}^3)$. E.g., starting from (2.337) we have

$$U^\dagger(\boldsymbol{a})|\boldsymbol{x}\rangle = |\boldsymbol{x} - \boldsymbol{a}\rangle \tag{2.358}$$

and therefore

$$U^\dagger(\boldsymbol{a})F_h^{\boldsymbol{x}}(M)U(\boldsymbol{a}) = \int_M d^3\boldsymbol{y} \int_{\mathbb{R}^3} d^3\boldsymbol{x}\, h(\boldsymbol{x} - \boldsymbol{y})|\boldsymbol{x} - \boldsymbol{a}\rangle\langle \boldsymbol{x} - \boldsymbol{a}| \tag{2.359}$$

$$= \int_{M-\boldsymbol{a}} d^3\boldsymbol{y} \int_{\mathbb{R}^3} d^3\boldsymbol{x}\, h(\boldsymbol{x} - \boldsymbol{y})|\boldsymbol{x}\rangle\langle \boldsymbol{x}| \tag{2.360}$$

$$= F_h^{\boldsymbol{x}}(M - \boldsymbol{a}). \tag{2.361}$$

In a similar way we can verify the validity of the other conditions in (2.325).

Importantly, the requirement that $h(\boldsymbol{x})$ has zero mean warrants that

$$\mathrm{Mean}_\psi(F_h^{\boldsymbol{x}}) = \langle \hat{\boldsymbol{x}}\rangle_\psi \tag{2.362}$$

$$= \mathrm{Mean}_\psi(E^{\boldsymbol{x}}). \tag{2.363}$$

Indeed, by the very definition of mean as first moment of the probability distribution we have

$$\mathrm{Mean}_\psi(F_h^{\boldsymbol{x}}) = \int_{\mathbb{R}^3} d^3\boldsymbol{x} \int_{\mathbb{R}^3} d^3\boldsymbol{y}\,\boldsymbol{x} h(\boldsymbol{x} - \boldsymbol{y})|\psi(\boldsymbol{y})|^2. \tag{2.364}$$

$$= \int_{\mathbb{R}^3} d^3\boldsymbol{x} \int_{\mathbb{R}^3} d^3\boldsymbol{y}(\boldsymbol{x} + \boldsymbol{y})h(\boldsymbol{x})|\psi(\boldsymbol{y})|^2 \tag{2.365}$$

$$= \int_{\mathbb{R}^3} d^3\boldsymbol{y}\,\boldsymbol{y}|\psi(\boldsymbol{y})|^2 \tag{2.366}$$

$$= \langle \hat{\boldsymbol{x}}\rangle_\psi. \tag{2.367}$$

The predictions for the second moment is however different, and a similar calculation leads to

$$\mathrm{Var}_\psi(F_h^{\boldsymbol{x}}) = \langle \hat{\boldsymbol{x}}^2\rangle_\psi - \langle \hat{\boldsymbol{x}}\rangle_\psi^2 + \mathrm{Var}(h) \tag{2.368}$$

$$= \mathrm{Var}_\psi(E^{\boldsymbol{x}}) + \mathrm{Var}(h). \tag{2.369}$$

The result can be seen as the effect of considering two independent classical random variables. Their independence leads to a distribution function given by the convolution as in (2.353), so that the variances are given by the sum as in (2.369). The probability density $h(\boldsymbol{x})$ describes the smearing associated to the considered measurement apparatus, which reflects its actual sensitivity and finite resolution, while the probability density $|\psi(y)|^2$ is the prediction associated to the projective position observable for the state ψ. We note in particular that the variance is no more expressed only by the mean value of the operator which can be used to evaluate the first moment, namely $\hat{x}$, and by the mean value of its square. Besides the variance for the standard sharp position measurement, a further contribution $\mathrm{Var}(h)$ appears, which is state independent and reflects the finite resolution of the apparatus used for the localization measurement. Note that the standard result is recovered in the limit of a sharply peaked probability density, corresponding to a pure state in the set of classical probability measures. Consider in particular the following family of Gaussian probability distributions, parametrized by the positive constant σ providing the width of the distribution

$$h_\sigma(x) = \left(\frac{1}{2\pi\sigma^2}\right)^{\frac{3}{2}} \mathrm{e}^{-\frac{1}{2}\frac{x^2}{\sigma^2}}. \tag{2.370}$$

We then have that

$$\begin{aligned}
\langle\psi|F^x_{h_\sigma}(M)\psi\rangle &= \int_M d^3y \int_{\mathbb{R}^3} d^3x \left(\frac{1}{2\pi\sigma^2}\right)^{\frac{3}{2}} \mathrm{e}^{-\frac{1}{2\sigma^2}(x-y)^2}|\psi(x)|^2 \\
&\xrightarrow{\sigma\to 0} \int_M d^3y |\psi(y)|^2,
\end{aligned} \tag{2.371}$$

so that in the limit of an infinite accuracy in the localization measurement of the apparatus exploited the POVM reduces to the standard PVM for the position observable.

Analogous results can obviously be obtained for the generalized momentum observable, asking for the corresponding covariance properties, namely

$$\begin{aligned}
U^\dagger(a)F^p(N)U(a) &= F^p(N) &&\forall a \in \mathbb{R}^3 \ \forall N \in \mathcal{B}(\mathbb{R}^3) \\
U^\dagger(q)F^p(N)U(q) &= F^p(N-q) &&\forall q \in \mathbb{R}^3 \ \forall N \in \mathcal{B}(\mathbb{R}^3) \\
U^\dagger(R)F^p(N)U(R) &= F^p(R^{-1}N) &&\forall R \in \mathrm{SO}(3) \ \forall N \in \mathcal{B}(\mathbb{R}^3).
\end{aligned} \tag{2.372}$$

2.3.5.2 Joint Position and Momentum Observables

A new and interesting situation appears when considering apparata performing both a measurement of the spatial location of a particle as well as of its momentum. As it is well known, no observable can be associated to such a measurement in the framework of standard textbook quantum mechanics, identifying observables with self-adjoint operators. Despite the fact that such measures are routinely performed, e.g. in particle physics when considering the track of particles in detection chambers. Let us consider the covariance equations for such an observable in the more general framework of POVMs. A position and momentum observable should be given by a

POVM $F^{x,p}$ defined on $\mathcal{B}(\mathbb{R}^3 \times \mathbb{R}^3)$ satisfying the following covariance equations under the action of translations, boosts and rotations respectively

$$U^\dagger(a)F^{x,p}(M \times N)U(a) = F^{x,p}(M - \mathbf{a} \times N) \quad \forall a \in \mathbb{R}^3 \; \forall M, N \in \mathcal{B}(\mathbb{R}^3)$$
$$U^\dagger(q)F^{x,p}(M \times N)U(q) = F^{x,p}(M \times N - q) \quad \forall q \in \mathbb{R}^3 \; \forall M, N \in \mathcal{B}(\mathbb{R}^3) \quad (2.373)$$
$$U^\dagger(R)F^{x,p}(M \times N)U(R) = F^{x,p}(R^{-1}M \times R^{-1}N) \quad \forall R \in \mathrm{SO}(3) \; \forall M, N \in \mathcal{B}(\mathbb{R}^3).$$

These covariance equations do not admit any solution within the set of PVM, while the general solution within the set of POVM takes the form [24]

$$F_S^{x,p}(M \times N) = \frac{1}{(2\pi\hbar)^3} \int_M d^3x \int_N d^3p \, W(x, p) S W^\dagger(x, p). \quad (2.374)$$

Here S is a trace class operator, positive, with trace equal to one and invariant under rotations

$$S \in \mathcal{T}(\mathcal{H}), \quad S \geq 0, \quad \mathrm{Tr}\{S\} = 1, \quad U^\dagger(R)SU(R) = S, \quad (2.375)$$

while $W(x, p)$ denote the unitary Weyl operators

$$W(x, p) = e^{-\frac{i}{\hbar}(x \cdot \hat{p} - \hat{x} \cdot p)} \quad (2.376)$$
$$= e^{-\frac{i}{\hbar}x \cdot \hat{p}} e^{+\frac{i}{\hbar}\hat{x} \cdot p} e^{+\frac{i}{2\hbar}x \cdot p} \quad (2.377)$$

built in terms of the canonical position and momentum operators $\hat{x}$ and $\hat{p}$, namely the generators of boosts and translations respectively, where in the second line we have used the Baker-Campbell-Hausdorff formula (2.387) recalled in Remark 2.12. The proof that $F_S^{x,p}$ is a POVM, as well as verification of the covariance equations (2.373), can be obtained as in Remark 2.11, further exploiting the properties of the operator S given in (2.375).

2.12 Baker-Campbell-Hausdorff Formula

Relations between the exponentials of non-commuting variables [33], and more generally between operator orderings [34], are of fundamental importance in both physics and mathematics. We recall here that the Baker-Campbell-Hausdorff formula [35–37] provides the operator Z that solves the equation

$$e^A e^B = e^Z \quad (2.378)$$

in the form

$$Z = A + B + \frac{1}{2}[A, B] + \frac{1}{2}\sum_{n \geq 2}\frac{1}{(n+1)!}([A, B]_n + [B, A]_n), \quad (2.379)$$

where the symbol $[\cdot, \cdot]_n$ denote nested commutators according to

$$[A, B]_n = [A, \ldots [A, [A, B]] \ldots], \tag{2.380}$$

where the operator A appears n times. It is often expressed as

$$e^A e^B = e^{A+B+\frac{1}{2}[A,B]+\frac{1}{12}([A,[A,B]]+[B,[B,A]])+\cdots}. \tag{2.381}$$

A dual formula to (2.378) is given by the Zassenhaus formula [33]

$$e^{A+B} = e^A e^B \prod_{n=2}^{\infty} e^{Z_n(A,B)}, \tag{2.382}$$

where $Z_n(A, B)$ is a functional of the operators A and B implicitly recursively defined by the relation

$$Z_n(A, B) = \frac{1}{n!} \left\{ \frac{d^n}{d\lambda^n} (e^{-\lambda^{n-1}Z_{n-1}} \cdots e^{-\lambda^2 Z_2} e^{-\lambda B} e^{-\lambda A} e^{\lambda(A+B)}) \right\}_{\lambda=0}. \tag{2.383}$$

The first terms can be determined from the Baker-Campbell-Hausdorff formula to be

$$Z_2 = -\frac{1}{2}[A, B], \tag{2.384}$$

as well as

$$Z_3 = \frac{1}{6}[A, [A, B]] - \frac{1}{3}[B, [B, A]]. \tag{2.385}$$

A special case of particular relevance in quantum mechanics is obtained when the commutator of the operators A and B is proportional to the identity

$$[A, B] = c\mathbb{1}, \tag{2.386}$$

with $c \in \mathbb{C}$, so that we are left with the identity

$$e^A e^B = e^{A+B+\frac{1}{2}c}. \tag{2.387}$$

This last identity is also known as Kermack-McCrae identity [38].

The connection with position and momentum observable, as well as the reason why such a joint observable can be expressed only in the formalism of POVM, where position observables appear as smeared versions of the usual position observable, and similarly for momentum, can be understood looking at the marginal observables. Starting from $F_S^{x,p}$ we can consider a measure of position irrespective of the momentum of the particle, thus coming to the marginal position observable

$$F_S^x(M) = F_S^{x,p}(M \times \mathbb{R}^3) \tag{2.388}$$

$$= \int_M d^3x \int_{\mathbb{R}^3} \frac{d^3p}{(2\pi\hbar)^3} e^{\frac{i}{\hbar}(z-z')\cdot p}$$

$$\times \int_{\mathbb{R}^3} d^3z \int_{\mathbb{R}^3} d^3z' e^{-\frac{i}{\hbar}x\cdot\hat{p}} |z\rangle\langle z|S|z'\rangle\langle z'|e^{+\frac{i}{\hbar}x\cdot\hat{p}} \tag{2.389}$$

$$= \int_M d^3x \int_{\mathbb{R}^3} d^3z \, |z+x\rangle\langle z|S|z\rangle\langle z+x| \tag{2.390}$$

$$= \int_M d^3y \int_{\mathbb{R}^3} d^3x \, h_{S^x}(x-y)|x\rangle\langle x|, \tag{2.391}$$

obtained by exploiting (2.377) and writing the position matrix elements of the operator S. In the expression for the marginal appears the function

$$h_{S^x}(x) = \langle x|S|x\rangle, \tag{2.392}$$

which thanks to the properties of the operator S detailed in (2.375) is a well-defined probability density invariant under rotations. On similar grounds the marginal momentum observable is given by

$$F_S^p(M) = F_S^{x,p}(\mathbb{R}^3 \times N) \tag{2.393}$$

$$= \int_N d^3q \int_{\mathbb{R}^3} d^3p \, |p\rangle\langle p-q|S|p-q\rangle\langle p| \tag{2.394}$$

$$= \int_N d^3q \int_{\mathbb{R}^3} d^3p \, h_{S^p}(p-q)|p\rangle\langle p|, \tag{2.395}$$

where again the function

$$h_{S^p}(p) = \langle p|S|p\rangle \tag{2.396}$$

is a well-defined probability density, which corresponds to the momentum probability density of a system described by the statistical operator S. A definite example is detailed in Remark 2.13. As it appears, the marginal observables are given by two POVM characterized by a smearing of the standard position and momentum observables, by means of the probability densities $h_{S^x}(x)$ and $h_{S^p}(p)$ respectively. The latter probability densities originate from the same positive trace class operator S of trace one, so that for each Cartesian direction $\mathrm{Var}(h_{S^x}) = \mathrm{Var}_S(E^x)$ as well as $\mathrm{Var}(h_{S^p}) = \mathrm{Var}_S(E^p)$, and therefore due to Heisenberg's uncertainty relations

$$\mathrm{Var}(h_{S^x})\,\mathrm{Var}(h_{S^p}) \geq \frac{\hbar^2}{4}. \tag{2.397}$$

It is exactly this finite resolution in the marginal measurements of position and momentum stemming from $F_S^{x,p}$ that allows for a joint measurement for position and momentum in quantum mechanics. Considering the product of the variance of the two marginals F_S^x and F_S^p in a given state ρ we have, exploiting (2.369) for F_S^x and the analogue identity for F_S^p,

$$\mathrm{Var}_\rho(F_S^x)\,\mathrm{Var}_\rho(F_S^p) = \ \mathrm{Var}_\rho(E^x)\,\mathrm{Var}_\rho(E^p) + \mathrm{Var}(h_{Sx})\,\mathrm{Var}(h_{Sp})$$
$$+ \mathrm{Var}_\rho(E^x)\,\mathrm{Var}(h_{Sp}) + \mathrm{Var}_\rho(E^p)\,\mathrm{Var}(h_{Sx}). \quad (2.398)$$

Due to Heisenberg's uncertainty relation the first two terms at the r.h.s. are each greater than $\hbar^2/4$. Indeed, E^x and E^p are the usual PVM observables for position and momentum, while the variances of h_{Sx} and h_{Sp} satisfy (2.397). As a matter of fact Heisenberg's principle is at work twice. Exploiting the same constraints the last two terms at the r.h.s. can be lower bounded by the quantity

$$\frac{\hbar^2}{4}\left[\frac{\mathrm{Var}_\rho(E^x)}{\mathrm{Var}(h_{Sx})} + \frac{\mathrm{Var}(h_{Sx})}{\mathrm{Var}_\rho(E^x)}\right], \quad (2.399)$$

bounded below by $\hbar^2/2$ since $x + 1/x \geqslant 2$. As a result we have the inequality

$$\mathrm{Var}_\rho(F_S^x)\,\mathrm{Var}_\rho(F_S^p) \geqslant \hbar^2, \quad (2.400)$$

which at variance with the standard Heisenberg's relation makes reference to measurements for position and momentum performed with the very same measurement apparatus. Thus if we consider the statistics of a measurement of position obtained from the marginal of a given joint position and momentum observable, and similarly for momentum, the product of the variances stays above the lower bound that we have for sharp position and sharp momentum observables, just because of the added unsharpness necessary in order to consider a joint measurement.

2.13 POVM for Position and Momentum

In order to consider a definite example, let us take S to be a pure state corresponding to a Gaussian wavepacket of width σ centered in zero, so as to ensure invariance under rotations. Inserting it in (2.374) and exploiting the fact that the Weyl operators transform the Gaussian by shifting its mean value of position and momentum, as follows from (2.376), we obtain

$$F^{x,p}(M \times N) = \frac{1}{(2\pi\hbar)^3}\int_M d^3x_0 \int_N d^3p_0\,|\psi_{x_0 p_0}\rangle\langle\psi_{x_0 p_0}|, \quad (2.401)$$

where

$$\langle \boldsymbol{x} | \psi_{\boldsymbol{x}_0, \boldsymbol{p}_0} \rangle = \left(\frac{1}{2\pi\sigma^2} \right)^{\frac{3}{4}} e^{-\frac{1}{4\sigma^2}(\boldsymbol{x}-\boldsymbol{x}_0)^2 + \frac{i}{\hbar} \boldsymbol{p}_0 \cdot (\boldsymbol{x}-\boldsymbol{x}_0)}. \qquad (2.402)$$

Note that this is a POVM and not a PVM just because the Gaussian vectors $\{\psi_{\boldsymbol{x}_0 \boldsymbol{p}_0}\}_{\boldsymbol{x}_0, \boldsymbol{p}_0 \in \mathbb{R}^3}$ are not orthogonal, but rather provide an overcomplete set. They correspond to coherent states as we shall discuss in Remark 5.11. Indeed, precursors of the notion of POVM have been used in quantum optics long before their mathematical formalization [39]. Verification of the normalization of the POVM is in this case equivalent to the proof that coherent states provide a nonorthogonal resolution of the identity. It is quite instructive to consider the associated marginal position and momentum observables, given respectively by the POVMs

$$F^{\boldsymbol{x}}(M) = \int_M d^3 \boldsymbol{x}_0 \int_{\mathbb{R}^3} d^3 \boldsymbol{x} \left(\frac{1}{2\pi\sigma^2} \right)^{\frac{3}{2}} e^{-\frac{1}{2\sigma^2}(\boldsymbol{x}-\boldsymbol{x}_0)^2} |\boldsymbol{x}\rangle\langle\boldsymbol{x}| \qquad (2.403)$$

and

$$F^{\boldsymbol{p}}(N) = \int_N d^3 \boldsymbol{p}_0 \int_{\mathbb{R}^3} d^3 \boldsymbol{p} \left(\frac{2\sigma^2}{\pi\hbar^2} \right)^{\frac{3}{2}} e^{-\frac{2\sigma^2}{\hbar^2}(\boldsymbol{p}-\boldsymbol{p}_0)^2} |\boldsymbol{p}\rangle\langle\boldsymbol{p}|. \qquad (2.404)$$

It is now clear that depending on the value of σ we can have more or less coarse-grained position and momentum observables. No limit on σ can however be taken in order to have a sharp observable for both position and momentum. In the limit $\sigma \to 0$ we would recover as in (2.371) the standard position observable, but the marginal for momentum would identically vanish, intuitively corresponding to a complete lack of information on momentum. The opposite would happen in the limit $\sigma \to \infty$.

References

1. G. Ludwig, *Foundations of quantum mechanics* (Springer-Verlag, New York, 1983)
2. M. Born, Z. Phys. **37**, 863 (1926). https://doi.org/10.1007/BF01397477
3. M. Reed, B. Simon, *I: Functional Analysis* (Academic Press, San Diego, 1980)
4. R. Schatten, *Norm ideals of completely continuous operators*. Ergebnisse der Mathematik und ihrer Grenzgebiete (Springer-Verlag, 1960)
5. I. Bengtsson, K. Życzkowski, *Geometry of Quantum States: An Introduction to Quantum Entanglement*, 2nd edn. (Cambridge University Press, 2017)
6. S. Wißmann, A. Karlsson, E.M. Laine, J. Piilo, H.-P. Breuer, Phys. Rev. A **86**, 062108 (2012). https://doi.org/10.1103/PhysRevA.86.062108

7. A.S. Holevo, *Quantum Systems, Channels, Information, Studies in Mathematical Physics*, vol. 16 (De Gruyter, Amsterdam, 2012)
8. E.T. Jaynes, Phys. Rev. **106**, 620 (1957). https://doi.org/10.1103/PhysRev.106.620
9. E.T. Jaynes, Phys. Rev. **108**, 171 (1957). https://doi.org/10.1103/PhysRev.108.171
10. R. Balian, *From microphysics to macrophysics. Vol. I* (Springer-Verlag, Berlin, 2007). https://doi.org/10.1007/978-3-540-45475-5
11. E. Schrödinger, Math. Proc. Cambridge Philos. Soc. **32**, 446 (1936). https://doi.org/10.1017/S0305004100019137
12. N. Gisin, Helv. Phys. Acta **62**, 363 (1989). https://doi.org/10.5169/seals-116034
13. L.P. Hughston, R. Jozsa, W.K. Wootters, Phys. Lett. A **183**, 14 (1993). https://doi.org/10.1016/0375-9601(93)90880-9
14. A.K. Kirkpatrick, Found. Phys. Lett. **19**, 95 (2006). https://doi.org/10.1007/s10702-006-1852-1
15. G. Cassinelli, E. De Vito, A. Levrero, J. Math. Anal. Appl. **210**, 472 (1997). https://doi.org/10.1006/jmaa.1997.5480
16. T. Heinosaari, M. Ziman, *The Mathematical Language of Quantum Theory* (Cambridge University Press, Cambridge, 2011)
17. A.S. Holevo, *Probabilistic and statistical aspects of quantum theory* (North-Holland, Amsterdam, 1982)
18. P. Busch, Phys. Rev. Lett. **91**, 120403 (2003). https://doi.org/10.1103/PhysRevLett.91.120403
19. C.M. Caves, C.A. Fuchs, K.K. Manne, J.M. Renes, Found. Phys. **34**, 193 (2004). https://doi.org/10.1023/B:FOOP.0000019581.00318.a5
20. E.B. Davies, *Quantum Theory of Open Systems* (Academic Press, London, 1976)
21. A.M. Gleason, J. Math. Mech. **6**, 885 (1957)
22. V. Moretti, Spectral Theory and Quantum Mechanics (Springer-Verlag. Berlin (2013). https://doi.org/10.1007/978-3-319-70706-8
23. P. Busch, J. Kiukas, P. Lahti, Math. Slovaca **60**, 665 (2010). https://doi.org/10.2478/s12175-010-0039-1
24. A.S. Holevo, *Statistical Structure of Quantum Theory, Lect. Not. Phys.*, vol. m 67 (Springer, Berlin, 2001). https://doi.org/10.1007/3-540-44998-1
25. P. Busch, M. Grabowski, P. Lahti, *Operational quantum physics, Lect. Not. Phys.*, vol. m 31 (Springer-Verlag, 1995). https://doi.org/10.1007/978-3-540-49239-9
26. M. Naimark, C. R. (Dokl.) Acad. Sci. URSS **41**, 359 (1943)
27. M. Naimark, Bull. (Izv.) Acad. Sci. URSS (Ser. Math.) **7**, 237 (1943)
28. G.M.D. Ariano, P. Perinotti, M.F. Sacchi, J. Opt. B: Quantum Semiclass. Opt. **6**, S487 (2004). https://doi.org/10.1088/1464-4266/6/6/005
29. B. Moran, S. Howard, D. Cochran, in *Excursions in Harmonic Analysis, Volume 2: The February Fourier Talks at the Norbert Wiener Center*, ed. by T.D. Andrews, R. Balan, J.J. Benedetto, W. Czaja, K.A. Okoudjou (Birkhäuser Boston, Boston, 2013), pp. 49–64. https://doi.org/10.1007/978-0-8176-8379-54
30. M.G.A. Paris, Int. J. Quantum Inf. **07**, 125 (2009). https://doi.org/10.1142/S0219749909004839
31. J. Levy-Leblond, J. Math. Phys. **4**, 776 (1963). https://doi.org/10.1063/1.1724319
32. G. Cassinelli, E. Vito, P.J. Lahti, A. Levrero, *The Theory of Symmetry Actions in Quantum Mechanics*. Lect. Not. Phys. (Springer, Berlin, 2004). https://doi.org/10.1007/b99455
33. R.M. Wilcox, J. Math. Phys. **8**, 962 (1967). https://doi.org/10.1063/1.1705306
34. L. Ferialdi, Phys. Rev. D **107**, 105010 (2023). https://doi.org/10.1103/PhysRevD.107.105010
35. J. Campbell, Proc. London Math. Soc. **28**, 381 (1897)
36. H. Baker, Proc. London Math. Soc. **3**, 24 (1905)
37. F. Hausdorff, Ber. Verh. Saechs. Akad. Wiss. Leipzig **58**, 19 (1906)
38. W.O. Kermack, W.H. McCrea, Proc. Edinburgh Math. Soc. **2**, 220 (1931). https://doi.org/10.1017/S0013091500007781
39. M.O. Scully, M.S. Zubairy, *Quantum Optics* (Cambridge University Press, Cambridge, 1997)

Maps

3

Abstract

A crucial aspect in the formulation of quantum theory is the description of state transformations and correspondingly, in a dual way, of observable transformations. We address in detail their structure and interpretation, building on the notion of map. Relevant transformations include time evolutions due to the free dynamics of the system, the coupling with other degrees of freedom acting as an environment, as well as the interaction with a measurement apparatus. The general characterization of these maps puts into evidence the notion of complete positivity, related to non-commutativity and entanglement, reflecting at the level of a single system the quantum features arising in the description of composite systems.

3.1 Introduction

In Chap. 2 we have worked out, starting from the premises of Chap. 1, the general notion of state for a quantum mechanical system, namely statistical operators, and characterized the corresponding convex set. Exploiting the general notion of duality relation in a statistical theory, we have further put into evidence the convex set of effects as elementary observations, leading to POVMs as general notion of observables. Transformations of the set of states into itself play a key role in the development of the theory. The same goes for transformations of the set of elementary observations. As already seen in Sect. 2.2.6, the behavior of these transformations on convex sets fully determines their extension to the containing linear set, so that it is of interest to study the notion of map as linear transformation between spaces of operators. The non-commutativity of these spaces brings with itself the introduction of the mathematical notion of complete positivity, which plays an important role in the whole theory of open quantum systems, and is very helpful in providing a structural characterization of maps. As we will see in Chap. 4, this notion is further

B. Vacchini, *Open Quantum Systems*, Graduate Texts in Physics,
https://doi.org/10.1007/978-3-031-58218-9_3

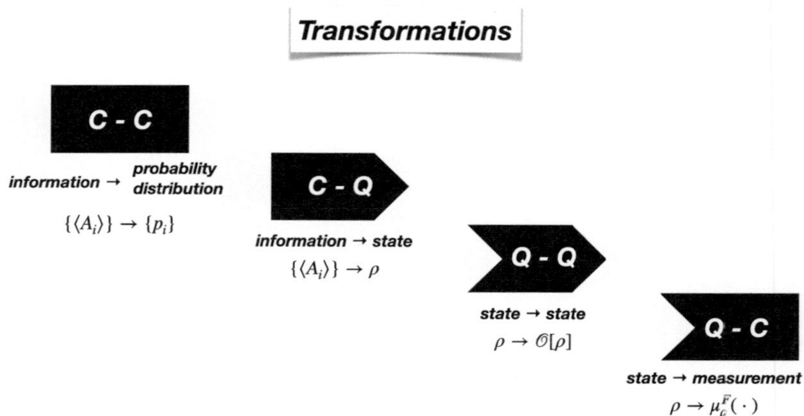

Fig. 3.1 Visual symbolic representation of the different possible kinds of transformations between classical and quantum objects. In a classical-classical transformation we can encode the information about some measurement in a classical state, that is a probability distribution. We can also encode the same information in a quantum system, which does provide a very different information carrier. We thus implement a classical-quantum transformation. A quantum-quantum transformation sends a quantum state in another state or in a subcollection and corresponds to the action of an operation $\mathcal{O}$, that we shall describe in Sect. 3.2.2. The assignment of measurement outcomes to a state, as can be obtained with the POVMs introduced in Sect. 2.3.1, provides a quantum-classical transformation

tightly connected to the tensor product structure in which composite systems have to be described in quantum mechanics, and therefore to the notion of entanglement. The treatment of state transformations is necessary in order to discuss sequence of measurements, that ask for knowledge not only of the statistics of outcomes for a given observable, but also of the way in which the state has been modified to extract this information.

This will lead us to introduce the important notion of operation. The noise associated to these transformations will be further discussed by means of suitable quantifiers. Along this path we can better understand and describe measurement transformations, which can always be interpreted as coming from the interaction with other quantum degrees of freedom, thus letting the notion of instrument emerge. In the case of a finite-dimensional Hilbert space we can further introduce many different convenient representations for these quantum maps, useful for their evaluation and for the assessment of different properties. The different kinds of relevant transformations are symbolically depicted in Fig. 3.1. See also [1] for a nice compact overview.

3.2 Transformations of States and Observables

In order to understand the statistical structure of quantum mechanics, we have considered as simplest statistical experiments those depicted in Fig. 2.1, in which a single particle is prepared and affects a registration apparatus. In this setting, the

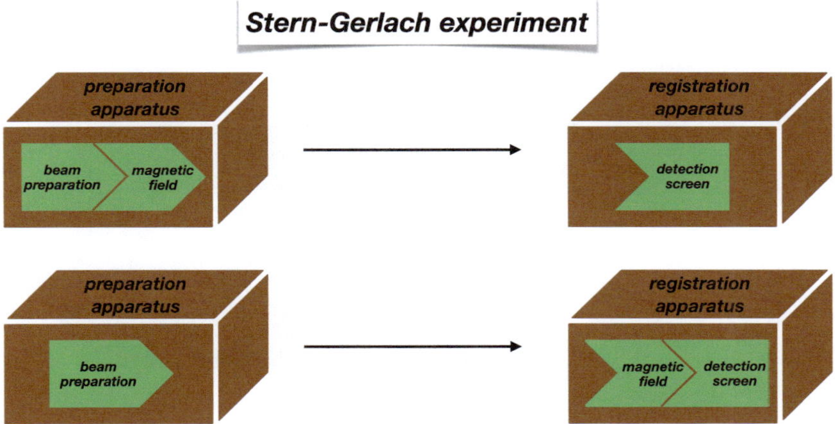

Fig. 3.2 Different subdivision into steps of the Stern-Gerlach experiment. The three basic steps consisting of beam preparation, interaction with the inhomogeneous magnetic field, and final detection on the screen are grouped in different but equivalent ways to define a preparation and a registration stage. The different steps are depicted according to the symbols introduced in Fig. 3.1 to distinguish classical-quantum, quantum-quantum and quantum-classical transformations

preparation procedure based on a classical input delivers a quantum output, namely a state $\rho \in S(\mathcal{H})$, while the registration procedure described by the POVM F with outcome space $(\Omega, \mathcal{F})$ accepts a quantum input and produces a classical output $\mu_\rho^F \in \mathrm{Prob}(\Omega)$, corresponding to the probability distribution for the outcomes of the considered observable.

In general, a measurement can be performed in different steps, however its description can always be reduced to this scheme by including intermediate transformations either on the side of the preparation or of the registration. Note that if many intermediate steps can be put into evidence, such a reduction can generally be performed in different ways. Think for example of the previously considered Stern-Gerlach experiment. The experimenter prepares a beam of particles, then spatially separates different spin components along a given axis thanks to an inhomogeneous magnetic field, finally a measurement of position is performed by detection on a screen. The action of the magnetic field, described by a unitary dynamical evolution, can be seen as part of the state preparation representing a transformation on the space of states, thus corresponding to the Schrödinger picture, or equivalently it can be seen as part of the registration, determining a transformation on the space of observables and thus corresponding to the Heisenberg picture. The two alternative viewpoints are depicted in Fig. 3.2.

Given this situation it immediately appears that it is of interest to characterize the possible transformations of states and of observables, which leads to a description of the intermediate steps in the finalization of a measurement procedure. We will call operations such transformations, which accept a quantum input and produce a quantum output. These transformations of relevant spaces of operators into them-

Fig. 3.3 We recollect, using the notation of Fig. 3.1 to denote quantum-quantum transformations, different kinds of relevant transformations of quantum states

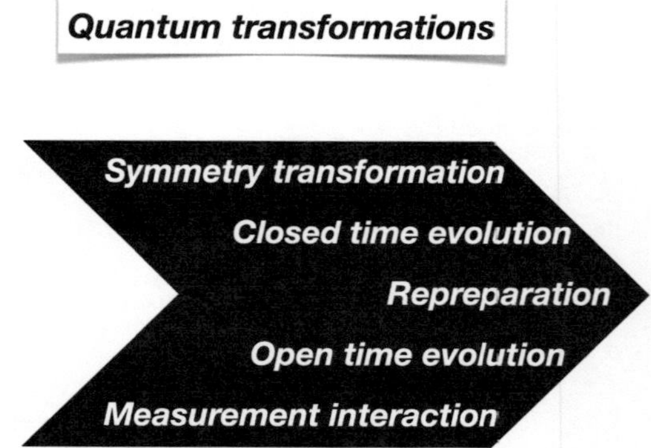

selves are obviously of great importance, and some of the most relevant are recalled in Fig. 3.3.

Each transformation can be seen in a dual way, exchanging the role of states and observables. They allow to describe not only intermediate steps in a given measurement setting, but also symmetry operations, input-output transformations over a finite time, so-called one-shot transformations, as well as a general time-dependent dynamics. Such transformations are moreover necessary if besides a single measurement, leading from a quantum state to a classical probability distribution, we want to describe repeated measurements of the same or of different observables, leading to conditional probabilities. Indeed, in order to perform two measurements one after the other we need to know, besides the statistics of the first measurement, the state transformed as a consequence of it, so as to apply to this new state the second measurement. In such a way, we can obtain conditional probabilities. By iterating this procedure we can also think about performing a high number of subsequent measurements of the same observable, so that in the limit of arbitrarily small time intervals in between these subsequent measurements a description of measurement continuous in time can be obtained, thus recovering in a suitable framework the notion of trajectory also in quantum mechanics [2–4].

Let us first focus on transformations of the state space $S(\mathcal{H}) \subset \mathcal{T}(\mathcal{H})$. Such maps have to respect the convex structure of the state space, and according to the procedure detailed in the proof of Theorem 2.3 they can be uniquely extended to the whole linear space $\mathcal{T}(\mathcal{H})$. This extension allows to exploit useful duality relations. In a general transformation an ensemble of quantum states will be sent to a sub-ensemble. Consider for example the situation in which in the preparation the experimenter selects, out of an unpolarized beam of particles, those which are polarized along a given axis, thus picking up only a fraction of the beam. Such state transformations that we have called operations will thus send elements of $S(\mathcal{H})$ in the space $\tilde{S}(\mathcal{H}) = \{\rho \in \mathcal{T}(\mathcal{H}), \rho \geqslant 0, \mathrm{Tr}\,\rho \leqslant 1\} \subset \mathcal{T}(\mathcal{H})$ of subcollections or sub-normalized statistical operators. Denoting by O such an operation, it is natural to ask this map to be: linear, that is $O[\alpha T + \beta S] = \alpha O[T] + \beta O[S]$ for all $S, T \in \mathcal{T}(\mathcal{H})$ and $\alpha, \beta \in \mathbb{C}$; positive, namely $O[T] \geqslant 0$ for $T \geqslant 0$; trace-nonincreasing, that is $\mathrm{Tr}\{O[T]\} \leqslant \mathrm{Tr}\{T\}$ for all positive $T \in \mathcal{T}(\mathcal{H})$. The latter condition reflects the fact that a state can be sent to a subcollection. It turns out, however, that the requirement of being positive is not sufficient, since it allows for the following unphysical situation. Suppose we have an operation acting on $\mathcal{T}(\mathcal{H})$, and we consider our system described on $\mathcal{H}$ together with another system described on a Hilbert space $\mathbb{C}^n$, with $n \in \mathbb{N}$. We extend our map to $\mathcal{T}(\mathcal{H} \otimes \mathbb{C}^n)$ by letting it act in a trivial way on the other degrees of freedom, namely we consider the map $O \otimes \mathbb{1}_{\mathcal{M}_n}(\mathbb{C})$. It might happen, as we shall see by means of example in Sect. 4.3.3, that the latter map is not positive, so that states are no more sent to states. Such a behavior, coming about just due to the non-commutativity of the space of operators, is strictly related to the notion of entanglement, that we shall briefly discuss in Sect. 4.3, and it calls for a more stringent requirement on the transformation maps known as complete positivity. The property of complete positivity entered the stage of quantum mechanics in the 70's, when its relevance was discovered in connection with studies on the quantum theory of measurement [5–8].

3.2.1 Complete Positivity

Let us now introduce the notion of complete positivity. We first recall that, as discussed in Sect. 2.2.2, the space of bounded operators $\mathcal{B}(\mathcal{H})$ is the dual of the space of trace class operators $\mathcal{T}(\mathcal{H})$, i.e. $\mathcal{B}(\mathcal{H}) = \mathcal{T}'(\mathcal{H})$ according to the duality formula

$$\mathrm{Tr} : \mathcal{B}(\mathcal{H}) \times \mathcal{T}(\mathcal{H}) \rightarrow \mathbb{C} \tag{3.1}$$
$$(B, T) \mapsto \mathrm{Tr}\{B^{\dagger}T\},$$

with $B \in \mathcal{B}(\mathcal{H})$ and $T \in \mathcal{T}(\mathcal{H})$. Given a map Φ acting on states

$$\Phi : \mathcal{T}(\mathcal{H}) \rightarrow \mathcal{T}(\mathcal{H}) \tag{3.2}$$

the duality relation allows to introduce the dual or adjoint map acting on the dual space

$$\Phi' : \mathcal{B}(\mathcal{H}) \rightarrow \mathcal{B}(\mathcal{H}), \tag{3.3}$$

whose action is defined asking

$$\mathrm{Tr}\{B\Phi[T]\} = \mathrm{Tr}\{\Phi'[B]T\} \tag{3.4}$$

for all $B \in \mathcal{B}(\mathcal{H})$ and $T \in \mathcal{T}(\mathcal{H})$. Note that from the important relation

$$\mathrm{Tr}\{\Phi[T]\} = \mathrm{Tr}\{\mathbb{1}\Phi[T]\} \tag{3.5}$$
$$= \mathrm{Tr}\{\Phi'[\mathbb{1}]T\} \tag{3.6}$$

for all $T \in \mathcal{T}(\mathcal{H})$ we immediately have that if Φ is trace-nonincreasing, then $\Phi'[\mathbb{1}]$ is an effect, that is to say $0 \leqslant \Phi'[\mathbb{1}] \leqslant \mathbb{1}$, as can be immediately verified taking as trace class operators T one-dimensional projections. In particular, trace preservation of Φ corresponds to the fact that Φ' is unit-preserving, namely

$$\mathrm{Tr}\{\Phi[T]\} = \mathrm{Tr}\{T\} \quad \Leftrightarrow \quad \Phi'[\mathbb{1}] = \mathbb{1}. \tag{3.7}$$

As simple but paradigmatic example consider the map

$$\Phi(t)[\rho] = U(t)\rho U^{\dagger}(t), \tag{3.8}$$

with $\{U(t)\}_{t \in \mathbb{R}}$ a group of unitary operators, which gives the time evolution in Schrödinger picture. Its adjoint is easily seen to be given by

$$\Phi'(t)[X] = U^{\dagger}(t)X U(t), \tag{3.9}$$

corresponding to the time evolution in Heisenberg picture. We notice that the notion of dual map allows to introduce a notion of time evolution for the observables even if the dynamics is no more unitary.

We are now in the position to define an important property of linear maps known as complete positivity. We say that a map Φ'

$$\Phi' : \mathcal{B}(\mathcal{H}) \to \mathcal{B}(\mathcal{H}) \tag{3.10}$$
$$B \mapsto \Phi'[B]$$

is completely positive if the following equivalent conditions hold:

1. For any $n \in \mathbb{N}$ the map

$$\Phi' \otimes \mathbb{1}_{\mathcal{M}_n(\mathbb{C})} : \mathcal{B}(\mathcal{H} \otimes \mathbb{C}^n) \to \mathcal{B}(\mathcal{H} \otimes \mathbb{C}^n) \tag{3.11}$$
$$B \otimes C \mapsto \Phi'[B] \otimes C,$$

with $B \in \mathcal{B}(\mathcal{H})$ and $C \in \mathcal{M}_n(\mathbb{C})$, sends positive operators to positive operators
2. For any $n \in \mathbb{N}$ the inequality

$$\sum_{i,j=1}^{n} \langle \psi_i | \Phi'[B_i^{\dagger} B_j] | \psi_j \rangle \geqslant 0 \qquad \forall \{\psi_i\} \subset \mathcal{H}, \forall \{B_i\} \subset \mathcal{B}(\mathcal{H}) \tag{3.12}$$

holds.

Note that due to the relation $(\Phi \otimes \mathbb{1})' = \Phi' \otimes \mathbb{1}$ between adjoint maps, which can be checked starting from the duality relation (3.4), the requirement of complete positivity can be equivalently set on the map Φ itself rather than its adjoint, that is asking the maps

$$\Phi \otimes \mathbb{1}_{M_n(\mathbb{C})} : \mathcal{T}(\mathcal{H} \otimes \mathbb{C}^n) \to \mathcal{T}(\mathcal{H} \otimes \mathbb{C}^n) \tag{3.13}$$

to be positive for all $n \in \mathbb{N}$, or up to the dimension of $\mathcal{H}$ in the finite-dimensional case, as we shall see in Sect. 4.4.1. Making reference to (3.13) the condition of complete positivity can be rephrased asking that the map extended in the trivial way to a tensor product structure remains positive, as further discussed in Sect. 4.4.

In order to show the equivalence of the two conditions (3.11) and (3.12) we recall that $\mathcal{H} \otimes \mathbb{C}^n = \bigoplus_{i=1}^n \mathcal{H}$, so that a pure state in $\mathcal{H} \otimes \mathbb{C}^n$ can be represented as

$$\Psi = \begin{pmatrix} \psi_1 \\ \vdots \\ \psi_n \end{pmatrix}, \tag{3.14}$$

with $\psi_i \in \mathcal{H}$, and at the same time an element of $\mathcal{B}(\mathcal{H} \otimes \mathbb{C}^n)$ can be identified as a $n \times n$ matrix with operator entries, that is $\mathcal{B}(\mathcal{H} \otimes \mathbb{C}^n) = \mathcal{M}_n(\mathcal{B}(\mathcal{H}))$, according to

$$B = \sum_{k,l=1}^n B_{kl} \otimes |e_k\rangle\langle e_l|, \tag{3.15}$$

with $\{B_{kl}\} \subset \mathcal{B}(\mathcal{H})$ and $\{e_i\}_{i=1}^n$ a basis of vectors in $\mathbb{C}^n$. A positive operator A in $\mathcal{B}(\mathcal{H} \otimes \mathbb{C}^n)$ can therefore be written in the form

$$A = B^\dagger B \tag{3.16}$$

$$= \sum_{r=1}^n \sum_{k,l=1}^n B_{rk}^\dagger B_{rl} \otimes |e_k\rangle\langle e_l|, \tag{3.17}$$

or possibly as a sum of such terms, and in matrix form, using the Kronecker product and making reference to the chosen basis

$$A = \sum_{r=1}^n \begin{pmatrix} B_{r1}^\dagger B_{r1} & \dots & B_{r1}^\dagger B_{rn} \\ \vdots & \ddots & \vdots \\ B_{rn}^\dagger B_{r1} & \dots & B_{rn}^\dagger B_{rn} \end{pmatrix} \tag{3.18}$$

$$= \sum_{r=1}^n \boldsymbol{B}_r^\dagger \otimes \boldsymbol{B}_r \tag{3.19}$$

with $\boldsymbol{B}_r = (B_{r1}, \dots, B_{rn})$ a vector of operators and $\otimes$ here denotes the outer product of two vectors. Positivity of the map $\Phi' \otimes \mathbb{1}_{M_n(\mathbb{C})}$ on $\mathcal{B}(\mathcal{H} \otimes \mathbb{C}^n)$ is then expressed as

$$\langle\Psi|(\Phi'\otimes\mathbb{1}_{\mathcal{M}_n(\mathbb{C})})\sum_{r=1}^{n}\boldsymbol{B}_r^\dagger\otimes\boldsymbol{B}_r\Psi\rangle\geqslant 0 \tag{3.20}$$

for any $\Psi\in\mathcal{H}\otimes\mathbb{C}^n$ and for all $\{B_{ri}\}\subset\mathcal{B}(\mathcal{H})$, which can be equivalently written

$$\sum_{i,j=1}^{n}\langle\psi_i|\Phi'[B_i^\dagger B_j]\psi_j\rangle\geqslant 0 \tag{3.21}$$

$\forall n\in\mathbb{N}$ as well as $\forall\{\psi_i\}\subset\mathcal{H}$ and $\forall\{B_i\}\subset\mathcal{B}(\mathcal{H})$, since validity of (3.20) for the case in which r takes on a single value is necessary but also sufficient to ensure the condition for the generic case. We have thus proven the equivalence between (3.11) and (3.12).

A simple but important fact is that any unitary transformation

$$\mathcal{U}[\rho]=U\rho U^\dagger, \tag{3.22}$$

is not only positive, but also completely positive. Indeed, condition (3.12) now corresponds to positivity of the quantity

$$\sum_{i,j=1}^{n}\langle\psi_i|\Phi'[B_i^\dagger B_j]\psi_j\rangle = \sum_{i,j=1}^{n}\langle\psi_i|U^\dagger(B_i^\dagger B_j)U\psi_j\rangle \tag{3.23}$$

$$= \sum_{i,j=1}^{n}\langle\psi_i|(U^\dagger B_i^\dagger U)(U^\dagger B_j U)\psi_j\rangle \tag{3.24}$$

$$= \left\|\sum_{j=1}^{n}(U^\dagger B_j U)\psi_j\right\|_{\mathcal{H}}^2, \tag{3.25}$$

for any choice of vectors in $\mathcal{H}$ and operators in $\mathcal{B}(\mathcal{H})$, which is obviously verified. According to the very definition of complete positivity, the dual map which according to (3.4) reads

$$\mathcal{U}'[\rho]=U^\dagger\rho U, \tag{3.26}$$

so that

$$\mathcal{U}'[\rho]=\mathcal{U}[\rho]^\dagger \tag{3.27}$$
$$=\mathcal{U}^{-1}[\rho] \tag{3.28}$$

is also completely positive.

3.1 Partial Trace Operation

The partial trace Tr_E is the unique linear assignment

$$\mathrm{Tr}_E[\cdot] : \mathcal{T}(\mathcal{H}_S \otimes \mathcal{H}_E) \to \mathcal{T}(\mathcal{H}_S) \qquad (3.29)$$
$$T \mapsto \mathrm{Tr}_E[T]$$

such that

$$\mathrm{Tr}_{S \otimes E}\{T(A_S \otimes \mathbb{1}_E)\} = \mathrm{Tr}_S\{\mathrm{Tr}_E[T]A_S\} \qquad (3.30)$$

$\forall A_S \in \mathcal{B}(\mathcal{H}_S)$ and $\forall T \in \mathcal{T}(\mathcal{H}_S \otimes \mathcal{H}_E)$. The partial trace thus associates to any state in $\mathcal{H}_S \otimes \mathcal{H}_E$ the unique system state in $\mathcal{H}_S$ which correctly reproduces all the local observations, namely those on the system only. The constraint (3.30) implies in particular for factorized operators $T = T_S \otimes T_E \in \mathcal{T}(\mathcal{H}_S \otimes \mathcal{H}_E)$ the chain of equalities

$$\mathrm{Tr}_S\{\mathrm{Tr}_E[T_S \otimes T_E]A_S\} = \mathrm{Tr}_{S \otimes E}\{(T_S \otimes T_E)(A_S \otimes \mathbb{1}_E)\} \qquad (3.31)$$
$$= \mathrm{Tr}_{S \otimes E}\{T_S A_S \otimes T_E\} \qquad (3.32)$$
$$= \mathrm{Tr}_S\{T_S A_S\}\,\mathrm{Tr}_E\{T_E\}, \qquad (3.33)$$

where the last one follows from the standard definition of trace over the Hilbert space $\mathcal{H}_S \otimes \mathcal{H}_E$. Since this relation is valid for any A_S, for factorized states the partial trace reads

$$\mathrm{Tr}_E[T_S \otimes T_E] = \mathrm{Tr}_E\{T_E\}T_S. \qquad (3.34)$$

Exploiting this result together with the relation

$$T = \sum_{j,k}\sum_{m,n}\langle\psi_j \otimes \varphi_k|T\psi_m \otimes \varphi_n\rangle|\psi_j\rangle\langle\psi_m| \otimes |\varphi_k\rangle\langle\varphi_n|, \qquad (3.35)$$

which expresses in a unique way the operator T in the operator bases $\{|\psi_j\rangle\langle\psi_m|\}$ and $\{|\varphi_k\rangle\langle\varphi_n|\}$ obtained from the c.o.n.s. $\{\psi_i\}$ and $\{\varphi_n\}$ in $\mathcal{H}_S$ and $\mathcal{H}_E$ respectively, together with the identity

$$\mathrm{Tr}_E[|\psi_j\rangle\langle\psi_m| \otimes |\varphi_k\rangle\langle\varphi_n|] = |\psi_j\rangle\langle\psi_m|\delta_{k,n} \qquad (3.36)$$

following from (3.34), we finally come to

$$\mathrm{Tr}_E[T] = \sum_{j,k,m}\langle\psi_j \otimes \varphi_k|T\psi_m \otimes \varphi_k\rangle|\psi_j\rangle\langle\psi_m| \qquad (3.37)$$

$$= \sum_k \langle \varphi_k | T \varphi_k \rangle, \tag{3.38}$$

which provides a direct way to evaluate the partial trace. Note that, as it should be, the result is independent from the choice of basis in $\mathcal{H}_E$. The partial trace is thus obtained by summing over the diagonal matrix elements of the operator with respect to an arbitrary basis in $\mathcal{H}_E$. Note that the partial trace typically leads from pure to mixed states, as we shall discuss in Sect. 4.3. For the sake of clarity, whenever confusion might arise, we will denote the partial trace over the degrees of freedom of the Hilbert space $\mathcal{H}_E$ as $\mathrm{Tr}_E[\cdot]$, and the trace of an operator on the Hilbert space $\mathcal{H}_E$ in the sense of (2.38) as $\mathrm{Tr}_E\{\cdot\}$ or simply $\mathrm{Tr}_E \cdot$.

Another fundamental example of completely positive trace-preserving map, acting between different operator spaces, is the partial trace introduced in Remark 3.1. We now show that the partial trace operation $\mathrm{Tr}_E[\cdot]$ is completely positive. Indeed, according to (3.38) we have $\forall \{\psi_i\} \subset \mathcal{H}_S$, $\forall \{T_i\} \subset \mathcal{T}(\mathcal{H}_S \otimes \mathcal{H}_E)$ and $\forall n \in \mathbb{N}$

$$\sum_{i,j=1}^{n} \langle \psi_i | \mathrm{Tr}_E[T_i^\dagger T_j] \psi_j \rangle = \sum_{i,j=1}^{n} \sum_k \langle \psi_i \otimes \varphi_k | T_i^\dagger T_j \psi_j \otimes \varphi_k \rangle \tag{3.39}$$

$$= \sum_k \left\| \sum_{j=1}^{n} T_j \psi_j \otimes \varphi_k \right\|^2. \tag{3.40}$$

Also in this case it is worth considering the dual map $\mathrm{Tr}'_E[\cdot]$, which will provide another example of completely positive map. Following (3.4) the map dual to the partial trace, namely

$$\mathrm{Tr}'_E[\cdot] : \mathcal{B}(\mathcal{H}_S) \to \mathcal{B}(\mathcal{H}_S \otimes \mathcal{H}_E) \tag{3.41}$$
$$B \mapsto \mathrm{Tr}'_E[B]$$

is obtained from the relation

$$\mathrm{Tr}_{S \otimes E}\{T \, \mathrm{Tr}'_E[B]\} = \mathrm{Tr}_S\{\mathrm{Tr}_E[T] B\}, \tag{3.42}$$

so that thanks to (3.30) we have the identification

$$\mathrm{Tr}'_E[B] = B \otimes \mathbb{1}_E. \tag{3.43}$$

The map dual to the partial trace thus sends an operator acting on the system degrees of freedom to an operator on the tensor product Hilbert space acting trivially on the environmental degrees of freedom. Note that according to condition (3.11) complete positivity is stable under composition. If Φ and Λ are completely positive, then both $\Phi \otimes \mathbb{1}$ and $\Lambda \otimes \mathbb{1}$ are positive, so that such is

$$(\Phi \otimes \mathbb{1}) \circ (\Lambda \otimes \mathbb{1}) = (\Phi \circ \Lambda) \otimes \mathbb{1} \tag{3.44}$$

and therefore $\Phi \circ \Lambda$ is completely positive. The same is true for linear combinations with positive coefficients, so that the set of completely positive maps is a positive cone. We stress furthermore that $\Phi \otimes \mathbb{1}$ is positive iff $\mathbb{1} \otimes \Phi$ is positive. A fundamental example of completely positive map is considered in Remark 3.2.

> ### 3.2 Basic Completely Positive Map
>
> As a fundamental example we consider a simple transformation built as follows. Consider an operator $S \in \mathcal{B}(\mathcal{H})$ and define
>
> $$O_S[X] = S X S^{\dagger}. \tag{3.45}$$
>
> Linearity is obvious, while regarding complete positivity for arbitrary n, any $T \in \mathcal{T}(\mathcal{H} \otimes \mathbb{C}^n)$ with $T \geqslant 0$, and any $\psi \in \mathcal{H} \otimes \mathbb{C}^n$ we have
>
> $$\langle \psi | (O_S \otimes \mathbb{1}_{M_n(\mathbb{C})})[T] \psi \rangle = \langle \psi | (S \otimes \mathbb{1}_{\mathbb{C}^n}) T (S \otimes \mathbb{1}_{\mathbb{C}^n})^{\dagger} \psi \rangle \tag{3.46}$$
> $$= \langle \tilde{\psi} | T \tilde{\psi} \rangle \tag{3.47}$$
> $$\geqslant 0, \tag{3.48}$$
>
> where we have denoted as $\tilde{\psi} \in \mathcal{H} \otimes \mathbb{C}^n$ the vector $(S \otimes \mathbb{1}_{\mathbb{C}^n})^{\dagger} \psi$. The adjoint map has the same form with S replaced by its adjoint, namely $O_S' = O_{S^{\dagger}}$.

3.2.2 Operations

We are now in the position to properly define an operation as a map

$$O : \mathcal{T}(\mathcal{H}) \to \mathcal{T}(\mathcal{H}) \tag{3.49}$$
$$T \mapsto O[T]$$

which is linear, completely positive and trace-nonincreasing, i.e. $\mathrm{Tr}\{O[T]\} \leqslant \mathrm{Tr}\{T\}$ for all $T \in \mathcal{T}(\mathcal{H})$, $T \geqslant 0$. Note that in general initial and final Hilbert spaces can be different. If in particular $\mathrm{Tr}\{O[T]\} = \mathrm{Tr}\{T\}$ for all $T \in \mathcal{T}(\mathcal{H})$, so that the map is trace-preserving the quantum operation is called a channel. An operation generally sends $\mathcal{S}(\mathcal{H})$ into $\widetilde{\mathcal{S}}(\mathcal{H})$, so that a statistical operator is sent to a subcollection, while a channel sends $\mathcal{S}(\mathcal{H})$ into $\mathcal{S}(\mathcal{H})$, so that statistical operators are transformed in statistical operators. As we have already seen, if O is trace-preserving, then the dual map O' preserves the identity, i.e. $O'[\mathbb{1}] = \mathbb{1}$, and in this case it is called unital. Since O is trace-nonincreasing, $O'[\mathbb{1}]$ is a positive operator between zero and one, that is an effect. Furthermore, as we have seen, the map O is completely positive iff such is O', so that complete positivity can equally well be verified on the map or its dual. Note finally that the space of operations or channels is closed under convex combinations. A paradigmatic example of operation is given in Remark 3.3. A very special example of operation is given by a unitary transformation of the form (3.22), whose complete positivity has already been proven. This operation is a channel, and is reversible, in the sense that $\Phi_U^{-1} = \Phi_{U^\dagger}$ is again a channel. Unitary transformations are the only reversible operations.

An operation of great relevance for the treatment of open quantum systems is the so-called assignment map. If $\mathcal{H}_S$ is the Hilbert space of the system of interest, given a Hilbert space $\mathcal{H}_E$ and a fixed statistical operator ρ_E in it we define the assignment map as follows

$$\mathcal{A}_{\rho_E} : \mathcal{T}(\mathcal{H}_S) \to \mathcal{T}(\mathcal{H}_S \otimes \mathcal{H}_E). \tag{3.50}$$
$$T \mapsto T \otimes \rho_E.$$

Complete positivity of this transformation is immediately verified making reference to the condition (3.11). Suppose to extend this map in the trivial way to another ancillary Hilbert space $\mathcal{H}_A$. For any positive operator $X \in \mathcal{T}(\mathcal{H}_S \otimes \mathcal{H}_A)$ we have

$$(\mathcal{A}_{\rho_E} \otimes \mathbb{1}_A)X = X \otimes \rho_E \tag{3.51}$$

which is still a positive operator. This operation is again in particular a channel, due to normalization of ρ_E. The dual map $\mathcal{A}'_{\rho_E}[\cdot]$ can be determined starting from the constraint (3.4), which in this case reads

$$\mathrm{Tr}_S\{T\mathcal{A}'_{\rho_E}[B]\} = \mathrm{Tr}_{S\otimes E}\{\mathcal{A}_{\rho_E}[T]B\} \tag{3.52}$$
$$= \mathrm{Tr}_{S\otimes E}\{(T \otimes \rho_E)B\} \tag{3.53}$$
$$= \mathrm{Tr}_S\{T\,\mathrm{Tr}_E[(\mathbb{1}_S \otimes \rho_E)B]\} \tag{3.54}$$

with arbitrary $B \in \mathcal{B}(\mathcal{H}_S \otimes \mathcal{H}_E)$ and $T \in \mathcal{T}(\mathcal{H}_S)$, leading to the following identification of the dual of the assignment map

$$\mathcal{A}'_{\rho_E} : \mathcal{B}(\mathcal{H}_S \otimes \mathcal{H}_E) \to \mathcal{B}(\mathcal{H}_S) \tag{3.55}$$
$$B \mapsto \mathrm{Tr}_E[(\mathbb{1}_S \otimes \rho_E)B].$$

This map is in particular unit-preserving, corresponding to the fact that $\mathcal{A}_{\rho_E}$ is a channel. As a result while the assignment map $\mathcal{A}_{\rho_E}$ sends a state in the Hilbert space

$\mathcal{H}_S$ of the system to a state in the larger Hilbert space $\mathcal{H}_S \otimes \mathcal{H}_E$ by taking the tensor product with a fixed state of the environment, the dual map assigns to an observable in the larger space an observable in the space of the system only, by averaging it with respect to the same fixed environmental state. We notice further that the map $\mathrm{Tr}_E \circ \mathcal{A}_{\rho_E}$ is a completely positive trace-preserving map acting as the identity on $\mathcal{T}(\mathcal{H}_S)$, so that the partial trace acts as left inverse of the assignment map

$$\mathrm{Tr}_E \circ \mathcal{A}_{\rho_E} : \mathcal{T}(\mathcal{H}_S) \to \mathcal{T}(\mathcal{H}_S) \tag{3.56}$$
$$T \mapsto T,$$

that is

$$\mathrm{Tr}_E \circ \mathcal{A}_{\rho_E} = \mathbb{1}_{\mathcal{T}(\mathcal{H}_S)}. \tag{3.57}$$

On the other hand, the map $\mathcal{A}_{\rho_E} \circ \mathrm{Tr}_E$ is a completely positive trace-preserving and idempotent map on $\mathcal{T}(\mathcal{H}_S \otimes \mathcal{H}_E)$

$$\mathcal{A}_{\rho_E} \circ \mathrm{Tr}_E : \mathcal{T}(\mathcal{H}_S \otimes \mathcal{H}_E) \to \mathcal{T}(\mathcal{H}_S \otimes \mathcal{H}_E) \tag{3.58}$$
$$X \mapsto \mathrm{Tr}_E\{T\} \otimes \rho_E,$$

that is

$$\mathcal{A}_{\rho_E} \circ \mathrm{Tr}_E = \mathcal{P}, \tag{3.59}$$

with $\mathcal{P}$ a projection. The map of (3.56) warrants (3.57) that will be used in Sect. 5.2 as compatibility condition to ensure the existence of a reduced dynamics, while the projection given by (3.58) corresponds to the standard projection that will be introduced in Sect. 5.5.1 to deal with the dynamics of a generic open quantum system. When it acts on a state in the bipartite Hilbert space $\mathcal{H}_S \otimes \mathcal{H}_E$, it sends it into a factorized state with the same first marginal and a fixed environmental marginal.

3.3 Basic Operation

Building on Remark 3.2, that is considering the completely positive map

$$O_S[X] = SXS^\dagger \tag{3.60}$$

with $S \in \mathcal{B}(\mathcal{H})$, we can obtain an operation by imposing the transformation to be trace-nonincreasing. This condition leads to

$$\mathrm{Tr}\{O_S[X]\} = \mathrm{Tr}\{S^\dagger SX\} \tag{3.61}$$
$$\leqslant \mathrm{Tr}\{X\}, \tag{3.62}$$

for arbitrary positive trace class operator X. The map O_S is therefore an operation iff $\langle \psi | S^\dagger S \psi \rangle \leqslant \|\psi\|^2$ for any state $\psi \in \mathcal{H}$. This condition can be equivalently stated as $\|S\| \leqslant 1$ or $S^\dagger S \leqslant \mathbb{1}$, so that the operator $S^\dagger S = O'_S[\mathbb{1}]$ is an effect. We immediately see that more in general the map $O[X] = \sum_k S_k X S_k^\dagger$ still gives an operation provided the operators $S_k^\dagger S_k$ are compatible effects, that is $\sum_k S_k^\dagger S_k \leqslant \mathbb{1}$.

3.2.3 Measurement and State Transformations

We now dwell on the physical meaning of operations in connection with conditional measurements. From the defining properties of an operation O it immediately follows that the operator $E = O'[\mathbb{1}]$ is positive and below the identity, so that it is an effect. An operation therefore uniquely identifies an effect. Vice versa, many distinct operations can lead to the same effect, so that to an effect we can associate a whole equivalence class of operations $[O]_E$, which are E compatible in the sense that $O'[\mathbb{1}] = E$ for all $O \in [O]_E$. This many to one relationship is associated to the meaning of operation as state transformation corresponding to the measurement of the elementary observable described by the effect E. There are in general many different ways to measure the same effect, corresponding to distinct apparata leading to different modifications of the system state, and they lead to the set of operations compatible with the given effect. Let us assume that the state of the system before the measurement is described by an operator $\rho \in \mathcal{S}(\mathcal{H})$, often called pre-measurement state. The probability to observe the event described by the effect E is then given by

$$P_\rho(E) = \mu_\rho^E \tag{3.63}$$
$$= \mathrm{Tr}\{\rho E\} \tag{3.64}$$
$$= \mathrm{Tr}\{\rho O'[\mathbb{1}]\} \tag{3.65}$$
$$= \mathrm{Tr}\{O[\rho]\}, \tag{3.66}$$

where O is any operation compatible with the given effect. As it appears, the probability for the realization of the effect is given by the trace of the subcollection $O[\rho]$, which provides the statistical weight associated to the event. Note that (3.66) can also be interpreted as the measurement of the identity after applying the operation O to the state ρ.

3.2.3.1 Sequence of Measurements

While as far as predictions on the measurement of E only are concerned, any operation in the equivalence class $[O]_E$ leads to the same result, this is no more true when we perform subsequently the measurement of another effect, let us call it F. Indeed, to perform this second measurement we need to know not only the statistics of the first measurement, but also the transformed state, so that the notion of operation now becomes essential. The post-measurement state conditioned on the occurrence of E is given by

$$\rho_{|E} = \frac{O[\rho]}{\text{Tr}\{O[\rho]\}}, \tag{3.67}$$

which is a suitably normalized statistical operator. We can now express the conditional probability for the occurrence of the effect F in the second measurement given the occurrence of E in the first measurement as

$$P_\rho(F|E) = \mu^F_{\rho_{|E}} \tag{3.68}$$
$$= \text{Tr}\{\rho_{|E} F\} \tag{3.69}$$
$$= \frac{\text{Tr}\{O[\rho]F\}}{\text{Tr}\{O[\rho]\}} \tag{3.70}$$
$$= \frac{\text{Tr}\{O[\rho]F\}}{P_\rho(E)}, \tag{3.71}$$

which also leads to the expression of the joint probability as

$$P_\rho(F, E) = P_\rho(F|E)P_\rho(E) \tag{3.72}$$
$$= \text{Tr}\{O[\rho]F\} \tag{3.73}$$
$$= \text{Tr}\{\rho O'[F]\}. \tag{3.74}$$

Note that the order in which measurements are performed is actually relevant. Indeed, denoting by O_E and O_F operations in the equivalence class compatible with the effects E and F respectively, we have the relations

$$P_\rho(F, E) = \text{Tr}\{O'_E[O'_F[\mathbb{1}]]\rho\} \tag{3.75}$$
$$= \text{Tr}\{O_F[O_E[\rho]]\}, \tag{3.76}$$

to be compared with

$$P_\rho(E, F) = \text{Tr}\{O'_F[O'_E[\mathbb{1}]]\rho\} \tag{3.77}$$
$$= \text{Tr}\{O_E[O_F[\rho]]\}. \tag{3.78}$$

As it appears, the second measurement actually depends on how the state has been transformed through the operation describing the first measurement. Different elements of the equivalence class $[O]_E$ lead to different results for $P_\rho(F, E)$. While the transformation on the state $O[\rho]$ corresponds to a Schrödinger picture, the transformation on the effect $O'[F]$ can be seen as a Heisenberg picture. As already discussed in Sect. 3.2 and depicted in Fig. 3.2 an experiment in which many steps are performed can be divided in different ways in a preparation and a registration part.

It is important to stress that while O is a linear map, the transformation which gives the state conditioned on the outcome

$$\rho \to \rho_{|E} = \frac{O[\rho]}{\mathrm{Tr}\{O[\rho]\}} \tag{3.79}$$

is no more linear since it involves an intermediate measurement. Indeed, considering a convex combination of states $\rho = p_1\rho_1 + p_2\rho_2$, with $p_{1,2} \geqslant 0$ and $p_1 + p_2 = 1$, we have

$$\rho_{|E} = \frac{p_1 O[\rho_1] + p_2 O[\rho_2]}{p_1 \,\mathrm{Tr}\{\rho_1 E\} + p_2 \,\mathrm{Tr}\{\rho_2 E\}} \tag{3.80}$$

$$= \frac{p_1 \,\mathrm{Tr}\{\rho_1 E\}}{\mathrm{Tr}\{\rho E\}} \frac{O[\rho_1]}{\mathrm{Tr}\{\rho_1 E\}} + \frac{p_2 \,\mathrm{Tr}\{\rho_2 E\}}{\mathrm{Tr}\{\rho E\}} \frac{O[\rho_2]}{\mathrm{Tr}\{\rho_2 E\}} \tag{3.81}$$

$$= p_{1|E}\rho_{1|E} + p_{2|E}\rho_{2|E}, \tag{3.82}$$

that is a convex combination of the conditioned states with new weights $p_{1|E}$ and $p_{2|E}$, corresponding to the probability of the first and second case conditioned on the outcome E.

It immediately appears that the formalism used to describe a sequence of two measurements allows to make predictions about the outcome of an arbitrary finite number of subsequent measurements. Indeed, starting from (3.76) and considering the measurement of the ordered sequence of effects $E_1, \ldots, E_n$ we have

$$P_\rho(E_n, \ldots, E_1) = \mathrm{Tr}\{O_{E_n}[\ldots O_{E_2}[O_{E_1}[\rho]]\ldots]\} \tag{3.83}$$

or equivalently

$$P_\rho(E_n, \ldots, E_1) = \mathrm{Tr}\{O'_{E_1}[\ldots O'_{E_{n-1}}[O'_{E_n}[\mathbb{1}]]\ldots]\rho\}, \tag{3.84}$$

corresponding again to Schrödinger and Heisenberg picture descriptions.

3.2.3.2 State Preparation

We now describe the connection between operations and state preparations. As discussed in Sect. 2.2.1, a possible way to perform the preparation of a state is to exploit apparata for the sharp measurement of a complete set of commuting observables, which can be identified with commuting PVMs. In this setting, let us denote by $A = (A_1, \ldots, A_k)$ the commuting self-adjoint operators representing the observables and by $a = (a_1, \ldots, a_k)$ the corresponding eigenvalues, which we suppose to be non-degenerate and that identify the eigenstates u_a. We thus have the spectral representation

$$A_i = \sum_a a_i P_a \tag{3.85}$$

where P_a is the one-dimensional projection corresponding to the state u_a

$$P_a = |u_a\rangle\langle u_a|. \tag{3.86}$$

We want to consider an operation O_a describing the statistics of the measurement of the effects in the range of our observable. According to (3.66) for any state ρ it has to satisfy the constraint

$$\mathrm{Tr}\{O_a[\rho]\} = \mathrm{Tr}\{\rho P_a\} \tag{3.87}$$
$$= \mu_\rho^{P_a} \tag{3.88}$$
$$= P_\rho(P_a), \tag{3.89}$$

where according to the fact that we are considering a PVM the operators $\{P_a\}$ are in particular projections. It is therefore natural to consider the Ansatz

$$O_a[\rho] = P_a \rho P_a, \tag{3.90}$$

which turns out to be a proper operation for any a, obviously compatible with the constraint (3.89). The effect associated to each of these operations $O'_a[\mathbb{1}] = P_a$ is as expected a projection. This kind of operations enjoy certain special properties which are known as repeatability and ideality, that will be further discussed in Remark 3.15. Repeatability means that if we repeat the measurement on the state selected according to a given outcome, with probability one we recover the same outcome, i.e.

$$\mu_{\rho|P_a}^{P_a} = \mathrm{Tr}[P_a \rho_{|P_a}] \tag{3.91}$$
$$= \mathrm{Tr}\left[P_a \frac{O_a[\rho]}{\mathrm{Tr}\,O_a[\rho]}\right] \tag{3.92}$$
$$= 1. \tag{3.93}$$

Ideality means that if an outcome occurs with probability one, then we can conclude that the initial state has been left unchanged by the measurement apparatus. More specifically, given

$$\mathrm{Tr}\{O_a[\rho]\} = 1 \tag{3.94}$$

for a given a, then

$$O_a[\rho] = \rho. \tag{3.95}$$

To prove this assume (3.94), so that according to (3.90) and idempotency of projections we have

$$\mathrm{Tr}\{(\mathbb{1} - P_a)\rho\} = 0. \tag{3.96}$$

We now observe that for any projection Q and positive operator T we have

$$\mathrm{Tr}\{QT\} = \mathrm{Tr}\left\{\left(Q\sqrt{T}\right)^{\dagger}\left(Q\sqrt{T}\right)\right\} \tag{3.97}$$

$$= \left\|Q\sqrt{T}\right\|_{2}^{2}, \tag{3.98}$$

where in (3.98) the Hilbert-Schmidt norm (2.50) appears, which can be zero only if its argument is the null operator. From $Q\sqrt{T} = 0$ we further conclude $\left(Q\sqrt{T}\right)^{\dagger} = 0$, hence $QT = TQ = 0$. As a result we have $(\mathbb{1} - P_a)\rho = \rho(\mathbb{1} - P_a) = 0$, and therefore $\rho P_a = P_a \rho = \rho$, which taking (3.90) and idempotency into account leads to (3.95). Note that the features of repeatability and ideality are related to the fact that the associated effects are actually projections, and the spectrum is discrete. The post-measurement state conditioned on the occurrence of the effect P_a is given according to the previous formulae by

$$\rho_{|a} = \frac{O_a[\rho]}{\mathrm{Tr}\{O_a[\rho]\}} \tag{3.99}$$

$$= \frac{P_a \rho P_a}{\mathrm{Tr}\{\rho P_a\}} \tag{3.100}$$

$$= P_a \tag{3.101}$$

$$= |u_a\rangle\langle u_a|, \tag{3.102}$$

namely the pure state corresponding to the unique eigenstate associated to the outcome a. The analogous situation in the presence of degeneracy is considered in Remark 3.4.

3.4 State Preparation in the Presence of Degeneracy

We now consider the most general situation in which the measurement is still associated to a collection of commuting PVM observables $\boldsymbol{B} = (B_1, \dots, B_r)$, but without assuming lack of degeneracy, so that denoting $\boldsymbol{b} = (b_1, \dots, b_r)$ we have the representation

$$B_i = \sum_{\boldsymbol{b}} b_i P_{\boldsymbol{b}} \tag{3.103}$$

but with $P_{\boldsymbol{b}}$ not necessarily one dimensional orthogonal projections

$$P_{\boldsymbol{b}} = \sum_s |u_{\boldsymbol{b},s}\rangle\langle u_{\boldsymbol{b},s}|. \tag{3.104}$$

One can immediately check that all previous statements still hold, in that they only rely on self-adjointness and idempotency of the projections, with the only exception of (3.102). The state obtained conditioning on a given outcome is now given by the mixed state

$$\rho_{|b} = \frac{O_b[\rho]}{\mathrm{Tr}\{O_b[\rho]\}} \tag{3.105}$$

$$= \frac{P_b \rho P_b}{\mathrm{Tr}\{\rho P_b\}}. \tag{3.106}$$

If no selection according to the outcome is performed, the state which a priori comes out of the measurement procedure is the mixture

$$\rho_{\mathrm{out}} = \sum_a \mu_\rho^{P_a} \rho_{|a} \tag{3.107}$$

$$= \sum_a \mu_\rho^{P_a} \frac{O_a[\rho]}{\mathrm{Tr}\{O_a[\rho]\}} \tag{3.108}$$

$$= \sum_a P_a \rho P_a, \tag{3.109}$$

that is a state diagonal in the basis common to the complete set of commuting observables. The a priori state obtained when performing no selection can be associated to a trace-preserving operation, as we shall further discuss in Sect. 3.2.7.

3.2.4 Mathematical Characterization of Operations

We will now provide a general characterization of completely positive maps, both in view of their structure and of their physical interpretation. This result will be accomplished by means of a few fundamental theorems. We first mention a basic result on the general structure of completely positive maps between algebras of operators [9], that provides the common root of Naimark and Kraus theorems, respectively Theorems 3.4 and 3.5.

Theorem 3.1 (Stinespring, 1955) *A normal linear map $\Phi' : \mathcal{B}(\mathcal{H}_S) \to \mathcal{B}(\mathcal{H}_S)$ is completely positive iff there exists a Hilbert space $\mathcal{H}_E$, a bounded operator $V : \mathcal{H}_S \to \mathcal{H}_S \otimes \mathcal{H}_E$ and a normal, unit-preserving $*$-homomorphism $\pi : \mathcal{B}(\mathcal{H}_S) \to$*

$\mathcal{B}(\mathcal{H}_S \otimes \mathcal{H}_E)$ such that $\Phi'[A_S] = V^\dagger \pi[A_S]V$ holds. π can be always realized as $\pi[A_S] = A_S \otimes \mathbb{1}_E$, so that we have the representation

$$\Phi'[A_S] = V^\dagger A_S \otimes \mathbb{1}_E V. \tag{3.110}$$

We do not provide a proof of the theorem, but recall that by definition a $*$-homomorphism π is a linear map from the algebra $\mathcal{B}(\mathcal{H}_S)$ to the larger algebra $\mathcal{B}(\mathcal{H}_S \otimes \mathcal{H}_E)$

$$\pi : \mathcal{B}(\mathcal{H}_S) \to \mathcal{B}(\mathcal{H}_S \otimes \mathcal{H}_E) \tag{3.111}$$
$$A_S \mapsto \pi[A_S]$$

that satisfies

$$\pi[A_S B_S] = \pi[A_S]\pi[B_S] \tag{3.112}$$
$$\pi[A_S^\dagger] = \pi[A_S]^\dagger \tag{3.113}$$
$$\pi[\mathbb{1}_S] = \mathbb{1}_S \otimes \mathbb{1}_E. \tag{3.114}$$

These properties imply that π is a positive map, and in particular a completely positive map. For a map that is normal according to the definition of Remark 3.5, this embedding in the larger algebra can always be written as $\pi(A_S) = A_S \otimes \mathbb{1}_E$, so that it is equivalent to the map dual to the partial trace considered in (3.43).

3.5 Normal Map

Normality is a regularity requirement [10], whose relevance already appeared in Sect. 2.2.7, implying that convergent monotone non-decreasing sequences of operators are transformed into convergent monotone non-decreasing sequences. It can be formulated as follows. A positive map

$$\mathcal{M}' : \mathcal{B}(\mathcal{H}_2) \to \mathcal{B}(\mathcal{H}_1) \tag{3.115}$$
$$B \mapsto \mathcal{M}[B]$$

is normal if $B_n \to B$, with $B_n \geqslant B_{n-1}$, implies $\mathcal{M}'[B_n] \geqslant \mathcal{M}'[B_{n-1}]$ and $\mathcal{M}'[B_n] \to \mathcal{M}'[B]$. The relevance of normality is due to the fact that it warrants the existence and equivalence of Heisenberg and Schrödinger picture, as implied by the following theorem:

Theorem 3.2 *If $\mathcal{M} : \mathcal{T}(\mathcal{H}_1) \to \mathcal{T}(\mathcal{H}_2)$ is a positive linear map, then its adjoint $\mathcal{M}' : \mathcal{B}(\mathcal{H}_2) \to \mathcal{B}(\mathcal{H}_1)$ is a linear normal positive map. Vice versa a linear normal positive map always admits a preadjoint, which is itself linear and positive and we have $0 \leqslant \mathcal{M}'[\mathbb{1}] \leqslant \mathbb{1}$ iff $0 \leqslant \text{Tr}\{\mathcal{M}[\rho]\} \leqslant \text{Tr}\,\rho \,\forall \rho \in \mathcal{S}(\mathcal{H})$.*

Theorem 3.1 has many important consequences. It is also called Stinespring dilation's theorem, because in providing the general characterization of a completely positive map it relies on involvement of another Hilbert space, together with the associated algebra of operators, thus working in a dilated setting. Together with this result Stinespring also proved that non-commutativity of the algebra, and therefore of the space of the observables, is a crucial ingredient for the emergence of complete positivity [9]. This fact is formalized by the following theorem [11].

Theorem 3.3 (Stinespring, 1955) *Let $\mathcal{A}$ and $\mathcal{B}$ be C^*-algebras. If either $\mathcal{A}$ or $\mathcal{B}$ is Abelian, then any positive map $\Phi : \mathcal{A} \to \mathcal{B}$ is completely positive.*

Proof The key starting point is the fact recalled in Remark 3.6 that any Abelian C^*-algebras can be realized as the space $C_\infty(\Omega)$ of continuous functions on a locally compact space vanishing at infinity, with the supremum norm, while assuming non-commutativity the C^*-algebra $\mathcal{B}$ can be taken as the space of bounded operators on a given Hilbert space, say $\mathcal{B}(\mathcal{H})$.

Let us first suppose $\mathcal{B}$ is Abelian. Then for any $\{A_i\}_{i=1,\dots,n} \subset \mathcal{A}$ and any $\{f_i\}_{i=1,\dots,n} \subset \mathcal{B}$, that is $f_i \in C_\infty(\Omega)$ we have

$$\left(\sum_{i,j=1}^{n} f_i^* \Phi[A_i^\dagger A_j] f_j \right)(\omega) = \sum_{i,j=1}^{n} f_i^*(\omega)\Phi[A_i^\dagger A_j](\omega)f_j(\omega) \tag{3.116}$$

$$= \Phi\left[\sum_{i,j=1}^{n} f_i^*(\omega)f_j(\omega)A_i^\dagger A_j \right](\omega) \tag{3.117}$$

$$= \Phi\left[\left(\sum_{i=1}^{n} f_i(\omega)A_i \right)^\dagger \left(\sum_{j=1}^{n} f_j(\omega)A_j \right) \right](\omega) \tag{3.118}$$

$$\geqslant 0 \tag{3.119}$$

where we have exploited both linearity and positivity of the map Φ. This map Φ is therefore completely positive in that it satisfies condition (3.12).

Let us now suppose that $\mathcal{A}$ is Abelian. For any $\{f_i\}_{i=1,\dots,n} \subset \mathcal{A}$ the expression $\Phi(f_i^* f_j)$ is an operator on $\mathcal{H}$ and in order to verify complete positivity according to (3.12) we have to show that

$$\sum_{i,j=1}^{n} \langle \psi_i | \Phi[f_i^* f_j] \psi_j \rangle \geqslant 0 \quad \forall \{f_i\}_{i=1,\dots,n} \subset \mathcal{A} \quad \forall \{\psi_i\}_{i=1,\dots,n} \subset \mathcal{H} \quad \forall n \in \mathbb{N}. \tag{3.120}$$

To this aim we note that the map $f \mapsto \mu_{ij}(f) = \langle \psi_i | \Phi[f]\psi_j \rangle$ for fixed i and j is a linear functional on the space $C_\infty(\Omega)$ so that by the Riesz-Markov theorem it can be identified with a regular measure, which thanks to the Radon-Nikodym theorem in particular can be expressed by means of a density given by a function $\mu_{ij}(\omega)$ in $L^1(\Omega)$ according to the expression

$$\mu_{ij}(f) = \int_{\Omega} d\omega \mu_{ij}(\omega) f(\omega). \qquad (3.121)$$

For any $\{v_i\}_{i=1,\dots,n} \subset \mathbb{C}$ we then have by linearity and thanks to positivity of Φ

$$\sum_{i,j=1}^{n} v_i^* \mu_{ij}(f^* f) v_j = \sum_{i,j=1}^{n} v_i^* \langle \psi_i | \Phi[f^* f] \psi_j \rangle v_j \qquad (3.122)$$

$$= \langle \left(\sum_{i=1}^{n} v_i \psi_i \right) | \Phi[f^* f] \left(\sum_{j=1}^{n} v_j \psi_j \right) \rangle \qquad (3.123)$$

$$\geqslant 0, \qquad (3.124)$$

so that the measure identified by the map $f \mapsto \sum_{i,j=1}^{n} v_i^* \mu_{ij}(f) v_j$ is positive and therefore associated to a positive density $\sum_{i,j=1}^{n} v_i^* \mu_{ij}(\omega) v_j \geqslant 0$ a.e. in Ω. We therefore have

$$\sum_{i,j=1}^{n} \langle \psi_i | \Phi[f_i^* f_j] \psi_j \rangle = \int_{\Omega} d\omega \sum_{i,j=1}^{n} \mu_{ij}(\omega)(f_i^* f_j)(\omega) \qquad (3.125)$$

$$= \int_{\Omega} d\omega \sum_{i,j=1}^{n} f_i^*(\omega) \mu_{ij}(\omega) f_j(\omega) \qquad (3.126)$$

$$\geqslant 0 \qquad (3.127)$$

which completes the proof. $\square$

As a first application of Stinespring's results we provide a general proof of Naimark's result Theorem 2.7, valid without restriction on the outcome space.

Theorem 3.4 (Naimark, 1943) *Let $F(\cdot)$ be a POVM in $\mathcal{H}_S$ with outcome space $(\Omega, \mathcal{F})$ so that $F : \mathcal{F} \to \mathcal{B}(\mathcal{H}_S)$. Then there exists a Hilbert space $\mathcal{H}_E$, a bounded operator $V : \mathcal{H}_S \to \mathcal{H}_S \otimes \mathcal{H}_E$ and a projection-valued measure $E(\cdot)$ on the same measure space $E : \mathcal{F} \to \mathcal{B}(\mathcal{H}_S \otimes \mathcal{H}_E)$ such that*

$$F(M) = V^{\dagger} E(M) V. \qquad (3.128)$$

Proof Let us first observe that the POVM defined on the measure space $(\Omega, \mathcal{F})$ induces in a natural way a map acting on the algebra $C_{\infty}(\Omega)$ of continuous functions on Ω with the supremum norm $\| \cdot \|_{\infty}$ as follows

$$\Phi_F : C_\infty(\Omega) \to \mathcal{B}(\mathcal{H}_S) \tag{3.129}$$

$$h \mapsto \Phi_F[h] = \int_\Omega h(x)\mathrm{d}F(x).$$

The operator $\Phi_F[h]$ with matrix elements

$$\langle\varphi|\Phi_F[h]\psi\rangle = \int_\Omega h(x)\mathrm{d}\langle\varphi|F(x)\psi\rangle \tag{3.130}$$

is bounded as follows from the inequality

$$\langle\varphi|\Phi_F[h]\varphi\rangle = \int_\Omega h(x)\mathrm{d}\langle\varphi|F(x)\varphi\rangle \tag{3.131}$$

$$\leqslant \|h\|_\infty \int_\Omega \mathrm{d}\langle\varphi|F(x)\varphi\rangle \tag{3.132}$$

$$= \|h\|_\infty \|\varphi\|^2 \tag{3.133}$$

valid $\forall \varphi \in \mathcal{H}_S$. As a result, to each function we associate its integral over the measure space with the given POVM. This map is obviously positive, and therefore, since the space $C_\infty(\Omega)$ is Abelian, thanks to Theorem 3.3 it is also completely positive. We can therefore exploit Stinespring result Theorem 3.1, which can be generally formulated with $\mathcal{B}(\mathcal{H}_S)$ replaced by an abstract C^*-algebra [9], to write the map in the form

$$\Phi_F[h] = V^\dagger \pi[h]V, \tag{3.134}$$

where π is a $*$-homomorphism $\pi : C_\infty(\Omega) \to \mathcal{B}(\mathcal{H}_S \otimes \mathcal{H}_E)$ between algebras. All such homomorphisms are of the form

$$\pi[h] = \int_\Omega h(x)\mathrm{d}E(x) \tag{3.135}$$

where E is a PVM on the measure space $(\Omega, \mathcal{F})$ taking values in $\mathcal{B}(\mathcal{H}_S \otimes \mathcal{H}_E)$ [12]. The defining properties of a $*$-homomorphism can be directly verified, and in particular we have

$$\langle\Phi|\pi[hf]\Psi\rangle = \int_\Omega h(x)f(x)\mathrm{d}\langle\Phi|E(x)\Psi\rangle \tag{3.136}$$

$$= \langle\Phi|\pi[h]\pi[f]\Psi\rangle, \tag{3.137}$$

just thanks to idempotency of the elements of the PVM. We have thus associated to the initial positive operator-valued measure F taking values in $\mathcal{B}(\mathcal{H}_S)$ a projection-valued measure E on the same measure space taking values in $\mathcal{B}(\mathcal{H}_S \otimes \mathcal{H}_E)$, such that (3.128) holds. $\qquad\square$

3.6 Algebras of Observables

The natural mathematical setting of the results considered in Sect. 3.2.4 is the framework of C^*-algebras, in which both classical and quantum observables can be considered. Without any pretension to be exhaustive, but hoping to help the reader in approaching the subject, we recall the basic definitions and relevant facts. We refer the reader to [11, 13] for a more appropriate treatment.

A set $\mathcal{A}$ is a C^*-algebra if the following conditions hold $\forall A, B \in \mathcal{A}$:

- $\mathcal{A}$ is a linear associative algebra on the field $\mathbb{C}$, that is a vector space with an associative product linear in both factors
- $\mathcal{A}$ is a normed space with norm $\| \cdot \|$, with respect to which the product is continuous $\|AB\| \leqslant \|A\|\|B\|$, so that in particular

$$\|A\| \geqslant 0, \quad \|A\| = 0 \ \text{ iff } \ A = 0$$

$$\|\lambda A\| = |\lambda|\|A\| \ \forall \lambda \in \mathbb{C}$$

$$\|A + B\| \leqslant \|A\| + \|B\|$$

- $\mathcal{A}$ is a Banach algebra, that is a complete space with respect to the topology defined by the norm $\| \cdot \|$
- $\mathcal{A}$ is a $*$-Banach algebra, that is we have an operation of involution denoted by $*$ such that

$$(A + B)^* = A^* + B^*$$
$$(\lambda A)^* = \lambda^* A^*$$
$$(AB)^* = B^* A^*$$
$$(A^*)^* = A$$

- C^* condition

$$\|A^* A\| = \|A\|^2$$

which implies in particular $\|A^*\| = \|A\|$.

Thanks to a result of Gelfand and Naimark, an Abelian C^*-algebra is isomorphic to the set of continuous functions on a measurable locally compact space X going to zero at infinity, equipped with the supremum norm, namely $C_\infty(X)$. A nonabelian C^*-algebra, that is a C^*-algebra whose elements do not commute among themselves, can always be represented as the space of

We now exploit Theorem 3.1 to obtain a most important structural characterization of completely positive maps. We consider for definiteness the case of a trace-preserving completely positive map Φ with equal initial and final space, however the result works as well in the case of different initial and final spaces, as well as considering general trace-nonincreasing maps, the only relevant feature being complete positivity.

Theorem 3.5 (Kraus, 1971) *Given a linear trace-preserving map*

$$\Phi : \mathcal{T}(\mathcal{H}_S) \to \mathcal{T}(\mathcal{H}_S) \tag{3.138}$$

$$\rho \mapsto \Phi[\rho]$$

the following statements are equivalent:

1. *Φ is completely positive according to either of the two equivalent formulations (3.11) and (3.12)*
2. *Φ can be written in Kraus form*

$$\Phi[\rho] = \sum_k A_k \rho A_k^\dagger \tag{3.139}$$

   where the sum is over a set which is at most denumerable, convergence is in the weak sense and $\sum_k A_k^\dagger A_k = \mathbb{1}$
3. *Φ can be expressed in the form*

$$\Phi[\rho] = \mathrm{Tr}_E[U\rho \otimes \rho_E U^\dagger] \tag{3.140}$$

   with ρ_E a state in a Hilbert space $\mathcal{H}_E$, which can be taken to be pure, and a unitary operator U on $\mathcal{H}_S \otimes \mathcal{H}_E$.

The expressions (3.139) and (3.140) are known as first and second representation theorem for a completely positive map [14]. In particular (3.139) is often called operator-sum representation for a completely positive map. The special feature of the first representation (3.139) is that it gives the generic structure of an arbitrary completely positive map in terms of operators acting in the Hilbert space of interest only. It is usually called Kraus representation of the map. The second representation (3.140), on the other hand, tells us that any completely positive transformation can be seen as originating from the interaction with an external system, upon taking an initially factorized state between system and environment. Each completely positive dynamics can therefore be obtained taking the partial trace with respect to a unitary evolution in a larger Hilbert space. The non-uniqueness of this construction reflects itself in the non-uniqueness of the Kraus representation, which we will address in Theorem 3.6 and Remark 3.7.

Proof We will prove the theorem showing that the following chain of implications holds $1 \Rightarrow 3 \Rightarrow 2 \Rightarrow 1$, so that all statements are equivalent.

Let us first show that $1 \Rightarrow 3$. By definition Φ is completely positive iff Φ' : $\mathcal{B}(\mathcal{H}_S) \rightarrow \mathcal{B}(\mathcal{H}_S)$ has this property. We now use Stinespring's theorem adapted to the case in which Φ' is a normal map, so that (3.110) holds. In this case for a completely positive map Φ' on the algebra $\mathcal{B}(\mathcal{H}_S)$ there exists a Hilbert space $\mathcal{H}_E$ and a bounded operator $V : \mathcal{H}_S \rightarrow \mathcal{H}_S \otimes \mathcal{H}_E$ such that

$$\Phi'[A_S] = V^\dagger (A_S \otimes \mathbb{1}_E) V, \tag{3.141}$$

while identity preservation implies $V^\dagger V = \mathbb{1}_S$. For fixed $\psi \in \mathcal{H}_E$ of norm one we can define an operator U by means of the expression

$$U(\varphi \otimes \psi) = V\varphi \qquad \forall \varphi \in \mathcal{H}_S, \tag{3.142}$$

which turns out to be an isometry thanks to the property of V and the fact that ψ is of norm one, indeed $\forall \varphi, \varphi' \in \mathcal{H}_S$ we have

$$\langle U(\varphi' \otimes \psi) | U(\varphi \otimes \psi) \rangle_{S \otimes E} = \langle V\varphi' | V\varphi \rangle_{S \otimes E} \tag{3.143}$$
$$= \langle \varphi' | V^\dagger V\varphi \rangle_S \tag{3.144}$$
$$= \langle \varphi' | \varphi \rangle_S \tag{3.145}$$
$$= \langle \varphi' \otimes \psi | \varphi \otimes \psi \rangle_{S \otimes E}. \tag{3.146}$$

The operator U can therefore be extended to define a unitary operator on the whole $\mathcal{H}_S \otimes \mathcal{H}_E$. With these results at hand we can identify the map Φ starting from Stinespring's theorem (3.141) and the duality relation (3.4), so that for arbitrary $B_S \in \mathcal{B}(\mathcal{H}_S)$ and $\rho \in \mathcal{T}(\mathcal{H}_S)$ we have

$$\text{Tr}_S\{\Phi[\rho] B_S\} = \text{Tr}_S\{\rho \Phi'[B_S]\} \tag{3.147}$$
$$= \text{Tr}_S\{\rho V^\dagger (B_S \otimes \mathbb{1}_E) V\}. \tag{3.148}$$

We now evaluate the trace over $\mathcal{H}_S$ with an arbitrary basis $\{\varphi_j\}$, so that using (3.142) we get

$$\text{Tr}_S\{\Phi[\rho] B_S\} = \sum_j \langle \varphi_j | \rho V^\dagger (B_S \otimes \mathbb{1}_E) V\varphi_j \rangle_S \tag{3.149}$$
$$= \sum_j \langle V\rho\varphi_j | (B_S \otimes \mathbb{1}_E) V\varphi_j \rangle_{S \otimes E} \tag{3.150}$$

$$= \sum_j \langle U(\rho\varphi_j \otimes \psi)|(B_S \otimes \mathbb{1}_E)U(\varphi_j \otimes \psi)\rangle_{S\otimes E} \tag{3.151}$$

$$= \sum_j \langle \varphi_j \otimes \psi|(\rho \otimes \mathbb{1}_E)U^\dagger(B_S \otimes \mathbb{1}_E)U(\varphi_j \otimes \psi)\rangle_{S\otimes E}. \tag{3.152}$$

This expression can be finally written as a trace over $\mathcal{H}_S \otimes \mathcal{H}_E$, by introducing the orthogonal projection $P_\psi = |\psi\rangle\langle\psi|$ in $\mathcal{H}_E$, coming to

$$\mathrm{Tr}_S\{\Phi[\rho]B_S\} = \mathrm{Tr}_{S\otimes E}\{(\rho \otimes \mathbb{1}_E)U^\dagger(B_S \otimes \mathbb{1}_E)U(\mathbb{1}_S \otimes P_\psi)\} \tag{3.153}$$

$$= \mathrm{Tr}_{S\otimes E}\{(B_S \otimes \mathbb{1}_E)U(\rho \otimes P_\psi)U^\dagger\} \tag{3.154}$$

$$= \mathrm{Tr}_S\{B_S \, \mathrm{Tr}_E[U\rho \otimes P_\psi U^\dagger]\}, \tag{3.155}$$

implying the representation

$$\Phi[\rho] = \mathrm{Tr}_E[U\rho \otimes P_\psi U^\dagger] \tag{3.156}$$

corresponding to (3.140) when ρ_E is identified with the pure state P_ψ.

Note that according to Sect. 4.3.1 the fact that the state ρ_E in (3.140) can be taken to be pure is warranted by the very notion of purification. Indeed, given an arbitrary state ρ_E we can consider for it a purification of the form (4.20)

$$\mathrm{Tr}_A[P_\Psi] = \rho_E, \tag{3.157}$$

with Ψ a pure state in $\mathcal{H}_E \otimes \mathcal{H}_A$, so that (3.140) becomes

$$\Phi[\rho] = \mathrm{Tr}_E[\mathrm{Tr}_A[U \otimes \mathbb{1}_A \rho \otimes P_\Psi U^\dagger \otimes \mathbb{1}_A]]. \tag{3.158}$$

The implication $3 \Rightarrow 2$ is now easily obtained by considering an orthogonal resolution for ρ_E, namely $\rho_E = \sum_\alpha \lambda_\alpha |\psi_\alpha\rangle\langle\psi_\alpha|$ and using the same orthonormal basis to evaluate the trace

$$\Phi[\rho] = \mathrm{Tr}_E[U\rho \otimes \rho_E U^\dagger] \tag{3.159}$$

$$= \sum_{\alpha\beta} \left(\sqrt{\lambda_\alpha}\langle\psi_\beta|U\psi_\alpha\rangle\right) \rho \left(\sqrt{\lambda_\alpha}\langle\psi_\beta|U\psi_\alpha\rangle\right)^\dagger \tag{3.160}$$

$$= \sum_K W_K \rho W_K^\dagger \tag{3.161}$$

with $K = \alpha, \beta$ a multi-index, and the operators W_K defined on $\mathcal{H}_S$ via

$$W_K = \sqrt{\lambda_\alpha}\langle\psi_\beta|U\psi_\alpha\rangle, \tag{3.162}$$

which further satisfy

$$\sum_K W_K^\dagger W_K = \mathbb{1}. \tag{3.163}$$

The last implication $2 \Rightarrow 1$ can be proven noting that the Kraus representation (3.139) immediately implies complete positivity according to (3.11). Indeed, starting from the Kraus representation (3.139) we have

$$\Phi \otimes \mathbb{1}_{\mathcal{M}_n(\mathbb{C})}[T] = \sum_K (W_K \otimes \mathbb{1}_{\mathbb{C}^n}) T (W_K \otimes \mathbb{1}_{\mathbb{C}^n})^\dagger, \tag{3.164}$$

so that $\Phi \otimes \mathbb{1}_{\mathcal{M}_n(\mathbb{C})}$ sends positive operators into positive operators for all n. $\qquad\square$

We have thus obtained a fundamental result according to which any operation admits a representation in the form

$$O[\rho] = \sum_k A_k \rho A_k^\dagger \tag{3.165}$$

with the constraint $\sum_k A_k^\dagger A_k \leqslant \mathbb{1}$, which warrants the property of being trace-nonincreasing. The set of operators $\{A_k\}$ is often called a collection of Kraus operators for the map O. Any completely positive map can therefore be written in the form (3.165) and it is trace-preserving, namely a channel, iff

$$\sum_k A_k^\dagger A_k = \mathbb{1}_S. \tag{3.166}$$

If the map is unit-preserving it is called unital, and this happens iff

$$\sum_k A_k A_k^\dagger = \mathbb{1}_S, \tag{3.167}$$

a condition warranted by unitary Kraus operators. A unital channel is therefore a bistochastic map, in that both the map and its dual preserve the identity operator. If the Hilbert space $\mathcal{H}_S$ of the system has dimension d, it is always possible to use d^2 or less operators in the sum. An important feature of the so-called Kraus form (3.139) is its non uniqueness, which also appears from the equivalent representation (3.140). This non uniqueness is put into evidence in Remark 3.7 and Remark 3.8, and can be characterized as follows.

3.7 Nonuniqueness of the Kraus Representation

Let us consider the following simple completely positive trace-preserving map defined on $\mathcal{H}_S = \mathbb{C}^d$

$$\mathcal{A} : \mathcal{T}(\mathcal{H}_S) \to \mathcal{T}(\mathcal{H}_S) \tag{3.168}$$

$$T \mapsto \mathcal{A}[T] = \mathrm{Tr}\{T\}\frac{\mathbb{1}_{\mathbb{C}^d}}{d},$$

which sends any trace class operator to a multiple of the identity operator determined by its trace, so that in particular all states are sent to the maximally mixed state. Consider any basis $\{\varphi_i\}_{i=1,\dots,d}$ in $\mathcal{H}_S$. We can then consider the rank one operators $E_{ij} = |\varphi_i\rangle\langle\varphi_j|$, which provide a canonical basis in the set $\mathcal{M}_d(\mathbb{C})$ of $d \times d$ complex matrices. For any $T \in \mathcal{T}(\mathcal{H}_S)$ exploiting the completeness relation (2.185) and the very definition of trace (2.38) we have

$$\sum_{i,j=1}^{d} E_{ij} T E_{ij}^{\dagger} = \sum_{i,j=1}^{d} |\varphi_i\rangle\langle\varphi_j|T\varphi_j\rangle\langle\varphi_i| \tag{3.169}$$

$$= \sum_{j=1}^{d} \langle\varphi_j|T\varphi_j\rangle \sum_{i=1}^{d} |\varphi_i\rangle\langle\varphi_i| \tag{3.170}$$

$$= \mathrm{Tr}\{T\}\mathbb{1}_{\mathbb{C}^d}.$$

We therefore have the Kraus representation

$$\mathcal{A}[T] = \frac{1}{d} \sum_{i,j=1}^{d} E_{ij} T E_{ij}^{\dagger}. \tag{3.171}$$

Actually any basis $\{\tau_\alpha\}_{\alpha=1,\dots,n^2}$ in the linear set $\mathcal{M}_d(\mathbb{C})$ of $d \times d$ complex matrices obeys the identity

$$\sum_{\alpha} \tau_\alpha T \tau_\alpha^{\dagger} = \mathrm{Tr}\{T\}\mathbb{1}_{\mathbb{C}^d}, \tag{3.172}$$

further discussed in Remark 3.19, and therefore leads to a different Kraus representation

$$\mathcal{A}[T] = \frac{1}{d} \sum_{\alpha=1}^{d^2} \tau_\alpha T \tau_\alpha^{\dagger}. \tag{3.173}$$

of the completely positive trace-preserving map considered in (3.168).

Theorem 3.6 *Two sets of Kraus operators* $\{A_1, \ldots, A_N\}$ *and* $\{B_1, \ldots, B_M\}$ *define the same completely positive transformation, that is*

$$\sum_{r=1}^{N} A_r T A_r^\dagger = \sum_{s=1}^{M} B_s T B_s^\dagger \tag{3.174}$$

for any $T \in \mathcal{T}(\mathcal{H}_S)$, *iff*

$$A_r = \sum_{s=1}^{M} u_{rs} B_s \tag{3.175}$$

where the complex numbers $\{u_{rs}\}$ *satisfy*

$$\sum_{r=1}^{N} u_{rs}^* u_{rs'} = \delta_{ss'}. \tag{3.176}$$

Proof The *if* part follows upon substitution. Using (3.175) and (3.176) we obtain

$$\sum_{r=1}^{N} A_r T A_r^\dagger = \sum_{r=1}^{N} \sum_{s,s'=1}^{M} u_{rs} B_s T u_{rs'}^* B_{s'}^\dagger \tag{3.177}$$

$$= \sum_{s=1}^{M} B_s T B_s^\dagger \tag{3.178}$$

for any $T \in \mathcal{T}(\mathcal{H}_S)$.

For the *only if* part we assume for any trace class operator T the identity (3.174). An equivalent requirement is to ask (3.174) with $T = |\psi\rangle\langle\psi|, \forall \psi \in \mathcal{H}_S$. If we start from the two sets of Kraus operators we obtain the following equivalent representations of a positive operator

$$W = \sum_{r=1}^{N} A_r |\psi\rangle\langle\psi| A_r^\dagger = \sum_{s=1}^{M} B_s |\psi\rangle\langle\psi| B_s^\dagger. \tag{3.179}$$

Upon defining the norm one vectors

$$\psi_r^A = \frac{A_r \psi}{\|A_r \psi\|} \tag{3.180}$$

and

$$\psi_s^B = \frac{B_s \psi}{\|B_s \psi\|} \tag{3.181}$$

we obtain, further exploiting the spectral theorem for W, the equivalent expressions

$$W = \sum_{i=1}^{d} w_i |\varphi_i\rangle\langle\varphi_i| \tag{3.182}$$

$$= \sum_{r=1}^{N} \mu_r^A |\psi_r^A\rangle\langle\psi_r^A| \tag{3.183}$$

$$= \sum_{s=1}^{M} \mu_s^B |\psi_s^B\rangle\langle\psi_s^B| \tag{3.184}$$

with $\{w_i\}$ the positive eigenvalues of W and $\{\varphi_i\}$ the corresponding orthogonal eigenvectors, while the weights in the other decompositions are defined according to (3.180) and (3.181) as

$$\mu_r^A = \|A_r\psi\|^2 \tag{3.185}$$

and

$$\mu_s^B = \|B_s\psi\|^2 \tag{3.186}$$

respectively. Making reference to Theorem 2.2 we are led to introduce the set of coefficients

$$v_{ri}^A = \sqrt{\frac{\mu_r^A}{w_i}} \langle\varphi_i|\psi_r^A\rangle \tag{3.187}$$

$$v_{si}^B = \sqrt{\frac{\mu_s^B}{w_i}} \langle\varphi_i|\psi_s^B\rangle \tag{3.188}$$

satisfying

$$\sum_{r=1}^{N} v_{ri}^A (v_{rj}^A)^* = \delta_{ij} \tag{3.189}$$

as well as

$$\sum_{s=1}^{M} v_{si}^B (v_{sj}^B)^* = \delta_{ij}, \tag{3.190}$$

that is the matrix v^A has d orthogonal columns of length N. It can thus be extended to an $N \times N$ unitary matrix, which we will denote in the same way and such that

$$\sum_{i=1}^{N} v_{ri}^A (v_{r'i}^A)^* = \delta_{rr'}. \tag{3.191}$$

In the same way we extend the matrix v^B to a unitary matrix fulfilling

$$\sum_{i=1}^{M} v_{si}^B (v_{s'i}^B)^* = \delta_{ss'}. \tag{3.192}$$

These relations allow to express the vectors in (3.183) and (3.184) respectively in terms of the orthogonal set $\{\varphi_i\}$. We have indeed

$$\psi_r^A = \sum_{i=1}^{N} \sqrt{\frac{w_i}{\mu_r^A}} \, v_{ri}^A \varphi_i \tag{3.193}$$

and similarly

$$\psi_s^B = \sum_{i=1}^{M} \sqrt{\frac{w_i}{\mu_s^B}} \, v_{si}^B \varphi_i, \tag{3.194}$$

where the sums have been extended upon defining $w_i = 0$ for $i > d$. We can now define the coefficients

$$u_{rs} = \sum_{i=1}^{M} v_{ri}^A (v_{si}^B)^*, \tag{3.195}$$

where without loss of generality we have supposed $M \geqslant N$. According to (3.181) and (3.186) we have the identity

$$\sum_{s=1}^{M} u_{rs} B_s \psi = \sum_{s=1}^{M} u_{rs} \sqrt{\mu_s^B} \, \psi_s^B, \tag{3.196}$$

which thanks to (3.194) implies

$$\sum_{s=1}^{M} u_{rs} B_s \psi = \sum_{s=1}^{M} \sum_{i=1}^{M} u_{rs} \sqrt{w_i} \, v_{si}^B \varphi_i. \tag{3.197}$$

We now replace the matrix u_{rs} with the defining expression (3.195), so that exploiting (3.191) we obtain

$$\sum_{s=1}^{M} u_{rs} B_s \psi = \sum_{i=1}^{M} v_{ri}^A \sqrt{w_i}\, \varphi_i.$$

(3.198)

As a last step we use the expression (3.187) for the coefficients v_{ri}^A, thus obtaining the relation

$$\sum_{s=1}^{M} u_{rs} B_s \psi = A_r \psi.$$

(3.199)

The validity of (3.199) for any $\psi \in \mathcal{H}_S$ in turn implies (3.175). Thanks to (3.191) and (3.192) we finally have

$$\sum_{r=1}^{N} u_{rs}^* u_{rs'} = \sum_{r=1}^{N} \sum_{i,j=1}^{M} (v_{ri}^A)^* v_{si}^B v_{rj}^A (v_{s'j}^B)^*$$

(3.200)

$$= \sum_{i=1}^{M} v_{si}^B (v_{s'i}^B)^*$$

(3.201)

$$= \delta_{ss'},$$

(3.202)

namely (3.176). $\square$

3.8 Decompositions of Qubit Channels

We consider a quantum system described in $\mathbb{C}^2$, often called qubit because of the relevance of this system in quantum information. A class of completely positive trace-preserving transformations for this system, also called channels as discussed in Sect. 3.2.2, is given by the expression

$$\mathcal{E}[\rho] = p\frac{1}{2}\mathbb{1}_{\mathbb{C}^2} + (1-p)\rho$$

(3.203)

where the positive parameter $0 \leqslant p \leqslant 1$ determines how close the transformed state gets to the maximal mixed or depolarized state $\frac{1}{2}\mathbb{1}_{\mathbb{C}^2}$, hence the name depolarizing channel for this transformation. To put into evidence linearity of the transformation (3.203) we express it as

$$\mathcal{E}[\rho] = p\frac{1}{2}\mathbb{1}_{\mathbb{C}^2}\,\mathrm{Tr}\{\rho\} + (1-p)\rho. \tag{3.204}$$

In view of Remark 3.7 we immediately obtain the representation

$$\mathcal{E}[\rho] = \frac{p}{2}\sum_{\alpha=1}^{4}\tau_\alpha\rho\tau_\alpha^\dagger + (1-p)\rho, \tag{3.205}$$

where $\{\tau_\alpha\}_{\alpha=1,\dots,4}$ is any basis in the linear set $M_2(\mathbb{C})$ of 2×2 complex matrices. From (3.205) one immediately identifies a set $\{B_s\}_{s=1}^{5}$ of Kraus operators given by $\left\{\sqrt{\frac{p}{2}}\tau_1,\,\dots,\,\sqrt{\frac{p}{2}}\tau_4,\,\sqrt{1-p}\,\mathbb{1}_{\mathbb{C}^2}\right\}$. A standard choice of basis is obtained considering the identity and the Pauli matrices, suitably normalized by a factor $1/\sqrt{2}$ as discussed in Sect. 3.3.2, leading to the expression

$$\mathcal{E}[\rho] = p\frac{1}{4}(\mathbb{1}_{\mathbb{C}^2}\rho\mathbb{1}_{\mathbb{C}^2} + \sigma_x\rho\sigma_x + \sigma_y\rho\sigma_y + \sigma_z\rho\sigma_z) + (1-p)\rho \tag{3.206}$$

$$= p\frac{1}{4}(\sigma_x\rho\sigma_x + \sigma_y\rho\sigma_y + \sigma_z\rho\sigma_z) + \left(1 - p\frac{3}{4}\right)\rho. \tag{3.207}$$

This expression immediately suggests another set $\{A_r\}_{r=1}^{4}$ of Kraus operators, namely $\left\{\frac{\sqrt{p}}{2}\sigma_x,\,\frac{\sqrt{p}}{2}\sigma_y,\,\frac{\sqrt{p}}{2}\sigma_z,\,\frac{\sqrt{4-3p}}{2}\mathbb{1}_{\mathbb{C}^2}\right\}$. The matrix $\{u_{rs}\}$ connecting the two set of Kraus operators according to (3.175) in this case has the following non-zero elements: $u_{11} = u_{22} = u_{33} = 1$, $u_{44} = \sqrt{p/(4-3p)}$, $u_{45} = 2\sqrt{(1-p)/(4-3p)}$.

3.2.5 Information Loss and Distinguishability Quantifiers

In Sect. 3.2.3 we have put into evidence how operations have a natural interpretation as transformations of states corresponding to the measurement of an elementary observable described by an effect. The effect is uniquely associated to the operation. The trace of the transformed state gives the probability for detection of the event associated to the given effect. In the case in which the associated effect is the identity, operations can describe the repreparation of a state, that is to say a new target state obtained from the incoming one, and they correspond to channels. In general, operations describe the transformation of a state in subcollections, corresponding to conditional state measurements and preparations.

In this perspective, we now show how to put into evidence the measurement character of state transformations such as operations. Let us first stress the fact that the measurement character is related to irreversibility. The very notion of irreversibility deserves an explanation, since it is used in many contexts with different facets. For example, we speak about logical irreversibility if there is no one-to-one correspondence between input and output in a logic circuit, while thermodynamic irreversibility is usually associated to detection of irreversible entropy production. In the present framework, measurement irreversibility is due to the fact that in quantum mechanics no measurement can be perfectly undone on all states, which would imply its description by means of a unitary transformation. According to the discussion in Sect. 3.2.1, it appears natural to identify physical transformations with completely positive trace-preserving maps. Importantly, it can be shown that a completely positive trace-preserving map admits an inverse iff it is a unitary transformation. The result can be proven directly or referring to the celebrated Wigner's theorem on symmetry transformations [15], together with the fact that transformations described by anti-unitary operators do not provide a channel. Indeed, they are related to the transposition map, which as we shall see in Sect. 4.3.3 is not completely positive.

In order to characterize the impact of a map, and in particular its measurement character, we naturally have to compare states before and after the action of the map. To this aim we have to introduce a way to compare quantum states. As discussed in Sect. 1.2, quantum states have the same role of probability distributions in a classical statistical description. Distinguishability quantifiers between quantum states can therefore be introduced relying on their classical counterparts. They characterize the difference in the statistical predictions obtained from the two states. A general class of distinguishability quantifiers is given by so-called divergences, both in classical and quantum setting [16]. These quantities are widely studied in the fields of quantum statistics and quantum information. We will here introduce two most prominent divergences, namely the trace distance and the quantum relative entropy. We will further briefly discuss a smoothened version of the quantum relative entropy, namely the quantum Jensen-Shannon divergence.

We still have to consider the strategy according to which states have to be compared. Two possible schemes can be considered, as depicted in Fig. 3.4. Given a completely positive trace-preserving map Φ, it can appear natural to consider the state before and after the transformation, namely ρ_{in} and $\rho_{\text{out}} = \Phi[\rho_{\text{in}}]$. Their comparison should characterize the disturbance of the map. Such a disturbance is however not related to the measurement character of the transformation. For example, a unitary channel can act in such a way that the support of ρ_{out} is orthogonal to the support of ρ_{in}. Given that orthogonal states can be perfectly discriminated [17], a distinguishability quantifier would reach its maximum value, despite the fact that the transformation is fully coherent and no measurement transformation is involved. Within such a scheme, we can rather estimate how the state has been changed, so that the value of

$$\mathcal{D}(\rho_{\text{in}}, \Phi[\rho_{\text{in}}]), \tag{3.208}$$

where $\mathcal{D}$ denotes the considered divergence or distinguishability quantifier, in the dependence on ρ_{in} is typically associated to a notion of disturbance.

An alternative strategy consists in comparing how distinct states are affected by the transformation, so that we consider

$$\mathcal{D}(\Phi[\rho_{\text{in}}^1], \Phi[\rho_{\text{in}}^2]), \tag{3.209}$$

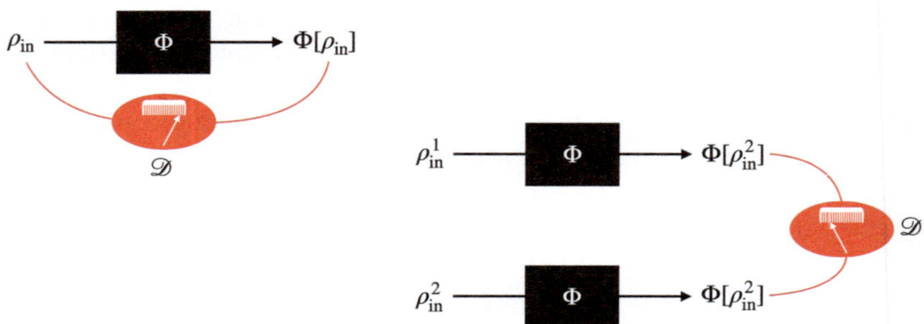

Fig. 3.4 Schematic representation of the different possible strategies to assess the effect of an operation on quantum states. In the first strategy we compare the state before and after the action of the map to estimate its disturbance character. In the second strategy we compare the action of the map on two distinct states to estimate its noise character. In both cases the comparison is performed by means of a suitable distinguishability quantifier

in the dependence on $\rho_{\mathrm{in}}^{1,2}$. This approach is well suited to quantify the noise that a completely positive trace-preserving map induces on the states, affecting their information content, provided $\mathcal{D}$ is insensitive to unitary transformations, which is indeed generally the case. It therefore does quantify the effect of the measurement transformation described by Φ. The use of the word noise, as something which hides the signal, is here justified by the fact that distinguishability quantifiers are contractive under the action of completely positive trace-preserving maps. The distinguishability between two states diminishes, so that the measurement transformation has the effect of noise with respect to the signal identified with the information allowing to distinguish them. A unitary transformation is in this sense noiseless, while the identity map is both noiseless and disturbance-free. A natural quantity to estimate the measurement disturbance of an operation on a given pair of states is therefore given by

$$M_O^{\mathcal{D}}(\rho_1, \rho_2) = \mathcal{D}(\rho_1, \rho_2) - \mathcal{D}(O[\rho_1], O[\rho_2]), \qquad (3.210)$$

with $\mathcal{D}$ the considered distinguishability quantifier.

This viewpoint about the characterization of the disturbance of a transformation, together with the notion of divergence or distinguishability quantifier, will turn out to be a convenient way to define a non-Markovian dynamics in the quantum framework, provided the behavior of the distinguishability is analyzed in time, as we shall discuss in Chap. 7.

3.2.5.1 Trace Distance

On the state space $\mathcal{S}(\mathcal{H})$ we can introduce a distinguishability quantifier starting from the trace norm $\|\cdot\|_1$ defined in (2.46) as follows

$$D(\rho_1, \rho_2) = \frac{1}{2}\|\rho_1 - \rho_2\|_1, \tag{3.211}$$

where $\rho_1, \rho_2 \in \mathcal{S}(\mathcal{H})$. Given its expression in terms of the trace norm, this quantity is a distance, hence the name trace distance, so that in particular it obeys the triangle inequality

$$D(\rho, \sigma) \leqslant D(\rho, \tau) + D(\tau, \sigma) \tag{3.212}$$

for any triple of states in $\mathcal{S}(\mathcal{H})$. In particular, since $\|\sigma\|_1 = 1$ for any state $\sigma \in \mathcal{S}(\mathcal{H})$, from

$$\|\rho_1 - \rho_2\|_1 \leqslant \|\rho_1\|_1 + \|\rho_2\|_1 \tag{3.213}$$

we have that the prefactor in (3.211) is such that

$$0 \leqslant D(\rho_1, \rho_2) \leqslant 1. \tag{3.214}$$

Since $\|\cdot\|_1$ is a norm, we have $D(\rho_1, \rho_2) = 0$ iff $\rho_1 = \rho_2$, while exploiting the Hahn-Jordan decomposition (2.213) we have that $D = 1$ iff ρ_1 and ρ_2 have orthogonal support. Explicit expressions for the trace distance are introduced in Remark 3.9 for the general case, and in Remark 3.10 for qubits.

3.9 Expressions for the Trace Distance

According to the very definition of trace norm (2.46) the trace distance between two states ρ_1 and ρ_2 is given by

$$D(\rho_1, \rho_2) = \frac{1}{2} \operatorname{Tr} |\rho_1 - \rho_2|. \tag{3.215}$$

Since the operator $\rho_1 - \rho_2$ is self-adjoint, according to (2.67) of Theorem 2.1 we therefore have

$$D(\rho_1, \rho_2) = \frac{1}{2} \sum_i |\lambda_i|, \tag{3.216}$$

where $\{\lambda_i\}$ are the eigenvalues of the trace class operator $\rho_1 - \rho_2$. According to the Hahn-Jordan decomposition for a self-adjoint operator considered in

(2.213) we have

$$\rho_1 - \rho_2 = N_+ - N_-, \tag{3.217}$$

where N_+ and N_- are positive operators with orthogonal support. Moreover due to the fact that $\rho_1 - \rho_2$ is traceless, that is $\mathrm{Tr}\{\rho_1 - \rho_2\} = 0$, we further have

$$\mathrm{Tr}\{N_+\} = \mathrm{Tr}\{N_-\}. \tag{3.218}$$

Recalling (2.216) we thus obtain the alternative expression

$$D(\rho_1, \rho_2) = \frac{1}{2}\,\mathrm{Tr}\,|N_+ - N_-| \tag{3.219}$$

$$= \frac{1}{2}\,\mathrm{Tr}\{N_+ + N_-\} \tag{3.220}$$

leading to

$$D(\rho_1, \rho_2) = \mathrm{Tr}\{N_\pm\}. \tag{3.221}$$

If we now take the orthogonal projection P_+ on the eigenspace of N_+, and we exploit orthogonality of the supports of N_+ and N_-

$$\mathrm{Tr}\{P_+(\rho_1 - \rho_2)\} = \mathrm{Tr}\{P_+(N_+ - N_-)\} \tag{3.222}$$

$$= \mathrm{Tr}\{P_+ N_+\} \tag{3.223}$$

$$= \mathrm{Tr}\{N_+\}, \tag{3.224}$$

we have

$$D(\rho_1, \rho_2) = \mathrm{Tr}\{P_+(\rho_1 - \rho_2)\}. \tag{3.225}$$

On the other hand, for any other orthogonal projection P we have

$$\mathrm{Tr}\{P(\rho_1 - \rho_2)\} = \mathrm{Tr}\{P(N_+ - N_-)\} \tag{3.226}$$

$$\leqslant \mathrm{Tr}\{P N_+\} \tag{3.227}$$

$$\leqslant \mathrm{Tr}\{N_+\}, \tag{3.228}$$

so that we obtain the important equality

$$D(\rho_1, \rho_2) = \max_{P \in \mathcal{P}(\mathcal{H})} \mathrm{Tr}\{P(\rho_1 - \rho_2)\}, \tag{3.229}$$

where the maximum is taken over the set of all orthogonal projections. This expression directly connects the value of the trace distance to experimentally observable quantities.

To convey some insight on the meaning of the trace distance we mention two special cases. For two pure states P_φ and P_ψ it can be shown that the trace distance reads [16]

$$D(P_\varphi, P_\psi) = \sqrt{1 - |\langle \varphi | \psi \rangle|^2}, \tag{3.230}$$

while for a finite-dimensional Hilbert space the trace distance between an arbitray state and the maximally mixed state obeys the inequality

$$D\left(\rho, \frac{\mathbb{1}_{\mathbb{C}^n}}{n}\right) \geq 1 - \frac{\mathrm{rank}(\rho)}{n}, \tag{3.231}$$

where $\mathrm{rank}(\rho)$ denotes the rank of the statistical operator ρ, that is the dimension of its support.

The most important property of the trace distance for our purposes is its contractivity under the action of completely positive trace-nonincreasing maps [18].

Theorem 3.7 (Ruskai, 1994) *Given a completely positive trace nonincreasing map O the inequality*

$$D(O[\rho_1], O[\rho_2]) \leq D(\rho_1, \rho_2) \tag{3.232}$$

holds $\forall \rho_1, \rho_2 \in \mathcal{S}(\mathcal{H})$, *and the equality sign is only attained for unitary transformations.*

Proof Let us consider a completely positive trace-nonincreasing map O. Due to linearity of the map, using the expression (3.211) of the trace distance and the Hahn-Jordan decomposition (3.217) we have

$$D(O(\rho_1), O(\rho_2)) = \frac{1}{2} \|O(N_+ - N_-)\|_1, \tag{3.233}$$

so that thanks to the triangle inequality and the expression (2.46) for the trace norm

$$D(O(\rho_1), O(\rho_2)) \leq \frac{1}{2} \|O(N_+)\|_1 + \frac{1}{2} \|O(N_-)\|_1 \tag{3.234}$$

$$= \frac{1}{2} \mathrm{Tr} \, |O(N_+)| + \frac{1}{2} \mathrm{Tr} \, |O(N_-)|. \tag{3.235}$$

We now exploit positivity of N_+ and N_-, together with the fact that O is completely positive and trace-nonincreasing, thus coming to

$$D(O(\rho_1), O(\rho_2)) \leqslant \frac{1}{2}\,\text{Tr}\{N_+\} + \frac{1}{2}\,\text{Tr}\{N_-\}. \tag{3.236}$$

Thanks to (3.221) we finally obtain

$$D(O(\rho_1), O(\rho_2)) \leqslant D(\rho_1, \rho_2). \tag{3.237}$$

Suppose now to consider a unitary transformation, that is a channel of the form (3.22). Importantly such a transformation is invertible and the inverse (3.28) is also completely positive, we thus have for any pair of states the chain of inequalities

$$D(\rho_1, \rho_2) = D(\mathcal{U}^{-1}\mathcal{U}[\rho_1], \mathcal{U}^{-1}\mathcal{U}[\rho_2]) \leqslant D(\mathcal{U}[\rho_1], \mathcal{U}[\rho_2]) \leqslant D(\rho_1, \rho_2), \tag{3.238}$$

leading to the desired equality. The only if comes from the fact that according to (3.215) the trace distance only depends on the modulus of the eigenvalues of $\rho_1 - \rho_2$, and unitary transformations are the only completely positive transformations leaving it invariant. $\qquad\square$

3.10 Trace Distance and Distinguishability

The meaning of the trace distance as distinguishability quantifier is best understood considering its role in assessing how well two states can be discriminated. Suppose to devise a preparation procedure that leads with equal probability to prepare the states ρ_1 and ρ_2. We want to determine the probability to successfully identify the actually prepared state by performing a single measurement. This paradigmatic situation is known as one-shot two-state discrimination problem [19,20]. Let us assume to measure an arbitrary projection P associating to the outcome 1 the state ρ_1, and to the outcome 0 the state ρ_2. According to the law of total probability, on average our success probability is given by

$$P_{\text{success}} = p(P = 1|\rho_1)p(\rho_1) + p(P = 0|\rho_2)p(\rho_2) \tag{3.239}$$

where $p(\rho_i)$ denotes the probability to prepare the state ρ_i. Given that the states are equiprobable and that the conditional probability to observe the event associated to the projection P if the state is ρ_1 is given according to quantum mechanics by

$$p(P = 1|\rho_1) = \text{Tr}\{P\rho_1\}, \tag{3.240}$$

we obtain

$$P_{\text{success}} = \frac{1}{2}[1 + \text{Tr}\{P(\rho_1 - \rho_2)\}]. \tag{3.241}$$

The highest probability of success in discriminating among the two states is given by choosing the optimal projection, so that according to (3.229) we obtain

$$P_{\text{optimal}} = \frac{1}{2}(1 + D(\rho_1, \rho_2)). \tag{3.242}$$

The trace distance quantifies the bias in favor of a correct identification of the prepared state, which justifies its role as distinguishability quantifier.

Let us notice that strictly speaking complete positivity is not required on the map O, positivity is enough for (3.232) to hold.

We thus have the important result that the action of a completely positive trace-nonincreasing map does reduce the trace distance between states. In particular, the action of a trace-preserving operation, that is a channel, generally decreases the distance between states and therefore according to Remark 3.10 their distinguishability. We further notice that the contractivity property (3.232) generally enforces via (3.238) invariance of the distinguishability quantifier with respect to unitary transformations. Any divergence contractive under completely positive trace-preserving maps is therefore invariant under the action of a unitary transformation. In such a way the strategy considered in (3.209) is indeed appropriate to characterize transformations with a measurement character, at variance with (3.208).

3.11 Trace Distance Between Qubits

According to (2.79), as further discussed in Sect. 3.3.2 a state in $\mathcal{S}(\mathbb{C}^2)$ can be written as linear combination of the identity matrix and the Pauli matrices. Each statistical operator can then be identified with a Euclidean vector r of norm smaller or equal to one, often called Bloch vector. Recalling that the eigenvalues of the statistical operator in this notation take the form $\frac{1}{2}(1 \pm \|\boldsymbol{r}\|)$ we obtain from (3.211)

$$D(\rho, \sigma) = \frac{1}{2}\|\boldsymbol{r} - \boldsymbol{s}\|, \tag{3.243}$$

so that the trace distance is half the Euclidean distance between the Bloch vectors identifying the states. This expression thus allows for a simple geometrical characterization of the trace distance in the Bloch sphere.

A further convenient expression putting into evidence the different role of diagonal and off-diagonal matrix elements in a fixed basis, that is so-called populations and coherences, is obtained expressing the state ρ_1 in matrix form according to

$$\rho_1 = \begin{pmatrix} p_1 & c_1 \\ c_1^* & 1 - p_1 \end{pmatrix}, \tag{3.244}$$

with $0 \leqslant p_1 \leqslant 1$ and $c_1 \in \mathbb{C}$, $|c_1| \leqslant \sqrt{p_1(1 - p_1)}$, so as to warrant positivity and normalization. Using the same notation for ρ_2 and denoting $\delta p = p_1 - p_2$ as well as $\delta c = c_1 - c_2$, we obtain by direct calculation

$$D(\rho, \sigma) = \sqrt{\delta p^2 + |\delta c|^2}. \tag{3.245}$$

3.2.5.2 Quantum Relative Entropy

We now want to consider on $\mathcal{S}(\mathcal{H})$ other distinguishability quantifiers related to the von Neumann entropy defined in (2.102). We thus introduce the quantum relative entropy

$$S(\rho_1, \rho_2) = \begin{cases} \mathrm{Tr}\{\rho_1(\log \rho_1 - \log \rho_2)\} & \text{if } \mathrm{supp}\rho_1 \subseteq \mathrm{supp}\rho_2 \\ +\infty & \text{otherwise} \end{cases}, \tag{3.246}$$

where $\rho_1, \rho_2 \in \mathcal{S}(\mathcal{H})$. Thanks to Klein's inequality (2.122) obeyed by the von Neumann entropy, the relative entropy is always positive. In particular, we have $S(\rho_1, \rho_2) = 0$ iff $\rho_1 = \rho_2$, while it becomes infinite if the states are orthogonal, a point further discussed in Remark 3.13 and Remark 3.14. From the very expression if follows that the relative entropy is invariant under unitary transformations. At variance with the trace distance (3.211), this quantity, though always positive, is generally unbounded and does not satisfy the triangle inequality.

Also for the quantum relative entropy a crucial property is the contractivity under the action of completely positive trace-nonincreasing maps [21], related to a notion of distinguishability discussed in Remark 3.12. The proof of this fact relies on the following two properties. First, the relative entropy is invariant under the action of the assignment map (3.50)

$$S(\mathcal{A}_{\rho_E}[\rho_1], \mathcal{A}_{\rho_E}[\rho_2]) = S(\rho_1, \rho_2), \tag{3.247}$$

with ρ_E positive operator with unit trace acting on $\mathcal{H}_E$, for any pair $\rho_1, \rho_2 \in \mathcal{S}(\mathcal{H})$. This follows directly from the definition (3.246) recalling that $\log(\rho \otimes T_E) = \log \rho \otimes \mathbb{1}_E + \mathbb{1}_S \otimes \log T_E$. Most importantly, as a consequence of a deep property of the von Neumann entropy known as strong subadditivity [22], the relative entropy is a contraction with respect to the partial trace operation (3.29), so that

$$S(\mathrm{Tr}_E[\sigma_1], \mathrm{Tr}_E[\sigma_2]) \leqslant S(\sigma_1, \sigma_2) \tag{3.248}$$

for any pair $\sigma_1, \sigma_2 \in \mathcal{S}(\mathcal{H} \otimes \mathcal{H}_E)$.

3.12 Quantum Relative Entropy and Distinguishability

The meaning of quantum relative entropy as distinguishability quantifier is best understood considering that it arises as quantum counterpart of the classical relative entropy, which appears in quantifying the capability to distinguishing two probability distributions. Let us consider first two commuting statistical operators, ρ and σ. They share the same eigenvectors and their eigenvalues identify two probability distributions which we denote as $\{\rho_i\}_i$ and $\{\sigma_i\}_i$ respectively. According to (3.246) we have

$$S(\rho, \sigma) = \begin{cases} \sum_i \rho_i \log\left(\frac{\rho_i}{\sigma_i}\right) & \text{if } \mathrm{supp}\rho \subseteq \mathrm{supp}\sigma \\ +\infty & \text{otherwise} \end{cases}. \tag{3.249}$$

This expression is known as classical relative entropy, or Kullback-Leibler divergence, between the two distributions $\{\rho_i\}_i$ and $\{\sigma_i\}_i$. Now, suppose to sample events according to the frequencies $\{\sigma_i\}_i$. The relative entropy allows to quantify the probability to erroneously conclude, by sampling N events, the observation of another probability distribution, identified by the frequencies $\{\rho_i\}_i$. This probability of error for large N behaves as

$$P = \mathrm{e}^{-NS(\rho, \sigma)}. \tag{3.250}$$

An analog interpretation as distinguishability quantifier holds for an arbitrary pair of statistical operators, with the sampling corresponding to measurements performed on N copies of the considered state [16].

Theorem 3.8 (Lindblad, 1975) *Given a completely positive trace-nonincreasing map O the inequality*

$$S(O[\rho_1], O[\rho_2]) \leqslant S(\rho_1, \rho_2) \tag{3.251}$$

holds $\forall \rho_1, \rho_2 \in S(\mathcal{H})$, and the equality sign is only attained for unitary transformations.

Proof For any completely positive trace-nonincreasing map O we can rely on the theorems by Stinespring and Kraus, Theorems 3.1 and 3.5 respectively. In particular, according to (3.140) we have the representation

$$O[\rho] = \mathrm{Tr}_E[\mathcal{U}[\mathcal{A}_{T_E}[\rho_1]]] \tag{3.252}$$

where $\mathcal{U}$ is defined as in (3.22) with U unitary operator on $\mathcal{H} \otimes \mathcal{H}_E$ and T_E a positive trace class operator acting on $\mathcal{H}_E$, with trace smaller than one, so that (3.247) is replaced by

$$S(\mathcal{A}_{T_E}[\rho_1], \mathcal{A}_{T_E}[\rho_2]) \leqslant S(\rho_1, \rho_2). \tag{3.253}$$

Exploiting the representation (3.252), together with contractivity under the partial trace (3.248), as well as invariance under unitary transformations and assignment map (3.247), together with (3.253), we thus obtain

$$\begin{aligned}
S(O[\rho_1], O[\rho_2]) &= S(\mathrm{Tr}_E[\mathcal{U}[\mathcal{A}_{T_E}[\rho_1]]], \mathrm{Tr}_E[\mathcal{U}[\mathcal{A}_{T_E}[\rho_2]]]) & (3.254)\\
&\leqslant S(\mathcal{U}[\mathcal{A}_{T_E}[\rho_1]], \mathcal{U}[\mathcal{A}_{T_E}[\rho_2]]) & (3.255)\\
&= S(\rho_1, \rho_2). & (3.256)
\end{aligned}$$

$\square$

Let us notice that also for the case of the relative entropy complete positivity is not required on the map O, positivity is enough for (3.251) to hold, as was shown in [23] along a different line of proof.

3.13 Quantum Relative Entropy Between Qubits

For the case of $\mathcal{H} = \mathbb{C}^2$, according to (2.79) we can identify the two states by means of two Euclidean vectors of norm one, r and s, associated to ρ and σ respectively. A direct calculation leads to

$$S(\rho, \sigma) = -h(\|r\|)$$

$$
-\frac{1}{2}\left(1 + \frac{\boldsymbol{r}\cdot\boldsymbol{s}}{\|\boldsymbol{s}\|}\right)\log\left(\frac{1+\|\boldsymbol{s}\|}{2}\right)
$$

$$
-\frac{1}{2}\left(1 - \frac{\boldsymbol{r}\cdot\boldsymbol{s}}{\|\boldsymbol{s}\|}\right)\log\left(\frac{1-\|\boldsymbol{s}\|}{2}\right),
\tag{3.257}
$$

where we have introduced the function

$$
h(x) = -\frac{1+x}{2}\log\left(\frac{1+x}{2}\right) - \frac{1-x}{2}\log\left(\frac{1-x}{2}\right),
\tag{3.258}
$$

corresponding to the Shannon entropy (2.101) of the distribution $\left\{\frac{1+x}{2}, \frac{1-x}{2}\right\}$. This function is monotonically decreasing, concave and even, with range between zero and one. Note that the last term of (3.257) is divergent for $\|\boldsymbol{s}\| \to 1$, corresponding to purity of the state σ, unless the states coincide.

3.14 Quantum Jensen-Shannon Divergence

The very definition of quantum relative entropy (3.246) implies that its value can become infinite, even comparing states in a finite-dimensional setting, as shown in Remark 3.13. In order to compare different pairs of states it is more convenient to consider other entropic distinguishability quantifiers, that is divergences whose expression involves the von Neumann entropy but taking on a finite range of values. To this aim we consider the quantum Jensen-Shannon divergence defined in terms of the von Neumann entropy (2.102) as [16]

$$
J(\rho,\sigma) = S\left(\frac{\rho+\sigma}{2}\right) - \frac{1}{2}S(\sigma) - \frac{1}{2}S(\rho).
\tag{3.259}
$$

We can verify by direct calculation that it can be equivalently written

$$
J(\rho,\sigma) = \frac{1}{2}\left[S\left(\rho, \frac{\rho+\sigma}{2}\right) + S\left(\sigma, \frac{\rho+\sigma}{2}\right)\right],
\tag{3.260}
$$

so that from Theorem 3.8 we immediately have the inequality

$$
J(O[\rho_1], O[\rho_2]) \leqslant J(\rho_1, \rho_2)
\tag{3.261}
$$

for any pair of states $\rho_1, \rho_2 \in \mathcal{S}(\mathcal{H})$, and any trace-nonincreasing operation O. From operator monotonicity of the logarithm we have

$$0 \leqslant J(\rho_1, \rho_2) \leqslant 1, \tag{3.262}$$

with $J(\rho_1, \rho_2) = 0$ iff $\rho_1 = \rho_2$, so that the Jensen-Shannon divergence provides a bounded entropic distinguishability quantifier. For the case of qubits relying on (3.258) we obtain in particular

$$J(\rho, \sigma) = h\left(\frac{\|\boldsymbol{r} + \boldsymbol{s}\|}{2}\right) - \frac{1}{2}[h(\|\boldsymbol{r}\|) + h(\|\boldsymbol{s}\|)]. \tag{3.263}$$

The connection to a distinguishability task for the Jensen-Shannon divergence is in the same spirit as discussed in Remark 3.12 for the relative entropy, with the difference that we are now comparing each probability distribution with an equal mixture of the other two, finally averaging over the results.

3.2.6   Measurement Disturbance

We now exemplify the previous discussion by considering an operation which transforms states by making them diagonal in a given basis. We introduce a basis $\{\varphi_k\}_k$ in $\mathcal{H}$, together with the associated one-dimensional orthogonal projections $\{P_k\}_k$, with $P_k = |\varphi_k\rangle\langle\varphi_k|$, and consider the completely positive transformation

$$\Delta : \mathcal{T}(\mathcal{H}_S) \rightarrow \mathcal{T}(\mathcal{H}_S) \tag{3.264}$$
$$\rho \mapsto \Delta[\rho] = \sum_k P_k \rho P_k,$$

which is in particular trace-preserving, so that Δ is a channel. As we shall see in Sect. 3.2.7.1 a channel with this structure is connected to a measurement á la von Neumann. We can now evaluate the disturbance of such a transformation for different pair of states

$$M_\Delta^{\mathcal{D}}(\rho_1, \rho_2) = \mathcal{D}(\rho_1, \rho_2) - \mathcal{D}(\Delta[\rho_1], \Delta[\rho_2]) \tag{3.265}$$

$$= \mathcal{D}(\rho_1, \rho_2) - \mathcal{D}\left(\sum_k P_k \,\mathrm{Tr}\,\{\rho_1 P_k\}, \sum_k P_k \,\mathrm{Tr}\,\{\rho_2 P_k\}\right) \tag{3.266}$$

according to the different distinguishability quantifiers introduced in (3.211), (3.246) and (3.259), namely trace distance, quantum relative entropy and Jensen-Shannon divergence respectively.

For the sake of convenience we consider the case $\mathcal{H} = \mathbb{C}^2$, so that we can exploit the explicit formulae already introduced. Considering the Bloch representation of (2.79) for a statistical operator in $\mathbb{C}^2$, the action of the dephasing channel (3.264) corresponds to the transformation $\boldsymbol{r} \to (\boldsymbol{r} \cdot \boldsymbol{k})\boldsymbol{k}$, where the norm one vector $\boldsymbol{k}$, $\|\boldsymbol{k}\| = 1$, identifies the basis $\{\varphi_k\}_{k=1,2}$ in $\mathbb{C}^2$. Relying on (3.243) we thus obtain

$$M_\Delta^D(\rho_1, \rho_2) = \frac{1}{2}(\|\boldsymbol{r}_1 - \boldsymbol{r}_2\| - |(\boldsymbol{r}_1 - \boldsymbol{r}_2) \cdot \boldsymbol{k}|). \tag{3.267}$$

where we can notice that the strength of the disturbance depends on the initial coherences, corresponding to different orientations of $\boldsymbol{r}_1$ and $\boldsymbol{r}_2$ with respect to the versor $\boldsymbol{k}$. This fact can be clearly put into evidence representing the states ρ_1 and ρ_2 as matrices in the basis $\{\varphi_k\}_{k=1,2}$. According to (3.245), we obtain in this notation for the trace distance the alternative expression

$$M_\Delta^D(\rho_1, \rho_2) = \sqrt{\delta p^2 + |\delta c|^2} - \sqrt{\delta p^2}, \tag{3.268}$$

where $\delta p = \langle \varphi_1 | \rho_1 \varphi_1 \rangle - \langle \varphi_1 | \rho_2 \varphi_1 \rangle$ and $\delta c = \langle \varphi_1 | \rho_1 \varphi_2 \rangle - \langle \varphi_1 | \rho_2 \varphi_2 \rangle$. The disturbance induced by Δ is due to the suppression of the coherences in the considered basis, hence the name dephasing. The strength of the disturbance reflects the relevance of the coherences in distinguishing the two incoming states ρ_1 and ρ_2. For the case of the relative entropy we rely on (3.257) and obtain

$$\begin{aligned}
M_\Delta^S(\rho_1, \rho_2) = {}& h(|\boldsymbol{r}_1 \cdot \boldsymbol{k}|) - h(\|\boldsymbol{r}_1\|) \\
& - \frac{1}{2}\left(1 + \frac{\boldsymbol{r}_1 \cdot \boldsymbol{r}_2}{\|\boldsymbol{r}_2\|}\right) \log\left(\frac{1 + \|\boldsymbol{r}_2\|}{2}\right) \\
& - \frac{1}{2}\left(1 - \frac{\boldsymbol{r}_1 \cdot \boldsymbol{r}_2}{\|\boldsymbol{r}_2\|}\right) \log\left(\frac{1 - \|\boldsymbol{r}_2\|}{2}\right) \\
& + \frac{1}{2}\left(1 + \frac{(\boldsymbol{r}_1 \cdot \boldsymbol{k})(\boldsymbol{r}_2 \cdot \boldsymbol{k})}{|\boldsymbol{r}_2 \cdot \boldsymbol{k}|}\right) \log\left(\frac{1 + |\boldsymbol{r}_2 \cdot \boldsymbol{k}|}{2}\right) \\
& + \frac{1}{2}\left(1 - \frac{(\boldsymbol{r}_1 \cdot \boldsymbol{k})(\boldsymbol{r}_2 \cdot \boldsymbol{k})}{|\boldsymbol{r}_2 \cdot \boldsymbol{k}|}\right) \log\left(\frac{1 - |\boldsymbol{r}_2 \cdot \boldsymbol{k}|}{2}\right).
\end{aligned} \tag{3.269}$$

Finally, we consider the estimate of the disturbance given by the Jensen-Shannon divergence introduced in Remark 3.14. Thanks to (3.263) we obtain the compact expression

$$\begin{aligned}
M_\Delta^J(\rho_1, \rho_2) = {}& h(\|\boldsymbol{r}_1 + \boldsymbol{r}_2\|/2) - h(|(\boldsymbol{r}_1 + \boldsymbol{r}_2) \cdot \boldsymbol{k}|/2) \\
& - \frac{h(\|\boldsymbol{r}_1\|) - h(|\boldsymbol{r}_1 \cdot \boldsymbol{k}|)}{2} - \frac{h(\|\boldsymbol{r}_2\|) - h(|\boldsymbol{r}_2 \cdot \boldsymbol{k}|)}{2}.
\end{aligned} \tag{3.270}$$

In both (3.269) and (3.270) it clearly appears that the quantification of the disturbance depends on the misalignment between $\boldsymbol{r}_1$ and $\boldsymbol{k}$, as well as $\boldsymbol{r}_2$ and $\boldsymbol{k}$, corresponding to the existence of coherences in the considered basis. A simple visualization of the behavior of the three quantifiers of measurement disturbance is given in Fig. 3.5, where they are compared taking as initial pair the maximally mixed state and a pure

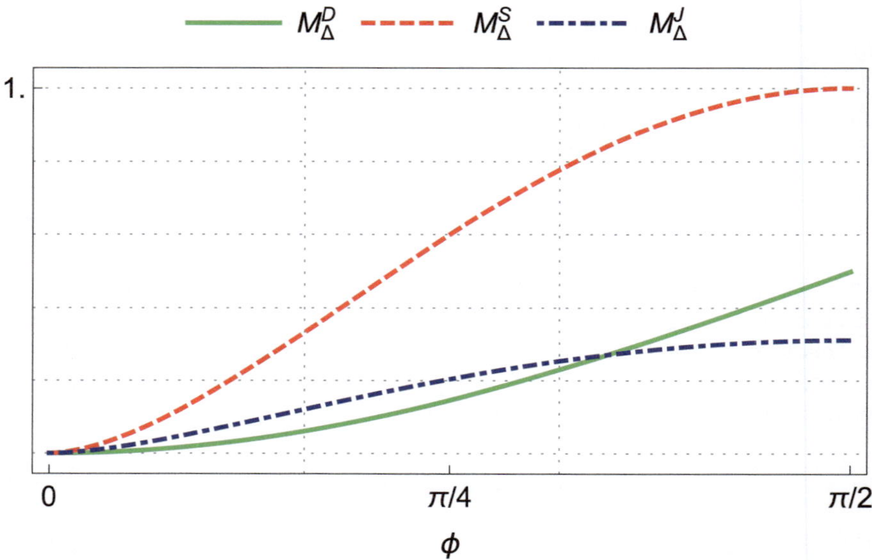

Fig. 3.5 Behavior of the measurement disturbance quantifiers M_Δ^D, M_Δ^S and M_Δ^J built in terms of trace distance, quantum relative entropy and quantum Jensen-Shannon divergence respectively. The quantifiers are evaluated for the case of a dephasing channel acting on a qubit, taking as initial pair the maximally mixed state and a pure state identified by a Bloch vector forming an angle ϕ with the dephasing direction. The disturbance is higher the larger the angle between the dephasing direction and the considered state

state. For this simple case they only depend on the angle between the Bloch vector identifying the pure state and the direction defined by the dephasing channel.

3.2.7 Instruments

We have seen in Sect. 3.2.3 that to the measurement of an elementary dichotomic observable described by an effect we can associate, in a way which is not unique, a transformation on the space of states known as operation. This operation describes the modification of the state taking place due to the measurement transformation, whose consequences have been described in more detail in Sect. 3.2.5. Moving from elementary observables described by effects, as characterized in Sect. 2.2.6, to the general notion of observable as a positive operator-valued measure, considered in Sect. 2.3.1, we are led to ask about the general properties of a state transformation which reproduces the statistics of the considered positive operator-valued measure. The relevant mathematical object describing this situation is known as instrument.

Given a measure space $(\Omega, \mathcal{F})$ we call instrument or operation-valued measure a map

$$\mathcal{I} : \mathcal{F} \rightarrow O(\mathcal{T}(\mathcal{H})), \tag{3.271}$$
$$M \mapsto \mathcal{I}(M)$$

where $O(\mathcal{T}(\mathcal{H}))$ denotes the set of operations acting on $\mathcal{T}(\mathcal{H})$, such that

1. $\mathcal{I}(M)$ is an operation $\forall M \in \mathcal{F}$
2. $\mathrm{Tr}\{\mathcal{I}(\Omega)[T]\} = \mathrm{Tr}\{T\}\ \forall T \in \mathcal{T}(\mathcal{H})$
3. $\mathcal{I}\left(\bigcup_i M_i\right) = \sum_i \mathcal{I}(M_i)$ in the weak topology for any sequence of disjoint sets in $\mathcal{F}$, i.e. $M_i \cap M_j = \emptyset$ for $i \neq j$.

The basic idea is that an instrument gives the subcollection obtained after performing a measurement on the incoming state and selecting the outcomes contained in the considered interval. The first condition tells us that $\mathcal{I}(M)$ is a completely positive trace-nonincreasing map sending states to subcollections for any interval $M \in \mathcal{F}$. The second condition ensures that if no selection is made on the basis of the obtained outcomes, the output is a normalized state. It thus ensures the proper normalization. The third condition warrants σ-additivity with respect to the dependence on the interval, thus justifying the name operation-valued measure. An instrument uniquely identifies a positive operator-valued measure as follows

$$F(M) = \mathcal{I}(M)'[\mathbb{1}], \tag{3.272}$$

indeed we can directly check that the map

$$F : \mathcal{F} \rightarrow \mathcal{E}(\mathcal{H}) \tag{3.273}$$
$$M \mapsto F(M) = \mathcal{I}(M)'[\mathbb{1}]$$

is a positive operator-valued measure. Given a state $\rho \in \mathcal{S}(\mathcal{H})$ the probability to obtain an outcome in M for the observable F is given in accordance with (3.63) by the formula

$$P_\rho(M) = \mu_\rho^F(M) \tag{3.274}$$
$$= \mathrm{Tr}\{\rho F(M)\} \tag{3.275}$$
$$= \mathrm{Tr}\{\rho \mathcal{I}(M)'[\mathbb{1}]\} \tag{3.276}$$
$$= \mathrm{Tr}\{\mathcal{I}(M)[\rho]\}. \tag{3.277}$$

In particular, (3.277) implies that the probability for outcomes in a definite interval M is given by the weight of the corresponding subcollection $\mathcal{I}(M)[\rho]$. In the same spirit of (3.67) the instrument also provides the new state conditioned on the occurrence of an outcome in M, namely

$$\rho(M) = \frac{\mathcal{I}(M)[\rho]}{\mathrm{Tr}\{\mathcal{I}(M)[\rho]\}}. \tag{3.278}$$

Note that as discussed in Sect. 3.2.3.1 the map associating to the pre-measurement state a new state conditioned on the measurement outcome $\rho \to \rho(M)$ is not linear. Moreover, it is not σ-additive in its dependence on the outcome M. If we consider the special case in which M becomes the whole space Ω, we have $P_\rho(\Omega) = 1$, so that the transformed state becomes

$$\rho(\Omega) = \mathcal{I}(\Omega)[\rho], \tag{3.279}$$

which is often called the a priori state. We recall that $\mathcal{I}(\Omega)$ is a quantum channel, so that it is trace-preserving. Equation (3.279) provides the state obtained after the interaction with the measurement apparatus described by the considered instrument, when no selection based on the outcome has been made. An instrument thus provides both the statistics of the outcomes and the transformed state for a general measurement described by a positive operator-valued measure. As discussed in Sect. 3.2.3.1 we are therefore able to consider subsequent measurements leading to conditional probabilities. To this aim we consider another positive operator-valued measure G on the outcome space $(\Gamma, \mathcal{G})$. Given an input ρ and an outcome $M \in \mathcal{F}$ for the POVM F, the conditional probability to obtain in a subsequent measurement an outcome $N \in G$ for the positive operator-valued measure G is given according to (3.274) and (3.278) by

$$P_\rho(N|M) = \mu^G_{\rho(M)}(N) \tag{3.280}$$

$$= \mathrm{Tr}\{G(N)\rho(M)\} \tag{3.281}$$

$$= \mathrm{Tr}\left\{G(N)\frac{\mathcal{I}(M)[\rho]}{\mathrm{Tr}\{\mathcal{I}(M)[\rho]\}}\right\} \tag{3.282}$$

$$= \frac{\mathrm{Tr}\{G(N)\mathcal{I}(M)[\rho]\}}{P_\rho(M)}. \tag{3.283}$$

Note that $P_\rho(N|M)$ denotes the conditional probability of obtaining first the outcome M for the POVM denoted with F, and subsequently the outcome N for the POVM denoted with G. The order is indeed relevant since in general $P_\rho(N|M) \neq P_\rho(M|N)$. Importantly, this comes from the fact that any measurement in quantum mechanics affects the state of the system, so that it will in general have an influence on any subsequent measurement. According to (3.274) and (3.280) the joint probability distribution is then given by

$$P_\rho(N, M) = \mathrm{Tr}\{G(N)\mathcal{I}(M)[\rho]\}, \tag{3.284}$$

and exploiting (3.4) it can be equivalently expressed in terms of the dual of the instrument

$$P_\rho(N, M) = \text{Tr}\{\rho \mathcal{I}(M)'[G(N)]\}. \tag{3.285}$$

We can now consider an instrument $\mathcal{K}$ compatible with the positive operator-valued measure G in the sense that

$$\text{Tr}\{\rho G(N)\} = \text{Tr}\{\mathcal{K}(N)[\rho]\}, \tag{3.286}$$

for any interval $N \in \mathcal{G}$ and any state $\rho \in \mathcal{S}(\mathcal{H})$, so that it correctly reproduces the statistics of the measurement, implying in particular

$$G(N) = \mathcal{K}(N)'[\mathbb{1}]. \tag{3.287}$$

This instrument describes a possible apparatus for the measurement of the POVM G. Combining (3.285) and (3.287) the joint probability distribution can be written

$$P_\rho(N, M) = \text{Tr}\{\rho \mathcal{I}(M)'[\mathcal{K}(N)'[\mathbb{1}]]\}, \tag{3.288}$$

which can be interpreted as a Heisenberg picture in the sense that we are evaluating the trace of the statistical operator with the effect $E = \mathcal{I}(M)'[\mathcal{K}(N)'[\mathbb{1}]]$ which keeps into account the POVMs that have been measured, their sequence and the apparatus used for the measurement. Exploiting the duality relation (3.4) we can express the very same joint probability distribution in the form

$$P_\rho(N, M) = \text{Tr}\{\mathcal{K}(N)[\mathcal{I}(M)[\rho]]\}, \tag{3.289}$$

corresponding to a Schrödinger picture in that the probability is given by the trace of the subcollection $\mathcal{K}(N)[\mathcal{I}(M)[\rho]]$ obtained by transforming the state of the system in the dependence on the measurements performed, their sequence and the apparata as described by the associated instruments. Equations (3.288) and (3.289) are the natural generalization of (3.75) and (3.76) respectively. These equivalent expressions provide a probability distribution on the rectangles $M \times N \in \Omega \times \Gamma$, from which we can construct a well-defined probability on the product of the σ-algebras $\mathcal{F} \otimes \mathcal{G}$. We have thus built a new instrument, defined on the measure space $(\Omega \times \Gamma, \mathcal{F} \otimes \mathcal{G})$, given by the composition of the two original instruments $\mathcal{K} \circ \mathcal{I}$, which describes two subsequent measurements

$$P_\rho(N, M) = \text{Tr}\{\mathcal{K} \circ \mathcal{I}(N \times M)[\rho]\}. \tag{3.290}$$

Most importantly, in such a way we have put multiple subsequent measurements on the same formal footing as a single measurement. This formalism opens the way for the description of measurements continuous in time in quantum mechanics [2–4]. A notion of trajectory for a system can therefore be recovered also in quantum mechanics, even though in a more refined description, taking into account the unavoidable uncertainty and disturbance associated to obtaining information from a measurement, as also discussed in Sect. 3.2.6.

Similarly to the case of operations, while an instrument as we have seen uniquely determines a positive operator-valued measure, the inverse connection is many to one, in that many instruments, leading to different state transformations, provide the same statistics of the outcomes and therefore the same positive operator-valued measure. To each positive operator-valued measure F corresponds a whole equivalence class of instruments, which we denote as $[\mathcal{I}]_F$. All elements of $[\mathcal{I}]_F$ are compatible with the POVM F in the sense of (3.274). The existence of these equivalence classes corresponds to the fact that the statistics of outcomes for the same observable can be obtained affecting the state in different ways, corresponding to different registration procedures. There are many apparata all measuring the position of an electron, and each of them differently affects the state of the electron when such a position measurement is performed. In turn this has an influence on a subsequent measurement on the electron, e.g. a characterization of its momentum distribution.

3.2.7.1   Von Neumann Instruments

An important class of instruments is given by the so-called von Neumann-Lüders instruments [24]. Their introduction provides a natural extension of the discussion on the connection between operations and state preparations in Sect. 3.2.3.2. We consider an observable identified by a self-adjoint operator A. Importantly, we assume a discrete spectrum, though allowing for degeneracy. According to the spectral theorem we have the representation

$$A = \sum_k a_k P_{a_k}. \tag{3.291}$$

At variance with (3.85) the projections P_{a_k} are not necessarily one-dimensional, since we allow for degeneracy of the eigenvalue a_k, fixed by a single real number. As discussed in Sect. 2.2 to each self-adjoint operator is uniquely associated a PVM [25]. For the operator of (3.291) according to (2.6) it takes the form

$$E^A(M) = \sum_{\{k \mid a_k \in M\}} P_{a_k}. \tag{3.292}$$

An instrument compatible with this projection-valued measure can be immediately constructed exploiting idempotency of the projection operators. Building on (3.90) we define

$$\mathcal{M}(\{a_k\})[\rho] = P_{a_k} \rho P_{a_k}, \tag{3.293}$$

while for an arbitrary interval M we set

$$\mathcal{M}(M)[\rho] = \sum_{\{k|a_k \in M\}} \mathcal{M}(\{a_k\})[\rho], \qquad (3.294)$$

so as to comply with σ-additivity. The defining properties of an instrument as introduced in (3.271) can be easily verified. An instrument constructed in this way is called von Neumann-Lüders instrument, and in particular if the P_{a_k} are one-dimensional projections, corresponding to lack of degeneracy, simply von Neumann instrument. According to (3.278) for this type of instruments the state conditioned on an outcome in M reads

$$\rho(M) = \frac{\sum_{\{k|a_k \in M\}} P_{a_k} \rho P_{a_k}}{\mathrm{Tr}\left\{ \sum_{\{k|a_k \in M\}} P_{a_k} \rho \right\}}. \qquad (3.295)$$

If we consider an outcome corresponding to one of the possible eigenvalues we have

$$\rho(\{a_k\}) = \frac{P_{a_k} \rho P_{a_k}}{\mathrm{Tr}\{P_{a_k} \rho\}}, \qquad (3.296)$$

and therefore in the absence of degeneracy, denoting as u_{a_k} the eigenvector relative to the eigenvalue a_k, the output state

$$\rho(\{a_k\}) = |u_{a_k}\rangle\langle u_{a_k}|, \qquad (3.297)$$

in accordance with (3.102) and independently of the purity of the incoming state ρ. According to (3.279) if no selection based on the outcome is performed we obtain

$$\rho(\Omega) = \sum_k P_{a_k} \rho P_{a_k} \qquad (3.298)$$

as in (3.109). The transformed state is thus diagonal in the basis determined by the considered observable. The effect of a von Neumann-Lüders instrument is therefore to remove the coherences of the input state in the associated basis. The transformation considered in Sect. 3.2.6 can therefore be interpreted as the channel associated to a von Neumann instrument. As detailed in Remark 3.15, von Neumann-Lüders instruments are repeatable and ideal. It is an important fact that any instrument which satisfies these two properties must be of the von Neumann-Lüders form, which implies in particular the pure point spectrum of the associated observable. Observables with a continuous spectrum therefore cannot be measured with arbitrary precision. Moreover, they do not admit ideal measurements. The perturbation induced by the measurement necessarily affects the state in a non-trivial way. In this sense, discrete and continuous spectra indeed do have crucially different features.

3.15 Repeatability and Ideality

The class of von Neumann-Lüders instruments has two important features, which actually single them out from all other instruments, namely repeatability and ideality. These properties have already been put into evidence in Sect. 3.2.3.2 in the framework of an operation describing a state preparation. Indeed, for any fixed outcome an instrument provides an operation, while if no selection is made it describes the preparation of a new output state starting from the input one. Repeatability corresponds to the fact that the state transformed by the instrument according to a certain outcome leads to the very same outcome if the instrument is applied once again. This condition can be expressed asking, for an arbitrary input state ρ,

$$\mathcal{M}(\{a_j\})[\mathcal{M}(\{a_k\})[\rho]] = \delta_{jk}\mathcal{M}(\{a_k\})[\rho] \qquad (3.299)$$

or equivalently in terms of the dual of the instrument

$$\mathcal{M}'(\{a_k\})[\mathcal{M}'(\{a_j\})[\mathbb{1}]] = \delta_{jk}\mathcal{M}'(\{a_k\})[\mathbb{1}]. \qquad (3.300)$$

The equivalence between (3.91) and (3.299) is directly verified. Denoting with F the positive operator-valued measure uniquely associated to the instrument according to (3.272) and recalling the expression (3.278) of the conditioned state we verify thanks to either (3.299) or (3.300)

$$\mathrm{Tr}\{\rho(M)F(M)\} = 1, \qquad (3.301)$$

expressing the fact that the conditional state relative to a certain outcome gives with probability one the same outcome in a subsequent measurement.

Ideality in turn corresponds to minimality in the perturbation of the state in the measurement, namely

$$\mathrm{Tr}\{\mathcal{M}(\{a_k\})[\rho]\} = 1 \qquad (3.302)$$

implies

$$\mathcal{M}(\{a_k\})[\rho] = \rho. \qquad (3.303)$$

States leading to an outcome that is certain are not modified. The proof is the same considered in Sect. 3.2.3.2 to prove the equivalence of (3.94) and (3.95), since for any fixed $\{a_k\}$ the map $\mathcal{M}(\{a_k\})$ is an operation, and relies on idempotency of orthogonal projections.

An important advancement in introducing instruments is the possibility to perform on an equal footing single and repeated measurements. It is therefore of special interest to consider the subsequent application of two von Neumann-Lüders instruments. We already know that their composition can be described as a single instrument on an outcome space given by the Cartesian product of the outcome spaces as in (3.290). The question is whether the class of von Neumann-Lüders instruments is stable under such a composition law. To this aim we consider two such instruments $\mathcal{M}^A$ and $\mathcal{M}^B$, related to two projective measurements corresponding to the observables

$$A = \sum_k a_k P_{a_k} \tag{3.304}$$

and

$$B = \sum_l b_l Q_{b_l} \tag{3.305}$$

respectively. If we apply in sequence the two instruments, the subcollection obtained from the input state ρ selecting first an outcome in M and then an outcome in N reads

$$\mathcal{M}^B(N)[\mathcal{M}^A(M)[\rho]] = \sum_{\{l|b_l \in N\}} \sum_{\{k|a_k \in M\}} Q_{b_l} P_{a_k} \rho P_{a_k} Q_{b_l} \tag{3.306}$$

$$= \mathcal{M}^B \circ \mathcal{M}^A(N \times M)[\rho], \tag{3.307}$$

where in the last line we have used the same notation as in (3.290). The composite instrument $\mathcal{M}^B \circ \mathcal{M}^A$ is however no more of the von Neumann-Lüders type, unless the two observables commute, i.e. $[P_{a_k}, Q_{b_l}] = 0$ for all l and all k, so that $[A, B] = 0$. This fact is best seen noting that the uniquely associated POVM

$$(\mathcal{M}^B \circ \mathcal{M}^A)'(N \times M)[\mathbb{1}] = \mathcal{M}^{A'}(M)\left[\mathcal{M}^{B'}(N)[\mathbb{1}]\right] \tag{3.308}$$

$$= \sum_{\{l|b_l \in N\}} \sum_{\{k|a_k \in M\}} P_{a_k} Q_{b_l} P_{a_k} \tag{3.309}$$

is no more a PVM. Indeed the operators $P_{a_k} Q_{b_l} P_{a_k}$ are positive and below the identity, so that $P_{a_k} Q_{b_l} P_{a_k} \in \mathcal{E}(\mathcal{H})$, but they are not idempotent, so that $P_{a_k} Q_{b_l} P_{a_k} \notin \mathcal{P}(\mathcal{H})$ unless A and B commute. The joint probability for an outcome in M for the observable A followed by an outcome in N for the observable B, given the input state ρ, is then written as

$$P_\rho(N, M) = \mathrm{Tr}\{\mathcal{M}^B(N)[\mathcal{M}^A(M)[\rho]]\} \tag{3.310}$$

$$= \sum_{\{l|b_l \in N\}} \sum_{\{k|a_k \in M\}} \mathrm{Tr}\{Q_{b_l} P_{a_k} \rho P_{a_k} Q_{b_l}\}. \tag{3.311}$$

Expression (3.311) can be extended to an arbitrary finite sequence of measurements and is known as Wigner's formula [24,26,27].

If we restrict to a discrete outcome space $\{a_1, a_2, \ldots\}$, also for the general case of a positive operator-valued measure F there is a simple way to associate to it an instrument compatible in the sense of (3.274). We simply set

$$\mathcal{I}(\{a_k\})[\rho] = \sqrt{F(\{a_k\})}\,\rho\,\sqrt{F(\{a_k\})}, \tag{3.312}$$

and extend by σ-additivity, thus defining a proper instrument. Compatibility follows from

$$\mathrm{Tr}\{\mathcal{I}(M)[\rho]\} = \mathrm{Tr}\left\{\sum_{\{k\,|\,a_k\in M\}} F(\{a_k\})\rho\right\} \tag{3.313}$$

$$= \mathrm{Tr}\{F(M)\rho\}, \tag{3.314}$$

where in (3.313) we have used the cyclic property of the trace operation and in (3.314) σ-additivity of the POVM.

3.2.7.2 General Instrument

We now provide the generic expression of an instrument, which encompasses the situation in which the outcome space is not necessarily discrete, exemplified in Remark 3.16 for the case of a joint measurement of position and momentum. Let Ω be a measurable space, μ a positive measure on it, and consider an operator-valued measurable function

$$V : \Omega \to \mathcal{B}(\mathcal{H}) \tag{3.315}$$

$$\omega \mapsto V(\omega)$$

satisfying

$$\int_\Omega \mathrm{d}\mu(\omega) V^\dagger(\omega) V(\omega) = \mathbb{1}. \tag{3.316}$$

The expression

$$\mathcal{I}(M)[\rho] = \int_M \mathrm{d}\mu(\omega) V(\omega) \rho V^\dagger(\omega) \tag{3.317}$$

defines an instrument with associated positive operator-valued measure

$$F(M) = \mathcal{I}(M)'[\mathbb{1}] \tag{3.318}$$

$$= \int_M \mathrm{d}\mu(\omega) V^\dagger(\omega) V(\omega), \tag{3.319}$$

well-defined and normalized thanks to σ-additivity of the measure μ and the normalization condition (3.316). For any state ρ the probability for an outcome in the set M according to (3.276) is then

$$P_\rho(M) = \text{Tr}\left\{\int_M d\mu(\omega) V^\dagger(\omega) V(\omega)\rho\right\}, \qquad (3.320)$$

and the state obtained by selecting an outcome in M reads

$$\rho(M) = \frac{\int_M d\mu(\omega) V(\omega)\rho V^\dagger(\omega)}{\text{Tr}\left\{\rho \int_M d\mu(\omega) V^\dagger(\omega) V(\omega)\right\}}, \qquad (3.321)$$

according to (3.278). As already put into evidence considering (3.109), as well as (3.279) and (3.298), given a transformation with a measurement character it is natural to consider the output state obtained if no selection is performed on the measurement, obtaining the so-called a priori state that for a general instrument reads

$$\rho(\Omega) = \int_\Omega d\mu(\omega) V(\omega)\rho V^\dagger(\omega). \qquad (3.322)$$

If we consider the complementary situation in which rather than the whole space we select as outcome an infinitesimal region $d\omega$ around the point $\omega \in \Omega$, we are lead to consider the so-called a posteriori state

$$\rho(\omega) = \frac{\mathcal{I}(d\omega)[\rho]}{\text{Tr}\{\mathcal{I}(d\omega)[\rho]\}}, \qquad (3.323)$$

which is the state we can associate to the system upon performing a measurement and obtaining the outcome ω. Note that while $\rho(M)$ in (3.321) is a function defined on $\mathcal{F}$, the a posteriori state is a function defined on Ω. It is a random variable depending on the outcomes and distributed according to the probability distribution $P(\omega) = \text{Tr}\{\mathcal{I}(\omega)[\rho]\}$. The a priori state is then given by the expectation value of the a posteriori state with respect to this probability measure

$$\mathcal{I}(\Omega)[\rho] = \int_\Omega \rho(\omega) P(d\omega). \qquad (3.324)$$

More generally, the subcollection corresponding to a given outcome has the expression

$$\mathcal{I}(M)[\rho] = \int_M \rho(\omega) P(d\omega). \qquad (3.325)$$

It is clear that this example includes the previously considered cases by taking measures with a purely discrete support.

3.16 Instrument for Joint Position and Momentum Measurement

We now exploit the results of Sect. 2.3.5 to provide an explicit expression
of an instrument related to the joint measurement of position and momentum
considered in Remark 2.13. To this aim let us recall (3.312) and consider the
identity

$$\sqrt{UAU^\dagger} = U\sqrt{A}U^\dagger, \tag{3.326}$$

valid for positive A and unitary U. We now start from (2.374) where S is a
positive operator and the Weyl operators (2.376) are unitary, and consider the
case in which S is the one-dimensional projection P_ψ associated to a Gaussian
state ψ centered in the origin, so as to satisfy the requirement (2.375) of
rotational invariance. We thus have $\sqrt{P_\psi} = P_\psi$ as well as

$$\langle x | W(x_0, p_0)\psi \rangle = \langle x | \psi_{x_0 p_0} \rangle, \tag{3.327}$$

with ψ_{x_0, p_0} as in (2.402). Indeed, according to (2.377) the Weyl operators
induce a translation in both position and momentum, thus leading to a Gaussian
wavepacket centered in x_0, p_0. In analogy with (3.312) we thus introduce the
expression

$$\mathcal{I}^{x,p}(M \times N)[\rho] = \frac{1}{(2\pi\hbar)^3} \int_M d^3x_0 \int_N d^3p_0 \, |\psi_{x_0 p_0}\rangle\langle\psi_{x_0 p_0}|\rho\psi_{x_0 p_0}\rangle\langle\psi_{x_0 p_0}|,$$
$$\tag{3.328}$$

which is an instrument associated to the POVM introduced in (2.401)

$$\mathcal{I}^{x,p}(M \times N)'[\mathbb{1}] = \frac{1}{(2\pi\hbar)^3} \int_M d^3x_0 \int_N d^3p_0 \, |\psi_{x_0 p_0}\rangle\langle\psi_{x_0 p_0}| \tag{3.329}$$
$$= F^{x,p}(M \times N). \tag{3.330}$$

In the considered example the positive measure μ is simply proportional to
the Lebesgue measure on the phase space, while the operator-valued measure
(3.315) associates to each point in the phase space the Gaussian wavepacket
with width σ centered in it.

3.2.8 Measurement Models

Thus far we have considered two levels for the description of a measurement in
quantum mechanics. In Sect. 2.3.1 we concentrated on describing the statistics of
the outcomes, showing that this can be generally obtained by means of a POVM.
In Sect. 3.2.7 we moved one step further considering the associated transforma-

Measurement process

POVM	Instrument	Measurement model
Q → C	Q → Q	Q×Q → Q
F	$\mathcal{I}$	$\mathfrak{M}$
statistics	statistics + transformed state	statistics + transformed state + measurement apparatus
	$F \to [\mathcal{I}]_F$	$\mathcal{I} \to [\mathfrak{M}]_{\mathcal{I}}$

Fig. 3.6 Scheme of the different steps taken in the description of the measurement process

tion induced on the state, thus coming to the notion of instrument. On top of the statistics, an instrument provides via (3.278) the state obtained from the incoming one by performing a selection based on the outcome. To any instrument one uniquely associates a positive operator-valued measure whose statistics is compatible with it according to (3.274), while the inverse relation is many to one, in analogy to what happens for the relationship between operations and effects discussed in Sect. 3.2.3. Indeed, the statistics of the same POVM F is obtained from a whole equivalence class of instruments $[\mathcal{I}]_F$, that however affect the state in different ways. We now make a further step in the description of the measurement, considering a third level, which following [15] we call measurement model and denote with the letter $\mathfrak{M}$. The measurement transformation on the state is now described in terms of the unitary interaction with another quantum system, namely the measurement apparatus from which the information on the outcome is extracted. Also in this case to a given instrument $\mathcal{I}$ corresponds a whole equivalence class of measurement schemes $[\mathfrak{M}]_{\mathcal{I}}$ compatible with it. These different levels of description of the measurement procedure are schematized in Fig. 3.6. The construction of a measurement scheme corresponding to an instrument can be obtained considering suitable representations of the operations in its range, based on dilations, as discussed in Sect. 3.2.4. We will now formulate the statement in its general form, referring to a theorem by Ozawa [28], and provide a proof for the special case in which the Hilbert space $\mathcal{H}$ of the system has finite dimension n, and the outcome space is also finite-dimensional.

Theorem 3.9 (Ozawa, 1984) *Let $\mathcal{I}$ be an instrument on the outcome space $(\Omega, \mathcal{F})$, for a system described in the Hilbert space $\mathcal{H}$. Then there exist a Hilbert space $\mathcal{H}_M$, a projection-valued measure E on $\mathcal{H}_M$ over the same outcome space, a statistical operator $\sigma_M \in \mathcal{T}(\mathcal{H}_M)$ and a unitary transformation U on $\mathcal{H} \otimes \mathcal{H}_M$ such that the representation*

$$\mathcal{I}(M)[\rho] = \mathrm{Tr}_{\mathcal{H}_M}\{U(\rho \otimes \sigma_M)U^{\dagger}(\mathbb{1}_{\mathcal{H}} \otimes E(M))\} \quad \forall M \in \mathcal{F} \quad (3.331)$$

holds for any state ρ of the system.

Proof We suppose dim $\mathcal{H} = n$ and the cardinality of Ω finite. For any fixed outcome in Ω according to (3.139) we have an operator-sum representation of the operation associated to it by the given instrument $\mathcal{I}$

$$\mathcal{I}(\{k\})[\rho] = \sum_{i \in I_k} A_i \rho A_i^\dagger, \tag{3.332}$$

where the index i runs over a set I_k with at most n^2 elements, corresponding to the cardinality of the representation of $\mathcal{I}(\{k\})$. The union of these expressions provides a representation for the channel $\mathcal{I}(\Omega)$ in terms of a finite set of Kraus operators

$$\mathcal{I}(\Omega)[\rho] = \sum_{i=1}^{N} A_i \rho A_i^\dagger, \tag{3.333}$$

with $N = \sum_k \sum_{i \in I_k}$. We now consider a dilation of the channel $\mathcal{I}(\Omega)$ as in (3.140) of Theorem 3.5. We thus introduce a Hilbert space $\mathcal{H}_M$, whose dimension can be taken to be N, and a pure state $P_\eta = |\eta\rangle\langle\eta|$ in $\mathcal{H}_M$ leading to the representation

$$\mathcal{I}(\Omega)[\rho] = \mathrm{Tr}_{\mathcal{H}_M}\{U(\rho \otimes P_\eta)U^\dagger\}. \tag{3.334}$$

In our construction we still have to introduce the observable for the apparatus degrees of freedom, which is often called pointer basis, recalling the fact that this should correspond to the reading of a pointer on the apparatus. To this aim we first observe that any basis, say $\{\varphi_i\}_{i=1,\ldots,N}$, in $\mathcal{H}_M$ induces a set of Kraus operators as shown in Theorem 3.5 by obtaining (3.139) from (3.140). We thus have the alternative representation

$$\mathcal{I}(\Omega)[\rho] = \sum_{i=1}^{N} B_i \rho B_i^\dagger, \tag{3.335}$$

with the operators $\{B_i\}$ defined as $B_i = \langle\varphi_i|U\eta\rangle$, so that their matrix elements read

$$\langle\Phi|B_i\Psi\rangle = \langle\Phi \otimes \varphi_i|U(\Psi \otimes \eta)\rangle \qquad \forall \Phi, \Psi \in \mathcal{H}. \tag{3.336}$$

As shown in Theorem 3.6 the two representations, which in particular have the same number of Kraus operators, are related by N^2 complex numbers $\{u_{ik}\}$ according to (3.175) as follows

$$A_i = \sum_{k=1}^{N} u_{ik} B_k. \tag{3.337}$$

Thanks to the constraint (3.176) these coefficients allow to introduce a new orthonormal system $\{\tilde{\varphi}_i\}_{i=1,\ldots,N}$ in $\mathcal{H}_M$ according to

$$\tilde{\varphi}_j = \sum_{k=1}^{N} u_{jk}^* \varphi_k. \tag{3.338}$$

In turn, any orthogonal basis allows to define a POVM, which is in particular a PVM, upon considering the associated one-dimensional projections

$$\tilde{E}(\{i\}) = |\tilde{\varphi}_i\rangle\langle\tilde{\varphi}_i|. \tag{3.339}$$

For any fixed outcome $\{i\}$ we can consider as in (2.286) the following expression of indirect measurement

$$\mathrm{Tr}_{\mathcal{H}_M}\{P_\eta U^\dagger(\mathbb{1}_{\mathcal{H}} \otimes \tilde{E}(\{i\}))U\}, \tag{3.340}$$

and build in terms of it the instrument $\mathcal{I}$. Thanks to (3.338) and (3.339) we have in fact, for any state ρ in $\mathcal{H}$ and each of the N possible values of i, the identity

$$\mathrm{Tr}_{\mathcal{H}_M}\{U(\rho \otimes P_\eta)U^\dagger(\mathbb{1}_{\mathcal{H}} \otimes \tilde{E}(\{i\}))\} = \langle\tilde{\varphi}_i|U\eta\rangle\rho\langle U\eta|\tilde{\varphi}_i\rangle \tag{3.341}$$

$$= \sum_{r,s=1}^{N} u_{ir}\langle\varphi_r|U\eta\rangle\rho u_{is}^*\langle U\eta|\varphi_s\rangle. \tag{3.342}$$

Recalling the definition (3.336) of the operators appearing in the Kraus representation (3.335), as well as their relationship with the alternative Kraus representation (3.333), we finally obtain

$$\mathrm{Tr}_{\mathcal{H}_M}\{U(\rho \otimes P_\eta)U^\dagger(\mathbb{1}_{\mathcal{H}} \otimes \tilde{E}(\{i\}))\} = \sum_{r=1}^{N} u_{ir} B_r \rho \sum_{s=1}^{N} u_{is}^* B_s^\dagger \tag{3.343}$$

$$= A_i \rho A_i^\dagger. \tag{3.344}$$

The desired representation of the instrument is obtained by suitably collecting the indexes in groups of cardinality I_k

$$\mathcal{I}(\{k\})[\rho] = \mathrm{Tr}_{\mathcal{H}_M}\left\{U(\rho \otimes P_\eta)U^\dagger\left(\mathbb{1}_{\mathcal{H}} \otimes \sum_{i\in I_k}\tilde{E}(\{i\})\right)\right\} \tag{3.345}$$

$$= \mathrm{Tr}_{\mathcal{H}_M}\{U(\rho \otimes P_\eta)U^\dagger(\mathbb{1}_{\mathcal{H}} \otimes E(\{k\}))\}, \tag{3.346}$$

recalling that $E(\{k\}) = \sum_{i\in I_k}\tilde{E}(\{i\})$ still is a projection, and finally extending by σ-additivity. $\qquad\square$

A measurement model $\mathfrak{M}$ can thus be identified with a quadruple $(\mathcal{H}_M, U, \sigma_M, G)$, where $\mathcal{H}_M$ can be seen as the Hilbert space of the measurement apparatus, U is a unitary operator on $\mathcal{H} \otimes \mathcal{H}_M$ coupling system and apparatus, σ_M the state in which the apparatus is initialized, and G an observable for the apparatus, which can be taken to be projection-valued. Recalling the compatibility of a POVM with an instrument as expressed by (3.274), we say that an instrument $\mathcal{I}$ is compatible with a measurement model $\mathfrak{M}$ identified by the quadruple $(\mathcal{H}_M, U, \sigma_M, G)$ if

$$\mathcal{I}(M)[\rho] = \mathrm{Tr}_{\mathcal{H}_M}\{U(\rho \otimes \sigma_M)U^\dagger(\mathbb{1}_\mathcal{H} \otimes G(M))\} \qquad (3.347)$$

holds for any interval $M \in \mathcal{F}$ and any system state ρ. A measurement model leading to a von Neumann instrument is considered in Remark 3.17.

3.17 Von Neumann Measurement

We now consider an example of measurement model leading to a von Neumann instrument as described in Sect. 3.2.7.1 and therefore compatible with a sharp observable. We consider in $\mathcal{H}$ a sharp observable with discrete and non-degenerate spectrum, so that according to (2.252) we can express it by means of the projection operators

$$E(\{k\}) = |\varphi_k\rangle\langle\varphi_k|, \qquad (3.348)$$

with $\{\varphi_k\}$ a basis in $\mathcal{H}$. Take as apparatus space $\mathcal{H}_M$ a Hilbert space with the same dimensionality of $\mathcal{H}$ and consider here a basis $\{\psi_i\}$ which allows to construct the projective observable for the apparatus

$$\tilde{E}(\{i\}) = |\psi_i\rangle\langle\psi_i|. \qquad (3.349)$$

We suppose to initialize the apparatus in the state $P_{\psi_1} = |\psi_1\rangle\langle\psi_1|$, so that we still only have to specify the unitary interaction. This is obtained by considering a unitary extension of the isometry

$$V(\varphi_k \otimes \psi_1) = \varphi_k \otimes \psi_k, \qquad (3.350)$$

which expresses the fact that if the system starts in an eigenstate of the considered observable, after the coupling the state of the apparatus is characterized by a label keeping track of this eigenstate. Note that this is exactly what happens in the Stern-Gerlach experiment described in Sect. 2.3.1.1, with the crucial difference that here the pointer basis $\{\psi_k\}$ is composed of vectors with orthogonal support. The quadruple $(\mathcal{H}_M, V, P_{\psi_1}, \tilde{E})$ provides a measurement model compatible with a von Neumann instrument and with the statistics of the outcomes of the system observable E. Recall first that a generic initial state of the system can be expressed as

$$\varphi = \sum_k c_k \varphi_k, \tag{3.351}$$

where the complex coefficients c_k satisfy $\sum_k |c_k|^2 = 1$. According to (2.245) the statistics of the outcomes for the observable E are then given by the so-called Born probabilities

$$\mu_{P_\varphi}^E(\{k\}) = |c_k|^2. \tag{3.352}$$

Due to (3.350) we further have

$$V(P_\varphi \otimes P_{\psi_1})V^\dagger = \sum_{j,k} c_j c_k^* |\varphi_j\rangle\langle\varphi_k| \otimes |\psi_j\rangle\langle\psi_k| \tag{3.353}$$

so that the instrument determined by the introduced measurement model according to (3.331) takes the form

$$\mathcal{I}(\{k\})[P_\varphi] = \mathrm{Tr}_{\mathcal{H}_M}\{V(P_\varphi \otimes P_{\psi_1})V^\dagger(\mathbb{1}_\mathcal{H} \otimes \tilde{E}(\{k\}))\} \tag{3.354}$$

$$= |c_k|^2 |\varphi_k\rangle\langle\varphi_k|, \tag{3.355}$$

where in the last line we have used (3.349). This instrument is of the von Neumann type (3.293) as becomes manifest when writing it in the form

$$\mathcal{I}(\{k\})[P_\varphi] = E(\{k\})P_\varphi E(\{k\}). \tag{3.356}$$

The statistics of the outcomes associated to it according to (3.274) is then given by

$$\mu_\varphi^E(\{k\}) = \mathrm{Tr}_\mathcal{H}\{\mathcal{I}(\{k\})[P_\varphi]\} \tag{3.357}$$

$$= \mathrm{Tr}_\mathcal{H}\,\mathrm{Tr}_{\mathcal{H}_M}\{V(P_\varphi \otimes P_{\psi_1})V^\dagger(\mathbb{1}_\mathcal{H} \otimes \tilde{E}(\{k\}))\} \tag{3.358}$$

$$= |c_k|^2, \tag{3.359}$$

indeed compatible with the projective measurement considered as starting point. We recall that according to (3.279) each instrument uniquely identifies a channel, that is to say a completely positive trace-preserving transformation, obtained when no selection on the outcomes is made. For the considered von Neumann instrument the associated instrument is the dephasing map making the state diagonal in the given basis

$$\mathcal{I}(\Omega)[\rho] = \sum_k E(\{k\})\rho E(\{k\}), \tag{3.360}$$

whose measurement disturbance has been discussed in Sect. 3.2.6 for the case $\mathcal{H} = \mathbb{C}^2$.

3.3   Quantum Maps and Their Representation

In Sect. 3.2.4 we have introduced the fundamental characterization theorems for completely positive transformations. These theorems led to the operator-sum representation (3.139) and to the dilation representation (3.140). Despite the relevance of these results, they apply only to completely positive maps, and involve a certain freedom. Indeed, as stressed by Theorem 3.6 the Kraus operators as well as their cardinality are not uniquely fixed. At the same time, the dilation of a completely positive map can be obtained in many different ways. It is therefore convenient to consider other representations for maps on the space of operators, including transformations which are not necessarily completely positive. To this aim we will restrict to the finite-dimensional Hilbert space $\mathcal{H} = \mathbb{C}^n$ and provide three distinct ways to perform such a task. In Sect. 3.3.1 we will consider a map as a linear operator in a Hilbert space and consider its matrix representation in a suitable basis. We will then consider in Sect. 3.3.3 its representation in terms of bases of maps. For a particular choice of basis, that we will call χ-matrix representation, the property of complete positivity can be easily characterized. In Sect. 4.4, after having introduced the notion of entanglement in Sect. 4.3, we will finally state Choi's theorem on the characterization of completely positive maps, pointing to its connection with the χ-matrix and operator-sum representation and with the Choi-Jamiołkowski isomorphism identifying maps with operators in a larger space. In that framework we will further stress that an operator-sum representation also arises for general Hermiticity preserving maps. The relationship between these parametrizations is summarized in Fig. 3.7.

$$\Lambda \in \begin{cases} \mathcal{B}(\mathcal{B}(\mathbb{C}^n)) & \langle \cdot, \cdot \rangle_{\mathrm{HS}} \\[1.5ex] \mathfrak{L}(\mathcal{B}(\mathbb{C}^n), \mathcal{B}(\mathbb{C}^n)) & \langle \cdot, \cdot \rangle_{\mathfrak{L}} \\[1.5ex] \mathcal{B}(\mathbb{C}^n \otimes \mathbb{C}^n) & \langle \cdot, \cdot \rangle_{\mathcal{H} \otimes \mathcal{H}} \end{cases}$$

Fig. 3.7 Schematic summary of the different possible mathematical frameworks for the description of the map Λ discussed in Sects. 3.3.1, 3.3.3 and 4.4 respectively

3.3.1   Linear Operators on Hilbert-Schmidt Space

In a finite-dimensional Hilbert space $\mathcal{H} = \mathbb{C}^n$ the bounded operators can be identified with $n \times n$ matrices with complex entries, so that

$$\mathcal{B}(\mathbb{C}^n) = \mathcal{M}_n(\mathbb{C}). \tag{3.361}$$

The space $\mathcal{B}(\mathbb{C}^n)$ is not only a Banach space, but also a Hilbert space of dimension n^2, with scalar product given by the Hilbert-Schmidt expression (2.53). Any map acting on operators on $\mathbb{C}^n$ can thus be seen as a linear operator in a Hilbert space, and for a fixed basis accordingly identified with a matrix. In particular, a map Λ can be identified with a self-adjoint operator if

$$\langle \Lambda(\chi), \omega \rangle_{\mathrm{HS}} = \langle \chi, \Lambda(\omega) \rangle_{\mathrm{HS}} \quad \forall \chi, \omega \in \mathcal{B}(\mathbb{C}^n) \tag{3.362}$$

and more in particular with a positive operator if

$$\langle \omega, \Lambda(\omega) \rangle_{\mathrm{HS}} \geqslant 0 \quad \forall \omega \in \mathcal{B}(\mathbb{C}^n). \tag{3.363}$$

To obtain a matrix representation for the map Λ seen as linear operator in the Hilbert space $\mathcal{B}(\mathbb{C}^n)$ we first consider a basis in this space, whose elements are therefore themselves *operators* rather than *vectors*. We denote the basis as $\{\tau_\alpha\}_{\alpha=1,\dots,n^2}$, and its elements are orthonormal with respect to the Hilbert-Schmidt scalar product

$$\langle \tau_\alpha, \tau_\beta \rangle_{\mathrm{HS}} = \mathrm{Tr}\{\tau_\alpha^\dagger \tau_\beta\} = \delta_{\alpha\beta}. \tag{3.364}$$

Each operator $\omega \in \mathcal{B}(\mathbb{C}^n)$ can be identified in this basis as a vector with components

$$\omega_\alpha = \langle \tau_\alpha, \omega \rangle_{\mathrm{HS}} \tag{3.365}$$
$$= \mathrm{Tr}\{\tau_\alpha^\dagger \omega\}. \tag{3.366}$$

For arbitrary $\omega \in \mathcal{B}(\mathbb{C}^n)$ we then have

$$\langle \tau_\alpha, \Lambda(\omega) \rangle = \Big\langle \tau_\alpha, \Lambda \Big(\sum_{\beta=1}^{n^2} \tau_\beta \langle \tau_\beta, \omega \rangle_{\mathrm{HS}} \Big) \Big\rangle_{\mathrm{HS}} \tag{3.367}$$

$$= \sum_{\beta=1}^{n^2} \langle \tau_\alpha, \Lambda(\tau_\beta) \rangle_{\mathrm{HS}} \langle \tau_\beta, \omega \rangle_{\mathrm{HS}} \tag{3.368}$$

leading to the following identification for the matrix elements of Λ in the basis $\{\tau_\alpha\}$

$$\Lambda_{\alpha\beta} = \langle \tau_\alpha, \Lambda(\tau_\beta) \rangle_{\mathrm{HS}} \tag{3.369}$$

$$= \mathrm{Tr}\{\tau_\alpha^\dagger \Lambda(\tau_\beta)\}. \tag{3.370}$$

This representation is useful for the functional calculus, since map composition goes over to matrix multiplication. Indeed, considering the composition of two maps Λ and Γ exploiting (3.369) and (3.368) we obtain

$$(\Lambda \circ \Gamma)_{\alpha\beta} = \mathrm{Tr}\{\tau_\alpha^\dagger (\Lambda \circ \Gamma)(\tau_\beta)\} \tag{3.371}$$

$$= \mathrm{Tr}\{\tau_\alpha^\dagger \Lambda(\Gamma(\tau_\beta))\} \tag{3.372}$$

$$= \sum_{\alpha',\beta'=1}^{n^2} \mathrm{Tr}\{\tau_\alpha^\dagger \Lambda_{\alpha'\beta'} \, \mathrm{Tr}\{\tau_{\beta'}^\dagger \Gamma(\tau_\beta)\}\tau_{\alpha'}\}, \tag{3.373}$$

so that thanks to the orthogonality condition (3.364) we have

$$(\Lambda \circ \Gamma)_{\alpha\beta} = \sum_{\beta'=1}^{n^2} \Lambda_{\alpha\beta'} \Gamma_{\beta'\beta}. \tag{3.374}$$

We note that according to (3.362) and (3.363) Hermiticity and positivity of Λ as linear operator on the Hilbert space reflect themselves in Hermiticity and positivity of the matrix $\Lambda_{\alpha\beta}$, but these properties have no direct relation to the notion of Hermiticity preserving or positivity preserving for the map Λ, as we shall clarify in Sect. 3.3.3.

3.3.2 Standard Basis Representation

We now mention a few particularly relevant examples of basis in $\mathcal{B}(\mathbb{C}^n)$.

3.3.2.1 Canonical Basis

A natural choice of basis of operators in $\mathcal{B}(\mathbb{C}^n)$ is obtained once fixed a basis of vectors in $\mathbb{C}^n$, say $\{\varphi_i\}_{i=1,\ldots n}$. We then consider the operators

$$E_{ij} = |\varphi_i\rangle\langle\varphi_j|, \tag{3.375}$$

often simply denoted as $|i\rangle\langle j|$. The collection $\{E_{ij}\}_{i,j=1,\ldots n}$ then provides a so-called canonical basis, given by matrices with just one non-zero entry, equal to one. For the case of $\mathcal{B}(\mathbb{C}^2)$ we typically consider the basis of eigenvectors of the σ_z Pauli matrix, and the canonical basis simply reads

$$E_{11} = \begin{pmatrix} 1 & 0 \\ 0 & 0 \end{pmatrix}, \quad E_{12} = \begin{pmatrix} 0 & 1 \\ 0 & 0 \end{pmatrix}, \quad E_{21} = \begin{pmatrix} 0 & 0 \\ 1 & 0 \end{pmatrix}, \quad E_{22} = \begin{pmatrix} 0 & 0 \\ 0 & 1 \end{pmatrix}. \tag{3.376}$$

3.3.2.2 Unitary Basis

An alternative useful choice of basis is obtained considering a suitably normalized set of unitary operators. We consider the operators $\{U_{rs}\}_{r,s=1,\dots n}$ with

$$U_{rs} = \frac{1}{\sqrt{n}} \sum_{k=1}^{n} e^{-i\frac{2\pi}{n}sk} |\varphi_{k\ominus r}\rangle\langle\varphi_k|, \tag{3.377}$$

where the symbol $\ominus$ denotes the difference modulo n, and $\{\varphi_i\}_{i=1,\dots n}$ is a basis in $\mathbb{C}^n$. These operators are unitary, i.e. $U_{rs}^{\dagger}U_{rs} = U_{rs}U_{rs}^{\dagger} = \mathbb{1}_{\mathbb{C}^n}$, as follows from their very definition. They are further orthogonal according to (2.53) as follows from the identity

$$\sum_{k=1}^{n} e^{i\frac{2\pi}{n}(s-s')k} = n\delta_{s,s'}. \tag{3.378}$$

We can therefore consider the orthonormal basis $\{\frac{1}{\sqrt{n}}U_{rs}\}_{r,s=1,\dots n}$, which for the special case of $\mathcal{B}(\mathbb{C}^2)$ simply becomes

$$U_{11} = \frac{1}{\sqrt{2}}\begin{pmatrix} 0 & 1 \\ -1 & 0 \end{pmatrix}, \quad U_{12} = \frac{1}{\sqrt{2}}\begin{pmatrix} 0 & 1 \\ 1 & 0 \end{pmatrix}, \quad U_{21} = \frac{1}{\sqrt{2}}\begin{pmatrix} -1 & 0 \\ 0 & 1 \end{pmatrix}, \quad U_{22} = \frac{1}{\sqrt{2}}\begin{pmatrix} 1 & 0 \\ 0 & 1 \end{pmatrix}. \tag{3.379}$$

3.3.2.3 Bloch Basis

As a last example we consider the so-called Bloch basis, which we will denote as $\{\Sigma_i\}_{i=1,\dots,n^2}$. We take the first element of the basis proportional to the identity or maximally mixed state. The other elements are determined by the requirement of being Hermitian and traceless. We thus have $\Sigma_1 = \frac{1}{\sqrt{n}}\mathbb{1}_{\mathbb{C}^2}$, together with $\Sigma_i = \Sigma_i^{\dagger}$ and $\mathrm{Tr}\{\Sigma_i\} = 0$ for $i = 2, \dots, n^2$. The requirement of orthonormality further fully determines the elements of the basis. For the case of $\mathbb{C}^2$ the elements of the basis are proportional to the identity and the three Pauli matrices

$$\Sigma_1 = \frac{1}{\sqrt{2}}\begin{pmatrix} 1 & 0 \\ 0 & 1 \end{pmatrix}, \quad \Sigma_2 = \frac{1}{\sqrt{2}}\begin{pmatrix} 0 & 1 \\ 1 & 0 \end{pmatrix}, \quad \Sigma_3 = \frac{1}{\sqrt{2}}\begin{pmatrix} 0 & -i \\ i & 0 \end{pmatrix}, \quad \Sigma_4 = \frac{1}{\sqrt{2}}\begin{pmatrix} 1 & 0 \\ 0 & -1 \end{pmatrix}. \tag{3.380}$$

In this basis maps are naturally described as affine transformations, as shown in Remark 3.18.

3.18 Affine Map Representation

The Bloch basis introduced in Sect. 3.3.2.3 is very often used also because it allows for a nice visualization of the action of a map. A generic operator $\mathcal{B}(\mathbb{C}^n)$ can be written in the form

$$S = \sum_{i=1}^{n^2} s_i \Sigma_i, \tag{3.381}$$

where according to (3.366) the coefficients are given by $s_i = \mathrm{Tr}\{\Sigma_i S\}$. According to (3.370) the action of a map Λ on this operator can be expressed in terms of the matrix

$$\Lambda_{ij} = \mathrm{Tr}\{\Sigma_i \Lambda(\Sigma_j)\} \tag{3.382}$$

according to

$$\Lambda(S) = \sum_{i,j=1}^{n^2} \Lambda_{ij} \mathrm{Tr}\{\Sigma_j S\}\Sigma_i, \tag{3.383}$$

thus leading to the identification

$$\Lambda(S) = \sum_{i=1}^{n^2} s_i' \Sigma_i \tag{3.384}$$

with

$$s_i' = \sum_{j=1}^{n^2} \Lambda_{ij} s_j. \tag{3.385}$$

Relaxing the normalization condition, any operator in $\mathcal{B}(\mathbb{C}^n)$ can be expressed in terms of the operators $\sigma_i = \sqrt{n}\,\Sigma_i$. In particular, statistical operators can be written in the form

$$\rho = \frac{1}{n}(\mathbb{1}_{\mathbb{C}^n} + \boldsymbol{r} \cdot \boldsymbol{\sigma}), \tag{3.386}$$

with $\boldsymbol{\sigma} = (\sigma_2, \ldots, \sigma_n)$ and $\boldsymbol{r} \in \mathbb{R}^{n^2-1}$. The constraints on the real vector $\boldsymbol{r}$ in order to warrant positivity of ρ is given by $(-)^i p_i(\boldsymbol{r}) \geqslant 0$ for $i = 1, \ldots, n$, where $p_i(\boldsymbol{r})$ is the i-th coefficient in the characteristic polynomial of the matrix

ρ, namely $\det(\rho - \lambda \mathbb{1}_{\mathbb{C}^n})$. These constraints express Descartes' rule of signs to determine the number of positive roots of a polynomial. For the special case of $\mathbb{C}^2$ we recover the Bloch sphere considered in Sect. 2.2.3.1, and (3.386) reduces to (2.79). The operators appearing in the representation (3.386) can be identified with the Pauli matrices and the only relevant constraint becomes $\|r\| \leqslant 1$. In this case the state space is in one-to-one correspondence with the Euclidean sphere of radius 1. Using this representation a map acting on states and preserving their normalization and Hermiticity can be identified with a transformation of the form

$$r \to r'. \tag{3.387}$$

Hermiticity and trace preservation allow to write the matrix (3.382) in the form

$$\Lambda = \begin{pmatrix} 1 & \mathbf{0} \\ \mathbf{b}^T & T \end{pmatrix}, \tag{3.388}$$

with $\mathbf{b} \in \mathbb{R}^{n^2-1}$ and T an Hermitian matrix, so that (3.387) is given by the affine transformation

$$r \to r' = Tr + b \tag{3.389}$$

which is naturally visualized for the case of $\mathbb{C}^2$, as shown in Fig. 3.8. In the special case in which the map Λ is unital, the identity $\mathbb{1}_{\mathbb{C}^n}$ is an eigenoperator of the map with orthogonal complement the set of traceless operators, so that the affine transformation (3.389) is actually linear. In the visualization of Fig. 3.8 in this case the rotated ellipsoid would still be centered in the origin.

3.3.3 Map Basis Representation

We now take a different point of view and consider the map Λ, rather than as a linear operator in Hilbert space, as a linear transformation from $\mathcal{B}(\mathbb{C}^n)$ into itself, namely $\Lambda \in \mathcal{L}(\mathcal{B}(\mathbb{C}^n), \mathcal{B}(\mathbb{C}^n))$. The latter space is also a Hilbert space of dimension $n^2 \times n^2$, which can be written as a direct sum as

$$\mathcal{L}(\mathcal{B}(\mathbb{C}^n), \mathcal{B}(\mathbb{C}^n)) = \bigoplus_{k=1}^{n^2} \mathcal{B}(\mathbb{C}^n), \tag{3.390}$$

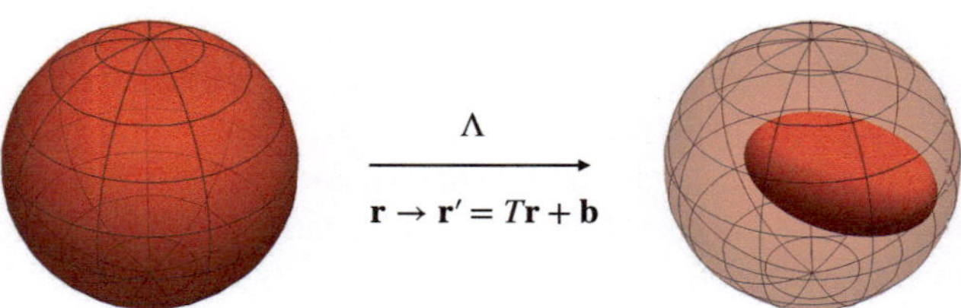

Fig. 3.8 Visualization of the action of a completely positive trace-preserving map in $\mathbb{C}^2$ on the Bloch sphere as an affine transformation of the Bloch vector. The transformation sends the sphere to a rotated and translated ellipsoid

and in it we have a scalar product constructed from the one in $\mathcal{B}(\mathbb{C}^n)$, namely for any arbitrary choice of basis $\{\tau_\gamma\}_{\gamma=1,\dots,n^2}$ in $\mathcal{B}(\mathbb{C}^n)$ we have

$$\langle \Lambda, \Gamma \rangle_{\mathfrak{L}} = \sum_{\gamma=1}^{n^2} \langle \Lambda(\tau_\gamma), \Gamma(\tau_\gamma) \rangle_{\mathrm{HS}} \tag{3.391}$$

$$= \sum_{\gamma=1}^{n^2} \mathrm{Tr}\{\Lambda(\tau_\gamma)^\dagger \Gamma(\tau_\gamma)\}. \tag{3.392}$$

In the space $\mathfrak{L}$ we can introduce two relevant collections of maps, namely the set $\{E_{\alpha\beta}\}_{\alpha,\beta=1,\dots n^2}$ defined as

$$E_{\alpha\beta}(\omega) = \tau_\alpha \, \mathrm{Tr}\{\tau_\beta^\dagger \omega\}, \tag{3.393}$$

and the set $\{F_{\alpha\beta}\}_{\alpha,\beta=1,\dots n^2}$ given by

$$F_{\alpha\beta}(\omega) = \tau_\alpha \omega \tau_\beta^\dagger. \tag{3.394}$$

3.19 Map Bases as Orthonormal Bases

We here show that the collection of maps $\{E_{\alpha\beta}\}_{\alpha,\beta=1,\dots n^2}$ and $\{F_{\alpha\beta}\}_{\alpha,\beta=1,\dots n^2}$ introduced in (3.393) and (3.394) respectively are indeed orthonormal bases.

We further consider in Remark 3.20 the matrices representing an arbitrary map in these bases. Recalling (3.393) and according to the very definition (3.391) of scalar product in $\mathfrak{L}$ we have

$$\langle E_{\alpha\beta}, E_{\alpha'\beta'}\rangle_{\mathfrak{L}} = \sum_{\gamma=1}^{n^2} \langle E_{\alpha\beta}(\tau_\gamma), E_{\alpha'\beta'}(\tau_\gamma)\rangle_{\mathrm{HS}} \tag{3.395}$$

$$= \sum_{\gamma=1}^{n^2} \mathrm{Tr}\{(\tau_\alpha \, \mathrm{Tr}\{\tau_\beta^\dagger \tau_\gamma\})^\dagger \tau_{\alpha'} \, \mathrm{Tr}\{\tau_{\beta'}^\dagger \tau_\gamma\}\} \tag{3.396}$$

$$= \sum_{\gamma=1}^{n^2} \mathrm{Tr}\{\tau_\gamma^\dagger \tau_\beta\} \, \mathrm{Tr}\{\tau_{\beta'}^\dagger \tau_\gamma\} \, \mathrm{Tr}\{\tau_\alpha^\dagger \tau_{\alpha'}\} \tag{3.397}$$

so that recalling (3.364) we are left with the orthonormality condition

$$\langle E_{\alpha\beta}, E_{\alpha'\beta'}\rangle_{\mathfrak{L}} = \delta_{\alpha\alpha'}\delta_{\beta\beta'}, \tag{3.398}$$

which implies linear independence of the elements of the collection $\{E_{\alpha\beta}\}_{\alpha,\beta=1,\ldots n^2}$, and therefore given the cardinality of the set also completeness. On a similar footing starting now from (3.394) we have

$$\langle F_{\alpha\beta}, F_{\alpha'\beta'}\rangle_{\mathfrak{L}} = \sum_{\gamma=1}^{n^2} \langle F_{\alpha\beta}(\tau_\gamma), F_{\alpha'\beta'}(\tau_\gamma)\rangle_{\mathrm{HS}} \tag{3.399}$$

$$= \sum_{\gamma=1}^{n^2} \mathrm{Tr}\{(\tau_\alpha \tau_\gamma \tau_\beta^\dagger)^\dagger \tau_{\alpha'} \tau_\gamma \tau_{\beta'}^\dagger\} \tag{3.400}$$

$$= \sum_{\gamma=1}^{n^2} \mathrm{Tr}\{\tau_\beta \tau_\gamma^\dagger \tau_\alpha^\dagger \tau_{\alpha'} \tau_\gamma \tau_{\beta'}^\dagger\}. \tag{3.401}$$

We now exploit the identity

$$\sum_{\gamma=1}^{n^2} \tau_\gamma^\dagger X \tau_\gamma = \mathbb{1}_{\mathbb{C}^n} \, \mathrm{Tr}\{X\}, \tag{3.402}$$

immediately verified for the canonical choice of basis (3.375) and easily proven to be invariant with respect to the choice of basis $\{\tau_\gamma\}_{\gamma=1,\ldots,n^2}$ in $\mathcal{B}(\mathbb{C}^n)$, so that we have

$$\langle F_{\alpha\beta}, F_{\alpha'\beta'}\rangle_{\mathfrak{L}} = \mathrm{Tr}\{\tau_{\beta}\tau_{\beta'}^{\dagger}\}\,\mathrm{Tr}\{\tau_{\alpha}^{\dagger}\tau_{\alpha'}\} \tag{3.403}$$

$$= \delta_{\alpha\alpha'}\delta_{\beta\beta'}, \tag{3.404}$$

that is orthonormality, and hence completeness. We note in passing that (3.402) implies the following representations for the identity operator

$$\mathbb{1}_{\mathbb{C}^n} = \frac{1}{n}\sum_{\gamma=1}^{n^2} \tau_{\gamma}^{\dagger}\tau_{\gamma}, \tag{3.405}$$

in analogy to the completeness relation (2.73).

3.20 Matrix Representation for Map Bases

We now consider an arbitrary map $\Lambda \in \mathfrak{L}(\mathcal{B}(\mathbb{C}^n), \mathcal{B}(\mathbb{C}^n))$ and determine the matrix of coefficients associated to it in the different bases. We use the notation

$$\Lambda = \sum_{\alpha,\beta=1}^{n^2} \langle E_{\alpha\beta}, \Lambda\rangle_{\mathfrak{L}}\, E_{\alpha\beta}, \tag{3.406}$$

so that according to the definition of $E_{\alpha\beta}$ (3.393) we have

$$\langle E_{\alpha\beta}, \Lambda\rangle_{\mathfrak{L}} = \sum_{\gamma=1}^{n^2} \langle E_{\alpha\beta}(\tau_{\gamma}), \Lambda(\tau_{\gamma})\rangle_{\mathrm{HS}} \tag{3.407}$$

$$= \sum_{\gamma=1}^{n^2} \mathrm{Tr}\{(\tau_{\alpha}\,\mathrm{Tr}\{\tau_{\beta}^{\dagger}\tau_{\gamma}\})^{\dagger}\Lambda(\tau_{\gamma})\}, \tag{3.408}$$

and thanks to (3.364) we come to the identification

$$\langle E_{\alpha\beta}, \Lambda\rangle_{\mathfrak{L}} = \mathrm{Tr}\{\tau_{\alpha}^{\dagger}\Lambda(\tau_{\beta})\} = \Lambda_{\alpha\beta}, \tag{3.409}$$

where the matrix $\Lambda_{\alpha\beta}$ has been encountered in (3.370) when representing Λ as an element of $\mathcal{B}(\mathbb{C}^n)$ and allows to write the action of the map on an arbitrary operator $\omega \in \mathcal{B}(\mathbb{C}^n)$ in the form

$$\Lambda(\omega) = \sum_{\alpha,\beta=1}^{n^2} \Lambda_{\alpha\beta} E_{\alpha\beta}(\omega). \tag{3.410}$$

Along the same lines we denote

$$\Lambda = \sum_{\alpha,\beta=1}^{n^2} \langle F_{\alpha\beta}, \Lambda \rangle_{\mathfrak{L}} \, F_{\alpha\beta} \tag{3.411}$$

$$= \sum_{\alpha,\beta=1}^{n^2} \Lambda'_{\alpha\beta} F_{\alpha\beta}, \tag{3.412}$$

and identify thanks to the definition of $F_{\alpha\beta}$ (3.394)

$$\Lambda'_{\alpha\beta} = \langle F_{\alpha\beta}, \Lambda \rangle_{\mathfrak{L}} \tag{3.413}$$

$$= \sum_{\gamma=1}^{n^2} \langle F_{\alpha\beta}(\tau_\gamma), \Lambda(\tau_\gamma) \rangle_{\mathrm{HS}} \tag{3.414}$$

$$= \sum_{\gamma=1}^{n^2} \mathrm{Tr}\{((\tau_\alpha \tau_\gamma \tau_\beta^\dagger)^\dagger \Lambda(\tau_\gamma)\}, \tag{3.415}$$

so that the relevant matrix of coefficients is given by

$$\Lambda'_{\alpha\beta} = \sum_{\gamma=1}^{n^2} \mathrm{Tr}\{\tau_\beta \tau_\gamma^\dagger \tau_\alpha^\dagger \Lambda(\tau_\gamma)\}. \tag{3.416}$$

Exploiting (3.410) and (3.393) we also have

$$\Lambda'_{\alpha\beta} = \sum_{\gamma=1}^{n^2} \mathrm{Tr}\left[\tau_\beta \tau_\gamma^\dagger \tau_\alpha^\dagger \sum_{\delta,\varepsilon=1}^{n^2} \Lambda_{\delta\varepsilon} \tau_\delta \, \mathrm{Tr}\{\tau_\varepsilon^\dagger \tau_\gamma\} \right], \tag{3.417}$$

which again due to (3.364) leads to the following connection between the matrices representing the map Λ in the two bases

$$\Lambda'_{\alpha\beta} = \sum_{\gamma,\delta=1}^{n^2} M_{\alpha\beta,\delta\gamma} \Lambda_{\delta\gamma}, \tag{3.418}$$

with

$$M_{\alpha\beta,\delta\gamma} = \mathrm{Tr}\{\tau_\alpha^\dagger \tau_\delta \tau_\beta \tau_\gamma^\dagger\}. \tag{3.419}$$

In Remark 3.19 it is explicitly shown that both collections provide orthonormal bases according to the scalar product (3.391). Each basis uniquely associates a matrix of coefficients to each map in $\mathcal{L}(\mathcal{B}(\mathbb{C}^n), \mathcal{B}(\mathbb{C}^n))$. We use the notation

$$\Lambda = \sum_{\alpha,\beta=1}^{n^2} \Lambda_{\alpha\beta} E_{\alpha\beta} \tag{3.420}$$

$$= \sum_{\alpha,\beta=1}^{n^2} \Lambda'_{\alpha\beta} F_{\alpha\beta}, \tag{3.421}$$

with $\Lambda_{\alpha\beta}$ and $\Lambda'_{\alpha\beta}$ derived in Remark 3.20 and given respectively by (3.409) and (3.416). These matrices are easily connected as in (3.418), but encode quite different information on the map Λ. The matrix $\Lambda_{\alpha\beta}$ has the advantage to translate map composition in matrix multiplication. Hermiticity and positivity of the matrix $\Lambda_{\alpha\beta}$ correspond to self-adjointness and positivity of Λ as linear operator in the sense of (3.362) and (3.363) respectively. These properties are however generally not relevant when considering the transformation of states and observables arising as a consequence of some measurement interaction or of time evolution. Instead, the matrix $\Lambda'_{\alpha\beta}$ related to the basis (3.394), that we shall call χ-matrix representation, directly provides important information on the map Λ. More precisely Λ is Hermiticity preserving iff $\Lambda'_{\alpha\beta}$ is Hermitian, and it is completely positive iff $\Lambda'_{\alpha\beta}$ is a positive matrix, as shown in Remark 3.21. We recall that a map is said to be Hermiticity preserving if self-adjoint operators are transformed into self-adjoint operators, namely

$$(\Lambda(\omega))^\dagger = \Lambda(\omega^\dagger) \qquad \forall \omega \in \mathcal{B}(\mathbb{C}^n), \tag{3.422}$$

while it is said to be positivity preserving if positive operators are transformed into positive operators, that is

$$\Lambda(\omega) \geqslant 0 \qquad \forall \omega \in \mathcal{B}(\mathbb{C}^n), \omega \geqslant 0. \tag{3.423}$$

For the sake of simplicity, a positivity preserving map is often simply termed positive map. To clarify and stress the difference between positivity as linear operator, as in (3.363), and positivity as a map, in the sense of being positivity preserving according to (3.423), we point to two simple examples. Consider for a system described in $\mathbb{C}^2$

the map

$$F_{zz}(\omega) = \frac{1}{2}\sigma_z \omega \sigma_z^\dagger, \tag{3.424}$$

completely positive by construction, but not corresponding to a positive linear operator, since

$$\langle \sigma_x, F_{zz}(\sigma_x)\rangle_{\text{HS}} = -1. \tag{3.425}$$

As a complementary example consider

$$E_{zz}(\omega) = \frac{1}{2}\sigma_z \, \text{Tr}\{\sigma_z^\dagger \omega\}. \tag{3.426}$$

This expression provides a positive linear operator since

$$\langle \omega, E_{zz}(\omega)\rangle_{\text{HS}} = \frac{1}{2}|\,\text{Tr}\{\omega^\dagger \sigma_z\}|^2. \tag{3.427}$$

We have however

$$E_{zz}(|+\rangle\langle+|) = \frac{1}{2}\sigma_z. \tag{3.428}$$

The map is therefore not positive, and in particular it cannot be completely positive.

3.21 Map Properties and Matrix Representation

We show how Hermiticity preservation and complete positivity of a map are univocally encoded in properties of the χ-matrix representation of the map. The proof of these facts will also provide us, in the finite-dimensional case, a simple criterion to verify complete positivity according to (3.139) of Theorem 3.5, and thus to obtain a Kraus representation.

We first prove the connection between Hermiticity preservation as given by the requirement (3.422) and Hermiticity of the matrix $\Lambda'_{\alpha\beta}$. Starting from (3.412) and recalling (3.394) we have

$$(\Lambda(\omega))^\dagger = \sum_{\alpha,\beta=1}^{n^2} (\Lambda'_{\alpha\beta})^* \tau_\beta \omega^\dagger \tau_\alpha^\dagger \tag{3.429}$$

$$= \sum_{\alpha,\beta=1}^{n^2} (\Lambda'_{\beta\alpha})^* \tau_\alpha \omega^\dagger \tau_\beta^\dagger, \tag{3.430}$$

so that indeed (3.422) is verified $\forall \omega \in \mathcal{B}(\mathbb{C}^n)$ iff $\Lambda'_{\alpha\beta} = (\Lambda'_{\beta\alpha})^*$, so that the matrix $\Lambda'_{\alpha\beta}$ is Hermitian.

To clarify the relationship between complete positivity of the map and positivity of the associated matrix in the χ-matrix representation, let us first suppose that $\Lambda'_{\alpha\beta}$ is non-negative. The matrix can then be decomposed as

$$\Lambda'_{\alpha\beta} = (U D U^\dagger)_{\alpha\beta} \tag{3.431}$$

$$= \sum_{\gamma=1}^{n^2} \lambda_\gamma u_{\alpha\gamma} u^*_{\beta\gamma}, \tag{3.432}$$

where U is a unitary matrix, with matrix elements $u_{\alpha\beta}$, and D a diagonal matrix with entries the positive eigenvalues λ_γ of $\Lambda'_{\alpha\beta}$. We thus obtain for the action of the map Λ the representation

$$\Lambda(\omega) = \sum_{\gamma=1}^{n^2} \lambda_\gamma \left(\sum_{\alpha=1}^{n^2} u_{\alpha\gamma} \tau_\alpha \right) \omega \left(\sum_{\beta=1}^{n^2} u^*_{\beta\gamma} \tau^\dagger_\beta \right) \tag{3.433}$$

$$= \sum_{\gamma=1}^{n^2} \lambda_\gamma \tilde{\tau}_\gamma \omega \tilde{\tau}^\dagger_\gamma. \tag{3.434}$$

where we have set

$$\tilde{\tau}_\gamma = \sum_{\alpha=1}^{n^2} u_{\alpha\gamma} \tau_\alpha. \tag{3.435}$$

Thanks to positivity of the coefficients $\{\lambda_\gamma\}_{\gamma=1,\dots,n^2}$ the expression (3.434) provides an operator-sum representation for the map Λ that can be recast in the form (3.139), thus warranting its complete positivity according to Theorem 3.5. In particular, the representation (3.434) is called a canonical Kraus operator-sum decomposition, since the Kraus operators $\{\tilde{\tau}_\gamma\}_{\gamma=1,\dots,n^2}$ provide an orthonormal basis. This representation is furthermore unique if the eigenvalues are non-degenerate. Importantly, the result relies on positivity of the eigenvalues which is equivalent to positivity of the matrix $\Lambda'_{\alpha\beta}$. Conversely, consider a completely positive map Λ. Thanks to Theorem 3.5 it admits a Kraus operator-sum decomposition of the form (3.139)

$$\Lambda(\omega) = \sum_{\gamma=1}^{n^2} A_\gamma \omega A^\dagger_\gamma. \tag{3.436}$$

We now express the Kraus operators A_γ in the basis $\{\tau_\alpha\}_{\alpha=1,\ldots,n^2}$ according to

$$A_\gamma = \sum_{\alpha=1}^{n^2} c_{\gamma\alpha} \tau_\alpha, \tag{3.437}$$

so that substituting in (3.436) we obtain

$$\Lambda(\omega) = \sum_{\gamma=1}^{n^2} \left(\sum_{\alpha=1}^{n^2} c_{\gamma\alpha} \tau_\alpha \right) \omega \left(\sum_{\beta=1}^{n^2} c_{\gamma\beta} \tau_\beta \right)^\dagger \tag{3.438}$$

$$= \sum_{\alpha,\beta=1}^{n^2} \left(\sum_{\gamma=1}^{n^2} c_{\gamma\alpha} c_{\gamma\beta}^* \right) \tau_\alpha \omega \tau_\beta^\dagger, \tag{3.439}$$

which according to (3.412) and (3.394) leads us to the identification

$$\Lambda'_{\alpha\beta} = \sum_{\gamma=1}^{n^2} c_{\gamma\alpha} c_{\gamma\beta}^*. \tag{3.440}$$

The obtained matrix is therefore positive, since

$$\sum_{\alpha,\beta=1}^{n^2} v_\alpha \Lambda'_{\alpha\beta} v_\beta^* \geq 0 \tag{3.441}$$

for any collection of vectors $\{v_\alpha\}_{\alpha=1,\ldots,n^2}$ in $\mathbb{C}^{n^2}$. We have thus shown that complete positivity of the map is equivalent to positivity of the matrix associated to it in the χ-matrix representation. This provides us also with a simple criterion for the verification of the property of complete positivity introduced in Sect. 3.2.1 for the finite-dimensional case. We further note that diagonalization of the associated matrix $\Lambda'_{\alpha\beta}$ also provides a way to construct a canonical Kraus operator-sum decomposition for the map.

It is worth stressing that for the canonical choice of basis (3.375) the matrices $\Lambda_{\alpha\beta}$ and $\Lambda'_{\alpha\beta}$ are related by a simple exchange of indexes. In this basis we make the identification

$$\tau_\alpha \rightarrow |e_i\rangle\langle e_j| \tag{3.442}$$

$$\Lambda = \begin{pmatrix} - & \boxtimes & - & \otimes \\ - & \boxtimes & - & \otimes \\ \boxminus & - & \ominus & - \\ \boxminus & - & \ominus & - \end{pmatrix} \rightarrow \Lambda' = \begin{pmatrix} - & \boxminus & - & \ominus \\ - & \boxminus & - & \ominus \\ \boxtimes & - & \otimes & - \\ \boxtimes & - & \otimes & - \end{pmatrix}$$

Fig. 3.9 Rearrangement of matrix elements connecting the matrices $\Lambda_{\alpha\beta}$ and $\Lambda'_{\alpha\beta}$ for the canonical choice of basis $\{E_{ij}\}$. The elements denoted by a dash are kept fixed

as well as

$$\tau_\beta \rightarrow |e_k\rangle\langle e_l|, \tag{3.443}$$

where the greek indexes run over n^2 values, and latin indexes correspondingly cover n values. We thus have

$$\Lambda_{\alpha\beta} \rightarrow \Lambda_{ij,kl}, \tag{3.444}$$

where according to the definition (3.409) we have

$$\Lambda_{ij,kl} = \mathrm{Tr}\{(|e_i\rangle\langle e_j|)^\dagger \Lambda(|e_k\rangle\langle e_l|)\} \tag{3.445}$$
$$= \langle e_i|\Lambda(|e_k\rangle\langle e_l|)|e_j\rangle. \tag{3.446}$$

In the same way we have

$$\Lambda'_{\alpha\beta} \rightarrow \Lambda'_{ij,kl}, \tag{3.447}$$

where relying on (3.416) we obtain

$$\Lambda'_{ij,kl} = \sum_{r,s=1}^{n} \mathrm{Tr}\{|e_k\rangle\langle e_l|(|e_r\rangle\langle e_s|)^\dagger(|e_i\rangle\langle e_j|)^\dagger \Lambda(|e_r\rangle\langle e_s|)\} \tag{3.448}$$
$$= \langle e_i|\Lambda(|e_j\rangle\langle e_l|)|e_k\rangle. \tag{3.449}$$

Comparing (3.446) and (3.449) we immediately realize that

$$\Lambda_{ij,kl} = \Lambda'_{ik,jl}, \tag{3.450}$$

that is they are simply related by changing the order of the last two indexes [29]. This relationship corresponds to a rearrangement of the matrix elements as depicted in Fig. 3.9.

References

1. A.S. Holevo, V. Giovannetti, Rep. Prog. Phys. **75**, 046001 (2012). https://doi.org/10.1088/0034-4885/75/4/046001
2. A. Barchielli, L. Lanz, G.M. Prosperi, Nuovo Cimento B **72**, 79 (1982). https://doi.org/10.1007/BF02721671
3. H. Carmichael, *An Open Systems Approach to Quantum Optics* (Springer, Berlin, 1993)
4. A. Barchielli, M. Gregoratti, *Quantum Trajectories and Measurements in Continuous Time.* Lecture Notes in Physics, vol. 782 (Springer, Berlin, 2009). https://doi.org/10.1007/978-3-642-01298-3
5. R. Haag, D. Kastler, J. Math. Phys. **5**, 848 (1964). https://doi.org/10.1063/1.1704187
6. K.-E. Hellwig, K. Kraus, Commun. Math. Phys. **11**, 214 (1969). https://doi.org/10.1007/BF01645807
7. K.-E. Hellwig, K. Kraus, Commun. Math. Phys. **16**, 142 (1970). https://doi.org/10.1007/BF01646620
8. K. Kraus, Ann. Phys. (N.Y.) **64**, 311 (1971). https://doi.org/10.1016/0003-4916(71)90108-4
9. W.F. Stinespring, Proc. Am. Math. Soc. **6**, 211 (1955). https://doi.org/10.1090/S0002-9939-1955-0069403-4
10. E.B. Davies, *Quantum Theory of Open Systems* (Academic Press, London, 1976)
11. M. Takesaki, *Theory of Operator Algebras I* (Springer, Berlin, 2002). https://doi.org/10.1007/978-1-4612-6188-9
12. V. Paulsen, *Completely Bounded Maps and Operator Algebras* (Cambridge University Press, 2003)
13. F. Strocchi, *An Introduction to the Mathematical Structure of Quantum Mechanics* (World Scientific, 2005). https://doi.org/10.1142/7038
14. K. Kraus, *States, Effects, and Operations.* Lecture Notes in Physics, vol. 190 (Springer, Berlin, 1983). https://doi.org/10.1007/3-540-12732-1
15. T. Heinosaari, M. Ziman, *The Mathematical Language of Quantum Theory* (Cambridge University Press, Cambridge, 2011)
16. I. Bengtsson, K. Życzkowski, *Geometry of Quantum States: An Introduction to Quantum Entanglement*, 2nd edn. (Cambridge University Press, 2017)
17. M. Nielsen, I. Chuang, *Quantum Computation and Quantum Information* (Cambridge University Press, Cambridge, 2000)
18. M.B. Ruskai, Rev. Math. Phys. **6**, 1147 (1994). https://doi.org/10.1142/S0129055X94000407
19. C.W. Helstrom, *Quantum Detection and Estimation Theory* (Academic Press, New York, 1976)
20. A.S. Holevo, *Probabilistic and Statistical Aspects of Quantum Theory* (North-Holland, Amsterdam, 1982)
21. G. Lindblad, Commun. Math. Phys. **40**, 147 (1975). https://doi.org/10.1007/BF01609396
22. E.H. Lieb, M.B. Ruskai, J. Math. Phys. **14**, 1938 (1973). https://doi.org/10.1063/1.1666274
23. A. Müller-Hermes, D. Reeb, Ann. Henri Poincaré **18**, 1777 (2017). https://doi.org/10.1007/s00023-017-0550-9
24. G. Lüders, Ann. Phys. **443**, 322 (1950). https://doi.org/10.1002/andp.19504430510
25. M. Reed, B. Simon, *I: Functional Analysis* (Academic Press, San Diego, 1980)
26. J. Schwinger, Proc. Natl. Acad. Sci. U.S.A. **45**, 1542 (1959). https://doi.org/10.1073/pnas.45.10.1542
27. B.R. Judd, G.W. Mackey (eds.), *The Collected Works of Eugene Paul Wigner* (Springer, New York, 1993). https://doi.org/10.1007/978-3-662-02781-3
28. M. Ozawa, J. Math. Phys. **25**, 79 (1984). https://doi.org/10.1063/1.526000
29. E.C.G. Sudarshan, P.M. Mathews, J. Rau, Phys. Rev. **121**, 920 (1961). https://doi.org/10.1103/PhysRev.121.920

Part II
Open Quantum Systems Theory

*Natürlich ist man nicht so einfältig zu denken,
dass solchermaßen zu erraten sei,
wie es auf der Welt wirklich zugeht.*
—*Erwin Schrödinger, Die gegenwärtige Situation in der
Quantenmechanik, Naturwiss. 23, 844 (1935)*

The aim of the Part II is to introduce basic concepts and models of open quantum system theory, putting emphasis on the master equation description and on the quantum features of the reduced dynamics, including effects related to multitime correlation functions. Complete positivity is considered in its relation to entanglement. Exactly solvables models are compared with results obtained in the weak-coupling approximation, and further studied in view of their memory effects according to a characterization based on the information exchange between system and environment. The so-called Lindblad master equation is considered in detail, together with more general time-local and memory-kernel master equations.

Composite Systems

4

Abstract

The treatment of open systems by necessity requires dealing with composite systems. In the framework of quantum probability the composition of systems has to be formulated in terms of the tensor product of the Hilbert spaces associated to each single system. This fact determines the kind of correlations arising in quantum mechanics, leading in particular to the emergence of entanglement. We discuss the notion of entanglement in a bipartite setting for pure states, and briefly point to the extension of the definition to mixed states. We address in particular the connection between entanglement and complete positivity of maps.

4.1 Introduction

The treatment of interacting systems calls for the very definition of the mathematical framework for the description of composite systems. In quantum mechanics the combination of two or more systems leads to consider as Hilbert space, in which all the relevant degrees of freedom are described, the tensor product of the Hilbert spaces associated to the single systems. While the relevant formalism for dealing with closed systems, that is perfectly isolated ones, has been introduced in Chap. 2, the investigation of general state transformations has already led us to consider how these transformations can arise as the consequence of an interaction with other quantum degrees of freedom in Chap. 3, so that the system has to be considered open. The focus was on transformations due to measurement. Indeed, the analysis of the interaction of a system with a measurement apparatus provides the archetype of an open system description. However, the perfect shielding of a system from all others can only be an abstraction, so that all systems are in principle open. The actual

B. Vacchini, *Open Quantum Systems*, Graduate Texts in Physics,
https://doi.org/10.1007/978-3-031-58218-9_4

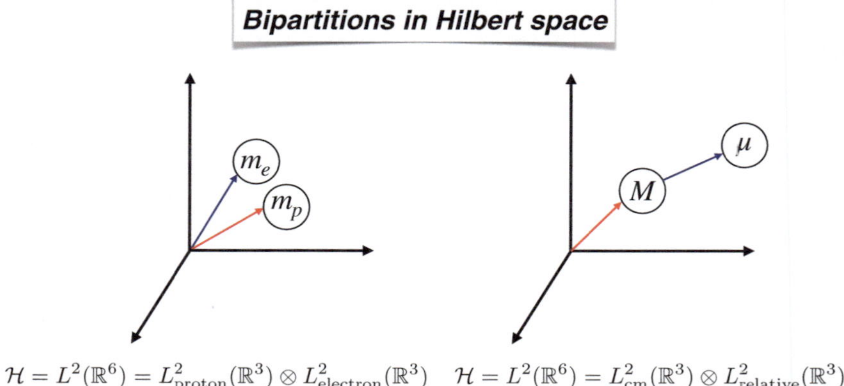

Fig. 4.1 Schematic representation of two different bipartitions of the Hilbert space in which the degrees of freedom of electron and proton in the hydrogen atom are described

relevance of this fact for a realistic and accurate description of experiments involving the system, however, depends on the situation.

As a first clarification of this aspect, we consider the very simple example of the quantum mechanical treatment of the hydrogen atom, as usually addressed in textbooks. The hydrogen atom as made up of a proton and an electron is indeed a composite system, yet we can focus directly on describing a charged particle in a Coulomb potential, without mentioning any underlying tensor product structure. The interacting degrees of freedom of proton and electron should be described on the tensor product Hilbert space $L^2_{\text{proton}}(\mathbb{R}^3) \otimes L^2_{\text{electron}}(\mathbb{R}^3)$. Importantly, however, we can consider as in the classical case an equivalent set of degrees of freedom, namely center of mass and relative coordinate, leading to the isomorphic space $L^2_{\text{cm}}(\mathbb{R}^3) \otimes L^2_{\text{relative}}(\mathbb{R}^3)$, as schematically depicted in Fig. 4.1. The latter degrees of freedom do not interact, so that provided the initial condition is factorized when expressed in this space, we can simply neglect the tensor product structure and deal with the electron as if it were a closed system, described by the relative coordinate. This basic example allows to put into evidence a few fundamental facts. In the first instance, while composing systems we take the tensor product of the individual Hilbert spaces, given a set of degrees of freedoms there might be many ways to decompose them in a tensor product structure of Hilbert spaces. These decompositions are physically equivalent, but might lead to descriptions with a quite different level of complexity. Moreover, the emergence of a manifestly simpler description relies on two conditions: the existence of a combination of non interacting degrees of freedom on the one hand, and the relevance of factorized initial conditions with respect to these degrees of freedom. Only for the latter kind of initial conditions the other degrees of freedom can be neglected in the course of the evolution, so that e.g. we can derive the spectrum of the hydrogen atom and introduce stationary states for the electronic configuration without bothering about the center of mass dynamics. The situation is different for example in the Stern-Gerlach experiment considered at the end of Sect. 2.3.1. In that setting considering an initially factorized condition between center of mass and spin degrees of freedom the interaction term forced us to keep into account the composite structure of the system. Most importantly, we have to keep in mind that the very definition of correlations depends on the adopted tensor product description, i.e. they cannot be associated to a collection of degrees of freedom without specifying the way in which they are analyzed.

In general, the interaction between the considered system and other degrees of freedom is unavoidable and influences in a non-trivial way its time evolution. At the same time, it cannot be described in full detail, that is considering the overall set of interacting degrees of freedom as closed system. This approach would result in being exceedingly complicated or computationally expensive, possibly providing a lot of irrelevant information. It is therefore often appropriate or even necessary to consider the observed degrees of freedom as an open system, in that they interact with other quantum degrees of freedom, typically called environment, and to look for their effective description. The environment is often also called reservoir or bath if it involves a large or infinite amount of degrees of freedom, typically at thermal equilibrium. In such an open quantum system description we are therefore naturally faced with the tensor product space $\mathcal{H}_S \otimes \mathcal{H}_E$. With $\mathcal{H}_S$ we denote the Hilbert space in which to describe the open system of interest, while $\mathcal{H}_E$ is the Hilbert space of the environmental degrees of freedom, whose influence on the system's dynamics we want to keep effectively into account. We are in particular faced with a fixed bipartition, determining the relevant correlations. As a first step in the description of open quantum systems we therefore discuss some basic features of states in a bipartite setting. We will put into evidence the notion of entanglement, arising just due to the introduction of tensor product structures, and its deep connection with the property of complete positivity of a map introduced in Sect. 3.2.1.

4.2 Reduced State

If the overall Hilbert space has a bipartite expression $\mathcal{H}_S \otimes \mathcal{H}_E$ any state $\rho \in \mathcal{S}(\mathcal{H}_S \otimes \mathcal{H}_E)$, namely any positive element of $\mathcal{T}(\mathcal{H}_S \otimes \mathcal{H}_E)$ with trace equal to one, uniquely determines two reduced states in $\mathcal{S}(\mathcal{H}_S)$ and $\mathcal{S}(\mathcal{H}_E)$. This happens in analogy to the classical framework, where a probability distribution on a measure space with a product structure fixes marginal distributions obtained integrating over a subset of degrees of freedom, e.g. the marginal position distribution obtained from the overall phase-space distribution involving both position and momentum for a massive particle in Remark 1.2. In the quantum case the reduced state for the reduced system is obtained by means of the partial trace introduced in Remark 3.1. If $\rho \in \mathcal{S}(\mathcal{H}_S \otimes \mathcal{H}_E)$ we have that the operator

$$\rho_S = \mathrm{Tr}_E[\rho], \tag{4.1}$$

with the partial trace defined by (3.38), so that it can be written as

$$\rho_S = \sum_k \langle \varphi_k | \rho \varphi_k \rangle, \tag{4.2}$$

with $\{\varphi_k\}$ any basis in $\mathcal{H}_E$, is indeed a statistical operator, that is $\rho_S \in \mathcal{S}(\mathcal{H}_S)$. Positivity of ρ_S immediately follows from the complete positivity of the partial trace operation proven in Remark 3.1. The state ρ_S introduced in (4.1) is called reduced state of the system. It is exactly the unique state that provides the correct statistics for any observation on the system only. In this respect, it is a key quantity in the description of an open quantum system. Indeed, for any observable of the form $A_S \otimes \mathbb{1}_E$ we have

$$\langle A \rangle_\rho = \mathrm{Tr}\{\rho(A_S \otimes \mathbb{1}_E)\} \tag{4.3}$$

$$= \sum_{i,k} \langle \psi_i \otimes \varphi_k | \rho(A_S \otimes \mathbb{1}_E)\psi_i \otimes \varphi_k \rangle \tag{4.4}$$

$$= \sum_i \langle \psi_i | \mathrm{Tr}_E[\rho]A_S \psi_i \rangle \tag{4.5}$$

$$= \mathrm{Tr}_S\{\rho_S A_S\}, \tag{4.6}$$

where we have introduced two arbitrary c.o.n.s. $\{\psi_i\}$ and $\{\varphi_k\}$ in $\mathcal{H}_S$ and $\mathcal{H}_E$ respectively, and used the very definition of trace and partial trace. In a symmetric way we can obtain the reduced state of the environment as

$$\rho_E = \mathrm{Tr}_S[\rho], \tag{4.7}$$

where we have considered the partial trace with respect to the degrees of freedom of the system, so that $\rho_E \in \mathcal{S}(\mathcal{H}_E)$.

4.3 Bipartite States and Entanglement

We have seen that a state in a Hilbert space with a tensor product structure uniquely identifies reduced states, which contain all the necessary information in order to predict the outcome of local observations, that is measurements involving only a subset of the degrees of freedom. We now point to the fact that states on a bipartite Hilbert space can have a rich structure in terms of correlations, featuring in particular correlations that cannot be obtained in a classical state, such as entanglement (English translation by Schrödinger himself [1] of the original term Verschränkung used in his seminal paper [2]). We provide some basic facts related to this concept relevant for the theory of open quantum systems, pointing in particular to its relationship with the notion of complete positivity of a map discussed in Sect. 3.2.1.

4.3.1   Schmidt Decomposition and Purifications

An important and useful property for the characterization of pure states in a bipartite Hilbert space is the fact that they can be expressed as sum of biorthogonal vectors. Let $\Psi \in \mathcal{H}_S \otimes \mathcal{H}_E$ be a norm one vector, that is a pure state in the bipartite space, then it can be written in the form

$$\Psi = \sum_{i=1}^{D} \sqrt{\lambda_i}\, \chi_S^i \otimes \chi_E^i \tag{4.8}$$

with $D = \min\{\dim \mathcal{H}_S, \dim \mathcal{H}_E\}$, where $\{\chi_S^i\}$ and $\{\chi_E^i\}$ are orthonormal bases in $\mathcal{H}_S$ and $\mathcal{H}_E$ respectively. The representation (4.8) is known as Schmidt decomposition. The coefficients $\sqrt{\lambda_i}$ are non-negative numbers called Schmidt coefficients, such that $\sum_{i=1}^{D} \lambda_i = \|\Psi\|$. The number of non-zero coefficients, the so-called Schmidt number, is uniquely determined by the vector Ψ. The name of this decomposition goes back to the seminal mathematical contribution on which it is based [3], where the singular value decomposition for operators appeared (see [4,5] and [6] for historical remarks and more details).

Proof Given any two orthonormal bases $\{\varphi_S^i\}$ and $\{\psi_E^i\}$ in $\mathcal{H}_S$ and $\mathcal{H}_E$ respectively, an arbitrary state can be written in the form

$$\Psi = \sum_{i=1}^{\dim \mathcal{H}_S} \sum_{j=1}^{\dim \mathcal{H}_E} c_{ij}\varphi_S^i \otimes \psi_E^j. \tag{4.9}$$

If $\dim \mathcal{H}_S \neq \dim \mathcal{H}_E$, we suppose without loss of generality $\dim \mathcal{H}_E > \dim \mathcal{H}_S$, and we can write

$$\Psi = \sum_{i=1}^{\dim \mathcal{H}_S} \varphi_S^i \otimes \eta_E^i \tag{4.10}$$

upon defining

$$\eta_E^i = \sum_{j=1}^{\dim \mathcal{H}_E} c_{ij}\psi_E^j. \tag{4.11}$$

The environmental vectors $\{\eta_E^i\}_{i=1,\ldots,\dim \mathcal{H}_S}$ can be expressed in terms of $\dim \mathcal{H}_S$ orthonormal environmental vectors, say $\{\varphi_E^j\}_{j=1,\ldots,\dim \mathcal{H}_S}$, according to

$$\eta_E^i = \sum_{j=1}^{\dim \mathcal{H}_S} a_{ij}\varphi_E^j. \tag{4.12}$$

We finally obtain

$$\Psi = \sum_{i,j=1}^{\dim \mathcal{H}_S} a_{ij} \varphi_S^i \otimes \varphi_E^j, \tag{4.13}$$

with $\{\varphi_S^i\}$ and $\{\varphi_E^j\}$ orthonormal basis in $\mathcal{H}_S$ and $\mathcal{H}_E$ respectively. According to the singular value decomposition theorem the operator defined via $a_{ij} = (A)_{ij}$ can be brought in the form

$$A = U\Lambda V \tag{4.14}$$

where U and V are unitary, while Λ is diagonal with non-negative entries $\left\{\sqrt{\lambda_k}\right\}$, determined by the eigenvalues $\{\lambda_k\}$ of the positive operator $A^\dagger A$. Using the notation $(U)_{ij} = u_{ij}$ and $(V)_{ij} = v_{ij}$ we have

$$\Psi = \sum_{i,j=1}^{\dim \mathcal{H}_S} \sum_{k,l}^{\dim \mathcal{H}_S} u_{ik}\delta_{kl}\sqrt{\lambda_k}v_{kj}\varphi_S^i \otimes \varphi_E^j, \tag{4.15}$$

so that upon defining two new bases $\{\chi_S^i\}_i$ and $\{\chi_E^i\}_i$ according to

$$\chi_S^k = \sum_i u_{ik}\varphi_S^i \tag{4.16}$$

$$\chi_E^k = \sum_i v_{kj}\varphi_E^j \tag{4.17}$$

we obtain the desired expression (4.8). $\square$

From the Schmidt decomposition (4.8) we also learn that any state can be obtained, actually in many ways, as partial trace from a pure state on a larger Hilbert space. To this aim we consider a state $\rho_S \in \mathcal{S}(\mathcal{H}_S)$, that according to (2.72) can be written in the form

$$\rho_S = \sum_i p_i |\varphi_S^i\rangle\langle\varphi_S^i|. \tag{4.18}$$

We now consider another Hilbert space $\mathcal{H}_E$ (with dimension at least $\dim \mathcal{H}_S$) and a basis $\{\psi_E^i\}$ in it. For any such basis the pure state

$$\Psi = \sum_{i=1} \sqrt{p_i}\varphi_S^i \otimes \psi_E^i \tag{4.19}$$

on $\mathcal{H}_S \otimes \mathcal{H}_E$ leads according to (4.1) to ρ_S as reduced state, that is

$$\mathrm{Tr}_E[|\Psi\rangle\langle\Psi|] = \rho_S. \tag{4.20}$$

A pure state Ψ such that (4.20) holds is called a purification of ρ_S. Note that for every unitary operator U_E on $\mathcal{H}_E$ the state $(\mathbb{1}_S \otimes U_E)\Psi$ is still a purification of ρ_S. The different possible purifications of a state ρ_S are related to its possible decompositions. We state this fact as a theorem, which actually provides an alternative formulation of Theorem 2.2, following [7]

Theorem 4.1 (Gisin, 1989) *A state $\rho \in \mathcal{S}(\mathcal{H}_S)$ with* $\dim \mathcal{H}_S = d$ *admits the representations*

$$\rho_S = \sum_{i=1}^{d} \lambda_i |\varphi_S^i\rangle \langle \varphi_S^i| \tag{4.21}$$

and

$$\rho_S = \sum_{k=1}^{M} \mu_k |\psi_S^k\rangle \langle \psi_S^k| \tag{4.22}$$

with $\{\varphi_S^i\}$ and $\{\psi_S^k\}$ sets of normalized vectors iff there exist a Hilbert space $\mathcal{H}_E$ and a norm one vector $\Psi \in \mathcal{H}_S \otimes \mathcal{H}_E$, as well as bases $\{\varphi_E^i\}$ and $\{\psi_E^k\}$ in $\mathcal{H}_E$ such that

$$\Psi = \sum_{i=1}^{d} \sqrt{\lambda_i}\, \varphi_S^i \otimes \varphi_E^i \tag{4.23}$$

$$= \sum_{k=1}^{M} \sqrt{\mu_k}\, \psi_S^k \otimes \psi_E^k. \tag{4.24}$$

The physical interpretation of this result is that there is a unique vector Ψ in the extended Hilbert space such that all demixtures can be seen to arise by a non-selective measurement, in the sense of the discussion in Sect. 3.2.7, of a suitable observable in $\mathcal{H}_E$ on this very state Ψ.

Proof The *if* part of the statement immediately follows by taking the partial trace of the expressions at the r.h.s. of (4.23) and (4.24), which thanks to orthonormality of the considered bases in $\mathcal{H}_E$ leads to the r.h.s. of (4.21) and (4.22) respectively. To prove the *only if* part we resort to Theorem 2.2. Suppose that the r.h.s. of (4.21) and (4.22) indeed represent the same state. Starting from (4.22) and considering a basis $\{\psi_E^k\}$ in $\mathcal{H}_E$ with $\dim \mathcal{H}_E \geqslant \dim \mathcal{H}_S$ we can introduce as in (4.19) the following purification of the state ρ_S

$$\Psi = \sum_{k=1}^{M} \sqrt{\mu_k}\, \psi_S^k \otimes \psi_E^k. \tag{4.25}$$

Furthermore according to Theorem 2.2 the relation (2.194) holds, so that in the present notation

$$\sqrt{\mu_k}\,\psi_S^k = \sum_{i=1}^{d} U_{ki}\sqrt{\lambda_i}\,\varphi_S^i. \tag{4.26}$$

We thus have

$$\Psi = \sum_{k=1}^{M}\sum_{i=1}^{d} U_{ki}\sqrt{\lambda_i}\,\varphi_S^i \otimes \psi_E^k \tag{4.27}$$

$$= \sum_{i=1}^{d}\sqrt{\lambda_i}\,\varphi_S^i \otimes \left(\sum_{k}^{M} U_{ki}\psi_E^k\right). \tag{4.28}$$

The vectors

$$\varphi_E^i = \sum_{k=1}^{M} U_{ki}\psi_E^k \tag{4.29}$$

also provide a basis $\{\varphi_E^i\}$ in $\mathcal{H}_E$, thanks to the fact that the complex numbers U_{ki} are elements of a unitary matrix. We thus obtain

$$\Psi = \sum_{i=1}^{d}\sqrt{\lambda_i}\,\varphi_S^i \otimes \varphi_E^i \tag{4.30}$$

as in (4.23). $\square$

4.3.2 Pure and Mixed Entangled States

Each pure state has a Schmidt decomposition in the form (4.8), characterized by an invariant which is the Schmidt number, that thus allows for a classification of these states. In particular, the Schmidt number fixes the number of factorized elements appearing in representing a given state as a superposition. It thus provides information on how far a pure state is from a factorized one. Indeed, if a state has Schmidt number equal to one it is said to be separable or factorized. This happens iff the state is of the form

$$\Psi = \chi_S \otimes \chi_E, \tag{4.31}$$

that is to say it is a product state. Note that a pure state in a bipartite system is separable iff its marginals are pure. For such a state one has

$$\langle A_S \otimes A_E \rangle_\Psi = \langle A_S \rangle_{\chi_S} \langle A_E \rangle_{\chi_E}, \tag{4.32}$$

so that the statistical outcomes of product observables are uncorrelated. The state does not contain any correlation. If the Schmidt number is larger than one the state is said to be entangled. The correlations in pure quantum states are therefore identified with entanglement, and states are either factorized or entangled. If in particular the Schmidt number is equal to its maximum value D and all Schmidt coefficients are equal, that is the state takes the form

$$\Psi = \frac{1}{\sqrt{D}} \sum_{k=1}^{D} \chi_S^k \otimes \chi_E^k, \tag{4.33}$$

where note that any phase can be absorbed in the definition of the basis elements, the state is said to be maximally entangled. In a given Hilbert space one can consider a basis of maximally entangled states, often called Bell basis, exemplified in Remark 4.1

4.1 Bell states in $\mathbb{C}^2 \otimes \mathbb{C}^2$

We consider the simple but telling case of $\mathcal{H}_S = \mathcal{H}_E = \mathbb{C}^2$, namely two qubits. A basis in this space can be obtained considering the tensor product of the elements of a basis in each space, e.g. the eigenvectors $\{\psi_+, \psi_-\}$ of the Pauli σ_z operator. We thus obtain the basis $\{\psi_+ \otimes \psi_+, \psi_- \otimes \psi_+, \psi_+ \otimes \psi_-, \psi_- \otimes \psi_-\}$, composed of joint eigenvectors of the commuting operators $\sigma_z \otimes \mathbb{1}_{\mathbb{C}^2}$ and $\mathbb{1}_{\mathbb{C}^2} \otimes \sigma_z$. It is usually called computational basis, due to its relevance in quantum information and computation. We now consider another pair of commuting observables on $\mathbb{C}^2 \otimes \mathbb{C}^2$, namely $\sigma_z \otimes \sigma_z$ and $\sigma_x \otimes \sigma_x$. Indeed, exploiting the relation (2.83) for the Pauli matrices, together with the useful identity

$$[A \otimes B, C \otimes D] = [A, C] \otimes BD + CA \otimes [B, D], \tag{4.34}$$

we have for any pair of indexes $i, j = 1, 2, 3$

$$[\sigma_i \otimes \sigma_i, \sigma_j \otimes \sigma_j] = 0. \tag{4.35}$$

The common eigenvectors of these operators are easily verified to be

$$\psi_e^\pm = \frac{1}{\sqrt{2}}(\psi_+ \otimes \psi_+ \pm \psi_- \otimes \psi_-) \tag{4.36}$$

$$\psi_o^\pm = \frac{1}{\sqrt{2}}(\psi_+ \otimes \psi_- \pm \psi_- \otimes \psi_+), \tag{4.37}$$

where the subscript denotes even or odd parity and is determined by the eigenvalue with respect to $\sigma_z \otimes \sigma_z$, while the superscript refers to the eigenvalue with respect to $\sigma_x \otimes \sigma_x$. Due to lack of degeneracy and thanks to (4.35) they are also eigenvectors of $\sigma_y \otimes \sigma_y$. The collection $\{\psi_e^+, \psi_e^-, \psi_o^+, \psi_o^-\}$ provides a basis known as Bell basis. The expressions (4.36) and (4.37) for the states $\psi_e^\pm$ and $\psi_o^\pm$ respectively already provide a Schmidt decomposition according to (4.8), showing in particular that they are maximally entangled complying with (4.33).

It is a simple but important fact that both reduced states (4.1) and (4.7) when determined by a pure state in a bipartite Hilbert space have the same set of eigenvalues, indeed from (4.8) we obtain

$$\rho_S = \mathrm{Tr}_E[|\Psi\rangle\langle\Psi|] = \sum_{i=1}^{D} \lambda_i |\chi_S^i\rangle\langle\chi_S^i| \tag{4.38}$$

and

$$\rho_E = \mathrm{Tr}_S[|\Psi\rangle\langle\Psi|] = \sum_{i=1}^{D} \lambda_i |\chi_E^i\rangle\langle\chi_E^i|. \tag{4.39}$$

In particular, both ρ_S and ρ_E are pure states iff the state Ψ is factorized, while they are proportional to the identity iff it is maximally entangled. These states share all the properties characterized by their probability distribution $\{\lambda_i\}$ only, in particular as discussed in Sect. 2.2.3 purity and von Neumann entropy

$$S(\rho_S) = S(\rho_E) = -\sum_{i=1}^{D} \lambda_i \log \lambda_i. \tag{4.40}$$

For a factorized pure state the von Neumann entropy of the associated reduced states is equal to zero, while for a maximally entangled state it takes the maximum value $\log D$. Maximal correlations in the total pure state imply for the reduced states a trivial statistics. The quantity

$$E(|\Psi\rangle) = -\sum_{i=1}^{D} \lambda_i \log \lambda_i, \tag{4.41}$$

is uniquely determined by the Schmidt coefficients of the state and therefore serves as a quantifier of its entanglement, thus deserving the name of entanglement entropy. For the special case of $\mathcal{H}_S = \mathcal{H}_E = \mathbb{C}^2$ the reduced states can be written in the form (2.79) and the entanglement entropy takes the expression

$$E(|\Psi\rangle) = h(\|\boldsymbol{r}\|), \tag{4.42}$$

where $\boldsymbol{r}$ is the Bloch vector of the reduced states and h the binary Shannon entropy introduced in (3.258). The entanglement entropy is thus the Shannon entropy of the squared Schmidt coefficients.

For the case of mixed states in a bipartite setting the classification of states in view of their correlation properties is much more involved. Also mixed states can be divided into two classes, namely separable and entangled states. A state $\rho \in \mathcal{S}(\mathcal{H}_S \otimes \mathcal{H}_E)$ is said to be separable if it can be expressed as a convex combination of product states, that is in the form

$$\rho = \sum_i p_i \rho_S^i \otimes \rho_E^i, \tag{4.43}$$

where p_i is a probability distribution, namely $p_i \geqslant 0$ and $\sum_i p_i = 1$, together with $\{\rho_S^i\} \subset \mathcal{S}(\mathcal{H}_S)$ and $\{\rho_E^i\} \subset \mathcal{S}(\mathcal{H}_E)$. In particular, within this class a state is said to be factorized if it can be expressed as a tensor product in the form

$$\rho = \rho_S \otimes \rho_E, \tag{4.44}$$

with $\rho_S \in \mathcal{S}(\mathcal{H}_S)$ and $\rho_E \in \mathcal{S}(\mathcal{H}_E)$. At variance with pure states not all separable states are factorized. As for pure states, all states that are not separable are said to be entangled. This means that all states that cannot be brought in the form (4.43) are entangled. Differently from pure states, there is generally no special relationship between the reduced states obtained from a mixed state. However, if one of them is pure we can conclude that ρ is factorized.

If a state is separable the statistical outcomes for a product observable take the form

$$\langle A_S \otimes A_E \rangle_\rho = \sum_i p_i \langle A_S \rangle_{\rho_S^i} \langle A_E \rangle_{\rho_E^i}, \tag{4.45}$$

so that they only exhibit correlations that can be realized in any statistical theory, and do not rely on specific features of quantum theory. A separable state of the form (4.43) can be obtained relying on the capability to prepare independently the statistical operators $\{\rho_S^i\}$ and $\{\rho_E^i\}$ and to mix the different factorized states. The only additional resource needed is therefore a random generator reproducing the probability distribution $\{p_i\}$. If the state is not separable, further resources are needed. This can be easily understood for the case of pure entangled states. Realizing an entangled state according to (4.8) calls for the capability to prepare coherent superpositions of the different factorized states appearing in the expression. While mixing states does not call for a new preparation apparatus, but only for a suitably alternated use of the available ones, preparing a superposition state indeed requires a new apparatus, leading a quantum system to exhibit typical quantum features. Think for example of an interference experiment. The experimental resources needed to make different sources coherently interfere are definitely not equivalent to those needed to activate them independently. Superposing is a typical quantum feature calling for new resources, at variance with mixing which is available in any statistical theory and reflected in the convex structure of the state space. For the case of pure states, the Schmidt decomposition (4.8) allows for a full characterization of entanglement, and the number of Schmidt coefficients provides a unified way to quantify it. When moving to mixed states the characterization of entangled states is a very difficult problem, as well as the quantification of the amount of their entanglement. We should be able to exclude representations of the form (4.43). As we shall discuss in Sect. 4.3.3, the characterization of entangled states is strictly related to the capability of distinguishing between positive and completely positive maps. Note moreover that the set of entangled states is not convex. As a crass counterexample consider the uniform mixture of the projections associated to the Bell states (4.36) and (4.37), which provides the separable maximally mixed state. Relevant classes of entangled states in $\mathbb{C}^2 \otimes \mathbb{C}^2$ are considered in Remark 4.2.

4.2 Bell-diagonal and Werner states

In the case $\mathcal{H}_S = \mathcal{H}_E = \mathbb{C}^2$ states that are diagonal in the Bell basis, that is of the form

$$\rho = p_e^+ P_{\psi_e^+} + p_e^- P_{\psi_e^-} + p_o^+ P_{\psi_o^+} + p_o^- P_{\psi_o^-}, \tag{4.46}$$

where the positive weights sum up to one, are called Bell-diagonal states. The matrix representing them in the computational basis is in X shape, that is, it has non-zero elements only on the main diagonal and the anti-diagonal

$$\rho = \frac{1}{2} \begin{pmatrix} p_e^+ + p_e^- & 0 & 0 & p_e^+ - p_e^- \\ 0 & p_o^+ + p_o^- & p_o^+ - p_o^- & 0 \\ 0 & p_o^+ - p_o^- & p_o^+ + p_o^- & 0 \\ p_e^+ - p_e^- & 0 & 0 & p_e^+ + p_e^- \end{pmatrix}. \tag{4.47}$$

These states are states with maximally mixed marginals, that is both associated reduced states ρ_S and ρ_E are proportional to the identity. For this special class of states, entanglement properties are known not only for the pure elements. It can be shown that the entanglement properties are determined by the highest weight. In particular, they are entangled if the highest weight is above 1/2 [8]. States with maximally mixed marginals can be conveniently written in the form

$$\rho = \frac{1}{4}\left(\mathbb{1}_{\mathbb{C}^2} \otimes \mathbb{1}_{\mathbb{C}^2} + \sum_{i=1}^{3} a_i \sigma_i \otimes \sigma_i\right), \tag{4.48}$$

so that the defining property is immediately verified since the Pauli matrices are traceless, upon the identification

$$a_1 = (p_o^+ - p_o^-) + (p_e^+ - p_e^-)$$
$$a_2 = (p_o^+ - p_o^-) - (p_e^+ - p_e^-), \tag{4.49}$$
$$a_3 = (p_e^+ + p_e^-) - (p_o^+ + p_o^-)$$

leading to

$$\rho = \frac{1}{4}\begin{pmatrix} 1+a_3 & 0 & 0 & a_1 - a_2 \\ 0 & 1-a_3 & a_1 + a_2 & 0 \\ 0 & a_1 + a_2 & 1-a_3 & 0 \\ a_1 - a_2 & 0 & 0 & 1+a_3 \end{pmatrix}. \tag{4.50}$$

The Bell states are recovered when the $\{a_i\}$ coefficients all have modulus one and exhibit an odd number of negative signs, namely

$$P_{\psi_o^\pm} = \frac{1}{2}\begin{pmatrix} 0 & 0 & 0 & 0 \\ 0 & 1 & \pm 1 & 0 \\ 0 & \pm 1 & 1 & 0 \\ 0 & 0 & 0 & 0 \end{pmatrix}, \quad P_{\psi_e^\pm} = \frac{1}{2}\begin{pmatrix} 1 & 0 & 0 & \pm 1 \\ 0 & 0 & 0 & 0 \\ 0 & 0 & 0 & 0 \\ \pm 1 & 0 & 0 & 1 \end{pmatrix}. \tag{4.51}$$

It is useful to recall that denoting the state in terms of 2×2 blocks

$$\rho = \begin{pmatrix} A & B \\ C & D \end{pmatrix} \tag{4.52}$$

the reduced state ρ_S is obtained by replacing each block of dimension $\dim\mathcal{H}_E$ by its trace, while ρ_E is given by the sum of the $\dim\mathcal{H}_S$ blocks on the diagonal, that is

$$\rho_S = \mathrm{Tr}_E[\rho] \tag{4.53}$$
$$= \begin{pmatrix} \mathrm{Tr}\,A & \mathrm{Tr}\,B \\ \mathrm{Tr}\,C & \mathrm{Tr}\,D \end{pmatrix}, \tag{4.54}$$

and

$$\rho_E = \text{Tr}_S[\rho] \tag{4.55}$$

$$= A + D. \tag{4.56}$$

An important class of states within the set of states with maximally mixed marginals is given by the so-called Werner states. These states are characterized by invariance under transformations of the form $U \otimes U$, where U is a unitary operator, and were introduced in [9] to show that, at variance with the case of pure states [10], for mixed states entanglement is not a sufficient condition for violation of the Bell inequality. With reference to (4.48) they can be written in the form

$$\rho_W(r) = \frac{1}{4}\left(\mathbb{1}_{\mathbb{C}^2} \otimes \mathbb{1}_{\mathbb{C}^2} - r \sum_{i=1}^{3} \sigma_i \otimes \sigma_i \right), \tag{4.57}$$

where r is a positive parameter lying between zero and one. According to (4.46), (4.48) and (4.49), they can also be expressed as

$$\rho_W(r) = r P_{\psi_o^-} + \frac{1-r}{4} \mathbb{1}_{\mathbb{C}^2} \otimes \mathbb{1}_{\mathbb{C}^2}. \tag{4.58}$$

These states are entangled for $r > 1/3$, but violate the Bell inequality only for $r > 1/\sqrt{2}$. Note that they are mixed unless $r = 1$.

4.3.3 Positive Maps and Entanglement Detection

The notion of entanglement has a deep connection with the notion of complete positivity. In particular, we can infer about entanglement or lack of entanglement of a state by acting on it with maps that are positive but not completely positive. If we consider a positive map Φ and a separable state $\rho \in \mathcal{S}(\mathcal{H}_S \otimes \mathcal{H}_E)$, then thanks to the definition of separability (4.43) we have

$$(\Phi \otimes \mathbb{1}_{\mathcal{T}(\mathcal{H}_E)})\rho = \sum_i p_i \Phi[\rho_S^i] \otimes \rho_E^i, \tag{4.59}$$

which is still a positive operator. Things can be different if one considers the action of a positive map on states which are not separable. As paradigmatic example we introduce the transposition map

$$\Phi^T[\rho] = \rho^T, \tag{4.60}$$

which for a given basis $\{e_i\}$ in $\mathcal{H}_S$ is defined as follows

$$\langle e_r | \rho^T e_s \rangle = \langle e_s | \rho e_r \rangle, \tag{4.61}$$

so that it transforms positive operators in positive operators. We can now consider a pure state in $\mathcal{H}_S \otimes \mathcal{H}_E$, which relying on (4.8) can be written in the form

$$P_\Psi = \sum_{i,j=1}^{D} \sqrt{p_i p_j} E_S^{ij} \otimes E_E^{ij}, \tag{4.62}$$

where we have introduced the rank-one operators

$$E_S^{ij} = |\chi_S^i\rangle\langle\chi_S^j| \tag{4.63}$$

as well as

$$E_E^{ij} = |\chi_E^i\rangle\langle\chi_E^j|. \tag{4.64}$$

We now consider the action of $\Phi^T \otimes \mathbb{1}_{\mathcal{T}(\mathcal{H}_E)}$, also known as partial transpose, on the state (4.62)

$$(\Phi^T \otimes \mathbb{1}_{\mathcal{T}(\mathcal{H}_E)}) P_\Psi = \sum_{i,j=1}^{D} \sqrt{p_i p_j} (E_S^{ij})^T \otimes E_E^{ij} \tag{4.65}$$

$$= \sum_{i,j=1}^{D} \sqrt{p_i p_j} E_S^{ji} \otimes E_E^{ij}. \tag{4.66}$$

The obtained operator is however not positive. This fact can be put into evidence exhibiting for it an eigenvector corresponding to a negative eigenvalue. As anticipated, this is possible provided the Schmidt number associated to the state Ψ is at least two, that is the state is entangled. In this case, considering two of the non-zero Schmidt coefficients, say p_r and p_s, it can be easily verified that the vector

$$\Theta = \frac{1}{\sqrt{2}} (\chi_S^r \otimes \chi_E^s - \chi_S^s \otimes \chi_E^r) \tag{4.67}$$

is such that

$$(\Phi^T \otimes \mathbb{1}_{\mathcal{T}(\mathcal{H}_E)}) P_\Psi \Theta = -\sqrt{p_r p_s}\,\Theta, \tag{4.68}$$

so that thanks to positivity of the Schmidt coefficients it is an eigenvector corresponding to a negative eigenvalue. This example shows in particular that though the definition of complete positivity considered in Sect. 3.2.1 involves the trivial extension of the map, obtained by considering the tensor product with the identity, its actual verification calls on the availability of entangled states, whose realization requires an interaction among the degrees of freedom associated to the two involved Hilbert spaces.

As already mentioned in Sect. 3.2, the trivial extension of a positive map to a tensor product structure of Hilbert spaces is thus generally no more positive, which leads to introduce the notion of complete positivity considered in Sect. 3.2.1. We have now seen that the difference between a completely positive and a positive map can be put into evidence only considering entangled states. In particular, according to the previous example, given a map Φ that is positive but not completely positive the condition

$$(\Phi \otimes \mathbb{1}_{\mathcal{T}(\mathcal{H}_E)})\rho \geqslant 0 \tag{4.69}$$

is a necessary condition for separability of ρ, while the condition

$$(\Phi \otimes \mathbb{1}_{\mathcal{T}(\mathcal{H}_E)})\rho \leqslant 0 \tag{4.70}$$

is a sufficient condition for entanglement of ρ. In a dual way entangled states can be used to distinguish completely positive from positive maps. In particular, we have the following theorem [11].

Theorem 4.2 (M. & P. & R. Horodecki, 1996) *Let $\rho \in \mathcal{S}(\mathcal{H}_S \otimes \mathcal{H}_E)$, then ρ is separable iff the operator $(\Phi \otimes \mathbb{1}_{\mathcal{T}(\mathcal{H}_E)})\rho$ is positive for any positive map Φ on $\mathcal{S}(\mathcal{H}_S)$.*

For the special case in which $\mathcal{H}_S \otimes \mathcal{H}_E$ coincides with $\mathbb{C}^2 \otimes \mathbb{C}^2$ or $\mathbb{C}^2 \otimes \mathbb{C}^3$, it is further enough to consider the action of the transposition Φ^T as defined in (4.60) to verify separability of a state [11].

In Sect. 3.2.4 we have seen thanks to Theorem 3.3 that the distinction between positivity and complete positivity appears only considering the non-commutativity of the space of observables. Here, thanks to Theorem 4.2 we learn that this distinction appears only considering entangled states.

4.4 Choi-Jamiołkowski Isomorphism

We now discuss an important characterization of completely positive maps in finite dimension introduced by Choi [12], which in particular connects complete positivity of a map with positivity of a matrix. It thus also provides a convenient way to establish complete positivity of a transformation, at the same way leading to a Kraus representation of the map. This analysis recovers the results obtained in Sect. 3.3.3 analyzing convenient ways to represent maps, and puts them in a deeper perspective.

4.4.1 Choi Matrix

We now consider a finite-dimensional Hilbert space $\mathcal{H} = \mathbb{C}^n$ and a linear map $\Lambda \in \mathcal{L}(\mathcal{M}_n(\mathbb{C}), \mathcal{M}_d(\mathbb{C}))$, sending operators acting on $\mathbb{C}^n$, namely according to the notation used in (3.361) elements of $\mathcal{M}_n(\mathbb{C})$, to operators acting on $\mathbb{C}^d$. We then have the following fundamental result.

Theorem 4.3 (Choi, 1972) *Consider the linear map*

$$\Lambda : \mathcal{M}_n(\mathbb{C}) \to \mathcal{M}_d(\mathbb{C}) \tag{4.71}$$
$$\rho \mapsto \Lambda(\rho)$$

the following statements are equivalent:

1. Λ is completely positive
2. Λ is n-positive, that is $\mathbb{1}_{\mathcal{M}_n(\mathbb{C})} \otimes \Lambda$ is a positive map
3. For any orthonormal basis $\{e_i\}_{i=1}^{n}$ in $\mathbb{C}^n$ the nd $\times$ nd square matrix

$$\mathfrak{C}_\Lambda = \begin{pmatrix} \Lambda(|e_1\rangle\langle e_1|) & \dots & \Lambda(|e_1\rangle\langle e_n|) \\ \vdots & \ddots & \vdots \\ \Lambda(|e_n\rangle\langle e_1|) & \dots & \Lambda(|e_n\rangle\langle e_n|) \end{pmatrix} \tag{4.72}$$

is positive.

The matrix $\mathfrak{C}_\Lambda$ appearing in (4.72) is called the Choi matrix associated to the map Λ [13]. The theorem tells us that complete positivity of a map is equivalent to the standard notion of positivity of a suitable square matrix with dimension equal to the product of the dimensions of initial and final Hilbert space. It further tells us that for a map acting on a Hilbert space of finite dimension in order to infer complete positivity we do not have to verify (3.13) for any natural number, but only for the given dimension.

Proof To show the equivalence of the statements we will prove that the chain of implications *1⇒2⇒3⇒1* holds.

The implication *1⇒2* is entailed by the very definition of complete positivity according to (3.13), which calls for *n*-positivity for arbitrary $n \in \mathbb{N}$.

To verify that *2⇒3* we consider a basis $\{e_i\}_{i=1}^{n}$ in $\mathbb{C}^n$. We define the operator

$$X = nP_\Upsilon, \tag{4.73}$$

where P_Υ is the projection on the maximally entangled stated obtained from this basis according to (4.33)

$$\Upsilon = \frac{1}{\sqrt{n}} \sum_{i=1}^{n} e_i \otimes e_i. \tag{4.74}$$

We thus have

$$X = \sum_{i,j=1}^{n} |e_i\rangle\langle e_j| \otimes |e_i\rangle\langle e_j|, \tag{4.75}$$

so that the Choi matrix can be written as

$$\mathfrak{C}_\Lambda = \sum_{i,j=1}^{n} |e_i\rangle\langle e_j| \otimes \Lambda[|e_i\rangle\langle e_j|] \tag{4.76}$$

$$= \mathbb{1}_{\mathcal{M}_n(\mathbb{C})} \otimes \Lambda[X]. \tag{4.77}$$

Positivity of X and n-positivity of Λ then imply positivity of the matrix $\mathfrak{C}_\Lambda$, as discussed in Sect. 3.2.1, and recalling that $\mathbb{1}_{\mathcal{M}_n(\mathbb{C})} \otimes \Lambda$ is a positive transformation iff such is $\Lambda \otimes \mathbb{1}_{\mathcal{M}_n(\mathbb{C})}$.

We now conclude the proof proving the implication $3 \Rightarrow 1$. Given that $\mathfrak{C}_\Lambda$ is a positive matrix we can diagonalize it, and using non normalized eigenvectors we obtain the representation

$$\mathfrak{C}_\Lambda = \sum_{j=1}^{nd} |v_j\rangle\langle v_j|, \tag{4.78}$$

with v_j vectors in $\mathbb{C}^n \otimes \mathbb{C}^d$. These vectors can be expressed with reference to a basis $\{e_j\}$ in $\mathbb{C}^n$ in the form

$$v_j = \sum_{k=1}^{n} e_k \otimes v_{jk}, \tag{4.79}$$

with $v_{jk} \in \mathbb{C}^d$. We can now define the collection of linear operators

$$V_j : \mathbb{C}^n \to \mathbb{C}^d \tag{4.80}$$

$$e_k \mapsto v_{jk}$$

which relate the considered basis in $\mathbb{C}^n$ to the vectors v_{jk}. Thanks to (4.78) and (4.79) we obtain the representation

$$\mathfrak{C}_\Lambda = \sum_{j=1}^{nd} \sum_{k,q=1}^{n} |e_k \otimes v_{jk}\rangle\langle e_q \otimes v_{jq}| \tag{4.81}$$

$$= \sum_{j=1}^{nd} \sum_{k,q=1}^{n} |e_k\rangle\langle e_q| \otimes |v_{jk}\rangle\langle v_{jq}|. \tag{4.82}$$

We now exploit the operators defined in (4.80) coming to

$$\mathfrak{C}_\Lambda = \sum_{j=1}^{nd} \sum_{k,q=1}^{n} |e_k\rangle\langle e_q| \otimes V_j|e_k\rangle\langle e_q|V_j^\dagger. \tag{4.83}$$

The comparison between (4.76) and (4.83) leads to the identification

$$\Lambda[|e_k\rangle\langle e_q|] = \sum_{j=1}^{nd} V_j|e_k\rangle\langle e_q|V_j^\dagger, \tag{4.84}$$

valid for all the n^2 linearly independent operators $E_{kq} = |e_k\rangle\langle e_q|$ introduced in (3.375) that provide a basis in $\mathcal{M}_n(\mathbb{C})$ as discussed in Sect. 3.3.2. We have thus obtained a Kraus representation for the map

$$\Lambda[T] = \sum_{j=1}^{nd} V_j T V_j^\dagger, \tag{4.85}$$

which implies its complete positivity. $\square$

The proof of the theorem insists on the deep connection between complete positivity and entanglement. The Choi matrix whose positivity is equivalent to complete positivity of the map is just obtained by the action of the extended map on a maximally entangled state. The proof further provides a constructive path to obtain a Kraus operator-sum representation of the map according to (3.139). We obtain in particular a canonical representation as in (3.434), since the Kraus operators $\{V_j\}$ are orthogonal. Relying on their definition (4.79) and on (4.80) we have in fact

$$\langle v_i|v_j\rangle = \sum_{k,q=1}^{n} \langle e_q \otimes v_{iq}|e_k \otimes v_{jk}\rangle \tag{4.86}$$

$$= \sum_{q=1}^{n} \langle v_{iq}|v_{jq}\rangle \tag{4.87}$$

$$= \sum_{q=1}^{n} \mathrm{Tr}\{V_j |e_q\rangle \langle e_q| V_i^{\dagger}\} \tag{4.88}$$

$$= \mathrm{Tr}\{V_j V_i^{\dagger}\}. \tag{4.89}$$

The orthogonality of the decomposition in (4.78) finally implies

$$\mathrm{Tr}\{V_j V_i^{\dagger}\} = \delta_{ij}. \tag{4.90}$$

Note that while the choice of basis in $\mathbb{C}^n$ introduced in the proof is irrelevant since states connected by the action of local unitary transformations have the same degree of entanglement, it might lead to different Kraus decompositions. We finally notice that due to its very expression the Choi matrix associated to a trace-preserving map has trace equal to the dimensionality of the input space, namely

$$\mathrm{Tr}\,\mathfrak{C}_\Lambda = n. \tag{4.91}$$

We now point to the relationship between the Choi matrix $\mathfrak{C}_\Lambda$ of a map Λ introduced in (4.72) and the matrix $\Lambda'_{\alpha\beta}$ defined in (3.416) associated to Λ in Sect. 3.3.3 when considering its so-called χ-matrix representation. We simply have to start from (3.416) and consider the canonical basis of operators introduced in Sect. 3.3.2.1, so that according to (3.447) and (3.449) we have

$$\Lambda'_{ij,kl} = \langle e_j \otimes e_i| \left(\sum_{r,s=1}^{n} |e_r\rangle \langle e_s| \otimes \Lambda[|e_r\rangle \langle e_s|] \right) |e_l \otimes e_k\rangle, \tag{4.92}$$

and therefore

$$\Lambda'_{ij,kl} = (\mathfrak{C}_\Lambda)_{ji,lk}. \tag{4.93}$$

The Choi matrix and the χ-matrix therefore coincide, up to a swap of indexes that does not affect the positivity property, upon taking in the Hilbert-Schmidt space the canonical basis of operators of Sect. 3.3.2.1 determined by the considered basis $\{e_i\}_{i=1}^{n}$ in the Hilbert space. The swap in the indexes is due to the fact that $\Lambda'_{ij,kl}$ arises considering the matrix elements of the map $\Lambda \otimes \mathbb{1}_{\mathcal{M}_n(\mathbb{C})}$ applied to a maximally entangled state, while the Choi matrix in the form (4.72) has been introduced rather considering the action of $\mathbb{1}_{\mathcal{M}_n(\mathbb{C})} \otimes \Lambda$ on a maximally entangled state, and expressing the tensor product $A \otimes B$ of two matrices

$$A = \begin{pmatrix} a_{11} & a_{12} \\ a_{21} & a_{22} \end{pmatrix}, \quad B = \begin{pmatrix} b_{11} & b_{12} \\ b_{21} & b_{22} \end{pmatrix} \tag{4.94}$$

by means of their Kronecker product

$$\begin{pmatrix} a_{11}B & a_{12}B \\ a_{21}B & a_{22}B \end{pmatrix}. \tag{4.95}$$

This irrelevant swap of indexes is sometimes avoided identifying anyhow the matrix of (4.72) with the expression

$$\mathfrak{C}_\Lambda = \Lambda \otimes \mathbb{1}_{\mathcal{M}_n(\mathbb{C})}[X]. \tag{4.96}$$

4.3 Transposition and Choi map

We here dwell on two basic examples of positive but not completely positive linear maps. We consider the map

$$\Phi^{\mathrm{Choi}}[\rho] = (n-1)\,\mathrm{Tr}\{\rho\}\mathbb{1}_{\mathbb{C}^n} - \rho \tag{4.97}$$

introduced in [12] to deal with different positivity orders, so that we will call it Choi map, and the transposition

$$\Phi^T[\rho] = \rho^T, \tag{4.98}$$

already introduced in (4.60). Both maps act on $\mathcal{M}_n(\mathbb{C})$. We now consider a basis $\{e_i\}_{i=1}^n$ in $\mathbb{C}^n$ and the associated operators defining as in Sect. 3.3.2.1 a canonical basis

$$E_{ij} = |e_i\rangle\langle e_j|. \tag{4.99}$$

Thanks to (3.402) we then have the representation

$$\Phi^{\mathrm{Choi}}[\rho] = (n-1)\sum_{ij} E_{ij}\rho E_{ij}^\dagger - \rho. \tag{4.100}$$

We also have

$$\Phi^T[\rho] = \sum_{ij} E_{ij}\rho E_{ij}^\dagger - \sum_{ij}\frac{1}{\sqrt{2}}(E_{ij} - E_{ji})\rho\frac{1}{\sqrt{2}}(E_{ij} - E_{ji})^\dagger, \tag{4.101}$$

as can be checked by direct inspection recalling (4.61). As already discussed positivity of the transposed map Φ^T follows from (4.61). To show positivity of Φ^{Choi} we notice that given a positive operator $T \in \mathcal{M}_n(\mathbb{C})$ we have the representation

$$T = \sum_{i=1}^n t_i |f_i\rangle\langle f_i| \qquad t_i \geqslant 0 \tag{4.102}$$

with $\{f_i\}_{i=1}^n$ a basis in $\mathbb{C}^n$. We thus have

$$\Phi^{\mathrm{Choi}}[T] = \sum_{i=1}^n \left[(n-1) \left(\sum_{k=1}^n t_k \right) - t_i \right] |f_i\rangle\langle f_i| \qquad (4.103)$$

defining a positive operator since $\sum_{k=1}^n t_k \geqslant t_i$. The associated Choi matrices are however not positive. Taking the maximally entangled expression (4.73) the Choi matrix associated to the transposition map reads

$$\mathbb{1}_{\mathcal{M}_n(\mathbb{C})} \otimes \Phi^T[X] = \sum_{i,j=1}^n |e_i\rangle\langle e_j| \otimes \Phi^T[|e_i\rangle\langle e_j|] \qquad (4.104)$$

$$= \sum_{i,j=1}^n |e_i\rangle\langle e_j| \otimes |e_j\rangle\langle e_i| \qquad (4.105)$$

$$= \sum_{i,j=1}^n |e_i \otimes e_j\rangle\langle e_j \otimes e_i|. \qquad (4.106)$$

We can immediately verify that this matrix has n eigenvectors of the form $e_i \otimes e_i$ with eigenvalue 1, $n(n-1)/2$ eigenvectors of the form $(e_i \otimes e_j + e_j \otimes e_i)/\sqrt{2}$ for $i \neq j$ with eigenvalue 1, and $n(n-1)/2$ eigenvectors of the form $(e_i \otimes e_j - e_j \otimes e_i)/\sqrt{2}$ for $i \neq j$ with eigenvalue -1. Namely it is not positive. On the same footing we consider the matrix associated to the Choi map

$$\mathbb{1}_{\mathcal{M}_n(\mathbb{C})} \otimes \Phi^{\mathrm{Choi}}[X] = \sum_{i,j=1}^n |e_i\rangle\langle e_j| \otimes \Phi^{\mathrm{Choi}}[|e_i\rangle\langle e_j|] \qquad (4.107)$$

$$= (n-1)\mathbb{1}_{\mathbb{C}^n} \otimes \mathbb{1}_{\mathbb{C}^n} - X, \qquad (4.108)$$

where the matrix X has rank one. The only eigenvector with non zero eigenvalue is $\sum_{i=1}^n e_i \otimes e_i/\sqrt{n}$, with eigenvalue n. It follows that this vector is also an eigenvector of $\mathbb{1}_{\mathcal{M}_n(\mathbb{C})} \otimes \Phi^{\mathrm{Choi}}[X]$ with the negative eigenvalue -1. Importantly, we learn from (4.100) and (4.101) that these maps, though only positive and not completely positive as we have just shown, can be written as the difference of two completely positive maps. This follows from the fact that the r.h.s. of both (4.100) and (4.101) are written as difference of two maps in Kraus operator-sum representation. We will investigate the origin of this result in Sect. 4.4.2.

4.4.2 Isomorphism Between Maps and Operators

The connection between a map $\Lambda \in \mathcal{L}(\mathcal{M}_n(\mathbb{C}), \mathcal{M}_d(\mathbb{C}))$ and a $nd \times nd$ square matrix used in Theorem 4.3 of Sect. 4.4.1 actually has a deeper meaning and leads to an important identification between maps and operator spaces. Let us first recall that $nd \times nd$ square matrices can be identified with operators on the Hilbert space $\mathbb{C}^d \otimes \mathbb{C}^n$, that is according to (3.361) we have $\mathcal{B}(\mathbb{C}^d \otimes \mathbb{C}^n) = \mathcal{M}_{nd}(\mathbb{C})$. We can point to the existence of a one-to-one relationship between the space of maps and operators, known as Choi-Jamiołkowski isomorphism [13,14].

Theorem 4.4 (Choi & Jamiołkowski, 1972 & 1975) *The assignment*

$$\hat{\mathcal{J}} : \mathcal{L}(\mathcal{M}_n(\mathbb{C}), \mathcal{M}_d(\mathbb{C})) \rightarrow \mathcal{B}(\mathbb{C}^d \otimes \mathbb{C}^n) \tag{4.109}$$

$$\Lambda \mapsto \hat{\mathcal{J}}\Lambda$$

with

$$\hat{\mathcal{J}}\Lambda = \Lambda \otimes \mathbb{1}_{\mathcal{M}_n(\mathbb{C})}[P_\Upsilon], \tag{4.110}$$

where P_Υ is the projection on a maximally entangled state in $\mathbb{C}^n \otimes \mathbb{C}^n$, establishes an isomorphism between the vector space of linear maps $\mathcal{L}(\mathcal{M}_n(\mathbb{C}), \mathcal{M}_d(\mathbb{C}))$ and the space of linear operators $\mathcal{B}(\mathbb{C}^d \otimes \mathbb{C}^n)$. The inverse assignment is defined as

$$\check{\mathcal{J}} : \mathcal{B}(\mathbb{C}^d \otimes \mathbb{C}^n) \rightarrow \mathcal{L}(\mathcal{M}_n(\mathbb{C}), \mathcal{M}_d(\mathbb{C})) \tag{4.111}$$

$$\Psi \mapsto \check{\mathcal{J}}\Psi$$

with

$$\check{\mathcal{J}}\Psi[Y] = n \operatorname{Tr}_{\mathbb{C}^n}\{(\mathbb{1}_{\mathbb{C}^d} \otimes Y^T)\Psi\} \tag{4.112}$$

for any operator Y in $\mathcal{M}_n(\mathbb{C})$.

Proof We can immediately verify that (4.109) and (4.111) properly define linear maps. In particular, up to normalization, (4.109) coincides with the assignment used to define the Choi matrix. We have to prove that the two maps are one the inverse of the other, namely $\check{\mathcal{J}} \circ \hat{\mathcal{J}} = \mathbb{1}_{\mathcal{L}(\mathcal{M}_n(\mathbb{C}),\, \mathcal{M}_d(\mathbb{C}))}$ and $\hat{\mathcal{J}} \circ \check{\mathcal{J}} = \mathbb{1}_{\mathcal{B}(\mathbb{C}^d \otimes \mathbb{C}^n)}$. We first apply $\check{\mathcal{J}} \circ \hat{\mathcal{J}}$ to a generic map Λ and consider the action of the transformed map $(\check{\mathcal{J}} \circ \hat{\mathcal{J}})\Lambda$ on a generic operator in $\mathcal{M}_n(\mathbb{C})$. From the very definitions (4.109) and (4.111) we have

$$(\check{\mathcal{J}} \circ \hat{\mathcal{J}})\Lambda[Y] = n \operatorname{Tr}_{\mathbb{C}^n}\{(\mathbb{1}_{\mathbb{C}^d} \otimes Y^T)\Lambda \otimes \mathbb{1}_{\mathbb{C}^n}[P_\Upsilon]\}. \tag{4.113}$$

Using the explicit expression (4.74) for P_Υ we obtain

$$(\check{\mathcal{J}} \circ \hat{\mathcal{J}})\Lambda[Y] = \sum_{i,j=1}^{n} \Lambda[|e_i\rangle\langle e_j|] \operatorname{Tr}_{\mathbb{C}^n}\{Y^T |e_i\rangle\langle e_j|\} \tag{4.114}$$

$$= \sum_{i,j=1}^{n} \Lambda[|e_i\rangle\langle e_j|]\langle e_j|Y^T|e_i\rangle \tag{4.115}$$

$$= \sum_{i,j=1}^{n} \Lambda[|e_i\rangle\langle e_j|]\langle e_i|Y|e_j\rangle \tag{4.116}$$

$$= \Lambda\left[\sum_{i,j=1}^{n} |e_i\rangle\langle e_i|Y|e_j\rangle\langle e_j|\right], \tag{4.117}$$

where we have exploited the definition of transposed operator and the linearity of Λ. Evaluating the sums in (4.117) we obtain

$$(\check{\mathcal{J}} \circ \hat{\mathcal{J}})\Lambda[Y] = \Lambda[Y] \tag{4.118}$$

for any operator $Y \in \mathcal{M}_n(\mathbb{C})$ and any map $\Lambda \in \mathcal{L}(\mathcal{M}_n(\mathbb{C}), \mathcal{M}_d(\mathbb{C}))$, so that indeed $(\check{\mathcal{J}} \circ \hat{\mathcal{J}})$ acts as the identity on $\mathcal{L}(\mathcal{M}_n(\mathbb{C}), \mathcal{M}_d(\mathbb{C}))$. We now consider the behavior of $\hat{\mathcal{J}} \circ \check{\mathcal{J}}$. Considering an arbitrary element of $\mathcal{B}(\mathbb{C}^d \otimes \mathbb{C}^n)$ according to (4.109) and using again the explicit expression (4.74) for P_Υ we have

$$(\hat{\mathcal{J}} \circ \check{\mathcal{J}})\Psi = \check{\mathcal{J}}\Psi \otimes \mathbb{1}_{\mathcal{M}_n(\mathbb{C})}[P_\Upsilon]. \tag{4.119}$$

$$= \frac{1}{n} \sum_{i,j=1}^{n} \check{\mathcal{J}}\Psi[|e_i\rangle\langle e_j|] \otimes |e_i\rangle\langle e_j|. \tag{4.120}$$

We now use (4.111) leading to

$$(\hat{\mathcal{J}} \circ \check{\mathcal{J}})\Psi = \sum_{i,j=1}^{n} \operatorname{Tr}_{\mathbb{C}^n}\{(\mathbb{1}_{\mathbb{C}^d} \otimes |e_j\rangle\langle e_i|)\Psi\} \otimes |e_i\rangle\langle e_j| \tag{4.121}$$

$$= \sum_{i,j=1}^{n} \langle e_i|\Psi|e_j\rangle \otimes |e_i\rangle\langle e_j|, \tag{4.122}$$

so that we have

$$(\hat{\mathcal{J}} \circ \check{\mathcal{J}})\Psi = \Psi \tag{4.123}$$

for any operator $\Psi \in \mathcal{B}(\mathbb{C}^d \otimes \mathbb{C}^n)$, and indeed $(\check{\mathcal{J}} \circ \hat{\mathcal{J}})$ acts as the identity on this space. $\qquad\square$

$$\boxed{\textit{\textbf{Choi-Jamiołkowski isomorphism}}}$$

$$\mathcal{L}(\mathcal{M}_n(\mathbb{C}), \mathcal{M}_d(\mathbb{C})) \ni \Lambda : \mathcal{M}_n(\mathbb{C}) \to \mathcal{M}_d(\mathbb{C})$$

$$\Lambda \otimes \mathbb{1}_{\mathcal{M}_n(\mathbb{C})}[P_\Upsilon] = \hat{\mathcal{J}}\Lambda \xleftarrow{\hat{\mathcal{J}}} \Lambda$$

$$\Psi \xrightarrow{\check{\mathcal{J}}} (\check{\mathcal{J}}\Psi)[Y] = n\mathrm{Tr}_{\mathbb{C}^n}\{(\mathbb{1}_{\mathbb{C}^d} \otimes Y^T)\Psi\}$$

$$\mathcal{B}(\mathbb{C}^d \otimes \mathbb{C}^n) \ni \Psi : \mathbb{C}^d \otimes \mathbb{C}^n \to \mathbb{C}^d \otimes \mathbb{C}^n$$

Fig. 4.2 Schematic summary of the transformations involved in the Choi-Jamiołkowski isomorphism

The maps $\hat{\mathcal{J}}$ and $\check{\mathcal{J}}$ thus define a one-to-one relationship between maps and operators known as Choi-Jamiołkowski isomorphism, that we visualize in Fig. 4.2. Importantly, completely positive maps are identified with positive operators in a suitably higher dimensional space, and completely positive trace-preserving maps with statistical operators, according to (4.131). This was the heart of Choi's result Theorem 4.3. More generally, Hermiticity preserving maps, that is maps fulfilling the condition (3.422), are sent to Hermitian operators. Relying on this fact and along the same lines of the proof of Theorem 4.3, starting from diagonalization of the Hermitian operator $\Lambda \otimes \mathbb{1}_{\mathcal{M}_n(\mathbb{C})}[P_\Upsilon]$ we obtain for Λ an operator-sum representation of the form

$$\Lambda[T] = \sum_{j=1}^{nd} \gamma_j V_j T V_j^\dagger, \tag{4.124}$$

where the coefficients γ_j are real due to Hermiticity. In particular, upon sorting the coefficients in positive and negative ones we have

$$\Lambda[T] = \sum_{\{j|\gamma_j>0\}} |\gamma_j| V_j T V_j^\dagger - \sum_{\{j|\gamma_j<0\}} |\gamma_j| V_j T V_j^\dagger. \tag{4.125}$$

Any Hermiticity preserving transformation can therefore be expressed as the difference of two completely positive maps. This is in particular true for positive maps, as already put into evidence in Remark 4.3. This representations of the map is sometimes called pseudo-Kraus operator-sum representation. For the case of a Hermiticity preserving and trace-annihilating map, that is such that

$$\mathrm{Tr}\{\Lambda[T]\} = 0 \quad \forall T \tag{4.126}$$

we can further write

$$\Lambda[T] = \sum_{j=1}^{nd} \gamma_j \left[V_j T V_j^\dagger - \frac{1}{2}\{V_j^\dagger V_j, T\} \right]. \tag{4.127}$$

We will discuss the relevance of operator expressions of this form in Sect. 5.3.1. It is further useful to consider the marginals of the operator $\hat{\mathcal{J}}\Lambda$ associated to the map Λ by the isomorphism. The marginal with respect to the first Hilbert space $\mathbb{C}^d$ reads

$$\mathrm{Tr}_{\mathbb{C}^d}\{\hat{\mathcal{J}}\Lambda\} = \frac{1}{n} \sum_{i,j=1}^{n} \mathrm{Tr}_{\mathbb{C}^d}\{\Lambda[|e_i\rangle\langle e_j|]\}|e_i\rangle\langle e_j|, \tag{4.128}$$

so that

$$\mathrm{Tr}_{\mathbb{C}^d}\{\hat{\mathcal{J}}\Lambda\} = \frac{\mathbb{1}_{\mathbb{C}^n}}{n} \tag{4.129}$$

iff the map Λ is trace-preserving. The marginal with respect to the second Hilbert space $\mathbb{C}^n$ instead reads

$$\mathrm{Tr}_{\mathbb{C}^n}\{\hat{\mathcal{J}}\Lambda\} = \Lambda\left[\frac{\mathbb{1}_{\mathbb{C}^n}}{n}\right]. \tag{4.130}$$

These identities imply that completely positive trace-preserving maps are identified with positive operators with trace one, that is states in $\mathcal{S}(\mathbb{C}^d \otimes \mathbb{C}^n)$. Notice that this result relies on the different normalization of $\hat{\mathcal{J}}\Lambda$ with respect to the Choi matrix $\mathfrak{C}_\Lambda$. Indeed they differ by a factor equal to the dimension of the input space

$$\hat{\mathcal{J}}\Lambda = \frac{1}{n}\mathfrak{C}_\Lambda. \tag{4.131}$$

In this last equality we have used the identification (4.96). The marginals of these states are constrained to be the transform of the maximally mixed state with respect to the original map Λ, and the maximally mixed state itself. These results lead to the following equivalent conditions:

- Λ is hermiticity preserving iff $\hat{\mathcal{J}}\Lambda = (\hat{\mathcal{J}}\Lambda)^\dagger$
- Λ is completely positive iff $\hat{\mathcal{J}}\Lambda \geqslant 0$
- Λ is trace-preserving iff $\mathrm{Tr}_{\mathbb{C}^d}\{\hat{\mathcal{J}}\Lambda\} = \mathbb{1}_{\mathbb{C}^n}/n$
- Λ is trace-annihilating iff $\mathrm{Tr}_{\mathbb{C}^d}\{\hat{\mathcal{J}}\Lambda\} = 0$.

References

1. E. Schrödinger, Math. Proc. Camb. Philos. Soc. **32**, 446 (1936). https://doi.org/10.1017/S0305004100019137
2. E. Schrödinger, Naturwiss. **23**, 844 (1935). https://doi.org/10.1007/BF01491987
3. E. Schmidt, Math. Ann. **63**, 433 (1907). https://doi.org/10.1007/BF01449770
4. A. Pietsch, Math. Nachr. **283**, 6 (2010). https://doi.org/10.1002/mana.200910127
5. J.A. Miszczak, Int. J. Mod. Phys. C **22**, 897 (2011). https://doi.org/10.1142/S0129183111016683
6. A. Ekert, P.L. Knight, Am. J. Phys. **63**, 415 (1995). https://doi.org/10.1119/1.17904
7. N. Gisin, Helv. Phys. Acta **62**, 363 (1989). https://doi.org/10.5169/seals-116034
8. R. Horodecki, M. Horodecki, Phys. Rev. A **54**, 1838 (1996). https://doi.org/10.1103/PhysRevA.54.1838
9. R.F. Werner, Phys. Rev. A **40**, 4277 (1989). https://doi.org/10.1103/PhysRevA.40.4277
10. N. Gisin, Phys. Lett. A **154**, 201 (1991). https://doi.org/10.1016/0375-9601(91)90805-I
11. M. Horodecki, P. Horodecki, R. Horodecki, Phys. Lett. A **223**, 1 (1996). https://doi.org/10.1016/S0375-9601(96)00706-2
12. M.D. Choi, Can. J. Math. **24**, 520 (1972). https://doi.org/10.4153/CJM-1972-044-5
13. M.D. Choi, Linear Algebra Appl. **10**, 285 (1975). https://doi.org/10.1016/0024-3795(75)90075-0
14. A. Jamiolkowski, Rep. Math. Phys. **3**, 275 (1972). https://doi.org/10.1016/0034-4877(72)90011-0

Dynamical Evolution

5

Abstract

The dynamics of an open quantum system differs from a unitary evolution so that it can describe irreversible phenomena like dissipation and decoherence, with the latter a typical quantum effect. An important special class of open-system evolutions is given by quantum dynamical semigroups, whose mathematical structure is determined by the property of complete positivity and will be exposed in detail. To put into evidence the general features of a reduced dynamics we will further consider examples in which it can be exactly determined. The projection operator technique is finally introduced as general perturbative strategy for the derivation of master equations obeyed by an arbitrary reduced dynamics.

5.1 Introduction

We say that a quantum system with Hilbert space $\mathcal{H}_S$ is closed if its dynamics can be described by a reversible unitary evolution. If there is no explicit time dependence, this evolution is given by a group of transformations which thanks to Stone's theorem identifies a self-adjoint Hamiltonian H_S. The dynamics is then obtained as the solution of a Schrödinger equation determined by H_S

$$i\hbar \frac{d\psi(t)}{dt} = H_S \psi(t), \qquad H_S \in \mathcal{L}(\mathcal{H}_S), \ \ H_S = H_S^\dagger, \tag{5.1}$$

together with a suitable initial condition

$$\psi(t=0) = \psi_0, \quad \psi_0 \in \mathcal{H}_S, \quad \|\psi_0\| = 1. \tag{5.2}$$

© The Author(s), under exclusive license to Springer Nature Switzerland AG 2024

B. Vacchini, *Open Quantum Systems*, Graduate Texts in Physics,

https://doi.org/10.1007/978-3-031-58218-9_5

More generally, if the initial condition is given by a mixed state we have to solve the Liouville von Neumann equation

$$\frac{d\rho(t)}{dt} = -\frac{i}{\hbar}[H_S, \rho(t)], \qquad H_S \in \mathcal{L}(\mathcal{H}_S), \;\; H_S = H_S^\dagger, \tag{5.3}$$

which in particular preserves in time the purity (2.76) of the state, with an initial condition

$$\rho(t = 0) = \rho_0, \quad \rho_0 \in \mathcal{S}(\mathcal{H}_S). \tag{5.4}$$

The assumption of a reversible evolution, corresponding to perfect shielding of our system from the rest of the universe, can however only be an idealization. In general, the interaction with other quantum degrees of freedom acting as external environment cannot be neglected. This leads us to consider a bipartite Hilbert space $\mathcal{H}_S \otimes \mathcal{H}_E$ in which the system S and the environmental degrees of freedom E are described. Without loss of generality we can assume that the system $S + E$ is closed. Its dynamics is determined by a microscopic Hamiltonian of the form

$$H = H_S \otimes \mathbb{1}_E + \mathbb{1}_S \otimes H_E + V, \tag{5.5}$$

where H_S and H_E are the Hamiltonians describing the dynamics of S and E respectively when perfectly isolated. Due to the presence of the interaction term V the system is open. The dynamics of the system only, that is usually called reduced dynamics, will generally no more be given by a Schrödinger or Liouville von Neumann equation.

In this framework, pictorially visualized in Fig. 5.1, different dynamical effects appear, that cannot be described by a unitary evolution. On the one hand these effects can have a natural classical counterpart. For example the system can exchange energy with the environment, undergoing a dissipative evolution possibly admitting an equilibrium state. Also in this case, however, the quantum description imposes special contraints, related to the preservation in time of commutation relations between operators. On the other hand, completely new quantum dynamical effects such as decoherence can appear, that are connected to the behavior in time of typical quantum features of a state, such as its coherence. As we shall see, another interesting feature that can be put into evidence in the dynamics of an open system is a characterization of its memory properties. This feature is common to both classical and quantum case, but as for the case of dissipation, calls for different formulations in the two cases.

Historically, paradigmatic examples of open systems are given by the nuclear spin of an atom or a molecule immersed in a liquid or solid [1], a two-level atom coupled to a mode of the electromagnetic field in a cavity [2], or a particle in a harmonic potential undergoing Brownian motion due to interaction with a collection of harmonic oscillators [3]. For the sake of presentation of the conceptual aspects of the theory we will focus on a selection of physically relevant open systems, that allow for a detailed treatment and therefore to put into evidence the previously mentioned features, including the consideration of memory effects.

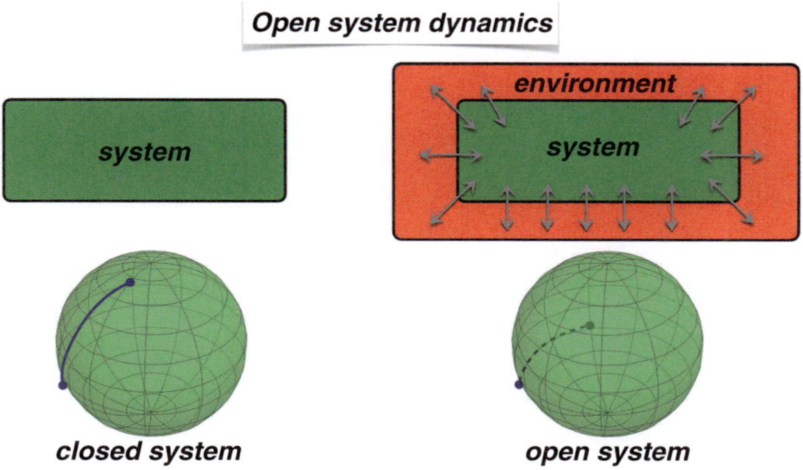

Open system dynamics

environment

system

system

closed system

open system

Fig. 5.1 Simple scheme of an open quantum system, namely a quantum system interacting with other quantum degrees of freedom, called environment, effectively taken into account. The resulting reduced dynamics for the system is no more unitary and exhibits news effects such as dissipation and decoherence. This fact is simply visualized for $\mathcal{H} = \mathbb{C}^2$. In this case the space of states $\mathcal{S}(\mathcal{H})$ can be identified with the Bloch sphere. The coherent dynamics of a closed dynamics can then be depicted as a trajectory on the surface of a sphere, while the trajectory of an open-system dynamics generally leads to the interior of the sphere

5.2 Reduced Dynamics

Given a state $\rho \in \mathcal{S}(\mathcal{H}_S \otimes \mathcal{H}_E)$ describing all the interacting degrees of freedom, as outlined in Sect. 4.2 we can obtain the reduced state ρ_S for the system and the reduced state ρ_E for the environment through the partial trace operation according to (4.1) and (4.7) respectively. Our aim is to describe the evolution of the system and the statistics of the possible measurements that can be performed on it. We are therefore only interested in the knowledge of $\rho_S(t)$ as a function of time. Given that our system is in interaction with an external environment, the question is whether we can eliminate the degrees of freedom of the environment to obtain effective closed equations of motion for the system only. For the sake of simplicity, and without loss of generality, we assume that the system $S + E$ is closed, so that the complete dynamics is unitary and determined by the Hamiltonian (5.5). If the state of system and environment at the initial time, that we take to be zero, is factorized

$$\rho_{SE}(0) = \rho_S(0) \otimes \rho_E, \tag{5.6}$$

with $\rho_S(0) \in \mathcal{S}(\mathcal{H}_S)$ and $\rho_E \in \mathcal{S}(\mathcal{H}_E)$, we can indeed describe the time dependence of the state $\rho_S(t)$ by the action of a map acting on $\rho_S(0)$, which turns out to be completely positive and trace-preserving. We have in fact

$$\rho_S(t) = \mathrm{Tr}_E \circ \mathcal{U}(t) \circ \mathcal{A}_{\rho_E}[\rho_S(0)], \tag{5.7}$$

where $\mathcal{A}_{\rho_E}$ denotes the assignment map introduced in (3.50), $\mathcal{U}(t)$ the unitary evolution determined by (5.5) in the notation of (3.22), and finally Tr_E denotes the partial trace defined in (3.29). Starting from (5.7) we introduce the collection of time-dependent maps

$$\Phi(t) = \mathrm{Tr}_E \circ \mathcal{U}(t) \circ \mathcal{A}_{\rho_E}, \tag{5.8}$$

such that

$$\rho_S(t) = \Phi(t)\rho_S(0), \tag{5.9}$$

and whose action is explicitly given by

$$\Phi(t)\rho_S(0) = \mathrm{Tr}_E[U(t)\rho_S(0) \otimes \rho_E U^\dagger(t)]. \tag{5.10}$$

The map $\Phi(t)$ is the composition of three maps that are completely positive and trace-preserving, as shown in Sects. 3.2.1 and 3.2.2, so that it preserves these properties. For fixed ρ_E we then have the commutative diagram

$$
\begin{array}{ccc}
\rho_S(0) \otimes \rho_E & \xrightarrow{\ \ \mathcal{U}(t)\ \ } & \mathcal{U}(t)[\rho_S(0) \otimes \rho_E] \\
{\scriptstyle \mathrm{Tr}_E}\downarrow & & \downarrow{\scriptstyle \mathrm{Tr}_E} \\
\rho_S(0) & \xrightarrow[\ \ \Phi(t)\ \ ]{} & \rho_S(t)
\end{array}
\quad ,
$$

which relies on the existence of the assignment map $\mathcal{A}_{\rho_E}$ that allows to draw the diagram

$$
\begin{array}{ccc}
\rho_S(0) & \xrightarrow{\ \ \Phi(t)\ \ } & \rho_S(t) \\
{\scriptstyle \mathcal{A}_{\rho_E}}\downarrow & & \uparrow{\scriptstyle \mathrm{Tr}_E} \\
\rho_S(0) \otimes \rho_E & \xrightarrow[\ \ \mathcal{U}(t)\ \ ]{} & \mathcal{U}(t)[\rho_S(0) \otimes \rho_E]
\end{array}
\quad .
$$

Importantly, the assignment map satisfies the compatibility condition

$$\mathrm{Tr}_E \circ \mathcal{A}_{\rho_E} = \mathbb{1}_{\mathcal{T}(\mathcal{H}_S)} \tag{5.11}$$

already put into evidence in (3.57) of Sect.3.2.2. As a result for fixed ρ_E the assignment (5.9) defines a linear completely positive trace-preserving map. Starting from (5.10) and considering a decomposition for ρ_E of the form (2.189)

$$\rho_E = \sum_i \lambda_i |\psi_i\rangle\langle\psi_i| \tag{5.12}$$

with $\lambda_i \geqslant 0$, $\sum_i \lambda_i = 1$ and $\{\psi_i\}$ a collection of normalized vectors in $\mathcal{H}_E$, together with a basis $\{\varphi_i\}$ in $\mathcal{H}_E$ to evaluate the partial trace, we obtain

$$\Phi(t)\sigma = \sum_{i,j} W_{i,j}(t)\sigma W_{i,j}(t)^\dagger \tag{5.13}$$

with $W_{i,j}(t)$ operators acting on $\mathcal{H}_S$ defined as

$$W_{i,j}(t) = \sqrt{\lambda_i}\langle\varphi_j|U(t)\varphi_i\rangle, \tag{5.14}$$

that further satisfy

$$\sum_{i,j} W_{i,j}(t)^\dagger W_{i,j}(t) = \mathbb{1}_{\mathcal{H}_S}, \tag{5.15}$$

corresponding to trace preservation.

It immediately appears that (5.10) and (5.13) provide the two possible representations of a completely positive map introduced respectively in (3.140) and (3.139) of Theorem 3.5. In particular, the operator-sum representation (5.13) puts into evidence that the time evolution of the system can be expressed in an intrinsic way by means of operators of the system only. The existence of the time-dependent maps $\{\Phi(t)\}_{t\geqslant 0}$, here warranted for initially factorized states, corresponds to the existence of a reduced dynamics, that is a description of the dynamics of the system only involving system operators and taking effectively into account the environmental influence. These maps are called quantum dynamical maps [4].

5.3 Quantum Dynamical Semigroups

In the case of closed systems, quantum dynamical maps are given by unitary transformations $\{\mathcal{U}(t)\}_{t\in\mathbb{R}}$ as in (3.8). The evolution is then in particular reversible, and in the absence of an explicit time dependence of the Hamiltonian is described by a one-parameter group of transformations, so that the composition law

$$\mathcal{U}(t+s) = \mathcal{U}(t) \circ \mathcal{U}(s) \qquad \forall t, s \in \mathbb{R} \tag{5.16}$$

holds. According to Stone's theorem [5] they take the form

$$\rho_S(t) = \mathcal{U}(t)\rho_S(0) \tag{5.17}$$

$$= \mathrm{e}^{-\frac{i}{\hbar}H_S t}\rho_S(0)\mathrm{e}^{+\frac{i}{\hbar}H_S t}, \tag{5.18}$$

and are determined by a self-adjoint Hamiltonian operator H_S. Note that unitary transformations are the only completely positive trace-preserving transformations that admit an inverse within the set of completely positive trace-preserving maps. More generally, we can consider an external time-dependent potential, leading to a collection of unitary transformations depending on two time indexes. Moving from closed to open systems, where the effect of an external environment has to be taken into account, reversibility is generally lost, while in keeping with the probabilistic interpretation any physical transformation still needs to be positive and trace-preserving. As we have seen in Sect. 5.2, if the considered evolution arises in an exact way from a microscopic Hamiltonian model the resulting dynamics in the presence of factorized initial conditions is in particular completely positive. In view of this fact, and of the relevance of complete positivity in the quantum context as discussed in Sects. 3.2.1 and 4.4, we are therefore interested in the possible expression of completely positive quantum dynamical maps describing a reduced dynamics in the sense of (5.9). An explicit general characterization of quantum dynamical maps is however only known for special cases. In particular, we can consider the simplest violation of reversibility, in which the collection of maps still obeys a composition law, but only forward in time, namely a semigroup of transformations

$$\Phi(t+s) = \Phi(t) \circ \Phi(s) \qquad \forall t, s \geqslant 0. \tag{5.19}$$

The validity of such a composition law is often associated with a notion of Markovian dynamics, as we shall better discuss in Chap. 7. We treat here the time-homogeneous Markovian case, in which the evolution only depends on the elapsed time, in analogy to (5.18), where the Hamiltonian was taken to be time independent. We are therefore led to define as quantum dynamical semigroup a collection $\{\Phi(t)\}_{t \geqslant 0}$ of quantum dynamical maps such that

1. $\Phi(t)$ is completely positive trace-preserving for any $t \geqslant 0$
2. $\Phi(t+s) = \Phi(t) \circ \Phi(s)$ for all $t, s \geqslant 0$
3. $\Phi(t) \to \mathbb{1}_{\mathcal{T}(\mathcal{H}_S)}$ for $t \to 0$.

In the case of an infinite-dimensional Hilbert space, the requirement $\Phi(t) \to \mathbb{1}_{\mathcal{T}(\mathcal{H}_S)}$ for $t \to 0$ is asked in the strong operator topology, and each $\Phi(t)$ is assumed to be continuous. As already exploited in going from (5.17) to (5.18), a one-parameter group of unitary operators according to Stone's theorem is expressed by means of a generator given by a self-adjoint operator. For a one-parameter semigroup of contraction operators, that is in the present case such that $\|\Phi(t)\rho\|_1 \leqslant \|\rho\|_1$ for any ρ, the Hille-Yosida theorem [6] characterizes the properties of the generator $\mathcal{L}$ such that the expression

$$\Phi(t) = e^{\mathcal{L}t} \tag{5.20}$$

holds, with $\mathcal{L}$ defined on a dense domain identified by existence of the limit

$$\mathcal{L} = \lim_{t \to 0^+} \frac{\Phi(t) - \mathbb{1}_{\mathcal{T}(\mathcal{H}_S)}}{t}, \tag{5.21}$$

and such that

$$\|(\mathcal{L} - \lambda \mathbb{1}_{\mathcal{T}(\mathcal{H}_S)})^{-1}\| \leqslant 1/\lambda \tag{5.22}$$

for all $\lambda > 0$. An alternative characterization of the generators of contraction semigroups is given by the Lumer-Phillips theorem, asking for dissipativity of the generator. These theorems provide kind of the analogue of Stone's theorem for more general evolutions [6–8]. Using the notation of (5.9) we have that for a quantum dynamical semigroup the time evolved state $\rho_S(t)$ obeys the differential equation

$$\frac{\mathrm{d}}{\mathrm{d}t}\rho_S(t) = \mathcal{L}\rho_S(t), \tag{5.23}$$

which is called quantum master equation. This name is generally used for an equation whose solution provides the time evolution of the statistical operator. The term master equation was introduced in a study on cosmic-ray showers calling for a statistical description [9]. The equation fixed the behavior of a probability distribution determining all the other distributions relevant for the considered problem, hence the name master. Master equations are widely used as rate equations in the classical physical and chemical literature. For the special case of a reversible evolution, so that (5.18) holds, corresponding to Stone's theorem, we recover the Liouville-von Neumann equation

$$\frac{\mathrm{d}}{\mathrm{d}t}\rho_S(t) = -\frac{i}{\hbar}[H_S, \rho_S(t)], \tag{5.24}$$

with H_S a self-adjoint operator, so that

$$\mathcal{L}\rho_S(t) = -\frac{i}{\hbar}[H_S, \rho_S(t)]. \tag{5.25}$$

The semigroup composition law (5.19) thus leads to a dynamics determined by a generator $\mathcal{L}$ according to (5.20) and obeying a master equation of the form (5.23). We still have to clarify the structure of the generator, apart from the special case of a unitary evolution already considered in (5.25) and corresponding to the limiting case of a closed dynamics. Furthermore, we have to identify the conditions under which such a composition law is physically motivated. As we shall see in Sect. 5.3.1, the requirement of complete positivity allows to provide a very useful characterization of the operator structure of generators of quantum dynamical semigroups. This result is known as GKSL theorem (see [10] for an interesting historical perspective on this result and [11] for a compact review on quantum dynamical semigroups), and provides a cornerstone in the theory of open quantum systems. It is worth mentioning that in the very same years in which these results appeared, investigations in quantum information related to these developments received important attention, see e.g. the pioneering work of Ingarden [12]. The proper definition of the conditions under which such a description is physically motivated is more involved, and will be explored in Sects. 5.4 and 6.2. The role of this kind of dynamics within the general framework of open quantum system dynamics will be further explored in Chap. 7. As a rule of thumb, a semigroup composition law appears in the presence of a clear separation of time-scales between environmental and reduced dynamics. Namely, when the correlation time of the environment, corresponding e.g. to the typical decay time of the correlation functions of suitable environmental observables, is much shorter than the relaxation time of the reduced system, in symbols

$$\tau_E \ll \tau_R. \tag{5.26}$$

It is important to stress that the results of Sect. 5.2 only warrant the existence of the quantum dynamical maps $\{\Phi(t)\}_{t \geqslant 0}$, and do not provide any explicit information on their structure or requirement on a particular composition law. It is the additional requirement (5.19) of a composition law forward in time that selects a special class of dynamics of physical interest and amenable to a complete mathematical characterization.

5.3.1 Structure of the Generator

The characterization of the structure of generators of quantum dynamical semigroups was obtained by Gorini, Kossakowski and Sudarshan [13] and Lindblad [14] and provides a landmark result.

Theorem 5.1 (Gorini et al. 1976; Lindblad 1976) *Let* $\dim \mathcal{H}_S = n$, *then a linear map* $\mathcal{L} : \mathcal{T}(\mathcal{H}_S) \mapsto \mathcal{T}(\mathcal{H}_S)$ *is the generator of a quantum dynamical semigroup iff it can be written in the form*

$$\mathcal{L}\rho = -\frac{i}{\hbar}[H, \rho] + \sum_{k=2}^{n^2} \gamma_k \left[L_k \rho L_k^\dagger - \frac{1}{2}\{L_k^\dagger L_k, \rho\} \right] \tag{5.27}$$

with $\gamma_k \geqslant 0$, $H = H^\dagger$, $\{L_k\} \subseteq \mathcal{B}(\mathcal{H}_S)$.

Note that according to standard usage we have introduced the anti-commutator between operators defined by

$$\{A, B\} = AB + BA. \tag{5.28}$$

For the sake of simplicity, we have stated the theorem considering the case of a finite-dimensional Hilbert space. The proof under this hypothesis was given by Gorini, Kossakowski and Sudarshan in [13], following seminal work by Kossakowski [15–17] considering a semigroup composition law as in (5.19), but not invoking complete positivity. As proven by Lindblad [14], the generator retains this form even if the Hilbert space is infinite-dimensional, provided the map $\mathcal{L}$ is norm continuous. This requirement implies that the Hamiltonian H, the operators L_k as well as $\sum_k \gamma_k L_k^\dagger L_k$ should be bounded. More generally, an expression of this form is a natural candidate in order to consider a master equation as in (5.23), admitting as solutions well-defined time evolutions of the state of the system. The relevance of the expression (5.27) for the generator can be hardly overestimated, providing a necessary and sufficient condition and therefore a complete characterization, as well as an exceedingly useful tool for phenomenological Ansatz. It is worth adding that it also appears in modifications of quantum mechanics, such as dynamical reduction models [18], providing a different formalization of measurements, so as to recover a classical picture for macroscopic objects. We now give a heuristic demonstration of the result, pointing to the basic ingredients necessary for the proof and suggesting possible extensions and physical interpretations.

Proof *Necessary condition.* We show following the argument used in [19] that the generator of a quantum dynamical semigroup must have the expression (5.27), often termed Lindblad or GKSL form. Due to (3.140) of Theorem 3.5 a completely positive trace-preserving map $\Phi(t)$ admits the operator-sum representation

$$\Phi(t)\rho = \sum_i A_i(t)\rho A_i^\dagger(t), \tag{5.29}$$

with $\sum_i A_i^\dagger(t) A_i(t) = \mathbb{1}_{\mathbb{C}^n}$. We now express the operators $A_i(t)$ in terms of the Bloch basis $\{\Sigma_i\}_{i=1,\dots,n^2}$ introduced in Sect. 3.3.2.3, whose first term is proportional to the identity, namely $\Sigma_1 = \mathbb{1}_{\mathbb{C}^n}/n$. We can write

$$A_i(t) = \sum_{k=1}^{n^2} v_{ik}(t)\Sigma_k, \tag{5.30}$$

so that we have

$$\Phi(t)\rho = \sum_{i,j=1}^{n^2} c_{ij}(t)\Sigma_i \rho \Sigma_j^\dagger \tag{5.31}$$

with $c_{ij}(t) = \sum_{k=1}^{n^2} v_{ki}(t) v_{kj}^*(t)$ a positive matrix. Thanks to the result by Hille and Yosida, the contraction semigroup $\Phi(t)$ identifies a generator via the relation

$$\mathcal{L}[\rho] = \lim_{t \to 0^+} \frac{\Phi(t) - \mathbb{1}_{M_n(\mathbb{C})}}{t}[\rho].$$
(5.32)

Starting from (5.31) we can write

$$\frac{\Phi(t) - \mathbb{1}_{M_n(\mathbb{C})}}{t} \rho = \frac{c_{11}(t) - n}{t} \Sigma_1 \rho \Sigma_1 + \sum_{i=2}^{n^2} \frac{c_{i1}(t)}{t} \Sigma_i \rho + \rho \sum_{i=2}^{n^2} \frac{c_{1i}(t)}{t} \Sigma_i^\dagger$$

$$+ \sum_{i,j=2}^{n^2} \frac{c_{ij}(t)}{t} \Sigma_i \rho \Sigma_j^\dagger,$$
(5.33)

so that due to the existence of the limit we set

$$\mathcal{L}[\rho] = a_{11} \rho + \sum_{i=2}^{n^2} a_i \Sigma_i \rho + \sum_{i=2}^{n^2} a_i^* \rho \Sigma_i^\dagger + \sum_{i,j=2}^{n^2} a_{ij} \Sigma_i \rho \Sigma_j^\dagger,$$
(5.34)

where we have defined

$$a_{11} = \lim_{t \to 0^+} \frac{1}{n} \frac{c_{11}(t) - n}{t}$$
(5.35)

$$a_i = \lim_{t \to 0^+} \frac{c_{i1}(t)}{t}$$
(5.36)

$$a_{ij} = \lim_{t \to 0^+} \frac{c_{ij}(t)}{t}$$
(5.37)

and exploited Hermiticity of the matrix $c_{ij}(t)$. We notice in particular that the matrix of coefficients $\{a_{kl}\}_{k,l=2,\dots,n^2}$ is positive, since it is obtained from the positive matrix $c_{ij}(t)$ by removing the first line and the first column. Asking for trace preservation of $\Phi(t)$, corresponding to the fact that $\mathcal{L}$ is trace-annihilating in the sense of (4.126), we come to

$$a_{11} \mathbb{1}_{\mathbb{C}^n} + \sum_{i=2}^{n^2} a_i \Sigma_i + \sum_{i=2}^{n^2} a_i^* \Sigma_i^\dagger + \sum_{i,j=2}^{n^2} a_{ij} \Sigma_j^\dagger \Sigma_i = 0.$$
(5.38)

Using the identification

$$K = \frac{a_{11}}{2} \mathbb{1}_{\mathbb{C}^n} + \sum_{i=2}^{n^2} a_i \Sigma_i,$$
(5.39)

this constraint leads to

$$K + K^\dagger = -\sum_{i,j=2}^{n^2} a_{ij} \Sigma_j^\dagger \Sigma_i. \tag{5.40}$$

Exploiting further the simple operator identity

$$K\rho + \rho K^\dagger = -\frac{i}{\hbar}\left[\frac{\hbar}{2i}(K^\dagger - K), \rho\right] + \frac{1}{2}\{K + K^\dagger, \rho\}, \tag{5.41}$$

the expression (5.34) of the generator finally becomes

$$\mathcal{L}[\rho] = -\frac{i}{\hbar}[H, \rho] + \sum_{i,j=2}^{n^2} a_{ij}\left[\Sigma_i \rho \Sigma_j^\dagger - \frac{1}{2}\{\Sigma_j^\dagger \Sigma_i, \rho\}\right], \tag{5.42}$$

where we have defined the Hermitian operator

$$H = \frac{\hbar}{2i}(K^\dagger - K). \tag{5.43}$$

In order to obtain the expression (5.27) we now exploit the crucial fact that the matrix of coefficients $\{a_{ij}\}_{i,j=2,\ldots,n^2}$, also known as Kossakowski matrix, is positive. We thus diagonalize it putting into evidence the positive eigenvalues. We have

$$A = U\Lambda U^\dagger \tag{5.44}$$

so that using the notation $(A)_{ij} = a_{ij}$, $(U)_{ik} = u_{ik}$ and $(\Lambda)_{ik} = \delta_{ik}\gamma_k$ we can define the new operators

$$L_k = \sum_{i=2}^{n^2} u_{ik} \Sigma_i, \tag{5.45}$$

and finally obtain (5.27). $\square$

A number of remarks are in order. The obtained expression of the generator (5.27) is often termed Lindbladian, while the name Liouvillian is typically used when the only non vanishing contribution is the commutator with a self-adjoint operator. It is often compactly written

$$\mathcal{L}[\rho] = -\frac{i}{\hbar}[H, \rho] + \mathcal{D}[\rho], \tag{5.46}$$

with $\mathcal{D}$ the dissipator defined as

$$\mathcal{D}[\rho] = \sum_{k=2}^{n^2} \gamma_k \left[L_k \rho L_k^\dagger - \frac{1}{2}\{L_k^\dagger L_k, \rho\} \right]. \tag{5.47}$$

The operators L_k are often called Lindblad operators. They typically describe the microscopic interaction events determining the reduced dynamics, e.g. a collision or the exchange of a quantum of energy, as we shall discuss in the examples of Sects. 5.3.2, 5.3.3, 5.4 and 6.5. The positive coefficients γ_k have the inverse dimension of time and are typically called decay or relaxation rates. The generator can thus be written, though as discussed in Remark 5.1 not in a unique way, as the sum of two contributions. The first term is a Liouvillian describing a coherent contribution to the dynamics determined by a self-adjoint operator, which is however not necessarily the Hamiltonian of the isolated system. This term alone would leave the spectrum of the state invariant, thus not modifying its purity and von Neumann entropy. The second term is the dissipator defined by (5.47) that describes both dissipation, namely energy exchange with the environment, and decoherence, that is loss of coherence taking place due to interaction with other degrees of freedom, as we shall discuss in Sect. 5.3.2. In particular, it can affect the purity of the state, leading from pure states to mixtures. It therefore generally provides an incoherent contribution to the dynamics. The dissipator (5.47) is given by the difference of two terms, often called gain and loss respectively. The names are given by analogy with standard usage in classical master equations, where they contribute to an increase or a decrease in a small region around the considered point of the probability density solution of the equation.

According to (5.23) the generator $\mathcal{L}$ determines the time evolution in the Schrödinger picture. The dual map associated to it via (3.4) reads

$$\mathcal{L}'[X] = +\frac{i}{\hbar}[H, X] + \sum_{k=2}^{n^2} \gamma_k \left[L_k^\dagger X L_k - \frac{1}{2}\{L_k^\dagger L_k, X\} \right], \tag{5.48}$$

and describes the evolution of observables in the Heisenberg picture. We recall that according to (3.7) trace preservation in the Schrödinger picture corresponds to identity preservation in the Heisenberg picture.

5.1 Alternative Expressions and Uniqueness of the Generator

We now put into evidence different possible alternative ways of writing a generator. We first observe that (5.27) admits the trivial rewriting

$$\mathcal{L}\rho = -\frac{i}{\hbar}[H, \rho] + \frac{1}{2}\sum_{k=2}^{n^2} \gamma_k([L_k, \rho L_k^\dagger] + [L_k\rho, L_k^\dagger]) \tag{5.49}$$

and similarly

$$\mathcal{L}'[X] = +\frac{i}{\hbar}[H, X] + \frac{1}{2}\sum_{k=2}^{n^2} \gamma_k (L_k^\dagger[X, L_k] + [L_k^\dagger, X]L_k). \tag{5.50}$$

More importantly, we point out that a generator in Lindblad form is determined by a self-adjoint operator and a completely positive, generally not trace-preserving, transformation,

$$\Psi[\rho] = \sum_{k=2}^{n^2} \gamma_k L_k \rho L_k^\dagger, \tag{5.51}$$

as follows

$$\mathcal{L}[\rho] = -\frac{i}{\hbar}[H, \rho] + \Psi[\rho] - \frac{1}{2}\{\Psi'[\mathbb{1}], \rho\}, \tag{5.52}$$

where Ψ' denotes again the dual map. For the dual generator we have correspondingly

$$\mathcal{L}'[X] = +\frac{i}{\hbar}[H, X] + \Psi'[X] - \frac{1}{2}\{\Psi'[\mathbb{1}], X\}. \tag{5.53}$$

Most importantly, it has to be noticed that the decomposition of a generator in a Hamiltonian and a dissipator as in (5.46) is not unique. Indeed, the expression (5.27) of $\mathcal{L}$ is invariant under the following transformations [20]:

1. $\sqrt{\gamma_k}L_k \to \sqrt{\gamma_k'}L_k' = \sum_r v_{kr}\sqrt{\gamma_r}L_r$ where the set of coefficients v_{kr} satisfies $\sum_{k=2}^{n^2} v_{kr}v_{ks}^* = \delta_{rs}$
2. $L_k \to L_k' = L_k + b_k$ together with $H \to H' = H + b + \frac{1}{2i\hbar}\sum_j \gamma_j(b_j^* L_j - b_j L_j^\dagger)$, with $b_k \in \mathbb{C}$ and $b \in \mathbb{R}$.

The first transformation comes from the freedom in the operator-sum representation of the completely positive map Ψ, as characterized by Theorem 3.6. The second transformation specifies the freedom in the division of the generator in two terms as in (5.46). It shows in particular that in finite dimension the Lindblad operators $\{L_k\}$ can be taken to be traceless, and this condition uniquely fixes H up to an additive constant. Interestingly this condition can also be understood as a minimization of the dissipative contribution to the generator in the splitting (5.46), with respect to a suitable average norm on the space of maps [21]. This unique splitting criterion has been suggested to have a thermodynamic meaning in [22].

Proof *Sufficient condition.* We now want to see that indeed a generator with the Lindblad form (5.27) leads to a completely positive map. The semigroup composition law directly follows from the exponential representation $\Phi(t) = e^{\mathcal{L}t}$, while preservation of Hermiticity and trace according to (3.422) and (4.126) is immediately checked from the master equation (5.23) associated to the generator. The crucial point is to prove that the obtained quantum dynamical map $\Phi(t)$ is completely positive. To this aim we provide an expression for the solution of the master equation (5.23) with $\mathcal{L}$ as in (5.27), which is of independent interest. Let us split the generator in two contributions as follows

$$\mathcal{L} = \mathcal{L}_R + \mathcal{L}_J, \tag{5.54}$$

where

$$\mathcal{L}_R[\rho] = -\frac{i}{\hbar}(H_{\text{eff}}\rho - \rho H_{\text{eff}}^\dagger) \tag{5.55}$$

with

$$H_{\text{eff}} = H - \frac{i}{2}\sum_{k=2}^{n^2} \gamma_k L_k^\dagger L_k \tag{5.56}$$

and

$$\mathcal{L}_J[\rho] = \sum_{k=2}^{n^2} \gamma_k L_k \rho L_k^\dagger. \tag{5.57}$$

The operator H_{eff}, often called effective Hamiltonian, is not self-adjoint and determines the contribution $\mathcal{L}_R$, where R stands for relaxation. The operator $\mathcal{L}_J$ describes a completely positive map, where J stands for jump. We note that the master equation (5.23) can be expressed as an equation for the map $\Phi(t)$ recalling that $\rho(t) = \Phi(t)\rho(0)$, so that it is equivalent to

$$\frac{\mathrm{d}}{\mathrm{d}t}\Phi(t) = \mathcal{L}\Phi(t), \tag{5.58}$$

with initial condition $\Phi(0) = \mathbb{1}$. We are thus led to consider the master equation

$$\frac{\mathrm{d}}{\mathrm{d}t}\mathcal{R}(t) = \mathcal{L}_R\mathcal{R}(t), \tag{5.59}$$

whose solution with initial condition $\mathcal{R}(0) = \mathbb{1}_{\mathcal{M}_n(\mathbb{C})}$ is easily checked to be

$$\mathcal{R}(t)[\rho] = e^{\mathcal{L}_R t}[\rho] \tag{5.60}$$

$$= e^{-\frac{i}{\hbar} H_{\text{eff}} t} \rho\, e^{+\frac{i}{\hbar} H_{\text{eff}}^{\dagger} t}, \tag{5.61}$$

and provides a semigroup of transformations that decrease the trace of the state, hence the term relaxation. In particular, (5.61) shows that $\mathcal{R}(t)$ is a completely positive transformation, since it is written in the Kraus form (3.139). With the aid of this map we can now write the solution of (5.23) according to Duhamel's formula as follows

$$\Phi(t) = \mathcal{R}(t) + \int_0^t d\tau\, \mathcal{R}(t - \tau)\mathcal{L}_J \Phi(\tau). \tag{5.62}$$

Denoting the convolution of time-dependent maps in analogy with (2.353) as

$$(\mathcal{A} \star \mathcal{B})(t) = \int_0^t d\tau\, \mathcal{A}(t - \tau)\mathcal{B}(\tau), \tag{5.63}$$

we can compactly write

$$\Phi(t) = \mathcal{R}(t) + (\mathcal{R} \star \mathcal{L}_J \Phi)(t). \tag{5.64}$$

The validity of (5.62) is verified by directly checking that $\rho_S(t) = \Phi(t)\rho_S(0)$ solves (5.23) for any initial condition. The advantage of (5.64) is that it can be rewritten in the form of a Dyson series

$$\Phi(t) = \mathcal{R}(t) + (\mathcal{R} \star \mathcal{L}_J \mathcal{R})(t) + (\mathcal{R} \star \mathcal{L}_J \mathcal{R} \star \mathcal{L}_J \mathcal{R})(t) + \dots, \tag{5.65}$$

which thanks to complete positivity of $\mathcal{R}(t)$ and $\mathcal{L}_J$ directly implies the same property for $\Phi(t)$, since it is here expressed as sum and composition of completely positive maps. $\qquad\square$

A number of further remarks are in order. Building on (5.65) we can explicitly write the general solution of (5.23) with $\mathcal{L}$ given by the Lindblad expression (5.27) as follows

$$\rho(t) = \mathcal{R}(t)\rho(0)$$
$$+ \sum_{n=1}^{\infty} \int_0^t dt_n \dots \int_0^{t_2} dt_1\, \mathcal{R}(t - t_n)\mathcal{L}_J \mathcal{R}(t_n - t_{n-1}) \dots \mathcal{L}_J \mathcal{R}(t_1)\rho(0). \tag{5.66}$$

The structure of this expression is further discussed in Remark 5.3. The map $\mathcal{R}(t)$ given in (5.61) describes a coherent time-dependent contribution decreasing the trace of the state. The map $\mathcal{L}_J$ instead describes an instantaneous transformation, which we will call jump. The expression (5.66), besides allowing to read some important properties of the solution such as Hermiticity and complete positivity, suggests its interpretation as a sum of contributions in a trajectory space. Each trajectory is characterized by the number of jumps and the time at which they take place, and is a trajectory in the space of operators. In between jumps, the state undergoes an effective dynamics. The solution is obtained by summing over all possible number of jumps and times at which they occur. Though this aspect will not be dealt with in the present work, it is mandatory to at least mention another very important physical interpretation of a dynamics described by a quantum dynamical semigroup. Indeed, the solution $\rho(t)$ of a master equation in GKSL form can always be expressed as an average over a set of stochastic trajectories, taking values in the space of vectors of the system's Hilbert space. Each trajectory now represents the evolution of a pure state conditioned on intermediate measurements on the systems. If the outcomes of the measurements are not recorded, the final state is obtained by averaging over these trajectories, with weights given by the probability of their occurrence. Importantly, this viewpoint provides a powerful numerical technique for the simulation of the master equation, the more convenient the higher the dimension of the system Hilbert space. Making reference e.g. to the situation in which the master equation describes an atom in a two-level approximation in interaction with the electromagnetic field, the single trajectories correspond to the evolution of the atom conditioned on the detection of photons exchanged with the field, and starting in a pure state. The measurements on the atom can be performed in many ways, corresponding to different detection schemes and therefore different noises affecting the system. The reduced dynamics of the system however, obtained when averaging over the noise and therefore over all measurement outcomes, is always the same. In this framework the stochastic trajectories correspond to the observation of the Bohr jumps between levels of the atom. This viewpoint was introduced in the seminal paper [23], see [24–26] for a detailed presentation and the connection to the proper underlying mathematical framework, as well as [27–31] for possible extensions of this viewpoint to open system dynamics that do not fall under the Markov semigroup description.

5.2 Laplace Transform

We collect here for the sake of convenience the definition and some relevant identities for the Laplace transform [32]. Given a function

$$g : \mathbb{R}_+ \to \mathbb{C} \tag{5.67}$$

we say that it admits Laplace transform if the integral

$$\hat{g}(u) = \int_0^{+\infty} dt\, g(t) \mathrm{e}^{-ut}, \tag{5.68}$$

defines an analytic function for $\Re u > b$, with b a real number, so that we have the assignment

$$g(t) \to \hat{g}(u). \tag{5.69}$$

This transform enjoys different properties that make it especially useful in dealing with differential and integro-differential equations. We have in particular for the Laplace transform of the derivative of the function

$$\frac{\mathrm{d}}{\mathrm{d}t} g(t) \to u\hat{g}(u) - g(0), \tag{5.70}$$

while the convolution goes over to multiplication

$$\int_0^{+\infty} \mathrm{d}\tau f(t-\tau)g(\tau) \to \hat{f}(u)\hat{g}(u). \tag{5.71}$$

We further put into evidence the useful relation

$$\int_0^t \mathrm{d}\tau g(\tau) \to \frac{\hat{g}(u)}{u}, \tag{5.72}$$

as well as the expression of the Laplace transform of some relevant functions, such as the constant function

$$\lambda \to \frac{\lambda}{u}, \tag{5.73}$$

and the exponential function

$$\mathrm{e}^{\lambda t} \to \frac{1}{u-\lambda}, \tag{5.74}$$

with $\lambda \in \mathbb{C}$. Thanks to linearity of the Laplace transform, more generally for a linear combination of exponentials we have

$$\sum_i a_i \mathrm{e}^{\lambda_i t} \to \sum_i a_i \frac{1}{u-\lambda_i}. \tag{5.75}$$

These formulae allow to evaluate the inverse Laplace transform of a function given by the ratio of polynomials, that by partial fraction decomposition can be brought to the form (5.75). We have in particular

$$\sinh(\lambda t) \to \frac{\lambda}{u^2-\lambda^2} \tag{5.76}$$

as well as

$$\cosh(\lambda t) \rightarrow \frac{u}{u^2 - \lambda^2}, \tag{5.77}$$

together with

$$\sin(\lambda t) \rightarrow \frac{\lambda}{u^2 + \lambda^2} \tag{5.78}$$

and

$$\cos(\lambda t) \rightarrow \frac{u}{u^2 + \lambda^2}. \tag{5.79}$$

The proof of the sufficient condition also provides a way to obtain a new class of generators, already suggested in [33], still identifying master equations whose solutions are given by completely positive dynamics. Indeed, we can generalize (5.27) to time-dependent operators and rates as follows

$$\begin{aligned} \mathcal{L}(t)\rho = &-\frac{i}{\hbar}[H(t), \rho] \\ &+ \sum_{k=2}^{n^2} \gamma_k(t) \left[L_k(t)\rho L_k(t)^\dagger - \frac{1}{2}\{L_k(t)^\dagger L_k(t), \rho\} \right], \end{aligned} \tag{5.80}$$

and consider the master equation

$$\frac{\mathrm{d}}{\mathrm{d}t}\rho_S(t) = \mathcal{L}(t)\rho_S(t). \tag{5.81}$$

Its solution takes the form

$$\rho_S(t) = \Phi(t, 0)\rho_S(0) \tag{5.82}$$

with

$$\Phi(t, s) = \overleftarrow{T} \exp\left(\int_s^t \mathrm{d}\tau\, \mathcal{L}(\tau) \right), \tag{5.83}$$

where the symbol $\overleftarrow{T}$ denotes chronological time-ordering, namely

$$\Phi(t, s) = \sum_{n=0}^{\infty} \frac{1}{n!} \int_{t_0}^t \mathrm{d}\tau_1 \ldots \int_{t_0}^t \mathrm{d}\tau_n\, \overleftarrow{T}[\mathcal{L}(\tau_1)\ldots\mathcal{L}(\tau_n)] \tag{5.84}$$

with

$$\overleftarrow{T}[\mathcal{L}(\sigma)\mathcal{L}(\tau)] = \begin{cases} \mathcal{L}(\sigma)\mathcal{L}(\tau) & \sigma \geqslant \tau \\ \mathcal{L}(\tau)\mathcal{L}(\sigma) & \sigma \leqslant \tau \end{cases}, \tag{5.85}$$

so that we obtain

$$\Phi(t,s) = \mathbb{1} + \int_s^t d\tau_1 \mathcal{L}(\tau_1) + \frac{1}{2}\int_s^t d\tau_2 \int_s^t d\tau_1 \, \overleftarrow{T}[\mathcal{L}(\tau_2)\mathcal{L}(\tau_1)] + \cdots \tag{5.86}$$

$$= \mathbb{1} + \int_s^t d\tau_1 \mathcal{L}(\tau_1) + \int_s^t d\tau_2 \int_s^{\tau_2} d\tau_1 \mathcal{L}(\tau_2)\mathcal{L}(\tau_1) + \cdots \tag{5.87}$$

We can now split the generator in two terms, in analogy to (5.54)

$$\mathcal{L}(t) = \mathcal{L}_R(t) + \mathcal{L}_J(t). \tag{5.88}$$

The crucial point is now to ask non-negativity of the rates $\gamma_k(t)$ at all times, so that the jump transformation

$$\mathcal{L}_J(t)[\rho] = \sum_{k=2}^{n^2} \gamma_k(t) L_k(t)\rho L_k^\dagger(t) \tag{5.89}$$

is still a completely positive map. The relaxing part $\mathcal{L}_R(t)$ leads to consider the time-local master equation

$$\frac{d}{dt}\mathcal{R}(t,s) = \mathcal{L}_R(t)\mathcal{R}(t,s), \tag{5.90}$$

whose solution with the constraint $\mathcal{R}(t,t) = \mathbb{1}_{\mathcal{M}_n(\mathbb{C})}$ is given for $t \geqslant s \geqslant 0$ by

$$\mathcal{R}(t,s) = \overleftarrow{T}\exp\left(\int_s^t d\tau \mathcal{L}_R(\tau)\right). \tag{5.91}$$

In particular, relying on the notation of (5.56) we introduce the non-Hermitian operator

$$H_{\mathrm{eff}}(t) = H(t) - \frac{i}{2}\sum_{k=2}^{n^2}\gamma_k(t)L_k^\dagger(t)L_k(t) \tag{5.92}$$

so that we have

$$\mathcal{R}(t,s)\rho = \overleftarrow{T}\exp\left(-\frac{i}{\hbar}\int_s^t d\tau\, H_{\mathrm{eff}}(\tau)\right)\rho\,\overrightarrow{T}\exp\left(+\frac{i}{\hbar}\int_s^t d\tau\, H_{\mathrm{eff}}(\tau)\right), \tag{5.93}$$

where in the rightmost operator anti-chronological time-ordering has been considered. This is a completely positive map satisfying the two-time composition law

$$\mathcal{R}(t, \tau) \circ \mathcal{R}(\tau, s) = \mathcal{R}(t, s) \qquad \forall t \geqslant \tau \geqslant s. \tag{5.94}$$

As a result we can still write a Dyson series for the time evolution (5.83), whose very structure ensures complete positivity of the time evolution. We have, upon setting $\mathcal{R}(t, 0) = \mathcal{R}(t)$,

$$\rho(t) = \mathcal{R}(t)\rho(0)$$
$$+ \sum_{n=1}^{\infty} \int_0^t dt_n \ldots \int_0^{t_2} dt_1 \mathcal{R}(t, t_n) \mathcal{L}_J(t_n) \mathcal{R}(t_n, t_{n-1}) \ldots \mathcal{L}_J(t_1) \mathcal{R}(t_1) \rho(0),$$
$$\tag{5.95}$$

and therefore a completely positive time evolution map characterized by two time indexes, satisfying the composition law

$$\Phi(t, \tau) \circ \Phi(\tau, s) = \Phi(t, s) \qquad \forall t \geqslant \tau \geqslant s, \tag{5.96}$$

again only valid forward in time.

5.3 Schwinger-Dyson Formula

The expression (5.66) is a special case of more general formulae providing expressions liable to a perturbation expansion for the exponential of the integral of the sum of two time-dependent operators, relevant for treatments in quantum mechanics, quantum field theory, statistical mechanics and open quantum system theory [34–37], related for the time-independent case to the Baker-Campbell-Hausdorff formula introduced in Remark 2.12.

We first consider the case in which the two operators are not time-dependent, so that the integration only implies time multiplication, namely we consider the expression

$$e^{(A+B)t}, \tag{5.97}$$

whose Laplace transform according to (5.68) reads

$$\frac{1}{u - (A + B)}. \tag{5.98}$$

We notice the formal identities

$$\frac{1}{u - (A + B)} = \frac{1}{u - A} + \frac{1}{u - (A + B)} B \frac{1}{u - A} \tag{5.99}$$

as well as

$$\frac{1}{u - (A + B)} = \frac{1}{u - A} \sum_{n=0}^{\infty} \left(B \frac{1}{u - A} \right)^n, \tag{5.100}$$

that follows from

$$\frac{1}{u - (A + B)} = \frac{1}{u - A} \frac{1}{1 - B \frac{1}{u-A}}. \tag{5.101}$$

Both (5.99) and (5.100) can be taken as operator identities, since ordering has been preserved. Denoting with

$$R_u(A) = (u - A)^{-1} \tag{5.102}$$

the resolvent of the operator A [5], these formulae are more conveniently written in the form

$$R_u(A + B) = R_u(A) + R_u(A + B) B R_u(A) \tag{5.103}$$

and

$$R_u(A + B) = R_u(A) \sum_{n=0}^{\infty} (B R_u(A))^n \tag{5.104}$$

respectively, where in the last line we have considered a so-called Neumann series.

Taking the inverse Laplace transform of (5.99) and (5.100) thanks to (5.71) and (5.75) we come to the expressions

$$e^{(A+B)t} = e^{At} + \int_0^t d\tau\, e^{(A+B)(t-\tau)} B e^{A\tau} \tag{5.105}$$

and

$$e^{(A+B)t} = e^{At} + \sum_{k=1}^{\infty} \int_0^t dt_k \ldots \int_0^{t_2} dt_1 e^{A(t-t_k)} B e^{A(t_k - t_{k-1})} \ldots B e^{At_1} \tag{5.106}$$

$$= e^{At} \overleftarrow{T} \exp\left(\int_0^t d\tau\, e^{-A\tau} B e^{A\tau} \right) \tag{5.107}$$

respectively, providing an alternative expression for (5.97) together with an explicit perturbation expansion. In particular, in (5.107) we have exploited the notion of time-ordering introduced in (5.85) to obtain a more compact expression. Notice that for a non-zero initial time we have correspondingly

$$e^{(A+B)(t-t_0)} = e^{At}\, \overleftarrow{T}\, \exp\left(\int_{t_0}^{t} d\tau\, e^{-A\tau} B e^{A\tau}\right) e^{-At_0}. \tag{5.108}$$

The previously introduced (5.66) is an instance of (5.106) with the identifications

$$A \to \mathcal{L}_R \tag{5.109}$$
$$B \to \mathcal{L}_J. \tag{5.110}$$

Alternatively if we consider the assignments

$$i\hbar A \to H_0 \tag{5.111}$$
$$i\hbar B \to V, \tag{5.112}$$

together with the notation

$$V_I(t) = e^{\frac{i}{\hbar} H_0 t} V e^{-\frac{i}{\hbar} H_0 t}, \tag{5.113}$$

where the index I denotes interaction picture, (5.108) provides the connection between the evolution operator in Schrödinger and interaction picture

$$e^{-\frac{i}{\hbar} H(t-t_0)} = e^{-\frac{i}{\hbar} H_0 t}\, \overleftarrow{T}\, \exp\left(-\frac{i}{\hbar} \int_{t_0}^{t} d\tau\, V_I(\tau)\right) e^{\frac{i}{\hbar} H_0 t_0}. \tag{5.114}$$

We now consider the situation in which the operators are truly time-dependent. Having in mind the case of a propagator forward in time we consider the chronologically time-ordered expression

$$\overleftarrow{T}\, \exp\left(\int_{t_0}^{t} d\tau\, (A(\tau) + B(\tau))\right). \tag{5.115}$$

We now rely on the previous result (5.107), obtained by means of the Laplace transform in the time-homogeneous case, and consider the Ansatz

$$\overleftarrow{T}\, e^{\int_{t_0}^{t} d\tau (A(\tau)+B(\tau))} = \overleftarrow{T}\, e^{\int_{t_0}^{t} d\tau A(\tau)}$$
$$\times \left\{ \overleftarrow{T}\, \exp\left(\int_{t_0}^{t} d\tau\, \overrightarrow{T}\, e^{-\int_{t_0}^{\tau} d\sigma A(\sigma)} B(\tau)\, \overleftarrow{T}\, e^{+\int_{t_0}^{\tau} d\sigma A(\sigma)}\right)\right\}. \tag{5.116}$$

Both sides of the equality have the same derivative and reduce to the identity at the initial time t_0, so that they coincide. Note the use of the anti-chronological time-ordering, defined as

$$\overrightarrow{T}\,[A(\sigma)A(\tau)] = \begin{cases} A(\sigma)A(\tau) & \sigma \leqslant \tau \\ A(\tau)A(\sigma) & \sigma \geqslant \tau \end{cases}, \tag{5.117}$$

so that

$$\overleftarrow{T}\,\mathrm{e}^{\int_{t_0}^{t}\mathrm{d}\tau\,A(\tau)}\,\overrightarrow{T}\,\mathrm{e}^{-\int_{t_0}^{t}\mathrm{d}\sigma\,A(\sigma)} = \mathbb{1}. \tag{5.118}$$

Similarly, relying on (5.105) and considering time dependence we come to the Ansatz

$$\overleftarrow{T}\,\exp\left(\int_{t_0}^{t}\mathrm{d}\tau\,(A(\tau)+B(\tau))\right) = \overleftarrow{T}\,\mathrm{e}^{\int_{t_0}^{t}\mathrm{d}\tau\,A(\tau)}$$
$$+ \int_{t_0}^{t}\mathrm{d}\tau\,\overleftarrow{T}\,\mathrm{e}^{\int_{\tau}^{t}\mathrm{d}\sigma\,(A(\sigma)+B(\sigma))}\,B(\tau)\,\overleftarrow{T}\,\mathrm{e}^{\int_{t_0}^{\tau}\mathrm{d}\sigma\,A(\sigma)}, \tag{5.119}$$

whose correctness can again be verified by explicitly considering the time derivative. These results recover (5.95) with the identifications

$$A \to \mathcal{L}_R(t) \tag{5.120}$$
$$B \to \mathcal{L}_J(t). \tag{5.121}$$

Another case of interest is obtained introducing a pair of complementary projections $\mathcal{P}$ and $\mathcal{Q}$, such that $\mathcal{P} + \mathcal{Q} = \mathbb{1}$, considering

$$A \to \mathcal{L}(t)\mathcal{Q} \tag{5.122}$$
$$B \to \mathcal{L}(t)\mathcal{P}, \tag{5.123}$$

where the time dependence might come from the interaction picture as in Sect. 5.5.2. In this case the time independent case according to (5.105) becomes

$$\mathrm{e}^{\mathcal{L}t} = \mathrm{e}^{\mathcal{L}\mathcal{Q}t} + \int_{0}^{t}\mathrm{d}\tau\,\mathrm{e}^{\mathcal{L}(t-\tau)}\mathcal{L}\mathcal{P}\mathrm{e}^{\mathcal{L}\mathcal{Q}\tau}, \tag{5.124}$$

while (5.119) becomes

$$\overleftarrow{T}\,e^{\int_{t_0}^{t}\,d\tau\,\mathcal{L}(\tau)} = \overleftarrow{T}\,e^{\int_{t_0}^{t}\,d\tau\,\mathcal{L}(\tau)Q}$$
$$\times\left\{\overleftarrow{T}\,\exp\left(\int_0^t\,d\tau\,\overrightarrow{T}\,e^{-\int_{t_0}^{\tau}\,d\sigma\,\mathcal{L}(\sigma)Q}\,\mathcal{L}(\tau)\mathcal{P}\,\overleftarrow{T}\,e^{+\int_{t_0}^{\tau}\,d\sigma\,\mathcal{L}(\sigma)Q}\right)\right\}.$$

$$(5.125)$$

5.3.2 Collisional Decoherence and Quantum Brownian Motion

We now introduce a master equation in Lindblad form to describe quantum decoherence. We will consider the paradigmatic example of decoherence for a massive test particle interacting through collisions with a background gas. In order to show the usefulness and the possible physical interpretation of the structure (5.27) of the generator, we will here follow a heuristic phenomenological route. The presented results can be further confirmed from a microscopic approach, that allows to provide an explicit expression for all the phenomenologically introduced quantities [38]. The word decoherence is used to describe the situation in which a quantum system loses the capability to show typical quantum effects, such as coherence or interference, due to the interaction with other quantum degrees of freedom, which are not accessible to the experimenter [39–42]. Here the interference fringes that can be observed at the output of an interferometer crossed by a massive particle, which according to quantum mechanics and at variance with the classical description can exhibit interference phenomena, are suppressed as a result of the disturbance due to the presence of a background gas.

The relevant Hilbert space for the system is $L^2(\mathbb{R}^3)$, while the environment is described as a gas of particles with a thermal distribution. A beam of massive particles is prepared, e.g. by evaporation from a oven, collimated and sent through an interferometer. We further assume that only a particle at a time can be found in the interferometer. The geometry and particular realization of the interferometer is not relevant in view of our general description of the disturbance effect due to the gas, though obviously affects the detailed description in the real experiments [43,44]. The schematic of an actually performed experiment is drawn in Fig. 5.2. The interaction with the background gas, which can be neglected only in the limit in which the experiment is performed in perfect vacuum, affects the dynamics of the otherwise free particle by means of collisions. Given that for a dilute gas these collisions can be considered independent, a Markovian memoryless description naturally applies. We thus introduce a generator in the form (5.27), starting from the given operator structure and providing a physically motivated Ansatz for the Hamiltonian and

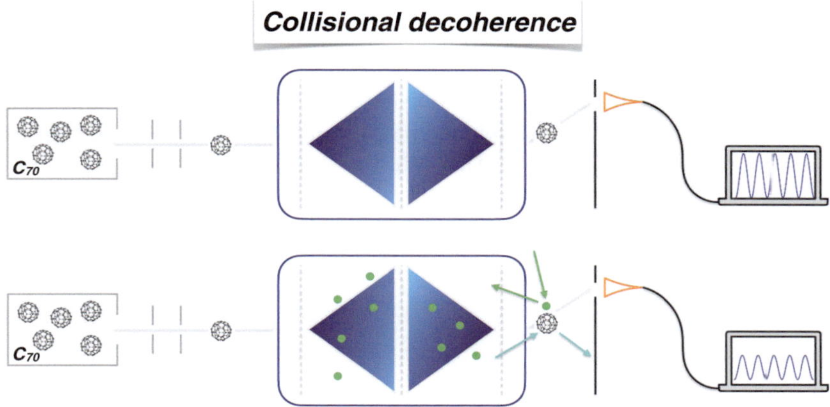

Fig. 5.2 Schematic representation of a three-grating interferometer for a massive particle in vacuum and in the presence of residual background gas, as used in [43]. The beam of massive particles leaves the source and is collimated by means of slits. The beam is so faint that the experiments are performed at the single particle level. The gratings prepare a coherent beam that is split, superposed and finally detected by integrating the signal reaching the screen. The effect of the collisions with background gas found in the interferometer is to disturb the coherent dynamics and reduce the visibility of the fringes. Above: visibility in the ideal situation of perfect vacuum; below: visibility in the presence of a dilute gas inducing collisions

the Lindblad operators appearing in it. We take as Hamiltonian contribution only a kinetic term, so that with reference to (5.27) we have the identification

$$H \rightarrow \frac{\hat{\boldsymbol{p}}^2}{2M}, \tag{5.126}$$

with $\hat{\boldsymbol{p}} = (\hat{p}_x, \hat{p}_y, \hat{p}_z)$ momentum operators and M mass of the particle. We further assume that the basic interaction mechanism with the environment is given by collisions in which the particle exchanges momentum with the gas. A microscopic interaction event thus leads to a change of the system momentum from $\boldsymbol{p}$ to $\boldsymbol{p} + \boldsymbol{q}$ according to the transformation

$$\langle \boldsymbol{p}|\rho_S|\boldsymbol{p}\rangle \rightarrow \langle \boldsymbol{p} + \boldsymbol{q}|\rho_S|\boldsymbol{p} + \boldsymbol{q}\rangle. \tag{5.127}$$

Using the generator of boosts introduced in (2.338) we can write

$$|\boldsymbol{p} + \boldsymbol{q}\rangle = U(\boldsymbol{q})|\boldsymbol{p}\rangle. \tag{5.128}$$

We thus identify the Lindblad operators characterizing the microscopic interaction events with unitary operators describing a momentum transfer, so that with reference to (5.27) we have the identification

$$L_k \to U(\boldsymbol{q}).$$ (5.129)

The momentum exchanged in a single collision is actually a random variable, so that we have to integrate over all possible momentum transfers weighted according to a suitable probability distribution $\mathcal{P}(\boldsymbol{q})$. We further quantify the number of collisions per unit interval of time by means of a rate Γ. We thus introduce the replacement

$$\sum_k \gamma_k \to \Gamma \int \mathrm{d}\boldsymbol{q}\, \mathcal{P}(\boldsymbol{q}).$$ (5.130)

Upon substitution in (5.27) we obtain the master equation

$$\frac{\mathrm{d}}{\mathrm{d}t}\rho_S(t) = -\frac{i}{\hbar}\left[\frac{\hat{\boldsymbol{p}}^2}{2M}, \rho_S(t)\right]$$
$$+ \Gamma \int \mathrm{d}\boldsymbol{q}\, \mathcal{P}(\boldsymbol{q})[U(\boldsymbol{q})\rho_S(t)U(\boldsymbol{q})^\dagger - \rho_S(t)].$$ (5.131)

In this expression the loss term, corresponding to the anti-commutator term in (5.27), is simply proportional to the identity since the Lindblad operators are unitary. The consequences of (5.131) for the position distribution of the massive particle are analyzed in Remark 5.4.

5.4 Space Localization and Momentum Spread

A master equation of the form (5.131) induces localization in position, which in turn suppresses spatial coherence. To justify this statement, we first rewrite the second term in the dissipator of (5.131) in a different but equivalent form. We consider the function

$$\mu(\boldsymbol{x}) = \int \frac{\mathrm{d}\boldsymbol{q}}{(2\pi\hbar)^{3/2}}\sqrt{\mathcal{P}(\boldsymbol{q})}\,\mathrm{e}^{\frac{i}{\hbar}\boldsymbol{q}\cdot\boldsymbol{x}},$$ (5.132)

well-defined and real thanks to positivity and isotropy of $\mathcal{P}(\boldsymbol{q})$, and satisfying

$$\int \mathrm{d}\boldsymbol{x}\,\mu^2(\boldsymbol{x}) = 1.$$ (5.133)

We then have for the expression of the gain term

$$\mathcal{E}[\rho] = \int \mathrm{d}\boldsymbol{q}\, \mathcal{P}(\boldsymbol{q})U(\boldsymbol{q})\rho U(\boldsymbol{q})^\dagger$$ (5.134)

in (5.131) the equivalent expressions

$$\mathcal{E}[\rho] = \int \int \mathrm{d}q\,\mathrm{d}k\,\delta(q-k)\sqrt{\mathcal{P}(q)}\,\mathrm{e}^{\frac{i}{\hbar}q\cdot\hat{x}}\rho\,\mathrm{e}^{-\frac{i}{\hbar}k\cdot\hat{x}}\sqrt{\mathcal{P}(k)} \tag{5.135}$$

$$= \int \frac{\mathrm{d}y}{(2\pi\hbar)^3}\int \int \mathrm{d}q\,\mathrm{d}k\sqrt{\mathcal{P}(q)}\,\mathrm{e}^{\frac{i}{\hbar}q\cdot(\hat{x}-y)}\rho\,\mathrm{e}^{-\frac{i}{\hbar}k\cdot(\hat{x}-y)}\sqrt{\mathcal{P}(k)} \tag{5.136}$$

$$= \int \mathrm{d}y\,\mu(\hat{x}-y)\rho\,\mu(\hat{x}-y)^{\dagger} \tag{5.137}$$

allowing to rewrite (5.131) in the form

$$\frac{\mathrm{d}}{\mathrm{d}t}\rho_S(t) = -\frac{i}{\hbar}\left[\frac{\hat{p}^2}{2M},\rho_S(t)\right]$$

$$+ \Gamma\left[\int \mathrm{d}y\,\mu(\hat{x}-y)\rho_S(t)\mu(\hat{x}-y) - \rho_S(t)\right]. \tag{5.138}$$

It can thus be seen that the effect of the randomly distributed momentum kicks
in the position representation corresponds to the transformation

$$\langle x|\rho_S|x\rangle \rightarrow \mu^2(x-y)\langle x|\rho_S|x\rangle, \tag{5.139}$$

that describes localization in space since the normalization condition (5.133)
corresponds to concentration of the function μ^2 in space.

In order to quantitatively understand how this master equation can describe decoherence effects, it is convenient to consider the matrix elements of (5.131) in the
position representation, thus coming to

$$\frac{\mathrm{d}}{\mathrm{d}t}\langle x|\rho_S(t)|y\rangle = \frac{i\hbar}{2M}(\Delta_x - \Delta_y)\langle x|\rho_S(t)|y\rangle$$

$$+ \Gamma\int \mathrm{d}q\,\mathcal{P}(q)\left[\mathrm{e}^{\frac{i}{\hbar}q\cdot(x-y)} - 1\right]\langle x|\rho_S(t)|y\rangle, \tag{5.140}$$

where we have used the explicit expression (2.338) for the boost generators. The
basic idea is that interference fringes arise due to the spatial coherence properties
of the beam, with modifications due to the collisional dynamics we would like to
investigate. We put this phenomenon into evidence in a simple way neglecting in the
first instance the coherent contribution due to the kinetic term. We can thus exploit the
fact that in the position basis the action of the dissipative part is simply multiplicative.
The exact treatment leading to the expression of the fringe visibility can be found in
Remark 5.5.

Introducing the function

$$\Phi_{\mathcal{P}}(x) = \int d\mathbf{q}\,\mathcal{P}(\mathbf{q})e^{\frac{i}{\hbar}\mathbf{q}\cdot x},$$

(5.141)

which is the Fourier transform of the probability distribution $\mathcal{P}(\mathbf{q})$, namely its characteristic function, we obtain a simple expression for the solution of the master equation in terms of the initial condition

$$\langle x|\rho_S(t)|y\rangle \approx e^{-\Gamma[1-\Phi_{\mathcal{P}}(x-y)]t}\langle x|\rho_S(0)|y\rangle,$$

(5.142)

where, as anticipated and for the sake of simplicity, we have neglected the contribution due to the free evolution. Corresponding to normalization of the probability distribution we have $\Phi_{\mathcal{P}}(0) = 1$, which warrants trace preservation and according to (5.142) implies in particular that the diagonal matrix elements in the position representations are not modified, so that indeed only off-diagonal matrix elements are affected by the dissipative contribution. For characteristic functions, discussed in Remark 6.1, we further have $|\Phi_{\mathcal{P}}(x)| \leqslant 1$, so that according to (5.142) off-diagonal matrix elements are suppressed exponentially in time, and therefore coherences are reduced due to interaction with the environmental degrees of freedom, here described by a gas. This is a typical decoherence effect, leading to reduction of visibility of the interference fringes. As discussed in Remark 5.5, the exact expression is given by (5.148) and only differs by an average of the decoherence effect over the longitudinal coordinate along the interferometer.

5.5 Exact Decoherence Dynamics

We here provide the exact solution of a master equation of the form

$$\frac{d}{dt}\langle x|\rho_S(t)|y\rangle = \left[\frac{i\hbar}{2M}(\Delta_x - \Delta_y) + f(x-y)\right]\langle x|\rho_S(t)|y\rangle,$$

(5.143)

to better justify the approximations leading to the simple expressions (5.142) and (5.154) for the evolution of the off-diagonal matrix elements. It can be expressed as a convolution integral via a suitable kernel with respect to the freely evolving state $\rho_S^0(t)$, namely [38,39]

$$\langle x|\rho_S(t)|y\rangle = \int\int \frac{d\kappa\,dz}{(2\pi\hbar)^3}e^{\frac{i}{\hbar}\kappa\cdot z}F(\kappa, x-y, t)\langle x-z|\rho_S^0(t)|y-z\rangle,$$

(5.144)

with

$$F(\kappa, \lambda, t) = \exp\left(\int_0^t d\tau f\left(\lambda - \frac{\kappa}{M}\tau\right)\right).$$

(5.145)

If the function F reduces to the identity we immediately recover the free solution. The solution of the free evolution equation, that is

$$\rho_S^0(t) = \mathrm{e}^{-\frac{i}{\hbar}\frac{\hat{P}^2}{2M}t}\rho_S(0)\mathrm{e}^{+\frac{i}{\hbar}\frac{\hat{P}^2}{2M}t}, \tag{5.146}$$

can be further explicitly expressed by means of the free propagator of a massive particle [45] according to

$$\langle x|\rho_S^0(t)|y\rangle = \left(\frac{M}{t}\right)^3 \int\int \frac{\mathrm{d}r\,\mathrm{d}s}{(2\pi\hbar)^3}\mathrm{e}^{\frac{i}{\hbar}\frac{M}{2t}[(x-r)^2-(y-s)^2]}\langle r|\rho_S(0)|s\rangle, \tag{5.147}$$

finally coming to

$$\langle x|\rho_S(t)|y\rangle = \left(\frac{M}{t}\right)^3 \int\int \frac{\mathrm{d}r\,\mathrm{d}s}{(2\pi\hbar)^3}\mathrm{e}^{\int_0^t \mathrm{d}\tau f(x-y-\frac{\tau}{t}[(x-r)-(y-s)])}$$

$$\times\,\mathrm{e}^{\frac{i}{\hbar}\frac{M}{2t}[(x-r)^2-(y-s)^2]}\langle r|\rho_S(0)|s\rangle. \tag{5.148}$$

The exact solution of (5.140) that refines the estimate (5.142) is recovered for the assignment

$$f(x) \to -\Gamma[1 - \Phi_{\mathcal{P}}(x)], \tag{5.149}$$

while the inclusion of the free evolution in the approximation (5.154) is obtained upon considering

$$f(x) \to -\frac{\Gamma}{6\hbar^2}x^2\langle q^2\rangle_{\mathcal{P}}. \tag{5.150}$$

In the exact expression, rather then the simple multiplicative action (5.142) or (5.154), the decoherence factor acts by convolution with the freely evolving solution over the length of the interferometer or equivalently the time of flight.

The evolution given by (5.142) can actually describe decoherence effects in different situations, provided that the basic interaction can be formulated in terms of random momentum transfers [46]. An interesting situation takes place when we can assume that the momentum transfers are small, so that we can expand the exponentials appearing in (5.140). In order to introduce the corresponding approximation in (5.142), we recall that the characteristic function of a probability distribution is also known as moment generating function, since the coefficients in its Taylor expansion are given by the moments of the probability distribution. Therefore, if we can assume that the momentum transfers are weak we have

$$\Phi_{\mathcal{P}}(\boldsymbol{x}) \approx 1 + \frac{i}{\hbar} \sum_{i=1}^{3} (\boldsymbol{x} - \boldsymbol{y})_i \langle q_i \rangle_{\mathcal{P}} \tag{5.151}$$

$$- \frac{1}{2\hbar^2} \sum_{i,j=1}^{3} (\boldsymbol{x} - \boldsymbol{y})_i (\boldsymbol{x} - \boldsymbol{y})_j \langle q_i q_j \rangle_{\mathcal{P}} + \ldots,$$

where we have introduced first and second moments of the probability distribution of transferred momenta. It is quite natural to assume isotropy, so that the first moment vanishes and we are left with

$$\Phi_{\mathcal{P}}(\boldsymbol{x}) \approx 1 - \frac{1}{2\hbar^2} \sum_{i,j=1}^{3} (\boldsymbol{x} - \boldsymbol{y})_i (\boldsymbol{x} - \boldsymbol{y})_j \delta_{ij} \langle q_i^2 \rangle_{\mathcal{P}} + \ldots \tag{5.152}$$

$$= 1 - \frac{1}{2\hbar^2} (\boldsymbol{x} - \boldsymbol{y})^2 \frac{1}{3} \langle \boldsymbol{q}^2 \rangle_{\mathcal{P}} + \ldots \tag{5.153}$$

leading to

$$\langle \boldsymbol{x} | \rho_S(t) | \boldsymbol{y} \rangle \approx e^{-\frac{\Gamma}{6\hbar^2} (\boldsymbol{x} - \boldsymbol{y})^2 \langle \boldsymbol{q}^2 \rangle_{\mathcal{P}} t} \langle \boldsymbol{x} | \rho_S(0) | \boldsymbol{y} \rangle. \tag{5.154}$$

The off-diagonal matrix elements are suppressed exponentially both with respect to elapsed interaction time and distance squared. This behavior was first put into evidence in a seminal paper on the subject, looking for a general model of interaction leading to the disappearance of typical quantum phenomena and therefore allowing for a classical description of the considered degrees of freedom [47]. Indeed, the theory of decoherence was initially mainly developed for the sake of characterizing the emergence of a classical description in the dynamics of a quantum system, typically on time-scales that are so short as to make quantum phenomena essentially unobservable [39,48].

5.6 Visibility Expression

We now derive a formula for the visibility of the interference fringes, in order to explicitly quantify the effect of the dissipative part of the master equation on it. For the sake of simplicity we consider the case of a double-slit experiment in the far-field approximation, so that the experimental setup is arranged as in Fig. 5.3. We thus consider a beam of particles that moves towards a grating perpendicular to its direction of propagation, with two identical slits of width δ separated by a distance $\boldsymbol{d}$, finally reaching after covering a distance L a detector where the interference fringes are observed. Denoting as ρ_S^{in} the state of the test particle after the passage through the collimation slits, the state prepared by the double-slit grating that provides the initial condition for the dynamical evolution map can be conveniently written as [49,50]

$$\rho_S(0) = 2\cos\left(\frac{\hat{\boldsymbol{p}} \cdot \boldsymbol{d}}{2\hbar}\right) \rho_S^{\text{in}} \cos\left(\frac{\hat{\boldsymbol{p}} \cdot \boldsymbol{d}}{2\hbar}\right), \tag{5.155}$$

where we recall that the operator $\mathrm{e}^{-\frac{i}{\hbar}\boldsymbol{d}\cdot\hat{\boldsymbol{p}}}$ according to (2.337) describes a translation by a vector $\boldsymbol{d}$ in position. With reference to the discussion in Sect. 3.2 we are considering both the collimation slits and the double-slit grating as part of the preparation procedure, so that we take (5.155) as initial condition.

We can now exploit covariance under space translations of the time evolution $\Phi(t)$ generated by (5.131), so that according to the covariance condition (2.323) discussed in Remark 2.9 we have

$$\Phi(t)[U(\boldsymbol{a})\rho U(\boldsymbol{a})^\dagger] = U(\boldsymbol{a})\Phi(t)[\rho]U(\boldsymbol{a})^\dagger, \tag{5.156}$$

with $U(\boldsymbol{a})$ the generator of translations introduced in (2.337). Further using (2.358) we obtain

$$\begin{aligned}
\langle \boldsymbol{x}|\Phi(t)[\rho_S(0)]|\boldsymbol{x}\rangle = {}& \langle \boldsymbol{x} - \tfrac{1}{2}\boldsymbol{d}|\Phi(t)[\rho_S^{\text{in}}]|\boldsymbol{x} - \tfrac{1}{2}\boldsymbol{d}\rangle \\
& + \langle \boldsymbol{x} + \tfrac{1}{2}\boldsymbol{d}|\Phi(t)[\rho_S^{\text{in}}]|\boldsymbol{x} + \tfrac{1}{2}\boldsymbol{d}\rangle \\
& + \langle \boldsymbol{x} - \tfrac{1}{2}\boldsymbol{d}|\Phi(t)[\rho_S^{\text{in}}]|\boldsymbol{x} + \tfrac{1}{2}\boldsymbol{d}\rangle + h.c.,
\end{aligned} \tag{5.157}$$

with t the time of flight, namely the time employed by the test particle to cross the interferometer and reach the detector. Upon defining

$$I(\boldsymbol{x}, \boldsymbol{y}) = \langle \boldsymbol{x}|\Phi(t)[\rho_S^{\text{in}}]|\boldsymbol{y}\rangle \tag{5.158}$$

we take as figure of merit for the visibility of the expected non-periodic signal the quantity

$$\mathcal{V} = \mathrm{Max}_x \frac{2\left|I\left(\boldsymbol{x} - \tfrac{1}{2}\boldsymbol{d}, \boldsymbol{x} + \tfrac{1}{2}\boldsymbol{d}\right)\right|}{I\left(\boldsymbol{x} - \tfrac{1}{2}\boldsymbol{d}, \boldsymbol{x} - \tfrac{1}{2}\boldsymbol{d}\right) + I\left(\boldsymbol{x} + \tfrac{1}{2}\boldsymbol{d}, \boldsymbol{x} + \tfrac{1}{2}\boldsymbol{d}\right)}. \tag{5.159}$$

As we shall see the quantity (5.159) takes the value 1 in the case of free dynamics, which is thus taken as reference value for the fringes visibility. According to (5.144) the position matrix elements of a state evolved according to the map $\Phi(t)$ can be expressed by convolution with the freely evolved state. We therefore observe in the first instance that according to (5.147) we have for an arbitrary initial state

$$\langle x - \tfrac{1}{2}d|\rho_S^0(t)|x + \tfrac{1}{2}d\rangle = \left(\frac{M}{t}\right)^3 e^{-\frac{i}{\hbar}\frac{M}{t}x\cdot d} \int\int \frac{d r d s}{(2\pi\hbar)^3} e^{\frac{i}{\hbar}\frac{M}{2t}(r^2-s^2)} e^{\frac{i}{\hbar}\frac{M}{2t}d\cdot(r+s)}$$

$$\times\ e^{-\frac{i}{\hbar}\frac{M}{t}x\cdot(r-s)}\langle r|\rho_S(0)|s\rangle. \tag{5.160}$$

This expression can be simplified if we are interested in the far-field approximation. For a signal with wavelength λ passing through slits of width δ and detected after a length L, the far-field approximation is expressed as $L \gg \delta^2/\lambda$. For the matter-wave experiment considered, assuming the test particle propagates along the interferometer with a constant longitudinal velocity p_z/M, according to the de Broglie relation we have $\lambda = \hbar/p_z$, while the time of flight is determined by the relation $L = p_z t/M$. The far-field approximation thus reads $\hbar t/M \gg \delta^2$. We suppose the slits far apart a distance $d = |d|$, so that the support of ρ_S^{in} is concentrated in a spatial region of the dimension of the slit, implying that the integrand can be neglected if $r^2 \gtrsim \delta^2$ or $s^2 \gtrsim \delta^2$. If the time of flight is long enough as to warrant the far-field approximation as well as $\hbar t/M \gg \delta d$ we are thus left with

$$\langle x - \tfrac{1}{2}d|\rho_S^0(t)|x + \tfrac{1}{2}d\rangle \approx \left(\frac{M}{t}\right)^3 e^{-\frac{i}{\hbar}\frac{M}{t}x\cdot d} \int\int \frac{d r d s}{(2\pi\hbar)^3}$$

$$\times\ e^{-\frac{i}{\hbar}\frac{M}{t}x\cdot(r-s)}\langle r|\rho_S(0)|s\rangle \tag{5.161}$$

$$= \left(\frac{M}{t}\right)^3 e^{-\frac{i}{\hbar}\frac{M}{t}x\cdot d}\langle Mx/t|\rho_S(0)|Mx/t\rangle, \tag{5.162}$$

where in the last line the diagonal matrix elements of the initial statistical operator in the momentum representation evaluated for the momentum Mx/t appear. An analogous calculation leads to

$$\langle x \pm \tfrac{1}{2}d|\rho_S^0(t)|x \pm \tfrac{1}{2}d\rangle \approx \left(\frac{M}{t}\right)^3 \langle Mx/t|\rho_S(0)|Mx/t\rangle. \tag{5.163}$$

We are now in the position to evaluate the matrix elements appearing in (5.159) in the far-field approximation. Inserting (5.162) and (5.163) into (5.144), with the identification

$$\rho_S(t) \to \Phi(t)[\rho_S^{\text{in}}], \tag{5.164}$$

we obtain respectively

$$\langle x - \tfrac{1}{2}d|\Phi(t)[\rho_S(0)]|x + \tfrac{1}{2}d\rangle = \left(\frac{M}{t}\right)^3 F\left(-\frac{M}{t}d, -d, t\right)\langle Mx/t|\rho_S^{\text{in}}|Mx/t\rangle \tag{5.165}$$

and

$$\langle x \pm \tfrac{1}{2}d|\Phi(t)[\rho_S(0)]|x \pm \tfrac{1}{2}d\rangle = \left(\frac{M}{t}\right)^3 F(0, 0, t)\langle Mx/t|\rho_S^{\text{in}}|Mx/t\rangle, \tag{5.166}$$

where we have further used the approximation $\langle M(x-r)/t|\rho_S^{\text{in}}|M(x-r)/t\rangle \approx \langle Mx/t|\rho_S^{\text{in}}|Mx/t\rangle$, corresponding to a flat momentum distribution, warranted by localization in position of ρ_S^{in}. The visibility can thus be expressed as the ratio

$$\mathcal{V} = \frac{\left|F\left(-\frac{M}{t}d, -d, t\right)\right|}{|F(0, 0, t)|}, \tag{5.167}$$

finally leading, using (5.145) and a suitable rescaling of the integration variable, to

$$\mathcal{V} = \exp\left(-t\,\Re\int_0^1 ds[f(0) - f(d(s - 1))]\right). \tag{5.168}$$

This formula generally puts into evidence a linear dependence on time in the argument of the exponential. For the case of interest we thus recover from (5.149) the expression

$$\mathcal{V} = \exp\left(-\Gamma\left(1 - \Re\int_0^1 ds\,\Phi_{\mathcal{P}}(d(s - 1))\right)t\right), \tag{5.169}$$

while (5.150) leads to

$$\mathcal{V} = \exp\left(-\Gamma\frac{d^2\langle q^2\rangle_{\mathcal{P}}}{18\hbar^2}t\right), \tag{5.170}$$

so that indeed the multiplicative factors in (5.142) and (5.154) provide a proper estimate of the decoherence effect.

We finally note that even though the master equation (5.131) has been obtained relying on the phenomenological analysis of a concrete physical system, for which also experiments have been performed, it suggests the general structure of a master

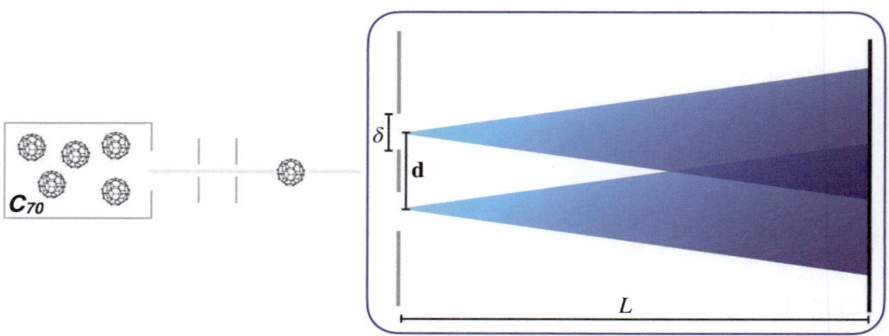

Fig. 5.3 Schematic representation of a double-slit experiment in the far-field approximation, for which the fringes visibility can be expressed as in (5.168)

equation describing a reduction of purity and therefore loss of coherence in some basis [51]. To clarify this fact we consider a master equation of the form

$$\frac{\mathrm{d}}{\mathrm{d}t}\rho_S(t) = -\frac{i}{\hbar}[H, \rho_S(t)] + \sum_k p_k[U_k\rho_S(t)U_k^\dagger - \rho_S(t)], \tag{5.171}$$

with U_k unitary operators and p_k a probability distribution. Equation (5.171) complies both with the Lindblad form (5.27) and with the more specific structure of (5.131). We can now consider the behavior in time of the purity defined in (2.76), which as discussed in Sect. 2.2.3 provides information on how far a state is from a mixed state. We have

$$\frac{\mathrm{d}}{\mathrm{d}t}\mathcal{P}(\rho_S(t)) = 2\,\mathrm{Tr}\left(\rho_S(t)\frac{\mathrm{d}}{\mathrm{d}t}\rho_S(t)\right), \tag{5.172}$$

so that exploiting the cyclic property of the trace operation together with normalization of the p_k we come to

$$\frac{\mathrm{d}}{\mathrm{d}t}\mathcal{P}(\rho_S(t)) = 2\sum_k p_k\,\mathrm{Tr}(\rho_S(t)U_k\rho_S(t)U_k^\dagger - \rho_S^2(t)). \tag{5.173}$$

We further exploit the identity

$$2\,\mathrm{Tr}(A^2 - AUAU^\dagger) = \|[A, U]\|_2^2 \tag{5.174}$$

valid for A self-adjoint and U unitary, where we have considered the Hilbert-Schmidt norm of (2.50). We thus finally obtain

$$\frac{\mathrm{d}}{\mathrm{d}t}\mathcal{P}(\rho_S(t)) = -\sum_k p_k \|[U_k, \rho_S(t)]\|_2^2, \tag{5.175}$$

so that for the dynamics corresponding to (5.171) the purity always decreases in time.

5.7 Double Commutator Master Equation

The behavior described in (5.154) can be obtained considering the matrix elements of the master equation

$$\frac{\mathrm{d}}{\mathrm{d}t}\rho_S(t) = -D[A, [A, \rho_S(t)]], \tag{5.176}$$

which is actually in the GKSL form (5.27) as follows from the identity

$$-\frac{1}{2}[A, [A, \rho]] = A\rho A - \frac{1}{2}\{A^2, \rho\}. \tag{5.177}$$

Using the identifications

$$A \to \frac{\hat{x}}{\hbar}, \tag{5.178}$$

as well as

$$D \to \frac{\Gamma}{6}\langle q^2 \rangle \mathcal{P}, \tag{5.179}$$

and including the free evolution term we have

$$\frac{\mathrm{d}}{\mathrm{d}t}\rho_S(t) = -\frac{i}{\hbar}\left[\frac{\hat{p}^2}{2M}, \rho_S(t)\right] - \frac{D}{\hbar^2}\sum_{i=1}^{3}[\hat{x}_i, [\hat{x}_i, \rho_S(t)]]. \tag{5.180}$$

The master equation (5.176) describes an exponential decay of the off-diagonal matrix elements of the statistical operator in the eigenbasis of the operator A, linear in time and quadratic in the difference of the eigenvalues. In the notation of (3.291) and assuming for the sake of simplicity lack of degeneracy we write

$$A = \sum_k a_k |u_{a_k}\rangle\langle u_{a_k}|, \tag{5.181}$$

leading to

$$\langle u_{a_r}|\rho_S(t)|u_{a_s}\rangle = e^{-D(a_r-a_s)^2 t}\langle u_{a_r}|\rho_S(0)|u_{a_s}\rangle. \tag{5.182}$$

This master equation ubiquitously appears in both microscopic and phenomenological models of decoherence, as we shall also see in (5.353) of Sect. 5.4.1. An interesting feature is the fact that in the presence of a complementary observable, acting as a derivation in the representation of the observable A, the master equation describes a diffusive behavior in the complementary basis. Considering in one dimension position and momentum we would get from (5.176), with the identification $A \rightarrow \hat{x}/\hbar$ as in (5.178),

$$\frac{d}{dt}\langle x|\rho_S(t)|y\rangle = -\frac{D}{\hbar^2}(x-y)^2\langle x|\rho_S(t)|y\rangle, \tag{5.183}$$

as well as

$$\frac{d}{dt}\langle p|\rho_S(t)|q\rangle = +D\left(\frac{\partial}{\partial p} - \frac{\partial}{\partial q}\right)^2\langle p|\rho_S(t)|q\rangle, \tag{5.184}$$

so that suppression of off-diagonal matrix elements in the position representation is associated, as a consequence of Heisenberg's uncertainty relation, to diffusion in the momentum representation.

5.3.3 Quantum Test Particle in a Gas

The master equation (5.131) in Lindblad form provides a satisfactory description of the decoherence effects in the dynamics of a massive test particle interacting via collisions with a dilute gas. As argued in Sect. 5.3.2, the phenomenon of decoherence is governed by the random momentum transfers and dominates for short times. On a longer time-scale, if the gas is at thermal equilibrium, thermalization of the test particle can take place. However, (5.131) does not keep properly into account the energy exchanged in the collisions. Indeed, considering the behavior of the mean value of the kinetic energy from (5.131) we would have

$$\frac{d}{dt}\langle\frac{\hat{p}^2}{2M}\rangle_t = \Gamma\int d\boldsymbol{q}\mathcal{P}(\boldsymbol{q})\frac{q^2}{2M} \tag{5.185}$$

where by definition

$$\langle\frac{\hat{p}^2}{2M}\rangle_t = \mathrm{Tr}\left\{\frac{\hat{p}^2}{2M}\rho_S(t)\right\}, \tag{5.186}$$

and we have used

$$U(\boldsymbol{q})^{\dagger}\hat{\boldsymbol{p}}U(\boldsymbol{q}) = \hat{\boldsymbol{p}} + \boldsymbol{q} \tag{5.187}$$

following from (2.338), together with isotropy of the probability density of random momentum transfers. The prediction for the behavior in time of the kinetic energy would therefore be a linear growth

$$\langle\frac{\hat{\boldsymbol{p}}^2}{2M}\rangle_t = \langle\frac{\hat{\boldsymbol{p}}^2}{2M}\rangle_0 + \langle\frac{\boldsymbol{q}^2}{2M}\rangle_{\mathcal{P}}\Gamma t, \tag{5.188}$$

leading to an unphysical infinite value. This fact can be explained by noting that the master equation (5.131) causes decoherence by inducing localization in position, which in turn implies an increase of the momentum spread, as also discussed at the end of Remark 5.7. Relying on (5.188) and the corresponding expression for the mean value of the momentum we have in fact

$$\mathrm{Var}_t(\boldsymbol{p}) = \langle\hat{\boldsymbol{p}}^2\rangle_t - \langle\hat{\boldsymbol{p}}\rangle_t^2 \tag{5.189}$$

$$= \mathrm{Var}_0(\boldsymbol{p}) + \langle\boldsymbol{q}^2\rangle_{\mathcal{P}}\Gamma t. \tag{5.190}$$

To obtain a realistic description of the energy exchange we first observe that the generator in the master equation (5.131) is characterized by covariance under space translations in the sense of (2.323). Indeed, writing (5.131) in the form

$$\frac{\mathrm{d}}{\mathrm{d}t}\rho_S(t) = \mathcal{L}[\rho_S(t)], \tag{5.191}$$

with

$$\mathcal{L}[\rho] = -\frac{i}{\hbar}\left[\frac{\hat{\boldsymbol{p}}^2}{2M}, \rho\right] + \Gamma\int \mathrm{d}\boldsymbol{q}\mathcal{P}(\boldsymbol{q})[U(\boldsymbol{q})\rho U(\boldsymbol{q})^{\dagger} - \rho], \tag{5.192}$$

similarly to (5.156) we have

$$\mathcal{L}[U(\boldsymbol{a})\rho U(\boldsymbol{a})^{\dagger}] = U(\boldsymbol{a})\mathcal{L}[\rho]U(\boldsymbol{a})^{\dagger}. \tag{5.193}$$

This symmetry has an important physical meaning, since it traces back to homogeneity of the gas. We now consider a modification of the generator $\mathcal{L}$ of (5.192) that preserves covariance under space translations [52,53], and admits as stationary solution the operator with the canonical structure

$$w(\hat{\boldsymbol{p}}) = \exp\left(-\beta\frac{\hat{\boldsymbol{p}}^2}{2M}\right), \tag{5.194}$$

with M the mass of the test particle and β the inverse temperature of the environmental gas. We first observe that the covariance property is still warranted if we consider a replacement of the unitary Lindblad operator $U(\boldsymbol{q})$ according to

$$U(\boldsymbol{q}) \to U(\boldsymbol{q})v(\boldsymbol{q}, \hat{\boldsymbol{p}}), \tag{5.195}$$

with $v(\boldsymbol{q}, \boldsymbol{p})$ a real function and we exploit this freedom to enforce the stationary solution (5.194). Modifying (5.192) according to (5.195) and using (5.187) we obtain

$$\mathcal{L}[w(\hat{\boldsymbol{p}})] = \Gamma \int \mathrm{d}\boldsymbol{q}\mathcal{P}(\boldsymbol{q})[U(\boldsymbol{q})w(\hat{\boldsymbol{p}})v(\boldsymbol{q}, \hat{\boldsymbol{p}})^2 U(\boldsymbol{q})^\dagger - w(\hat{\boldsymbol{p}})v(\boldsymbol{q}, \hat{\boldsymbol{p}})^2] \tag{5.196}$$

$$= \Gamma \int \mathrm{d}\boldsymbol{q}\mathcal{P}(\boldsymbol{q})[w(\hat{\boldsymbol{p}} - \boldsymbol{q})v(\boldsymbol{q}, \hat{\boldsymbol{p}} - \boldsymbol{q})^2 - w(\hat{\boldsymbol{p}})v(\boldsymbol{q}, \hat{\boldsymbol{p}})^2]. \tag{5.197}$$

If we consider v in exponential form

$$v(\boldsymbol{q}, \boldsymbol{p}) = \exp\left(-\frac{\beta}{4}E(\boldsymbol{q}, \boldsymbol{p})\right), \tag{5.198}$$

with

$$E(\boldsymbol{q}, \boldsymbol{p}) = \frac{(\boldsymbol{p} + \boldsymbol{q})^2}{2M} - \frac{\boldsymbol{p}^2}{2M} \tag{5.199}$$

$$= \frac{\boldsymbol{q}^2}{2M} + \frac{\boldsymbol{q} \cdot \boldsymbol{p}}{M} \tag{5.200}$$

the change of kinetic energy of the test particle as a consequence of a collision with random momentum transfer $\boldsymbol{q}$, we have that the integrand in (5.197) is an odd function of $\boldsymbol{q}$, so that $\mathcal{L}[w(\hat{\boldsymbol{p}})] = 0$. Keeping into account also energy exchanges thus allows to predict the physically relevant stationary solution. We now consider the obtained Lindblad master equation

$$\frac{\mathrm{d}}{\mathrm{d}t}\rho_S(t) = -\frac{i}{\hbar}\left[\frac{\hat{\boldsymbol{p}}^2}{2M}, \rho_S(t)\right] + \Gamma \int \mathrm{d}\boldsymbol{q}\mathcal{P}(\boldsymbol{q})$$
$$\times \left[U(\boldsymbol{q})\mathrm{e}^{-\frac{\beta}{4}E(\boldsymbol{q},\hat{\boldsymbol{p}})}\rho_S(t)\mathrm{e}^{-\frac{\beta}{4}E(\boldsymbol{q},\hat{\boldsymbol{p}})}U(\boldsymbol{q})^\dagger - \frac{1}{2}\left\{\rho_S(t), \mathrm{e}^{-\frac{\beta}{2}E(\boldsymbol{q},\hat{\boldsymbol{p}})}\right\}\right], \tag{5.201}$$

and study the behavior in time of the expectation value of the kinetic energy. The time derivative of the mean kinetic energy takes the compact form

$$\frac{\mathrm{d}}{\mathrm{d}t}\langle\frac{\hat{\boldsymbol{p}}^2}{2M}\rangle_t = \Gamma \int \mathrm{d}\boldsymbol{q}\mathcal{P}(\boldsymbol{q})\langle E(\boldsymbol{q}, \hat{\boldsymbol{p}})\mathrm{e}^{-\frac{\beta}{2}E(\boldsymbol{q},\hat{\boldsymbol{p}})}\rangle_t, \tag{5.202}$$

which generally leads to a complicated solution but allows for a simple explicit evaluation close to equilibrium, that is for energy exchanges small with respect to the thermal energy

$$\beta E(\boldsymbol{q}, \boldsymbol{p}) \ll 1. \tag{5.203}$$

In this case, as shown in Remark 5.8 the solution reads

$$\langle \frac{\hat{\boldsymbol{p}}^2}{2M} \rangle_t = \mathrm{e}^{-\eta t} \langle \frac{\hat{\boldsymbol{p}}^2}{2M} \rangle_0 + \frac{3}{2\beta}(1 - \mathrm{e}^{-\eta t}). \tag{5.204}$$

with η a positive constant defined by (5.210). We thus obtain on general grounds, close to thermal equilibrium, an exponential relaxation to the equipartition equilibrium value of the mean kinetic energy.

5.8 Thermalization

To obtain a general estimate of the behavior of (5.202) close to equilibrium we first recall the expression (5.200) of the energy exchange, so that (5.202) becomes

$$\frac{\mathrm{d}}{\mathrm{d}t}\langle \frac{\hat{\boldsymbol{p}}^2}{2M} \rangle_t = \Gamma \int \mathrm{d}\boldsymbol{q}\mathcal{P}(\boldsymbol{q})\mathrm{e}^{-\beta\frac{q^2}{4M}} \langle \left(\frac{q^2}{2M} + \frac{\boldsymbol{q} \cdot \hat{\boldsymbol{p}}}{M} \right) \mathrm{e}^{-\beta\frac{\boldsymbol{q}\cdot\hat{\boldsymbol{p}}}{2M}} \rangle_t. \tag{5.205}$$

This expression indeed recovers (5.185) in the infinite temperature limit $\beta \to 0$, and when (5.203) holds can be approximated as

$$\frac{\mathrm{d}}{\mathrm{d}t}\langle \frac{\hat{\boldsymbol{p}}^2}{2M} \rangle_t \approx \Gamma \int \mathrm{d}\boldsymbol{q}\mathcal{P}(\boldsymbol{q})\mathrm{e}^{-\beta\frac{q^2}{4M}} \langle \left(\frac{q^2}{2M} + \frac{\boldsymbol{q} \cdot \hat{\boldsymbol{p}}}{M} \right) \left(1 - \beta\frac{\boldsymbol{q} \cdot \hat{\boldsymbol{p}}}{2M} \right) \rangle_t. \tag{5.206}$$

Exploiting invariance under rotation of the momentum distribution $\mathcal{P}(\boldsymbol{q})$, that implies in particular

$$\int \mathrm{d}\boldsymbol{q}\mathcal{P}(\boldsymbol{q})\mathrm{e}^{-\beta\frac{q^2}{4M}}\boldsymbol{q} = 0, \tag{5.207}$$

as well as

$$\int \mathrm{d}\boldsymbol{q}\mathcal{P}(\boldsymbol{q})\mathrm{e}^{-\beta\frac{q^2}{4M}}q_i q_j = \delta_{ij}\frac{1}{3}\int \mathrm{d}\boldsymbol{q}\mathcal{P}(\boldsymbol{q})\mathrm{e}^{-\beta\frac{q^2}{4M}}q^2, \tag{5.208}$$

we are led to

$$\frac{\mathrm{d}}{\mathrm{d}t}\langle\frac{\hat{p}^2}{2M}\rangle_t = -\eta\left(\langle\frac{\hat{p}^2}{2M}\rangle_t - \frac{3}{2\beta}\right),$$

(5.209)

where we have defined

$$\eta = \beta\Gamma\frac{2}{3}\int \mathrm{d}q\,\mathcal{P}(q)\mathrm{e}^{-\beta\frac{q^2}{4M}}\frac{q^2}{M}.$$

(5.210)

The solution of (5.209) is then given by (5.204).

In the phenomenologically obtained master equation (5.201) a detailed connection to the microscopic interaction is required for the determination of the integration measure. For the case of a massive test particle interacting with a dilute gas of density n a microscopic approach leads to [38,53–56]

$$\frac{\mathrm{d}}{\mathrm{d}t}\rho_S(t) = -\frac{i}{\hbar}\left[\frac{\hat{p}^2}{2M}, \rho_S(t)\right] + \frac{2\pi}{\hbar}(2\pi\hbar)^3 n \int \mathrm{d}q\,|\tilde{V}(q)|^2$$
$$\times\left[U(q)\sqrt{S(q,\hat{p})}\rho_S(t)\sqrt{S(q,\hat{p})}U(q)^\dagger - \frac{1}{2}\left\{\rho_S(t), \sqrt{S(q,\hat{p})}\right\}\right],$$

(5.211)

where

$$S(q, p) = S(q, E(q, p))$$

(5.212)

is the dynamic structure factor introduced in Remark 5.9, that for a dilute gas of particles, so that we can assume Maxwell-Boltzmann statistics, takes the form [53]

$$S_{\mathrm{MB}}(q, p) = \sqrt{\frac{\beta m}{2\pi}}\frac{1}{|q|}\exp\left[-\frac{\beta}{8\,m}\frac{(2mE(q,p)+q^2)^2}{q^2}\right],$$

(5.213)

with m mass of the gas particles. In (5.201) the scattering cross section for collisions is described in the Born approximation, hence the appearance of the squared Fourier transform $|\tilde{V}(q)|^2$ of the interaction potential $V(x)$ between test and gas particles. The master equation (5.211) provide a quantum version of the classical linear Boltzmann equation, and was numerically studied in [57,58].

Starting from this expression we can consider the interesting limit in which the test particle is much more massive than the gas particles

$$\frac{m}{M} \ll 1, \tag{5.214}$$

so that small momentum transfers dominate, or equivalently small energy transfers, and we can expand the exponentials in (5.201) keeping terms up to second order in the position and momentum operators, coming after some algebra to the master equation

$$\frac{\mathrm{d}}{\mathrm{d}t}\rho_S(t) = -\frac{i}{\hbar}\left[\frac{\hat{\boldsymbol{p}}^2}{2M},\rho_S(t)\right] - \frac{D_{pp}}{\hbar^2}\sum_{i=1}^{3}[\hat{x}_i,[\hat{x}_i,\rho_S(t)]]$$

$$- \frac{i}{\hbar}\gamma\sum_{i=1}^{3}[\hat{x}_i,\{\hat{p}_i,\rho_S(t)\}] - \frac{D_{xx}}{\hbar^2}\sum_{i=1}^{3}[\hat{p}_i,[\hat{p}_i,\rho_S(t)]]. \tag{5.215}$$

The coefficients appearing in (5.215) take the expression

$$D_{pp} = \frac{2}{3}\frac{\pi^2 m^2}{\beta\hbar}\int \mathrm{d}\boldsymbol{q}\,|\tilde{V}(\boldsymbol{q})|^2|\boldsymbol{q}|\mathrm{e}^{-\beta\frac{q^2}{8m}}, \tag{5.216}$$

as well as

$$D_{xx} = \left(\frac{\beta\hbar}{4M}\right)^2 D_{pp}, \tag{5.217}$$

and

$$\gamma = \left(\frac{\beta}{2M}\right)D_{pp}. \tag{5.218}$$

The D_{pp} coefficient determines the strength of the decoherence effect in position, corresponding according to Remark 5.7 to momentum diffusion. The coefficient γ instead describes a friction effect, crucial for (5.201) to admit the operator (5.194) as stationary solution. This term appears together with another contribution with coefficient D_{xx}, that describes position diffusion, pointing to the fact that position and momentum diffusion appear together in the quantum mechanical description. In the high-temperature limit we recover the pure decoherence model of (5.180)

$$\frac{\mathrm{d}}{\mathrm{d}t}\rho_S(t) = -\frac{i}{\hbar}\left[\frac{\hat{\boldsymbol{p}}^2}{2M},\rho_S(t)\right] - \frac{D_{pp}}{\hbar^2}\sum_{i=1}^{3}[\hat{x}_i,[\hat{x}_i,\rho_S(t)]], \tag{5.219}$$

describing pure decoherence in position, with the identification

$$D \to D_{pp},\tag{5.220}$$

that is

$$\frac{\Gamma}{6}\langle q^2\rangle_{\mathcal{P}} \to \frac{2}{3}\frac{\pi^2 m^2}{\beta\hbar}\int \mathrm{d}\boldsymbol{q}\,|\tilde{V}(\boldsymbol{q})|^2|\boldsymbol{q}|\mathrm{e}^{-\beta\frac{q^2}{8m}}.\tag{5.221}$$

A master equation of the form (5.215) with $D_{xx}=0$, describing dissipation and decoherence in position, corresponding to momentum diffusion according to Remark 5.7, is known as Caldeira-Leggett master equation, since it was obtained considering the reduced dynamics of an harmonic oscillator interacting with a bosonic bath with an Ohmic spectral density [59], that is a spectral density growing linearly in frequency and without any cutoff. The notion of spectral density is discussed in Remark 5.14, within the study of the spin-boson-dephasing model of Sect. 5.4.1. A master equation of this form, that is with $D_{xx}=0$, is however not in Lindblad form, so that the obtained time evolution is not completely positive. More precisely, it is not even positive, so that states can be sent to non-positive operators [60]. Indeed, evaluating the Kossakowski matrix associated to the master equation (5.219) the condition for its positivity becomes [61,62]

$$D_{xx}D_{pp} - \frac{\hbar^2\gamma^2}{4} \geqslant 0,\tag{5.222}$$

that cannot be satisfied in the presence of friction if the momentum diffusion coefficient D_{xx} is equal to zero. It has to be noticed that in the present case complete positivity is warranted by the fact that (5.215) has been obtained by a limiting procedure from the Lindblad master equation (5.201). In particular, as follows from (5.217) and (5.218) the condition (5.222) is satisfied with the equal sign.

5.9 Dynamic Structure Factor

In Born approximation the energy dependent differential cross section describing scattering of a microscopic probe off a macroscopic sample as first obtained by van Hove in a seminal paper [63] can be expressed as

$$\frac{\mathrm{d}^2\sigma}{\mathrm{d}\Omega_{p'}\mathrm{d}E_{p'}} = (2\pi\hbar)^6\left(\frac{M}{2\pi\hbar^2}\right)^2\frac{p'}{p}|V(\boldsymbol{q})|^2\,S(\boldsymbol{q},E),\tag{5.223}$$

where $S(\boldsymbol{q},E)$ is the dynamic structure factor, p and p' are incoming and outgoing momenta of the scattered probe of mass M. This quantity is defined as the Fourier transform with respect to the energy and momentum transfer of the density-density autocorrelation function of the considered sample [64,65], namely

$$S(\boldsymbol{q}, E) = \frac{1}{2\pi\hbar} \int dt \int d\boldsymbol{x} \, e^{\frac{i}{\hbar}(Et - \boldsymbol{q} \cdot \boldsymbol{x})} G(\boldsymbol{x}, t) \tag{5.224}$$

where the time-dependent density-density autocorrelation function

$$G(\boldsymbol{x}, t) = \frac{1}{N} \int d\boldsymbol{y} \langle N(\boldsymbol{y}) N(\boldsymbol{x} + \boldsymbol{y}, t) \rangle \tag{5.225}$$

is expressed in terms of the particle density operator in the sample $N(\boldsymbol{x})$, with N the total number of particles in the sample. The dynamic structure factor thus contains information about the spectrum of spontaneous fluctuations of the system and it is of direct experimental access thanks to (5.223). If the sample is in a thermal state with inverse temperature β, or more precisely it obeys the β-KMS condition (6.69) discussed in Remark 6.2, the dynamic structure factor satisfies the detailed balance condition

$$S(\boldsymbol{q}, E) = e^{-\beta E} S(-\boldsymbol{q}, -E), \tag{5.226}$$

that is sufficient to warrant that the expression (5.194) is a stationary solution of (5.201).

In the considered physical model the dynamics was driven by collisions between massive particles. It is worth mentioning that the general idea of describing the interaction of a system with an environment as a sequence of *collisions*, generally interpreted as microscopic interaction events, can prove a convenient modelling to describe, provide insights and engineer system-environment models, see [66–70] for pioneering work in this direction and [71,72] for recent reviews.

5.4 Exactly Solvable Models

We will now consider microscopic interaction models that thanks to their simplicity and the presence of conservation laws allow for an analytic treatment. At least for a selection of environmental initial states, their reduced dynamics can be exactly evaluated and its dependence on features of the environment and of the coupling term is clearly spelled out. We will consider situations in which both decoherence and dissipative effects appear, with environments consisting of non-interacting collections of bosonic modes or spins. The knowledge of the time evolution will allow us to work out the dynamical map determining the reduced evolution, verify its complete positivity and obtain for it a Kraus representation. In particular, we will obtain the master equation associated to the exact dynamics and compare it with the generator of a quantum dynamical semigroup. This analysis will allow us to highlight different aspects of an open quantum system dynamics.

5.4.1 Spin-Boson Dephasing Model

We first consider a variant of the spin-boson model describing a pure decoherence effect induced in the system dynamics by the coupling with a collection of bosonic modes [73]. We will essentially follow [25] to obtain the exact solution, though alternative strategies can be considered, see e.g. [74]. This model, visualized in Fig. 5.4, will allow us to introduce a key quantity for the description of the linear coupling with a bosonic environment, namely the so-called spectral density. The Hamiltonian is of the form (5.5) with the identification

$$H_S = \frac{\hbar\omega_0}{2}\sigma_z, \tag{5.227}$$

so that the open system is a two-level system with energy gap $\hbar\omega_0$, while the Hamiltonian of the environment is given by

$$H_E = \sum_k \hbar\omega_k b_k^\dagger b_k, \tag{5.228}$$

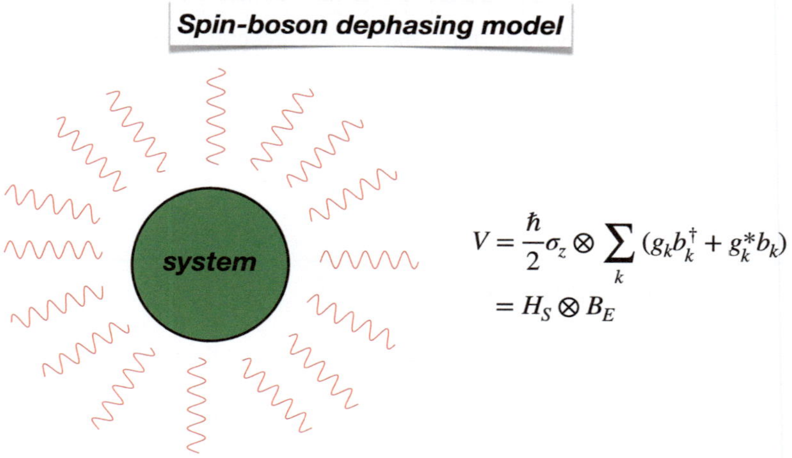

Fig. 5.4 Schematic representation of the spin-boson dephasing model, in which a two-level system is coupled to a collection of bosonic modes via an interaction term in which the system coupling operator commutes with the system Hamiltonian. The model is introduced in Sect. 5.4.1 and analyzed in Sect. 7.5.1.1 from the viewpoint of memory effects

where $b_k, b_k^\dagger$ denote respectively annihilation and creation operator of a mode labelled by the index k, for which we assume canonical commutation relations

$$[b_k, b_{k'}^\dagger] = \delta_{k,k'} \tag{5.229}$$

$$[b_k, b_{k'}] = 0 \tag{5.230}$$

$$[b_k^\dagger, b_{k'}^\dagger] = 0. \tag{5.231}$$

We do not make explicit here the range of the index k, since this discrete notation is used for the sake of simplicity to describe the interaction with a quantum field. To balance this shortcoming, we will consider a continuous limit for the relevant quantities at a later stage. The interaction term between the spin and the bosonic modes is taken to be of the form

$$V = \frac{\hbar}{2}\sigma_z \otimes X, \tag{5.232}$$

with

$$X = \sum_k (g_k b_k^\dagger + g_k^* b_k), \tag{5.233}$$

where g_k denotes the coupling strength to each mode, with the dimension of an inverse frequency, thus identifying a pure dephasing spin-boson model [73]. Indeed, the interaction term commutes with the free system Hamiltonian, which is therefore a constant of motion, implying time independence of the populations, that is the probabilities to be in the excited or ground state of the two-level system. Denoting with $|+\rangle$ and $|-\rangle$ the eigenvectors of the σ_z operator relative to the eigenvalues $+1$ and -1 respectively, and with ρ_S the reduced system operator defined in (4.1), the condition reads

$$\langle +|\rho_S(t)|+\rangle = \langle +|\rho_S(0)|+\rangle. \tag{5.234}$$

We note that the term dephasing is here used with the same significance as decoherence, though some authors use it only to denote loss of visibility or coherence not arising due to a quantum dynamical effect as in the present context, but rather due to a classical stochastic disturbance.

5.4.1.1 Unitary Evolution Operator

We work in the interaction picture with respect to the free evolution determined by $H_S + H_E$, so that exploiting

$$[\sigma_z, H_S] = 0, \tag{5.235}$$

together with

$$[b_k, H_E] = \hbar\omega_k b_k \tag{5.236}$$

and

$$[b_k^\dagger, H_E] = -\hbar\omega_k b_k^\dagger, \tag{5.237}$$

the interaction term becomes

$$V(t) = \frac{\hbar}{2}\sigma_z \otimes \sum_k (g_k b_k^\dagger e^{i\omega_k t} + g_k^* b_k e^{-i\omega_k t}). \tag{5.238}$$

The time evolution operator in interaction picture $\tilde{U}(t)$ is obtained as solution of the equation

$$\frac{\mathrm{d}}{\mathrm{d}t}\tilde{U}(t) = -\frac{i}{\hbar}V(t)\tilde{U}(t), \tag{5.239}$$

and reads

$$\tilde{U}(t) = \overleftarrow{T}\exp\left(-\frac{i}{\hbar}\int_0^t \mathrm{d}\tau\, V(\tau)\right), \tag{5.240}$$

where $\overleftarrow{T}$ denotes chronological time-ordering as defined in (5.85). Thanks to the fact that the commutator of the interaction operator $V(t)$ at different times is a $\mathbb{C}$-number, namely

$$[V(t), V(s)] = -\frac{i}{2}\sum_k \hbar^2 |g_k|^2 \sin[\omega_k(t-s)]\mathbb{1}_S, \tag{5.241}$$

as discussed in Remark 5.10 the time-ordering only contributes to an irrelevant phase and we can write

$$\tilde{U}(t) = e^{i\Phi(t)}U(t), \tag{5.242}$$

with

$$U(t) = \exp\left(-\frac{i}{\hbar}\int_0^t \mathrm{d}\tau\, V(\tau)\right) \tag{5.243}$$

and $\Phi(t)$ a time-dependent function. Given that

$$\int_0^t \mathrm{d}\tau\, V(\tau) = \sigma_z \otimes \left(\sum_k \hbar g_k e^{i\omega_k t/2}\frac{\sin(\omega_k t/2)}{\omega_k}b_k^\dagger + h.c.\right), \tag{5.244}$$

in view of the expression of the displacement operators considered in Remark 5.11 it is natural to introduce the time-dependent functions

$$\alpha_k(t) = -2ig_k e^{i\omega_k t/2}\frac{\sin(\omega_k t/2)}{\omega_k}, \tag{5.245}$$

thus coming to the expression

$$U(t) = \exp\left(\frac{1}{2}\sigma_z \otimes \sum_k [\alpha_k(t)b_k^\dagger - \alpha_k^*(t)b_k]\right). \tag{5.246}$$

5.10 Magnus Versus Dyson Expansion

We recall that given an operator equation of the form

$$\frac{\mathrm{d}}{\mathrm{d}t}U(t,t_0) = -\frac{i}{\hbar}H(t)U(t,t_0), \tag{5.247}$$

with initial condition $U(t_0,t_0) = \mathbb{1}$, we can consider two different perturbation expansions of the solution. The best-known result is due to Dyson [34] and relies on the introduction of a time-ordering operator, namely

$$U(t,t_0) = \overleftarrow{T}\exp\left(-\frac{i}{\hbar}\int_{t_0}^t \mathrm{d}\tau\,H(\tau)\right), \tag{5.248}$$

which can be expanded in the form

$$U(t,t_0) = \mathbb{1} + \sum_{n=1}^{\infty} S_n(t,t_0) \tag{5.249}$$

with

$$S_n(t,t_0) = \left(-\frac{i}{\hbar}\right)^n \frac{1}{n!}\int_{t_0}^t \mathrm{d}\tau_1\ldots\int_{t_0}^t \mathrm{d}\tau_n\,\overleftarrow{T}\,[H(\tau_1)\ldots H(\tau_n)], \tag{5.250}$$

according to the time-ordering introduced in (5.85). Another strategy was developed by Magnus [75] and builds on a representation of the solution in the form

$$U(t,t_0) = \exp\left(\sum_{n=1}^{\infty}\Omega_n(t,t_0)\right), \tag{5.251}$$

where the operators $\Omega_n(t, t_0)$ can be recursively determined and are given by time integrals of nested commutators. The first terms read

$$\Omega_1(t, t_0) = -\frac{i}{\hbar} \int_{t_0}^t \mathrm{d}\tau\, H(\tau) \tag{5.252}$$

$$\Omega_2(t, t_0) = \left(-\frac{i}{\hbar}\right)^2 \frac{1}{2!} \int_{t_0}^t \mathrm{d}\tau_1 \int_{t_0}^t \mathrm{d}\tau_2 [H(\tau_1), H(\tau_2)] \tag{5.253}$$

$$\Omega_3(t, t_0) = \left(-\frac{i}{\hbar}\right)^3 \frac{1}{3!} \int_{t_0}^t \mathrm{d}\tau_1 \int_{t_0}^t \mathrm{d}\tau_2 \int_{t_0}^t \mathrm{d}\tau_3 ([H(\tau_1), [H(\tau_2), H(\tau_3)]]$$
$$+ [H(\tau_3), [H(\tau_2), H(\tau_1)]]). \tag{5.254}$$

This expansion is related to the Baker-Campbell-Hausdorff formula introduced in Remark 2.12. Unitarity is here warranted at any order in the expansion, though the series is convergent only provided $\int_{t_0}^t \mathrm{d}\tau \|H(\tau)\| < \pi\hbar$. If the commutator of the operators $H(t)$ at different times is a $\mathbb{C}$-number, that is

$$[H(\tau_1), H(\tau_2)] = f(\tau_1, \tau_2)\mathbb{1}, \tag{5.255}$$

we have that $\Omega_n(t, t_0) = 0$ for $n \geqslant 3$ and we are left with

$$U(t, t_0) = \exp\left(-\frac{1}{2\hbar^2} \int_{t_0}^t \mathrm{d}\tau_1 \int_{t_0}^t \mathrm{d}\tau_2\, f(\tau_1, \tau_2)\right) \exp\left(-\frac{i}{\hbar} \int_{t_0}^t \mathrm{d}\tau\, H(\tau)\right).$$
$$\tag{5.256}$$

In order to gain knowledge about the operator $U(t)$ we evaluate its action on the states $|+\rangle \otimes |\psi\rangle$ and $|-\rangle \otimes |\psi\rangle$, with ψ an arbitrary state in $\mathcal{H}_E$ and $\{|+\rangle, |-\rangle\}$ the basis in $\mathbb{C}^2$ of eigenvectors of σ_z. The operator σ_z acts in a multiplicative way on its eigenvectors and we are left with

$$U(t)(|\pm\rangle \otimes |\psi\rangle) = \exp\left(\pm\frac{1}{2}\sum_k [\alpha_k(t)b_k^\dagger - \alpha_k^*(t)b_k]\right) |\pm\rangle \otimes |\psi\rangle. \tag{5.257}$$

We can now identify in (5.257) the expression of the displacement operator for the different modes, as introduced in (5.260) of Remark 5.11, so that we finally obtain

$$U(t)(|\pm\rangle \otimes |\psi\rangle) = |\pm\rangle \otimes \bigotimes_k D_k\left(\pm\frac{\alpha_k(t)}{2}\right) |\psi\rangle. \tag{5.258}$$

5.11 Displacement Operators

We here introduce the so-called displacement operators, widely used especially in quantum optics [76]. Consider the pair of bosonic annihilation and creation operators b and $b^\dagger$ obeying the canonical commutation relation

$$[b, b^\dagger] = \mathbb{1}. \tag{5.259}$$

For any $\alpha \in \mathbb{C}$ the operator

$$D(\alpha) = e^{\alpha b^\dagger - \alpha^* b} \tag{5.260}$$

is unitary, as can be verified using the Baker-Campbell-Hausdorff identity (2.387) that implies

$$D(\alpha) = e^{-\frac{1}{2}|\alpha|^2} e^{\alpha b^\dagger} e^{-\alpha^* b} \tag{5.261}$$

together with

$$D(\alpha) = e^{+\frac{1}{2}|\alpha|^2} e^{-\alpha^* b} e^{\alpha b^\dagger}. \tag{5.262}$$

In particular

$$D(\alpha)^\dagger = D(\alpha)^{-1} = D(-\alpha). \tag{5.263}$$

Such a unitary operator is called displacement operator with parameter α. The name comes from its action on the vacuum state $|0\rangle$ annihilated by b. Exploiting (5.261) we have

$$D(\alpha)|0\rangle = e^{-\frac{1}{2}|\alpha|^2} e^{\alpha b^\dagger} |0\rangle \tag{5.264}$$

$$= e^{-\frac{1}{2}|\alpha|^2} \sum_{n=0}^{\infty} \frac{\alpha^n}{n!} (b^\dagger)^n |0\rangle \tag{5.265}$$

$$= e^{-\frac{1}{2}|\alpha|^2} \sum_{n=0}^{\infty} \frac{\alpha^n}{\sqrt{n!}} |n\rangle, \tag{5.266}$$

so that

$$D(\alpha)|0\rangle = |\alpha\rangle, \tag{5.267}$$

where $|\alpha\rangle$ denotes the coherent state with parameter α, obtained by displacement of the vacuum. We have actually already encountered coherent states and displacement operators in Sect. 2.3.5.2. Indeed, displacement operators can be identified, for a suitable correspondence of the parameters, with the Weyl operators introduced in (2.376). We have in particular

$$D(\alpha) = W\left(\sqrt{2}\sigma\,\Re\alpha,\ \sqrt{2}\frac{\hbar}{\sigma}\,\Im\alpha\right), \tag{5.268}$$

where σ is the width of the Gaussian vacuum state $|0\rangle$. At the same time we have

$$|\alpha\rangle = |\psi_{x_0,p_0}\rangle, \tag{5.269}$$

where ψ_{x_0,p_0} denotes a Gaussian wavepacket of width σ and centred in $x_0 = \sqrt{2}\sigma\,\Re\alpha$ and $p_0 = \sqrt{2}(\hbar/\sigma)\,\Im\alpha$, according to the notation of (2.402). In this respect, coherent states are often termed the most classical states, in that they saturate the Heisenberg uncertainty relation. Coherent states are characterized by the property

$$b|\alpha\rangle = \alpha|\alpha\rangle, \tag{5.270}$$

and form a so-called overcomplete set since they generate the whole Hilbert space but are not orthogonal for different values of the complex number α. We have in particular

$$|\langle\alpha|\beta\rangle|^2 = e^{-|\alpha-\beta|^2}. \tag{5.271}$$

An important property of displacement operators relevant for the present treatment is their additive composition law

$$D(\alpha)D(\beta) = D(\alpha+\beta)e^{i\Im(\alpha\beta^*)}. \tag{5.272}$$

5.4.1.2 Quantum Dynamical Map

Thanks to knowledge of the unitary evolution operator we can obtain the expression of the reduced system state as a function of time. Due to conservation of the system energy leading to (5.234) we are only interested in the coherences or off-diagonal matrix elements. We therefore evaluate $\langle+|\rho_S(t)|-\rangle$ that, assuming an initially factorized state between system and environment, according to (5.9) and (5.10) reads

$$\langle +|\rho_S(t)|-\rangle = \langle +|\,\mathrm{Tr}_E\,[U(t)\rho_S(0) \otimes \rho_E U(t)^\dagger]|-\rangle \tag{5.273}$$

$$= \mathrm{Tr}_E\left\{\bigotimes_k D_k(\alpha_k(t))\rho_E\right\}\langle +|\rho_S(0)|-\rangle, \tag{5.274}$$

where we have used (5.258) and (5.272). According to standard usage it is convenient to write

$$\langle +|\rho_S(t)|-\rangle = e^{-\Gamma(t)}\langle +|\rho_S(0)|-\rangle, \tag{5.275}$$

upon defining

$$\Gamma(t) = -\log \mathrm{Tr}_E\left\{\bigotimes_k D_k(\alpha_k(t))\rho_E\right\}. \tag{5.276}$$

The effect on the system of the interaction with the environment is therefore a modification of its coherences by a time-dependent multiplicative factor, hence a decoherence effect. We now consider the explicit expression of the function $\Gamma(t)$, often called decoherence function, for the case of an environmental state of general interest. We assume a thermal bath of independent bosonic modes at inverse temperature $\beta = 1/(k_B T)$, namely

$$\rho_E = \frac{e^{-\beta H_E}}{Z_E}, \tag{5.277}$$

with H_E as in (5.228). This state is in factorized form

$$\rho_E = \bigotimes_k \rho_k^{\mathrm{th}}, \tag{5.278}$$

with

$$\rho^{\mathrm{th}} = \frac{e^{-\beta\hbar\omega b^\dagger b}}{\mathrm{Tr}\{e^{-\beta\hbar\omega b^\dagger b}\}}, \tag{5.279}$$

and the subscript refers to the considered mode, so that we are left with

$$\mathrm{Tr}_E\left\{\bigotimes_k D_k(\alpha_k(t))\rho_E\right\} = \prod_k \langle D_k(\alpha_k(t))\rangle_{\rho_k^{\mathrm{th}}}. \tag{5.280}$$

This relation actually provides the motivation for the choice of the logarithm in (5.276). We now exploit (5.309) proven in Remark 5.12, leading for the decoherence factor to

$$\Gamma(t) = \sum_k \frac{|\alpha_k(t)|^2}{2} \coth\left(\frac{\beta}{2}\hbar\omega_k\right),$$
(5.281)

and finally inserting the expression (5.245)

$$\Gamma(t) = \sum_k |g_k|^2 \frac{1 - \cos(\omega_k t)}{\omega_k^2} \coth\left(\frac{\beta}{2}\hbar\omega_k\right).$$
(5.282)

Note that according to this exact expression at short times the decay of the coherences is quadratic in time. Indeed, we have

$$\Gamma(t) \underset{t\to 0}{\approx} \frac{t^2}{2} \sum_k |g_k|^2 \coth\left(\frac{\beta}{2}\hbar\omega_k\right),$$
(5.283)

and therefore

$$\langle +|\rho_S(t)|-\rangle \approx \left[1 - \frac{t^2}{2} \sum_k |g_k|^2 \coth\left(\frac{\beta}{2}\hbar\omega_k\right)\right] \langle +|\rho_S(0)|-\rangle.$$
(5.284)

5.12 Expectations Over Thermal State

In order to obtain (5.281) we first prove the identity

$$\langle b^{\dagger q} b^q \rangle_{\text{th}} = q!(\langle b^\dagger b \rangle_{\text{th}})^q,$$
(5.285)

where $\langle \ldots \rangle_{\text{th}}$ denotes the expectation value with respect to the thermal state of (5.279). To this aim we write the expectation value of a generic observable as

$$\langle A \rangle_{\text{th}} = \sum_{n=0}^{\infty} p_n \langle n|A|n\rangle,$$
(5.286)

where $|n\rangle$ denotes an eigenvector of the number operator $b^\dagger b$, while the weights take the expression

$$p_n = (1 - w)w^n,$$
(5.287)

with

$$w = e^{-\beta\hbar\omega}.$$

(5.288)

In particular, for the number operator we recover the Bose-Einstein distribution

$$\langle b^\dagger b\rangle_{\text{th}} = (1-w)\sum_{n=0}^{\infty} nw^n$$

(5.289)

$$= \frac{w}{1-w}$$

(5.290)

$$= \frac{1}{e^{\beta\hbar\omega}-1}.$$

(5.291)

We now consider the identity

$$b^{\dagger q}b^q = b^\dagger b(b^\dagger b - 1)\dots(b^\dagger b - q + 1),$$

(5.292)

that is proven by induction, and insert it in (5.286) thus obtaining

$$\langle b^{\dagger q}b^q\rangle_{\text{th}} = (1-w)\sum_{n\geqslant q}^{\infty}\frac{n!}{(n-q)!}w^n$$

(5.293)

$$= (1-w)\sum_{n=0}^{\infty}\frac{(n+q)!}{n!}w^{n+q}$$

(5.294)

$$= q!w^q(1-w)\sum_{n=0}^{\infty}\frac{(n+q)\dots(1+q)}{n!}w^n$$

(5.295)

$$= q!\frac{w^q}{(1-w)^q},$$

(5.296)

where in the last equality we have used the formula

$$(1-x)^a = \sum_{n=0}^{\infty}\frac{a(a-1)\dots(a-n+1)}{n!}(-)^n x^n$$

(5.297)

with $a \to -(1+q)$. Comparing (5.290) and (5.296) we obtain (5.285).

We finally consider an operator given by a complex linear combination of creation and annihilation operators

$$A = \mu b + \nu b^\dagger.$$

(5.298)

Exploiting the Baker-Campbell-Hausdorff formula (2.387) we have

$$e^A = e^{\frac{1}{2}\mu\nu} e^{\nu b^\dagger} e^{\mu b} \tag{5.299}$$

and therefore

$$\langle e^A \rangle_{\text{th}} = e^{\frac{1}{2}\mu\nu} \langle e^{\nu b^\dagger} e^{\mu b} \rangle_{\text{th}} \tag{5.300}$$

$$= e^{\frac{1}{2}\mu\nu} \sum_{n,m=0}^{\infty} \frac{\nu^m \mu^n}{m!n!} \langle b^{\dagger\,m} b^n \rangle_{\text{th}} \tag{5.301}$$

$$= e^{\frac{1}{2}\mu\nu} \sum_{m=0}^{\infty} \frac{(\mu\nu)^m}{(m!)^2} \langle b^{\dagger\,m} b^m \rangle_{\text{th}} \tag{5.302}$$

so that exploiting (5.285) we have

$$\langle e^A \rangle_{\text{th}} = e^{\frac{1}{2}\mu\nu \langle \{b, b^\dagger\} \rangle_{\text{th}}}. \tag{5.303}$$

Relying on (5.291) we further have the identity

$$\langle \{b, b^\dagger\} \rangle_{\text{th}} = \coth\left(\frac{\beta}{2}\hbar\omega \right), \tag{5.304}$$

finally leading to

$$\langle e^A \rangle_{\text{th}} = e^{\frac{1}{2}\mu\nu \coth\left(\frac{\beta}{2}\hbar\omega\right)}. \tag{5.305}$$

We further notice that

$$\langle A^2 \rangle_{\text{th}} = \mu\nu \langle \{b, b^\dagger\} \rangle_{\text{th}}, \tag{5.306}$$

and therefore (5.305) also reads

$$\langle e^A \rangle_{\text{th}} = e^{\frac{1}{2}\langle A^2 \rangle_{\text{th}}}. \tag{5.307}$$

This result corresponds to the identity

$$\mathbb{E}[e^Y] = \exp\left(\frac{1}{2}\mathbb{E}[Y^2] \right) \tag{5.308}$$

valid for a classical random variable Y with Gaussian distribution and zero mean. Indeed, the state ρ^{th} is a Gaussian state in the sense that the associated position and momentum distributions are Gaussian. In particular, from equation (5.305) upon the substitution $\nu \to \alpha$ and $\mu \to -\alpha^*$ we recover for (5.260) the expression

$$\langle D(\alpha) \rangle_{\rho^{\mathrm{th}}} = \exp\left(-\frac{|\alpha|^2}{2}\coth\left(\frac{\beta}{2}\hbar\omega\right)\right), \tag{5.309}$$

directly leading to (5.281). From this identity we see that the expectation value of the displacement operator is a characteristic function in the language of probability theory, see also Remark 6.1.

To properly evaluate the decoherence function (5.282) we need an estimate of the relevant frequencies and of the strength of the coupling. To this aim we consider instead of a sum an integral over the modes by means of the formal replacement $\sum_k \to \int \mathrm{d}\omega\rho(\omega)$, with $\rho(\omega)$ a suitable density of modes. We further describe the relevance of the coupling to the different modes with a function $g_k \to g(\omega)$. These informations, making reference to both the environment and its coupling to the system, are encoded in a single non-negative function known as spectral density

$$J(\omega) = \rho(\omega)|g(\omega)|^2, \tag{5.310}$$

which characterizes the dynamics of a system interacting with a bosonic bath with a coupling term linear in the creation and annihilation operators as in (5.238). This treatment essentially recovers the result that would be obtained following a proper field treatment of the bosonic environment [77,78]. An example of spectral density together with its connection with the bath correlation function will be further discussed in Remark 5.14. We thus introduce the expression

$$\Gamma(t) = \int_0^\infty \mathrm{d}\omega\, J(\omega)\frac{1-\cos(\omega t)}{\omega^2}\coth\left(\frac{\beta}{2}\hbar\omega\right), \tag{5.311}$$

corresponding to (5.282) when considering the formal replacement

$$J(\omega) = \sum_k |g_k|^2 \delta(\omega - \omega_k). \tag{5.312}$$

According to (5.275) the function $\Gamma(t)$ describes how the system looses coherence as a result of the interaction with the environment. The connection between the function $\Gamma(t)$ and the bath correlation function is discussed in Remark 5.13. Note that as a general feature the function $\Gamma(t)$ is positive, so that coherences are suppressed in time according to the factor $\exp[-\Gamma(t)]$. Knowledge of the decoherence function fixes the expression of the dynamical map (5.9). Indeed, expressing the matrix elements of the initial statistical operator in the basis of eigenvectors of the σ_z operator

$$\rho_S(0) = \begin{pmatrix} \rho_{++}(0) & \rho_{+-}(0) \\ \rho_{-+}(0) & \rho_{--}(0) \end{pmatrix},$$
(5.313)

we have

$$\rho_S(t) = \begin{pmatrix} \rho_{++}(0) & \rho_{+-}(0)\mathfrak{E}(t) \\ \rho_{-+}(0)\mathfrak{E}(t) & \rho_{--}(0) \end{pmatrix},$$
(5.314)

with

$$\mathfrak{E}(t) = e^{-\Gamma(t)}$$
(5.315)

the function describing the effect of decoherence and identifying the map.

5.13 Bath Correlation Function

We here explicitly obtain the expression of the correlation function of the environmental operator appearing in the coupling term (5.232), in order to stress its role in determining the reduced dynamics. We define

$$C(t) = \langle X(t)X(0)\rangle_{\rho_E},$$
(5.316)

where $\langle\ldots\rangle_{\rho_E}$ denotes the expectation value with respect to the initial environmental state ρ_E, and $X(t)$ is the operator in interaction picture. Starting from the expression (5.233) we obtain, assuming $\mathrm{Tr}_E\{b_k\rho_E\} = 0$

$$C(t) = \sum_k |g_k|^2 (\langle b_k^\dagger b_k\rangle_{\rho_E} e^{i\omega_k t} + \langle b_k b_k^\dagger\rangle_{\rho_E} e^{-i\omega_k t}).$$
(5.317)

Note that the same correlation function would be relevant for a more general coupling of the form

$$V = \sum_k \hbar g_k A_S \otimes b_k^\dagger + \text{h.c.}$$
(5.318)

We can now use a standard rewriting of the product of two operators as a sum of their commutator and anti-commutator

$$AB = \frac{1}{2}\{A, B\} + \frac{1}{2}[A, B],$$
(5.319)

leading to

$$
C(t) = \frac{1}{2} \sum_k |g_k|^2 [\mathrm{e}^{i\omega_k t} \langle \{b_k, b_k^\dagger\} - [b_k, b_k^\dagger] \rangle_{\rho_E}
$$

$$
+ \mathrm{e}^{-i\omega_k t} \langle \{b_k, b_k^\dagger\} + [b_k, b_k^\dagger] \rangle_{\rho_E} \Big] \tag{5.320}
$$

$$
= \sum_k |g_k|^2 [\langle \{b_k, b_k^\dagger\} \rangle_{\rho_E} \cos(\omega_k t) - i \langle [b_k, b_k^\dagger] \rangle_{\rho_E} \sin(\omega_k t)] \tag{5.321}
$$

$$
= \sum_k |g_k|^2 [\langle \{b_k, b_k^\dagger\} \rangle_{\rho_E} \cos(\omega_k t) - i \sin(\omega_k t)]. \tag{5.322}
$$

For the case of a thermal state, exploiting (5.304) we come to

$$
C(t) = \sum_k |g_k|^2 \left(\coth\left(\frac{\beta}{2} \hbar \omega_k \right) \cos(\omega_k t) - i \sin(\omega_k t) \right). \tag{5.323}
$$

Making reference to the formal expression for the spectral density (5.312) we have in particular

$$
C(t) = \int_0^\infty \mathrm{d}\omega J(\omega) \left(\coth\left(\frac{\beta}{2} \hbar \omega \right) \cos(\omega t) - i \sin(\omega t) \right). \tag{5.324}
$$

Upon time integration of the correlation function we obtain

$$
\int_0^t \mathrm{d}\tau\, C(\tau) = \int_0^\infty \mathrm{d}\omega J(\omega) \left(\coth\left(\frac{\beta}{2} \hbar \omega \right) \frac{\sin(\omega t)}{\omega} + i \frac{\cos(\omega t) - 1}{\omega} \right) \tag{5.325}
$$

as well as

$$
\int_0^t \mathrm{d}\sigma \int_0^\sigma \mathrm{d}\tau\, C(\tau) = \int_0^\infty \mathrm{d}\omega J(\omega)
$$

$$
\times \left(\coth\left(\frac{\beta}{2} \hbar \omega \right) \frac{1 - \cos(\omega t)}{\omega^2} + i \frac{\sin(\omega t) - \omega t}{\omega^2} \right), \tag{5.326}
$$

so that we have the following relationship between decoherence function and environmental correlation function

$$
\Gamma(t) = \Re \int_0^t \mathrm{d}\sigma \int_0^\sigma \mathrm{d}\tau\, C(\tau), \tag{5.327}
$$

which provides an instance of fluctuation-dissipation relationship [65]. This expression corresponds to (5.311) and in the high-temperature case $\hbar\omega \ll k_B T$ takes the form

$$\Gamma(t) = \frac{2k_B T}{\hbar} \int_0^\infty d\omega J(\omega) \frac{1 - \cos(\omega t)}{\omega^3}. \tag{5.328}$$

The decoherence rate appearing in the master equation in terms of the bath correlation function then reads

$$\dot{\Gamma}(t) = \Re \int_0^t d\tau\, C(\tau), \tag{5.329}$$

which using the spectral density can be explicitly written as

$$\dot{\Gamma}(t) = \int_0^{+\infty} d\omega J(\omega) \coth\left(\frac{\beta\hbar\omega}{2}\right) \frac{\sin(\omega t)}{\omega}. \tag{5.330}$$

5.4.1.3 Kraus Decomposition and Master Equation

We now want to express the map identified by (5.314) in operator form, so as to check in particular its complete positivity. To this aim it is convenient to express it in the basis of maps introduced in (3.394), so that we have

$$\Phi(t)\rho_S(0) = \sum_{\alpha,\beta=1}^{4} \Phi'_{\alpha\beta}(t) F_{\alpha\beta}(\rho_S(0)). \tag{5.331}$$

The matrix $\Phi'_{\alpha\beta}$ is defined in (3.416), and its evaluation is particularly simple in the canonical operator basis defined in Sect. 3.3.2.1. Indeed, as discussed in Sect. 4.4.1 for this choice of basis it coincides with the Choi matrix. According to (4.93) we thus have

$$\Phi(t)\rho_S(0) = \sum_{i,j,k,l=1}^{2} \Phi'_{ij,kl}(t) F_{ij,kl}(\rho_S(0)) \tag{5.332}$$

$$= \sum_{i,j,k,l=1}^{2} (\mathcal{C}_{\Phi(t)})_{ji,lk} F_{ij,kl}(\rho_S(0)), \tag{5.333}$$

with

$$F_{ij,kl}(\rho_S(0)) = |i\rangle\langle j|\rho_S(0)|l\rangle\langle k|. \tag{5.334}$$

The canonical operator basis depends in turn on a choice of basis in the Hilbert space of the system, which we take again to be the set $\{|+\rangle, |-\rangle\}$ of eigenvectors of σ_z. From the very definition

$$\mathcal{C}_{\Phi(t)} = \begin{pmatrix} \Phi(t)[|+\rangle\langle+|] & \Phi(t)[|+\rangle\langle-|] \\ \Phi(t)[|-\rangle\langle+|] & \Phi(t)[|-\rangle\langle-|] \end{pmatrix}, \tag{5.335}$$

using (5.314) we come to

$$\mathcal{C}_{\Phi(t)} = \begin{pmatrix} 1 & 0 & 0 & \mathfrak{E}(t) \\ 0 & 0 & 0 & 0 \\ 0 & 0 & 0 & 0 \\ \mathfrak{E}(t) & 0 & 0 & 1 \end{pmatrix}. \tag{5.336}$$

The eigenvalues are determined to be $\{0, 0, 1 - \mathfrak{E}(t), 1 + \mathfrak{E}(t)\}$, so that recalling (5.315) they are indeed non-negative provided the decoherence function $\Gamma(t)$ is positive at all times, as generally true thanks to the form of its expression (5.311). According to (4.91) their sum is a constant equal to 2, namely the dimension of the space, since $\Phi(t)$ is trace-preserving at any time. Starting from (5.333) we can obtain an explicit Kraus representation for the map $\Phi(t)$ relying on the strategy described in Remark 3.21, writing

$$\mathcal{C}_{\Phi(t)} = U D U^\dagger, \tag{5.337}$$

with $D = \mathrm{diag}(0, 0, 1 - \mathfrak{E}(t), 1 + \mathfrak{E}(t))$ and U the unitary matrix obtained with the corresponding eigenvectors, which turn out to be time independent, namely

$$U = \begin{pmatrix} 0 & 0 & \frac{1}{\sqrt{2}} & \frac{1}{\sqrt{2}} \\ 0 & 1 & 0 & 0 \\ 1 & 0 & 0 & 0 \\ 0 & 0 & -\frac{1}{\sqrt{2}} & \frac{1}{\sqrt{2}} \end{pmatrix}. \tag{5.338}$$

Making reference to (3.433) we thus obtain

$$\Phi(t)\rho_S(0) = p_-(t)\sigma_z\rho_S(0)\sigma_z + p_+(t)\rho_S(0), \tag{5.339}$$

where we have introduced the positive numbers summing up to one

$$p_\pm(t) = \frac{1 \pm \mathfrak{E}(t)}{2}. \tag{5.340}$$

A map admitting such a Kraus representation is known as phase damping channel [79]. Noting that given a 2×2 matrix

$$M = \begin{pmatrix} a & b \\ c & d \end{pmatrix} \tag{5.341}$$

we have

$$\sigma_z M \sigma_z = \begin{pmatrix} a & -b \\ -c & d \end{pmatrix}, \tag{5.342}$$

it can be described saying that at time t with probability $p_+(t)$ the state has been left unchanged, and with the complementary probability the coherences are modified by a minus sign. It therefore corresponds to a random mixture of unitary transformations

$$\Phi(t)\rho_S(0) = \sum_{i=+,-} p_i(t) U_i \rho_S(0) U_i^\dagger, \tag{5.343}$$

with

$$U_+ = \mathbb{1}_{\mathbb{C}^2}, \qquad U_- = \sigma_z. \tag{5.344}$$

The Kraus representation, as discussed in Sect. 3.2.4, is not unique. Indeed, for the case at hand we can write

$$\Phi(t)\rho_S(0) = \sum_i K_i(t)\rho_S(0) K_i(t)^\dagger, \tag{5.345}$$

both with the set of Kraus operators

$$\{K_1(t), K_2(t)\} \rightarrow \left\{ \sqrt{\frac{1+\mathcal{E}(t)}{2}} \mathbb{1}_{\mathbb{C}^2}, \sqrt{\frac{1-\mathcal{E}(t)}{2}} \sigma_z \right\}, \tag{5.346}$$

corresponding to (5.339), with another set with the same cardinality

$$\{K_1(t), K_2(t)\} \rightarrow \left\{ \begin{pmatrix} 1 & 0 \\ 0 & \mathcal{E}(t) \end{pmatrix}, \begin{pmatrix} 0 & 0 \\ 0 & \sqrt{1-\mathcal{E}(t)^2} \end{pmatrix} \right\}, \tag{5.347}$$

as well as with an alternative set with a different cardinality

$$\{K_1(t), K_2(t), K_3(t)\} \rightarrow \left\{ \begin{pmatrix} \sqrt{1+\mathcal{E}(t)} & 0 \\ 0 & 0 \end{pmatrix}, \begin{pmatrix} 0 & 0 \\ 0 & \sqrt{1-\mathcal{E}(t)} \end{pmatrix}, \sqrt{\mathcal{E}(t)} \mathbb{1}_{\mathbb{C}^2} \right\}. \tag{5.348}$$

Starting from knowledge of the exact reduced dynamics we can obtain the master equation whose solution provides the evolved statistical operator of the system. The evolution is fixed by the equations

$$\frac{d}{dt}\langle+|\rho_S(t)|+\rangle = 0 \tag{5.349}$$

and

$$\frac{d}{dt}\langle+|\rho_S(t)|-\rangle = -\dot{\Gamma}(t)\langle+|\rho_S(t)|-\rangle, \tag{5.350}$$

following from energy conservation of the system and (5.275) respectively. Exploiting (5.342), given that $\dot{\Gamma}(t)$ is a real quantity, these equations can be recovered as matrix elements of the following operator equation

$$\frac{d}{dt}\rho_S(t) = -\frac{1}{2}\frac{\dot{\mathfrak{E}}(t)}{\mathfrak{E}(t)}[\sigma_z\rho_S(t)\sigma_z - \rho_S(t)], \tag{5.351}$$

which can be equivalently rewritten

$$\frac{d}{dt}\rho_S(t) = \frac{\dot{\Gamma}(t)}{2}[\sigma_z\rho_S(t)\sigma_z - \rho_S(t)], \tag{5.352}$$

or

$$\frac{d}{dt}\rho_S(t) = -\frac{\dot{\Gamma}(t)}{4}[\sigma_z, [\sigma_z, \rho_S(t)]], \tag{5.353}$$

complying with a typical form of master equation describing decoherence, namely (5.176), already discussed in Sect. 5.7. The master equation (5.352) has the same operator structure as (5.27), with a single self-adjoint Lindblad operator σ_z, so that the anti-commutator term becomes trivial. Recovery of the same operator structure is not surprising, since it was already found in (4.127) of Sect. 4.4.2 as a consequence of the requirement of trace and Hermiticity preservation. We have however the crucial difference that the constant coefficients of (5.27) are here replaced by a generally time-dependent function $\dot{\Gamma}(t)/2$. The connection between the function $\dot{\Gamma}(t)$ and the bath correlation function is discussed in Remark 5.13. A plot of the time derivative of the decoherence function appearing as coefficient in the master equation (5.352), is considered in Fig. 5.8 for the case of the spectral density (5.361), both at high and at low temperature. An exact semigroup composition law is only obtained if $\Gamma(t)$ can be approximated by a linear function of t, say $\Gamma(t) = 2\gamma t$, with γ a positive constant leading to the GKSL expression

$$\frac{d}{dt}\rho_S(t) = \gamma[\sigma_z\rho_S(t)\sigma_z - \rho_S(t)], \tag{5.354}$$

as discussed in Remark 5.14.

5.14 Spectral Density

The spectral density provides the central quantity allowing to determine the dynamics of a quantum system, described by either discrete or continuous variables, linearly coupled to a bosonic bath. As we have seen in Sect. 5.13, it determines the correlation functions of the bath operators appearing in the interaction term according to (5.324). This relation can be written in a more convenient way introducing an effective spectral density that includes in its expression the temperature dependent factor (5.304) [80,81]. We set in particular

$$J_{\text{eff}}^{\beta}(\omega) = J(|\omega|)\frac{\text{sgn}(\omega)}{2}\left[1 + \coth\left(\frac{\beta\hbar\omega}{2}\right)\right], \tag{5.355}$$

where

$$\text{sgn}(\omega) = \begin{cases} +1 & \omega > 0 \\ 0 & \omega = 0 \\ -1 & \omega < 0 \end{cases}, \tag{5.356}$$

so that (5.324) can be expressed as a Fourier transform

$$C(t) = \int_{-\infty}^{+\infty} d\omega\, J_{\text{eff}}^{\beta}(\omega)e^{-i\omega t}, \tag{5.357}$$

leading for its real and imaginary part respectively to

$$\Re C(t) = \frac{1}{2}\int_{-\infty}^{+\infty} d\omega\, J(\omega)\coth\left(\frac{\beta\hbar\omega}{2}\right)\cos(\omega t) \tag{5.358}$$

and to

$$\Im C(t) = -\frac{1}{2}\int_{-\infty}^{+\infty} d\omega\, J(\omega)\sin(\omega t). \tag{5.359}$$

The spectral density thus fixes the bath correlation function which in this framework fully describes the dynamics of the reduced system and thus determines the model. As we shall see also in Sect. 6.2, the two-point correlation functions of the bath operators appearing in the coupling generally play a crucial role in determining the reduced dynamics.

In particular, in the spin-boson decoherence model the key quantity is the decoherence function (5.276), that can be expressed as

$$\Gamma(t) = \int_{-\infty}^{+\infty} d\omega\, J_{\text{eff}}^{\beta}(\omega) \frac{1 - \cos(\omega t)}{\omega^2}. \tag{5.360}$$

As concrete example of spectral density, which further allows for an analytic treatment, we consider the reference expression [82–84]

$$J_{\text{Garg}}(\omega) = \kappa \omega_0^2 \frac{\eta \omega}{(\omega^2 - \omega_0^2)^2 + \eta^2 \omega^2}, \tag{5.361}$$

first introduced in [85], where the constant κ gives the coupling strength, and η provides the width of the dominant band of frequencies for the interaction, centred around the value ω_0. This expression describes the situation in which a narrow band of modes dominates the interaction with the system. It appears when the interaction between the system and the bosonic bath is mediated by a so-called reaction coordinate [83,86–88]. We note that this spectral density can be written as the difference between two Lorentzian functions with the same width, namely

$$J_{\text{Garg}}(\omega) = \frac{\kappa \omega_0^2}{2\Omega} \left(\frac{\frac{\eta}{2}}{(\omega - \Omega)^2 + \left(\frac{\eta}{2}\right)^2} - \frac{\frac{\eta}{2}}{(\omega + \Omega)^2 + \left(\frac{\eta}{2}\right)^2} \right), \tag{5.362}$$

where we have introduced the frequency

$$\Omega = \sqrt{\omega_0^2 - \left(\frac{\eta}{2}\right)^2}, \tag{5.363}$$

and we consider the so-called underdamped limit in which $\eta/2 < \omega_0$. The function (5.361) has four poles in correspondence to the values

$$\omega = \pm \left(\Omega \pm i\frac{\eta}{2} \right), \tag{5.364}$$

and in the vicinity of ω_0 the spectral density is well approximated by a single Lorentzian function, that is

$$J_{\text{Garg}}(\omega) \approx J_{\text{Lorentz}}(\omega) \tag{5.365}$$

with

$$J_{\text{Lorentz}}(\omega) = \frac{\kappa\omega_0}{2} \frac{\frac{\eta}{2}}{(\omega - \omega_0)^2 + \left(\frac{\eta}{2}\right)^2}, \tag{5.366}$$

as shown in Fig. 5.5.

For the spectral density of (5.361) we can evaluate the real part of the correlation function starting from the expression (5.357) by contour integration. To this aim we consider the poles of $J_{\text{Garg}}(\omega)$, given in (5.364), together with the representation

$$\coth\left(\frac{\beta\hbar\omega}{2}\right) = \frac{2}{\beta\hbar\omega} + \frac{1}{\beta\hbar} \sum_{n=1}^{+\infty} \frac{\omega}{\omega^2 + v_n^2}, \tag{5.367}$$

with

$$v_n = \frac{2\pi}{\beta\hbar} n \tag{5.368}$$

the so-called Matsubara frequencies, which puts into evidence an infinity of poles in correspondence to the values

$$\omega = \pm i \frac{2\pi}{\beta\hbar} n. \tag{5.369}$$

The formula (5.367) follows from the following representation formula for the hyperbolic cotangent [89]

$$\coth(\pi x) = \frac{1}{\pi x} + \frac{2}{\pi} \sum_{n=1}^{+\infty} \frac{x}{x^2 + n^2}. \tag{5.370}$$

The result reads [90]

$$\begin{aligned}
\Re C_{\text{Garg}}(t) ={}& \frac{\kappa\pi}{2\Omega} \frac{\sinh(\beta\hbar\Omega)}{\cosh(\beta\hbar\Omega) - \cos\left(\beta\hbar\frac{\eta}{2}\right)} \cos(\Omega t) e^{-\frac{\eta t}{2}} \\
&+ \frac{\kappa\pi}{2\Omega} \frac{\sinh\left(\beta\hbar\frac{\eta}{2}\right)}{\cosh(\beta\hbar\Omega) - \cos\left(\beta\hbar\frac{\eta}{2}\right)} \sin(\Omega t) e^{-\frac{\eta t}{2}} \\
&- \eta \frac{\kappa\pi}{\beta\hbar} \sum_{n=1}^{\infty} \frac{v_n e^{-v_n t}}{(v_n^2 + \omega_0^2)^2 - \eta^2 v_n^2}.
\end{aligned} \tag{5.371}$$

The contributions in the last term, given by a summation over the Matsubara frequencies, become relevant at low temperatures, namely when

$$\frac{2\pi k_B T}{\hbar} \ll \frac{\eta}{2}. \tag{5.372}$$

In a similar way we obtain for the temperature independent imaginary part of the correlation function

$$\Im C_{\mathrm{Garg}}(t) = \frac{\kappa\pi}{2\Omega} \sin(\Omega t)\,\mathrm{e}^{-\frac{\eta t}{2}}. \tag{5.373}$$

The behavior in time of the correlation function is plotted in Fig. 5.6 for the sake of example, considering high and low temperature regimes. The expression of the decoherence function corresponding to (5.361) is obtained relying on (5.327) and takes the form

$$\Gamma_{\mathrm{Garg}}(t) = \frac{\kappa\pi}{2\omega_0^4\Omega}\, \frac{\sinh(\beta\hbar\Omega)}{\cosh(\beta\hbar\Omega) - \cos\left(\frac{\beta\hbar\eta}{2}\right)}$$

$$\times \left\{\mathrm{e}^{-\frac{\eta t}{2}}\left[\left(\frac{\eta^2}{4} - \Omega^2\right)\cos(\Omega t) - \eta\Omega\sin(\Omega t)\right] + \frac{\eta}{2}\omega_0^2 t - \left(\frac{\eta^2}{4} - \Omega^2\right)\right\}$$

$$+ \frac{\kappa\pi}{2\omega_0^4\Omega}\, \frac{\sinh\left(\frac{\beta\hbar\eta}{2}\right)}{\cosh(\beta\hbar\Omega) - \cos\left(\frac{\beta\hbar\eta}{2}\right)}$$

$$\times \left\{\mathrm{e}^{-\frac{\eta t}{2}}\left[\left(\frac{\eta^2}{4} - \Omega^2\right)\sin(\Omega t) + \eta\Omega\cos(\Omega t)\right] + \Omega\omega_0^2 t + \eta\Omega\right\}$$

$$- \eta\frac{\kappa\pi}{\beta\hbar}\sum_{n=1}^{\infty}\frac{1}{\nu_n}\frac{\mathrm{e}^{-\nu_n t} + \nu_n t - 1}{(\nu_n^2 + \omega_0^2)^2 - \eta^2\nu_n^2}. \tag{5.374}$$

This function describes the suppression of the coherences according to (5.275), while its time derivative determines the coefficient in the master equation of the model (5.352). We consider their behavior in Figs. 5.7 and 5.8 respectively, in both high and low temperature limit. We stress in particular that in the long time limit the decoherence function $\Gamma(t)$ becomes approximately linear in time, so that in this regime the exact master equation (5.352) is well approximated by the GKSL expression (5.354) describing a Markovian dynamics.

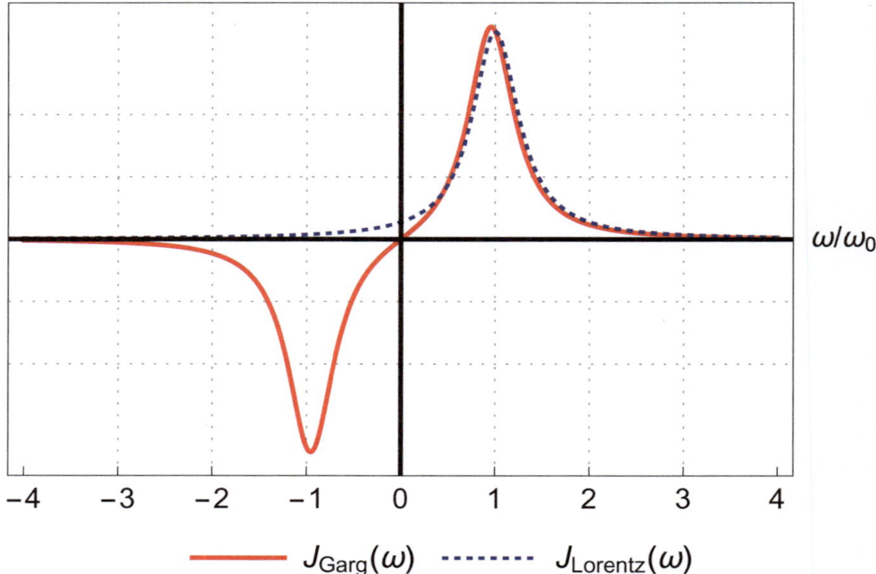

Fig. 5.5 Plot of the Garg spectral density $J_{\text{Garg}}(\omega)$ of (5.361), as well as of the Lorentzian function of (5.366) that well approximates it in the vicinity of the system frequency ω_0. The curves are plotted in the underdamped regime $\eta/\omega_0 = 0.6$

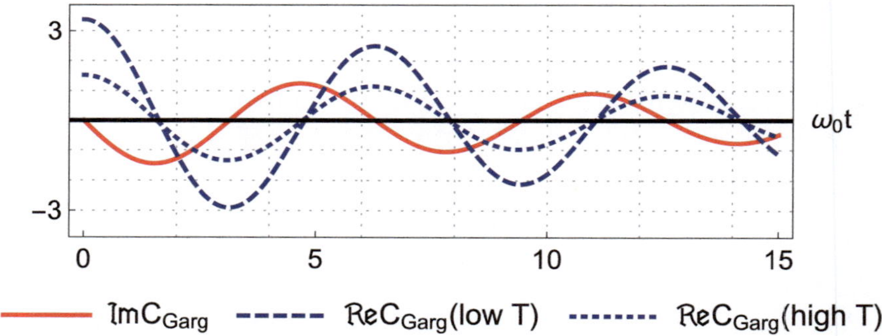

Fig. 5.6 Plot of $\Re C_{\text{Garg}}(t)$, $\Im C_{\text{Garg}}(t)$ for the spectral density of (5.361) in the underdamped regime $\eta/\omega_0 = 0.1$. The constant κ is set to one in arbitrary units. Both functions exhibit a decaying oscillating behavior. For the time dependent contribution $\Re C_{\text{Garg}}(t)$ we consider both high and low temperature (respectively $\beta\hbar\eta = 0.1$ and $\beta\hbar\eta = 10$), with larger oscillations at high temperature

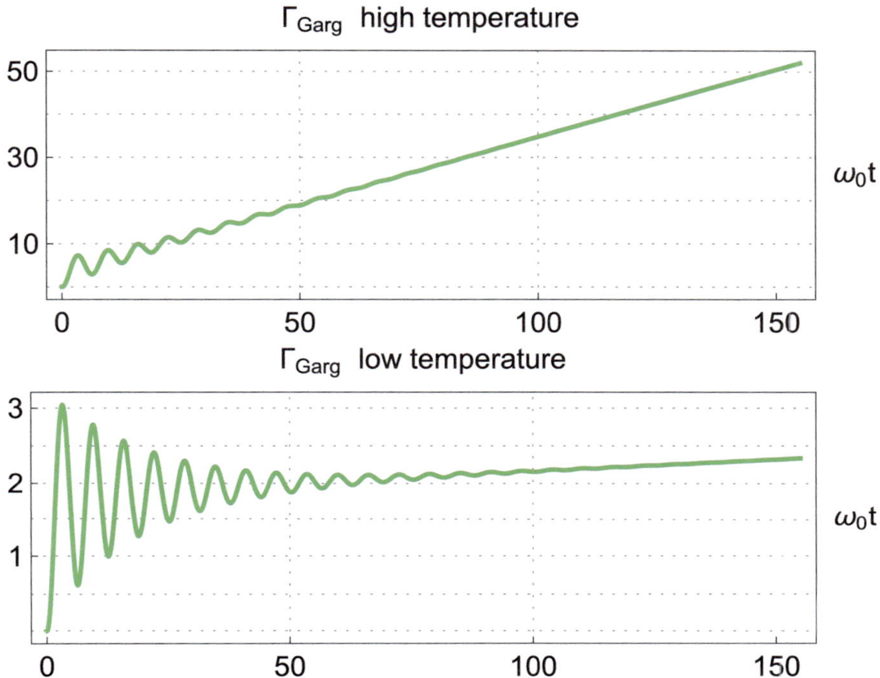

Fig. 5.7 Plot of the decoherence function $\Gamma_{\mathrm{Garg}}(t)$ for the spectral density of (5.361) in the under-damped regime, with the same choice of parameters as in Fig. 5.6. We consider the behavior in the short- and long-term region, for both high and low temperature. Note the approximately linear behavior for long times. At high temperatures the decoherence function $\Gamma_{\mathrm{Garg}}(t)$ gets larger, corresponding to a stronger decoherence effect

5.4.2 Hyperfine Couplings

The case of a central spin coupled to a spin bath, that we will consider here and in Sect. 5.4.3 in a setting that allows for a simplified analytic treatment, is of relevance in many physical systems. The description of noise and in particular decoherence effects plays a crucial role in spin-based quantum information processing systems, e.g. electron spins coupled via hyperfine interaction with an environment of nuclear spins, such as quantum dots or nitrogen vacancy center electron spins [91–93].

As visualized in Fig. 5.9 we now consider a two-level system interacting via a Heisenberg coupling with a collection of non-interacting spins, as considered for example in [94–96], but making reference to a situation in which an exact treatment can be put forward. The system Hamiltonian is still given by (5.227), while environment and couplings are described by

$$H_E + V = \sum_{k=1}^{N} \frac{\hbar \omega_k}{2} \sigma_z^k + \sum_{k=1}^{N} (J_x^k \sigma_x \otimes \sigma_x^k + J_y^k \sigma_y \otimes \sigma_y^k + J_z^k \sigma_z \otimes \sigma_z^k) \quad (5.375)$$

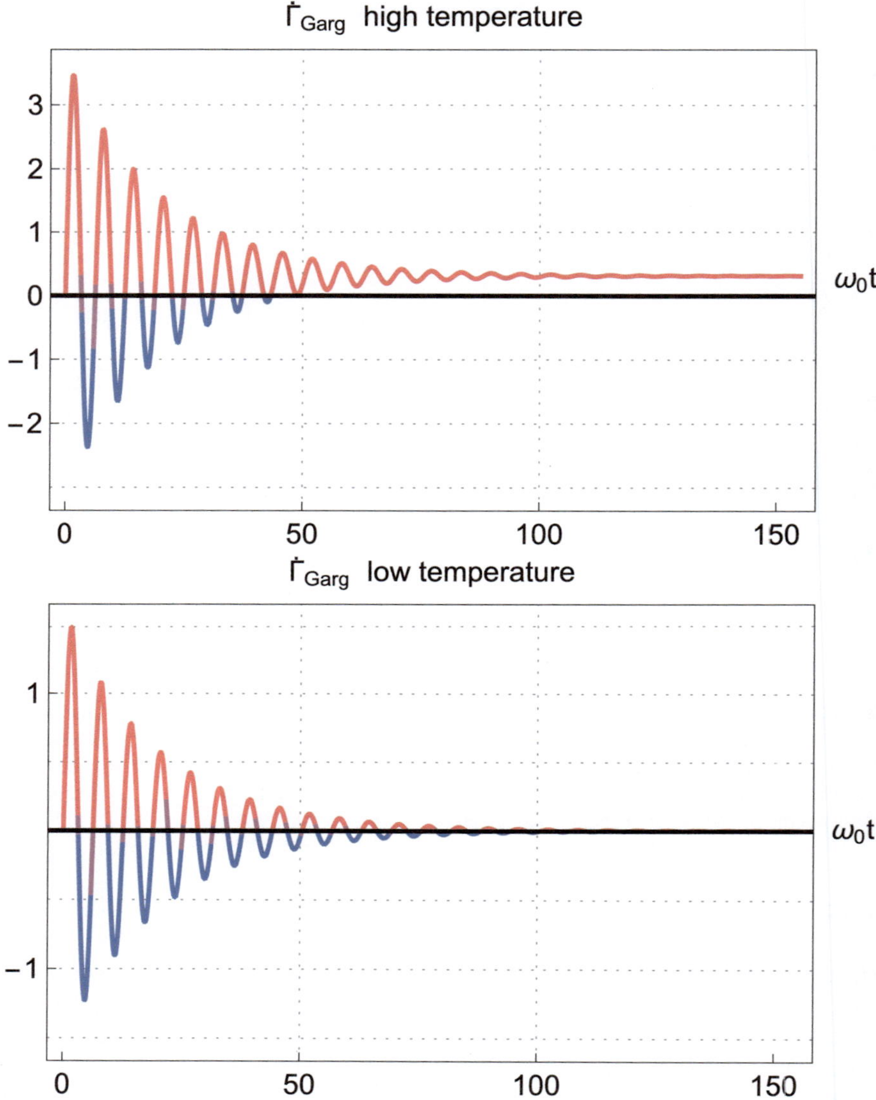

Fig. 5.8 Plot of the derivative of the decoherence function $\dot{\Gamma}_{\mathrm{Garg}}(t)$ for the spectral density of (5.361) in the under-damped regime, again for parameters as in Fig. 5.6. We consider the behavior in the short- and long-term region, for both high and low temperature. The function $\dot{\Gamma}_{\mathrm{Garg}}(t)$ identifies the multiplying coefficient in the master equation (5.352), whose positivity is granted only for long enough times. In particular, it becomes constant in the long-term region, leading to a master equation in GKSL form. The change in sign of the function $\dot{\Gamma}_{\mathrm{Garg}}(t)$ is put into evidence by a change of colour

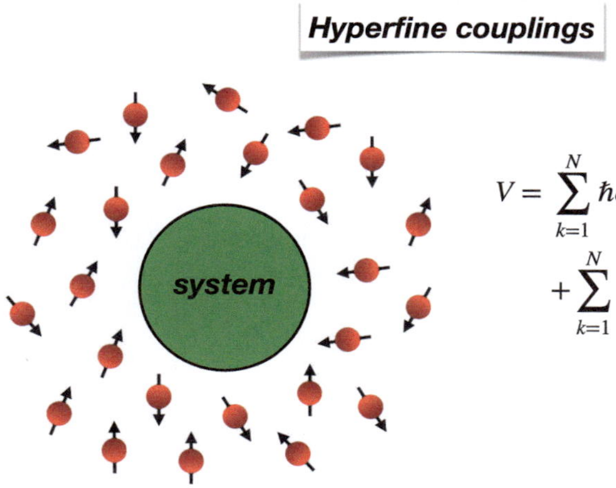

$$V = \sum_{k=1}^{N} \hbar\omega_\parallel^k \, |+\rangle\langle+| \otimes \sigma_z^k$$

$$+ \sum_{k=1}^{N} \hbar\omega_\perp^k (\sigma_+ \otimes \sigma_-^k + \sigma_- \otimes \sigma_+^k)$$

Fig. 5.9 Schematic representation of the central-spin model introduced in Sect. 5.4.2 and analyzed in Sect. 7.5.1.2 from the viewpoint of memory effects. A two-level system is coupled to a spin bath of two-dimensional environmental units. The coupling is of the Heisenberg type in the XXY configuration, namely the coupling constants in x and y directions are the same

where the operators acting in the space of the environment are defined according to

$$\sigma_i^k = \underbrace{\mathbb{1} \otimes \ldots \otimes \mathbb{1}}_{(k-1)\,\text{times}} \otimes \sigma_i \otimes \mathbb{1} \otimes \ldots, \tag{5.376}$$

so that the superscript clarifies on which spin the operator is acting. In order to consider an XXY model, that is a situation in which coupling constants are equal in the x and y direction, we set

$$J_x^k = J_y^k = \frac{\hbar\omega_\perp^k}{2} \tag{5.377}$$

as well as

$$J_z^k = \frac{\hbar\omega_\parallel^k}{2}. \tag{5.378}$$

It is further convenient, in order to obtain a simple expression later on in (5.387), to add and subtract a term of the form $\mathbb{1} \otimes \sum_{k=1}^{N} \frac{\hbar\omega_\parallel^k}{2} \sigma_z^k$ so as to define

$$H_E = \sum_{k=1}^{N} \left(\frac{\hbar\omega_k}{2} - \frac{\hbar\omega_\parallel^k}{2} \right) \sigma_z^k, \tag{5.379}$$

together with

$$V = \sum_{k=1}^{N} \hbar\omega_{\parallel}^{k} \left(\frac{\sigma_z + \mathbb{1}}{2} \right) \otimes \sigma_z^{k} + \sum_{k=1}^{N} \hbar\omega_{\perp}^{k} (\sigma_+ \otimes \sigma_-^{k} + \sigma_- \otimes \sigma_+^{k}), \qquad (5.380)$$

with σ_+ and σ_- defined as in (2.87) and (2.88).

In interaction picture with respect to the free Hamiltonians (5.227) and (5.379) we have

$$V(t) = |+\rangle\langle+| \otimes \sum_{k=1}^{N} \hbar\omega_{\parallel}^{k}\sigma_z^{k} + \sigma_+(t) \otimes \hbar\Sigma_-(t) + \sigma_-(t) \otimes \hbar\Sigma_+(t), \qquad (5.381)$$

where we have introduced the operators

$$\sigma_\pm(t) = \sigma_\pm e^{\pm i\omega_0 t} \qquad (5.382)$$

and

$$\Sigma_\pm(t) = \sum_{k=1}^{N} \omega_\perp^{k} \sigma_\pm^{k} e^{\pm i(\omega_k - \omega_\parallel^{k})t}, \qquad (5.383)$$

while $\{|+\rangle, |-\rangle\}$ denote the eigenvectors of σ_z.

5.4.2.1　Exact Reduced Dynamics and Quantum Dynamical Map

We now derive the exact reduced dynamics for the case in which the environment is initially in the state

$$|0\rangle = |-\rangle_1 \otimes |-\rangle_2 \otimes \ldots, \qquad (5.384)$$

corresponding to all spins in the ground state, which plays the role of vacuum or lowest energy state. As we shall see, for this initial environmental state the longitudinal interaction term in (5.375), characterized by the coupling (5.378), will only determine a coherent contribution to the dynamics. The relevance of the longitudinal coupling will be explored in Sect. 5.4.3. In accordance with this choice, following the strategy used in [25, 93, 97], we solve the Schrödinger equation

$$i\hbar\frac{\mathrm{d}}{\mathrm{d}t}|\psi_{SE}(t)\rangle = V(t)|\psi_{SE}(t)\rangle \qquad (5.385)$$

assuming for the whole system an initial pure state of the form

$$|\psi_{SE}(0)\rangle = (\mathsf{a}_-(0)|-\rangle + \mathsf{a}_+(0)|+\rangle) \otimes |0\rangle, \qquad (5.386)$$

with $a_{\pm}(0)$ complex amplitudes such that $|a_-(0)|^2 + |a_+(0)|^2 = 1$. We note that (5.381) implies

$$V(t)|-\rangle \otimes |0\rangle = 0, \tag{5.387}$$

and that the total number of excitations

$$\sigma_+\sigma_- + \sum_{k=1}^{N} \sigma_+^k \sigma_-^k \tag{5.388}$$

or equivalently the overall component along the z axis

$$\sigma_z + \sum_{k=1}^{N} \sigma_z^k \tag{5.389}$$

is a constant of the motion. We are therefore led for the evolved state to consider the Ansatz

$$|\psi_{SE}(t)\rangle = a_-(0)|-\rangle \otimes |0\rangle + a_+(t)|+\rangle \otimes |0\rangle + \sum_{k=1}^{N} a^k(t)|-\rangle \otimes |+\rangle_k, \tag{5.390}$$

where according to (5.376) we have defined

$$|+\rangle_k = \sigma_+^k|0\rangle \tag{5.391}$$
$$= |-\rangle_1 \otimes \ldots \otimes |-\rangle_{k-1} \otimes |+\rangle_k \otimes |-\rangle_{k+1} \ldots, \tag{5.392}$$

and the functions to be determined should satisfy the initial condition

$$\left\{ a_+(t), a_-(t), a^k(t) \right\} \xrightarrow{t \to 0} \{a_+(0), a_-(0), 0\}. \tag{5.393}$$

According to (5.381) we have

$$V(t)|+\rangle \otimes |0\rangle = -\sum_{k=1}^{N} \hbar\omega_{\parallel}^k|+\rangle \otimes |0\rangle + \sum_{k=1}^{N} \hbar\omega_{\perp}^k e^{+i(\omega_k - \omega_0 - \omega_{\parallel}^k)t}|-\rangle \otimes |+\rangle_k, \tag{5.394}$$

as well as

$$V(t)|-\rangle \otimes |+\rangle_k = \hbar\omega_{\perp}^k e^{-i(\omega_k - \omega_0 - \omega_{\parallel}^k)t}|+\rangle \otimes |0\rangle. \tag{5.395}$$

Exploiting (5.394) and (5.395) we obtain from (5.385) and the Ansatz equation (5.390) the following equations for the amplitudes

$$i\dot{\mathsf{a}}_+(t) = -\sum_{k=1}^{N}\omega_{\parallel}^{k}\mathsf{a}_+(t) + \sum_{k=1}^{N}\omega_{\perp}^{k}\mathrm{e}^{-i(\omega_k-\omega_0-\omega_{\parallel}^{k})t}\mathsf{a}^{k}(t) \tag{5.396}$$

and

$$i\dot{\mathsf{a}}^{k}(t) = \omega_{\perp}^{k}\mathrm{e}^{+i(\omega_k-\omega_0-\omega_{\parallel}^{k})t}\mathsf{a}_+(t). \tag{5.397}$$

The solution for $\mathsf{a}^{k}(t)$ with initial condition (5.393) reads

$$\mathsf{a}^{k}(t) = -i\omega_{\perp}^{k}\int_0^t \mathrm{d}\tau\, \mathrm{e}^{+i(\omega_k-\omega_0-\omega_{\parallel}^{k})\tau}\mathsf{a}_+(\tau), \tag{5.398}$$

so that for $\mathsf{a}_+(t)$ we obtain the equation

$$\dot{\mathsf{a}}_+(t) - i\sum_{k=1}^{N}\omega_{\parallel}^{k}\mathsf{a}_+(t) = -\sum_{k=1}^{N}(\omega_{\perp}^{k})^2\int_0^t \mathrm{d}\tau\, \mathrm{e}^{-i(\omega_k-\omega_0-\omega_{\parallel}^{k})(t-\tau)}\mathsf{a}_+(\tau). \tag{5.399}$$

It is now convenient to introduce the constant

$$\varpi = \sum_{k=1}^{N}\omega_{\parallel}^{k}, \tag{5.400}$$

together with the function

$$\mathsf{f}(t-s) = \sum_{k=1}^{N}(\omega_{\perp}^{k})^2\mathrm{e}^{-i(\omega_k-\omega_0-\omega_{\parallel}^{k})(t-s)} \tag{5.401}$$

$$= \langle 0|\Sigma_-(t)\Sigma_+(s)|0\rangle\mathrm{e}^{i\omega_0(t-s)} \tag{5.402}$$

$$= \mathrm{Tr}_E\{\Sigma_-(t)\Sigma_+(s)\rho_E\}\mathrm{e}^{i\omega_0(t-s)}, \tag{5.403}$$

which similarly to (5.316) captures the relevant correlation function for the description of the reduced dynamics. Note that, following (5.312) of the model in Sect. 5.4.1, we can introduce a spectral density

$$J(\omega) = \sum_{k=1}^{N}(\omega_{\perp}^{k})^2\delta(\omega + \omega_{\parallel}^{k} + \omega_0 - \omega_k), \tag{5.404}$$

such that in analogy to (5.357) and recalling that the environment is now considered to be at zero temperature we have

$$\mathsf{f}(t) = \int_{-\infty}^{+\infty} d\omega\, J(\omega) e^{-i\omega t}. \tag{5.405}$$

To keep a compact notation, we associate to any function $\mathsf{g}(t)$ of time a new function

$$g(t) = \mathsf{g}(t) e^{-i\varpi t}, \tag{5.406}$$

simply modified by multiplication with the phase $e^{-i\varpi t}$. With the definitions (5.401) and (5.406) we finally have

$$\dot{a}_+(t) = -\int_0^t d\tau f(t-\tau) a_+(\tau). \tag{5.407}$$

Denoting with $S(t)$ the solution of the integral equation

$$\dot{S}(t) = -\int_0^t d\tau f(t-\tau) S(\tau), \tag{5.408}$$

where the kernel is determined by the spectral density of the model according to (5.405), with initial condition

$$S(0) = 1, \tag{5.409}$$

we then have

$$a_+(t) = S(t) a_+(0). \tag{5.410}$$

Knowledge of the function $S(t)$ solution of (5.408) thanks to (5.390) and (5.398) completely determines the full dynamics, and therefore in particular the reduced dynamics of the system. Indeed, according to (5.10) and the results of Sect. 5.2 the reduced state of the system is given by

$$\Phi(t)\rho_S(0) = \mathrm{Tr}_E\left[|\psi_{SE}(t)\rangle\langle\psi_{SE}(t)|\right], \tag{5.411}$$

so that in the notation of (5.314) we have

$$\rho_S(t) = \begin{pmatrix} \rho_{++}(0)|S(t)|^2 & \rho_{+-}(0)S(t)e^{i\varpi t} \\ \rho_{-+}(0)S(t)^* e^{-i\varpi t} & \rho_{++}(0)(1-|S(t)|^2) + \rho_{--}(0) \end{pmatrix}. \tag{5.412}$$

Note that if all transversal couplings $\omega_\perp^k$ go to zero we recover $S(t) = 1$ and the dynamics reduces to a coherent unitary evolution.

5.4.2.2   Kraus Decomposition and Master Equation

Also in this case it is instructive to consider a Kraus representation of the map. Following the steps and the notation considered in Sect. 5.4.1.3, making reference to (5.412) we obtain for the Choi matrix the expression

$$
\mathfrak{C}_{\Phi(t)} = \begin{pmatrix} |S(t)|^2 & 0 & 0 & S(t)e^{i\varpi t} \\ 0 & 1-|S(t)|^2 & 0 & 0 \\ 0 & 0 & 0 & 0 \\ S(t)^*e^{-i\varpi t} & 0 & 0 & 1 \end{pmatrix},
\tag{5.413}
$$

whose eigenvalues $\{0, 0, 1 - |S(t)|^2, 1 + |S(t)|^2\}$ are positive provided $|S(t)| \leqslant 1$, as warranted by the expression (5.432) of Remark 5.15. The Choi matrix can be represented in the form (5.337) with

$$
D = \mathrm{diag}(0, 0, 1 - |S(t)|^2, 1 + |S(t)|^2)
\tag{5.414}
$$

and

$$
U = \frac{1}{\sqrt{1 + |S(t)|^2}} \begin{pmatrix} -\frac{S(t)}{|S(t)|}e^{i\varpi t} & 0 & 0 & S(t)e^{i\varpi t} \\ 0 & 0 & 1 & 0 \\ 0 & 1 & 0 & 0 \\ |S(t)| & 0 & 0 & 1 \end{pmatrix},
\tag{5.415}
$$

whose columns are the normalized eigenvectors of $\mathfrak{C}_{\Phi(t)}$ corresponding to the eigenvalues listed in (5.414). Also for this quantum dynamical map we make reference to the treatment in Remark 3.20, with the matrices appearing in (3.431) identified by (5.414) and (5.415). We consider as operator basis a canonical basis as in (3.375) of Sect. 3.3.2.1, obtained making reference to the eigenvectors of the Pauli σ_z operator. Given that only two eigenvalues are different from zero we are left with

$$
\Phi(t)\rho_S(0) = \sum_{i=1}^{2} K_i(t)\rho_S(0)K_i^\dagger(t),
\tag{5.416}
$$

where the Kraus operators read

$$
\{K_1(t), K_2(t)\} \rightarrow \left\{ \begin{pmatrix} 0 & 0 \\ \sqrt{1-|S(t)|^2} & 0 \end{pmatrix}, \begin{pmatrix} S(t)e^{i\varpi t} & 0 \\ 0 & 1 \end{pmatrix} \right\}.
\tag{5.417}
$$

According to (5.412) the dynamics of the two-level system is determined by the pair of equations

$$
\frac{\mathrm{d}}{\mathrm{d}t}\langle +|\rho_S(t)|+\rangle = 2\Re\frac{\dot{S}(t)}{S(t)}\langle +|\rho_S(t)|+\rangle
\tag{5.418}
$$

and

$$\frac{d}{dt}\langle+|\rho_S(t)|-\rangle = \left(i\varpi + \frac{\dot{S}(t)}{S(t)}\right)\langle+|\rho_S(t)|-\rangle, \tag{5.419}$$

that determine the evolution of the other matrix elements. In particular, due to trace preservation we have

$$\frac{d}{dt}\langle-|\rho_S(t)|-\rangle = -2\Re\frac{\dot{S}(t)}{S(t)}\langle+|\rho_S(t)|+\rangle. \tag{5.420}$$

For an arbitrary 2×2 matrix as in (5.341) we have the simple identity

$$\sigma_- M \sigma_+ = \begin{pmatrix} 0 & 0 \\ 0 & a \end{pmatrix}, \tag{5.421}$$

together with

$$\{\sigma_+\sigma_-, M\} = \begin{pmatrix} 2a & b \\ c & 0 \end{pmatrix} \tag{5.422}$$

and

$$[\sigma_+\sigma_-, M] = \begin{pmatrix} 0 & b \\ -c & 0 \end{pmatrix}, \tag{5.423}$$

so that (5.418), (5.419) and (5.420) are recovered as matrix elements of the master equation

$$\begin{aligned}
\frac{d}{dt}\rho_S(t) = {} & i\left(\varpi + \Im\frac{\dot{S}(t)}{S(t)}\right)[\sigma_+\sigma_-, \rho_S(t)] \\
& - 2\Re\frac{\dot{S}(t)}{S(t)}\left[\sigma_-\rho_S(t)\sigma_+ - \frac{1}{2}\{\sigma_+\sigma_-, \rho_S(t)\}\right].
\end{aligned} \tag{5.424}$$

5.15 Homogeneous Spin Bath

An explicit solution of the considered model requires the determination of the function $S(t)$. To this aim using the notation introduced in (5.63) for the convolution we write (5.408) in the compact form

$$\dot{S}(t) = -(f \star S)(t). \tag{5.425}$$

This equation is conveniently solved using the Laplace transform, defined in Remark 5.2, indeed exploiting (5.70) and (5.71) we obtain

$$uS(u) - S(0) = -\hat{f}(u)\hat{S}(u), \qquad (5.426)$$

which according to the initial condition (5.409) has solution

$$S(u) = \frac{1}{u + \hat{f}(u)}. \qquad (5.427)$$

The expression (5.401) for $f(t)$, further recalling (5.406), leads to

$$\hat{f}(u) = \sum_{k=1}^{N} \frac{(\omega_\perp^k)^2}{u - i\Omega^k}, \qquad (5.428)$$

where we have defined

$$\Omega^k = \omega_0 + \omega_\parallel^k - \omega_k - \varpi. \qquad (5.429)$$

An explicit solution allowing to put into evidence some general features is obtained considering uniform frequencies $\Omega^k \rightarrow \Omega$, as well as uniform rescaled couplings

$$\omega_\perp^k \rightarrow \frac{\omega_\perp}{\sqrt{N}}, \qquad (5.430)$$

where N denotes the number of modes, so that we obtain

$$S(u) = \frac{u - i\Omega}{u^2 - i\Omega u + \omega_\perp^2}, \qquad (5.431)$$

leading in the time domain thanks to (5.78) and (5.79) to the explicit solution

$$S(t) = e^{i\frac{\Omega}{2}t}\left[\cos(\chi t) - i\frac{\Omega}{2\chi}\sin(\chi t)\right], \qquad (5.432)$$

with

$$\chi = \sqrt{\left(\frac{\Omega}{2}\right)^2 + \omega_\perp^2}. \qquad (5.433)$$

In Fig. 5.10 we plot the behavior in time of populations and coherences as follows from (5.412) for the considered case. In particular, within this example

we can consider the explicit expression of the coefficients appearing in the
master equation describing the exact reduced dynamics. Relying on (5.432)
we obtain the coefficient

$$\Im \frac{\dot{S}(t)}{S(t)} = -\frac{\Omega}{2}\left(1-\left(\frac{\Omega}{2\chi}\right)^2\right)\frac{\sin^2(\chi t)}{\cos^2(\chi t)+\left(\frac{\Omega}{2\chi}\right)^2\sin^2(\chi t)} \tag{5.434}$$

relevant for the coherent part of the evolution, as well as

$$-2\Re\frac{\dot{S}(t)}{S(t)} = \chi\left(1-\left(\frac{\Omega}{2\chi}\right)^2\right)\frac{\sin(2\chi t)}{\cos^2(\chi t)+\left(\frac{\Omega}{2\chi}\right)^2\sin^2(\chi t)}, \tag{5.435}$$

which determines the dissipative part in (5.424). Note that in the presence of
coupling $\Omega/2\chi < 1$ according to (5.433). Both coefficients exhibit a periodic
time dependence, so that at variance with the model considered in Remark 5.14
they never become approximately constant. Moreover, the dissipative coefficient repeatedly takes on negative values. This feature is of importance in the
study of memory effects that we will consider in Chap. 7. Indeed, a further
example of spin-bath dynamics will be considered in Sect. 7.5.1.2, for the sake
of a detailed discussion of memory effects in the dynamics.

5.4.3 Spin-Star Decoherence Model

We now consider an example of exact reduced dynamics in which we can also have
access to the correlation properties of the total state. To this aim we consider a so-
called spin-star decoherence model, visualized in Fig. 5.11, describing the situation
in which a two-level system is coupled via a dephasing interaction with a collection
of surrounding two-level systems or spins. With respect to (5.375) only one of the
coupling constants, namely J_z^k defined in (5.378), is different from zero. This feature
allows to consider the unitary evolution operator for an arbitrary environmental state,
so that at variance with the situation considered in Sect. 5.4.2 this interaction term
will not simply lead to a coherent contribution to the dynamics, and to obtain explicit
expressions for an important class of environmental states.

The system Hamiltonian is still (5.227), while the bath Hamiltonian is given by

$$H_E = \sum_{k=1}^{N}\frac{\hbar\omega_k}{2}\sigma_z^k, \tag{5.436}$$

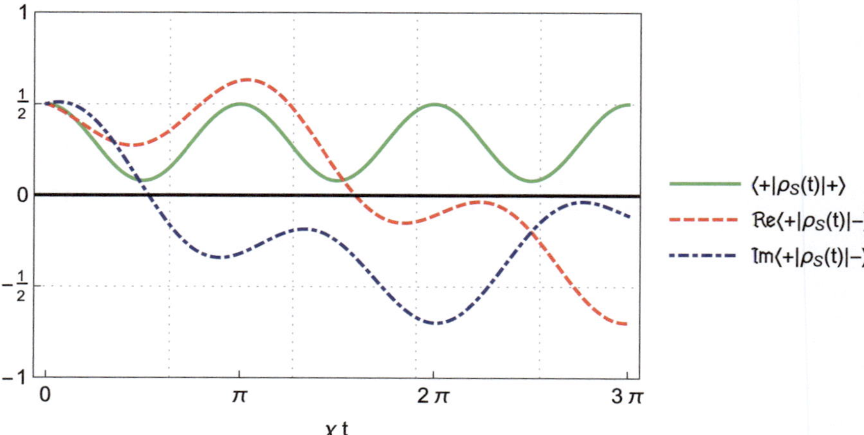

Fig. 5.10 Plot of the behavior in time of the population of the excited state, as well as of the real and the imaginary part of the coherence of the state for the model of Sect. 5.4.2. The initial condition is taken to be a balanced coherent superposition of excited and ground state. The coupling is fixed as $\Omega/2\chi = 0.4$

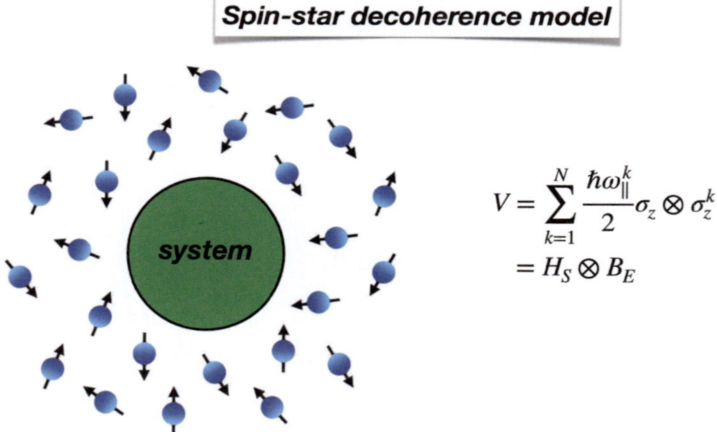

$$V = \sum_{k=1}^{N} \frac{\hbar\omega_{\parallel}^{k}}{2}\sigma_z \otimes \sigma_z^{k}$$
$$= H_S \otimes B_E$$

Fig. 5.11 The same system environment configuration as in Fig. 5.9, but with a dephasing interaction in which the system coupling operator commutes with its Hamiltonian. The model is introduced in Sect. 5.4.3 and analyzed in Sect. 7.5.1.3 from the viewpoint of memory effects

with operators denoted as in (5.376). The coupling term reads

$$V = \sum_{k=1}^{N} \frac{\hbar \omega_{\parallel}^{k}}{2} \sigma_z \otimes \sigma_z^k,$$
(5.437)

so that the operators σ_z and $\{\sigma_z^k\}_k$ are constants of motion.

5.4.3.1   Evolved Total State and Correlations

The coupling term is not affected by the interaction picture, so that the evolution operator takes the form

$$U(t) = \bigotimes_k U_k(t)$$
(5.438)

with

$$U_k(t) = e^{-i\omega_{\parallel}^k \sigma_z \otimes \sigma_z^k \frac{t}{2}}.$$
(5.439)

It is further convenient to write the evolution operator making a special choice of basis on the side of the system or of the environment. Expanding $U(t)$ in the basis of eigenvectors of σ_z we have

$$U(t) = |+\rangle\langle+| \bigotimes_k V_k(t) + |-\rangle\langle-| \bigotimes_k V_k(t)^{\dagger},$$
(5.440)

with

$$V_k(t) = e^{-i\omega_{\parallel}^k \sigma_z^k \frac{t}{2}},$$
(5.441)

providing an example of conditional transformation, in which the action on the environment depends on the state of the system, as we already observed in (5.257). On the other hand expanding $U(t)$ in the basis of eigenvectors of $\{\sigma_z^k\}_k$, which we denote as

$$|\{m_k\}\rangle = |m_1\rangle \otimes |m_2\rangle \otimes \dots,$$
(5.442)

with $m_k \in \{+, -\}$, we obtain

$$U(t) = \sum_{\{m_k\}} U_{\{m_k\}}(t) \otimes |\{m_k\}\rangle\langle\{m_k\}|,$$
(5.443)

with

$$U_{\{m_k\}}(t) = e^{-i\left(\sum_k \omega_{\parallel}^k m_k\right)\sigma_z \frac{t}{2}},$$
(5.444)

and where the sum is over all the 2^N possible configurations of environmental spins. The importance of the expression (5.443) lies in the fact that it allows to infer the structure of correlations in the total state. We will exploit this information in Sect. 7.5.1.3 to study the non-Markovian features of the model. Indeed, for the case in which the initial environmental state is diagonal in the basis (5.442), that is

$$\rho_E = \sum_{\{m_k\}} p(\{m_k\}) |\{m_k\}\rangle\langle\{m_k\}|, \tag{5.445}$$

with $p(\{m_k\})$ a probability distribution, namely $p(\{m_k\}) \geqslant 0$ and $\sum_{k=1}^{N} p(\{m_k\}) = 1$, thanks to (5.443) we have

$$\rho_{SE}(t) = \sum_{\{m_k\}} p(\{m_k\}) U_{\{m_k\}}(t)\rho_S(0) U_{\{m_k\}}(t)^\dagger \otimes |\{m_k\}\rangle\langle\{m_k\}|, \tag{5.446}$$

that is according to (4.43) a separable state only containing classical correlations with respect to the bipartition system-environment of the overall Hilbert space. With respect to the expression of a separable state (4.43) the weights are here given by $p(\{m_k\})$, the collection of system states by $U_{\{m_k\}}(t)\rho_S(0) U_{\{m_k\}}(t)^\dagger$, and the collection of environmental states by the pure states $|\{m_k\}\rangle\langle\{m_k\}|$. The expression (5.446) also leads to an explicit Kraus decomposition for the quantum dynamical map corresponding to a mixture of random unitaries as in (5.343), namely

$$\Phi(t)\rho_S(0) = \sum_{\{m_k\}} p(\{m_k\}) U_{\{m_k\}}(t)\rho_S(0) U_{\{m_k\}}(t)^\dagger. \tag{5.447}$$

In the standard case in which the environmental spins are initially uncorrelated, so that the initial state takes the form

$$\rho_{SE}(0) = \rho_S(0) \otimes \bigotimes_k \rho_E^k, \tag{5.448}$$

with ρ_E^k the initial state of the k-th environmental spin, e.g. considering (5.446) with $p(\{m_k\}) = \prod_k p_k(m_k)$, exploiting (5.440) we obtain for the total evolved state

$$\rho_{SE}(t) = \begin{pmatrix} \rho_{++}(0) \bigotimes_k V_k(t)\rho_E^k V_k(t)^\dagger & \rho_{+-}(0) \bigotimes_k V_k(t)\rho_E^k V_k(t) \\ \rho_{-+}(0) \bigotimes_k V_k(t)^\dagger \rho_E^k V_k(t)^\dagger & \rho_{--}(0) \bigotimes_k V_k(t)^\dagger \rho_E^k V_k(t) \end{pmatrix}. \tag{5.449}$$

5.4.3.2 Decoherence Dynamics

The expression (5.449) lends itself to obtain a compact expression for the reduced state, namely

$$\rho_S(t) = \begin{pmatrix} \rho_{++}(0) & \rho_{+-}(0)\mathfrak{D}(t) \\ \rho_{-+}(0)\mathfrak{D}(t)^* & \rho_{--}(0) \end{pmatrix}, \tag{5.450}$$

where exploiting (5.441) together with the identity

$$\mathrm{Tr}\{e^{-i\omega\sigma_z t}\rho\} = \cos(\omega t) - i\langle\sigma_z\rangle\sin(\omega t), \tag{5.451}$$

with

$$\langle\sigma_z\rangle = \mathrm{Tr}\{\sigma_z\rho\}, \tag{5.452}$$

we have introduced the decoherence factor

$$\mathfrak{D}(t) = \prod_{k=1}^{N}[\cos(\omega_{\parallel}^k t) - i\langle\sigma_z^k\rangle\sin(\omega_{\parallel}^k t)]. \tag{5.453}$$

Indeed, the expression (5.450) tells us that the effect of the environmental dynamics generally is a reduction of the coherence present in the initial state. Note in particular that different environmental states do lead to the very same reduced dynamics. More specifically, environmental states only differing for their amount of coherence in the eigenbasis of σ_z^k equally affect the system. Also in this case, as in (5.284), for $\langle\sigma_z^k\rangle = 0$ the exact solution for short times predicts a decay quadratic in time, namely

$$\mathfrak{D}(t) \underset{t\to 0}{\approx} 1 - \frac{t^2}{2}\sum_{k=1}^{N}(\omega_{\parallel}^k)^2, \tag{5.454}$$

so that

$$\langle+|\rho_S(t)|-\rangle \approx \left[1 - \frac{t^2}{2}\sum_{k=1}^{N}(\omega_{\parallel}^k)^2\right]\langle+|\rho_S(0)|-\rangle. \tag{5.455}$$

5.4.3.3   Master Equation

We can now proceed as in Sect. 5.4.1.3, obtaining the Choi matrix

$$\mathfrak{C}_{\Phi(t)} = \begin{pmatrix} 1 & 0 & 0 & \mathfrak{D}(t) \\ 0 & 0 & 0 & 0 \\ 0 & 0 & 0 & 0 \\ \mathfrak{D}(t)^* & 0 & 0 & 1 \end{pmatrix}, \tag{5.456}$$

with eigenvalues $\{0, 0, 1 - |\mathfrak{D}(t)|, 1 + |\mathfrak{D}(t)|\}$, so that the condition for complete positivity is

$$|\mathfrak{D}(t)| \leqslant 1, \tag{5.457}$$

always warranted by (5.453) thanks to

$$\langle \sigma_z^k \rangle^2 \leqslant 1. \tag{5.458}$$

Relevant choices of initial environmental states complying with (5.445) are given by the thermal and the maximally mixed state, formally corresponding to an infinite temperature state, leading respectively to

$$\langle \sigma_z^k \rangle = \mathrm{Tanh}\left(\frac{\beta \hbar \omega_k}{2} \right) \tag{5.459}$$

and

$$\langle \sigma_z^k \rangle = 0. \tag{5.460}$$

In particular, for the latter case considering homogeneous couplings we obtain

$$\mathfrak{D}(t) = \cos^N(\omega_\parallel t), \tag{5.461}$$

so that considering the same scaling for the couplings as in (5.430), for large N we have

$$\cos^N \left(\frac{\omega_\parallel}{\sqrt{N}} t \right) \underset{N \gg 1}{\approx} \exp\left(N \log\left(1 - \frac{\omega_\parallel^2 t^2}{2N} \right) \right), \tag{5.462}$$

and therefore

$$\mathfrak{D}(t) \approx \exp\left(-\frac{1}{2} \omega_\parallel^2 t^2 \right), \tag{5.463}$$

describing an exponential decay quadratic in time. At the other extreme, a zero temperature state, corresponding to $\langle \sigma_z^k \rangle = 1$, only induces a coherent modification to the dynamics, at variance with the spin-boson model considered in Sect. 5.4.1.

The expression (5.450) for the quantum dynamical map, which takes the same form as (5.314), leads to the evolution (5.349) and

$$\frac{\mathrm{d}}{\mathrm{d}t} \langle +|\rho_S(t)|- \rangle = \frac{\dot{\mathfrak{D}}(t)}{\mathfrak{D}(t)} \langle +|\rho_S(t)|- \rangle, \tag{5.464}$$

where at variance with (5.275) the multiplying factor is in general a complex quantity. Relying on the relations (5.342) and (5.423) we thus obtain for the master equation the expression

$$\frac{d}{dt}\rho_S(t) = i\Im\frac{\dot{\mathfrak{D}}(t)}{\mathfrak{D}(t)}[\sigma_+\sigma_-, \rho_S(t)]$$

$$-\frac{1}{2}\Re\frac{\dot{\mathfrak{D}}(t)}{\mathfrak{D}(t)}[\sigma_z\rho_S(t)\sigma_z - \rho_S(t)]. \tag{5.465}$$

Simple explicit examples of this kind of dynamics are considered in Remark 5.16.

5.16 Decoherence with Uniform and Random Couplings

To shed more light on the dynamics of the decoherence model of Sect. 5.4.3, we consider the situation of either uniformly or randomly coupled spins. Starting from the general expression (5.453) for the decoherence factor we obtain in particular

$$-\frac{1}{2}\Re\frac{\dot{\mathfrak{D}}(t)}{\mathfrak{D}(t)} = \sum_{k=1}^{N}\frac{\omega_\parallel^k}{4}\frac{\sin(2\omega_\parallel^k t)(1 - \langle\sigma_z^k\rangle^2)}{\cos^2(\omega_\parallel^k t) + \langle\sigma_z^k\rangle^2\sin^2(\omega_\parallel^k t)} \tag{5.466}$$

and

$$\Im\frac{\dot{\mathfrak{D}}(t)}{\mathfrak{D}(t)} = -\sum_{k=1}^{N}\omega_\parallel^k\frac{\langle\sigma_z^k\rangle}{\cos^2(\omega_\parallel^k t) + \langle\sigma_z^k\rangle^2\sin^2(\omega_\parallel^k t)}. \tag{5.467}$$

For the case in which the initial environmental state is the maximally mixed state, corresponding to an environment at infinite temperature, according to (5.460) we are left with

$$\frac{d}{dt}\rho_S(t) = \frac{1}{2}\sum_{k=1}^{N}\omega_\parallel^k\tan(\omega_\parallel^k t)[\sigma_z\rho_S(t)\sigma_z - \rho_S(t)]. \tag{5.468}$$

Considering again in the uniform case the scaling (5.430), in the limit of large N we are left with

$$\frac{d}{dt}\rho_S(t) \approx \frac{1}{2}\omega_\parallel^2 t[\sigma_z\rho_S(t)\sigma_z - \rho_S(t)], \tag{5.469}$$

corresponding to the approximated expression (5.463). The decoherence function in the complimentary situations of uniform and randomly distributed couplings is considered in Fig. 5.12, while in Fig. 5.13 we plot the corresponding time-dependent coefficient in the master equation (5.468).

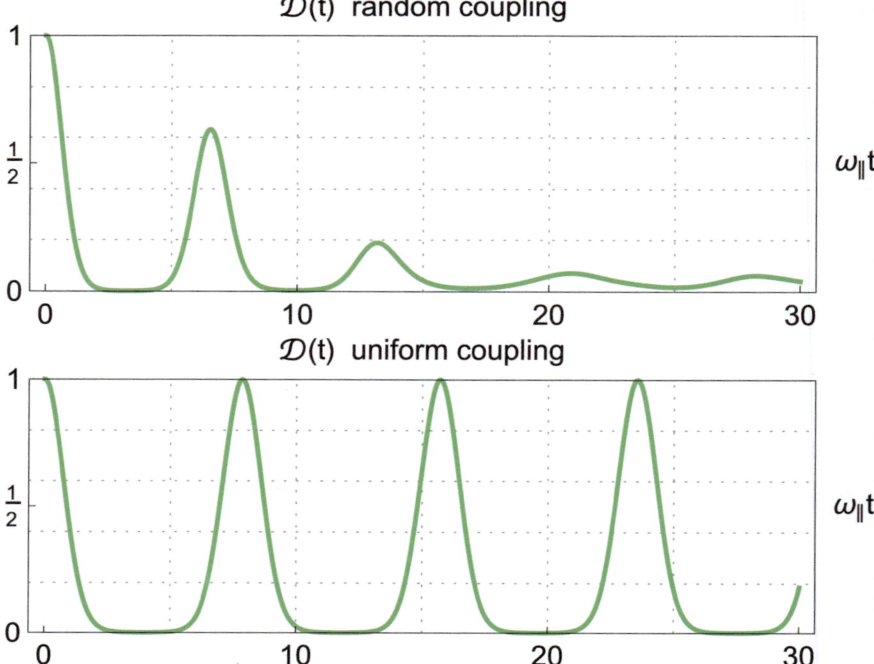

Fig. 5.12 Plot of the function $\mathcal{D}(t)$ that according to (5.450) determines the time dependence of the modulus of the coherence of the state in the spin-star decoherence model. In the top figure the bath is assumed to consist of $N = 25$ spins, with random couplings distributed around a reference value $\omega_\parallel = 0.4$ with variance 1×10^{-2} in arbitrary units. The environmental units are taken to be in the same state, so that $\langle \sigma_z^k \rangle$ is a k-independent constant set equal to 0.75, determining according to (5.459) the temperature of the thermal state. In the bottom figure we consider uniform coupling with the same parameters, namely $N = 25$ spins with $\omega_\parallel = 0.4$ in arbitrary units. The quadratic exponential decay corresponding to (5.469) and described by (5.463) clearly appears, together with the first revival taking place due to the finite dimension of the bath

5.5   Projection Operator Techniques

In the previous examples of open quantum system dynamics we have considered two somehow opposite perspectives. In Sect. 5.3 we have taken as reference the Lindblad form of the generator (5.27), which brings with itself the hypothesis of a semigroup evolution, and shaped it on the basis of phenomenological information on the interaction mechanism. At the other opposite, in Sect. 5.4 we have considered dynamics allowing for an exact treatment, obtaining in particular the exact evolution equation for the statistical operator of the system, which could then be compared with

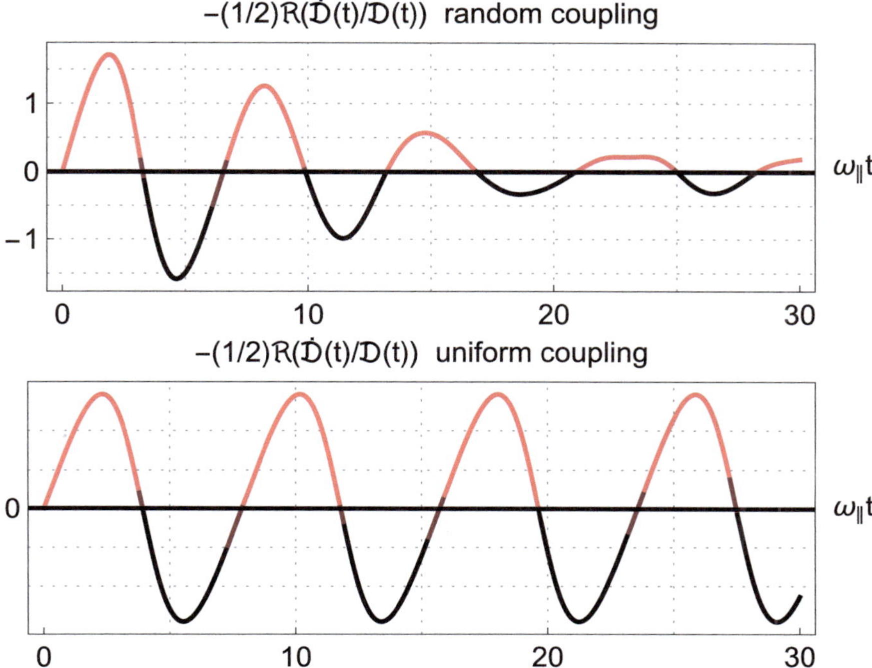

Fig. 5.13 Plot of the function $-(1/2)\Re(\dot{\mathfrak{D}}(t)/\mathfrak{D}(t))$ providing the coefficient in front of the dissipative contribution in the master equation (5.465), obeyed by the spin-star decoherence model. Top and bottom figures correspond to the same situation considered in Fig. 5.12. The change in sign of the coefficient is put into evidence by a change of colour

the semigroup evolution. In all these examples, complete positivity of the evolution map was granted from the phenomenological approach or from the exact microscopic treatment. In the general case, we are interested in obtaining information on the reduced evolution starting from the underlying microscopic dynamics, without being able to provide an exact solution to the problem. To this aim different strategies can be considered. Following [25] we will present here a technique, originally due to Nakajima and Zwanzig [98,99], which allows to obtain a representation for the exact evolution of the statistical operator describing all degrees of freedom, splitting it thanks to a projection operator in a so-called relevant and irrelevant part. For suitable choices of the projection operator we further obtain an exact master equation for the reduced statistical operator of the system. This equation is written in integro-differential form, but importantly allows for a perturbation expansion. Note however that the solutions of the approximated equations of motion are generally no more completely positive, so that it might be convenient to combine the perturbative expansion with other approximations, as we shall discuss in Sect. 6.2. The basic idea comes from non-equilibrium statistical mechanics. We have a complex system and try to obtain the relevant information by eliminating degrees of freedom by means of a projection operator, thus considering the dynamics of a subset of relevant variables, to be described in terms of effective master equations. In the present framework the relevant variables are those of the system, while in the general case they could be a collection of variables slowly varying with respect to others.

5.5.1   Standard and Correlated Projections

We define here as projection operator a map

$$\mathcal{P} : \mathcal{T}(\mathcal{H}_S \otimes \mathcal{H}_E) \to \mathcal{T}(\mathcal{H}_S \otimes \mathcal{H}_E) \tag{5.470}$$
$$\rho \mapsto \mathcal{P}\rho$$

with the following properties

1. $\mathcal{P}\rho \geqslant 0$ for $\rho \geqslant 0$
2. $\mathrm{Tr}\{\mathcal{P}\rho\} = \mathrm{Tr}\{\rho\}$
3. $\mathcal{P}^2 = \mathcal{P}$.

The map is therefore positive, trace-preserving and most importantly idempotent. The latter requirement is the defining property of a projection operator. Notice that orthogonal projections are a special class of projections that act in a Hilbert space, where we also have a notion of self-adjointness thanks to the presence of the scalar product. A crucial requirement to identify projection operators relevant for the determination of the reduced dynamics is given by the constraint

$$\mathrm{Tr}_E[\mathcal{P}\rho] = \rho_S, \tag{5.471}$$

telling us that knowledge of the projected state, that is the so-called relevant part, is enough to determine the system reduced state defined in (4.1). In other words, (5.471) implies that the assignment (5.470) sends a state on the bipartite Hilbert space $\mathcal{H}_S \otimes \mathcal{H}_E$ to another state in the same space preserving the first marginal. A useful and widely used expression for such a map is given by the so-called standard projection operator, depending on the the choice of a statistical operator σ_E on the space $\mathcal{H}_E$ and defined as

$$\mathcal{P}\rho = \mathrm{Tr}_E[\rho] \otimes \sigma_E \tag{5.472}$$

or equivalently according to the notation of (4.1)

$$\mathcal{P}\rho = \rho_S \otimes \sigma_E, \tag{5.473}$$

so that thanks to (3.59) it provides a completely positive map.

A general characterization of projection operators is given in Remark 5.17, where also correlated projection operators of the form

$$\mathcal{P}\rho = \sum_i \mathrm{Tr}_E[\Pi_i \rho \Pi_i] \otimes \frac{\Pi_i \sigma_E \Pi_i}{\mathrm{Tr}_E\{\Pi_i \sigma_E \Pi_i\}}, \tag{5.474}$$

are introduced. Here we still have $\sigma_E \in \mathcal{T}(\mathcal{H}_E)$, while the operators $\{\Pi_i\}$ denote an orthogonal resolution of the identity in $\mathcal{H}_E$ as in (2.73). As we shall see, any choice of projection in the sense of (5.470) and compatible with (5.471) leads to a representation for the exact total evolution, from which the reduced dynamics can be obtained. They however differ in the form of the evolution and in the information needed in order to characterize the state at the initial time.

5.17 Standard and Correlated Projections

We provide here a characterization of the structure of projection operators following [100, 101]. Given the bipartite structure of the considered Hilbert space $\mathcal{H}_S \otimes \mathcal{H}_E$, and the constraint to preserve the first marginal, it is natural to consider projection operators of the form

$$\mathcal{P} = \mathbb{1}_{\mathcal{T}(\mathcal{H}_S)} \otimes \Lambda, \tag{5.475}$$

with Λ a completely positive, trace-preserving and idempotent map on $\mathcal{T}(\mathcal{H}_E)$. The structure (5.475) is indeed a projection in the sense of (5.470) and warrants (5.471) in that the system degrees of freedom are unaffected by the transformation. For these maps we have the following representation theorem, in which we assume $\dim \mathcal{H}_E = n$.

Theorem 5.2 (Breuer, 2007) *A projection operator $\mathcal{P}$ on $\mathcal{T}(\mathcal{H}_S \otimes \mathcal{H}_E)$ of the form $\mathbb{1}_{\mathcal{T}(\mathcal{H}_S)} \otimes \Lambda$ with Λ completely positive and trace-preserving can be written as*

$$\mathcal{P}\rho = \sum_{i=1}^{n^2} \mathrm{Tr}_E[A_i \rho] \otimes B_i \tag{5.476}$$

with $\{A_i\}, \{B_i\}$ linearly independent sets of self-adjoint operators on $\mathcal{H}_E$ satisfying

1. $\mathrm{Tr}_E\{A_i B_j\} = \delta_{ij}$
2. $\sum_{i=1}^{n^2} \mathrm{Tr}_E\{B_i\} A_i = \mathbb{1}_E$
3. $\sum_{i=1}^{n^2} A_i^T \otimes B_i \geqslant 0.$

Proof We first consider a map $\mathcal{P}$ of the form (5.476) with $\{A_i\}$ and $\{B_i\}$ linearly independent and satisfying the above constraints. Idempotency can be verified thanks to *1*, while trace preservation follows from *2*. Hermiticity preservation in the sense of (3.422) is warranted by self-adjointness of the operators $\{A_i\}$ and $\{B_i\}$. The action of $\mathcal{P}$ on operators in tensor product form $\sigma_S \otimes \sigma_E$ is given by

$$\mathcal{P}\sigma_S \otimes \sigma_E = \sigma_S \otimes \sum_{i=1}^{n^2} \mathrm{Tr}_E\{A_i\sigma_E\}B_i, \tag{5.477}$$

so that indeed the map is of the form (5.475), with Λ defined as

$$\Lambda[\sigma_E] = \sum_{r=1}^{n^2} \mathrm{Tr}_E\{A_i\sigma_E\}B_i. \tag{5.478}$$

With $\{e_r\}_{r=1}^n$ an orthonormal basis in $\mathcal{H}_E$ the Choi matrix associated to the map Λ according to (4.76) reads

$$\mathfrak{C}_\Lambda = \sum_{r,s=1}^{n} |e_r\rangle\langle e_s| \otimes \Lambda[|e_r\rangle\langle e_s|] \tag{5.479}$$

$$= \sum_{r,s=1}^{n} \sum_{i=1}^{n^2} |e_r\rangle\langle e_s| \otimes \mathrm{Tr}_E\{A_i|e_r\rangle\langle e_s|\}B_i \tag{5.480}$$

$$= \sum_{i=1}^{n^2} \sum_{r,s=1}^{n} |e_r\rangle\langle e_s|A_i|e_r\rangle\langle e_s| \otimes B_i, \tag{5.481}$$

so that we have the identity

$$\mathfrak{C}_\Lambda = \sum_{i=1}^{n^2} A_i^T \otimes B_i, \tag{5.482}$$

warranting thanks to *3* complete positivity of Λ and therefore in particular positivity of $\mathcal{P}$. Note that in general different sets of self-adjoint operators $\{A_i\}$ and $\{B_i\}$ can represent the same map Λ.

We now consider a completely positive trace-preserving and idempotent map Λ and verify that it can be represented in the form (5.476). We first note that in a finite-dimensional Hilbert space an idempotent operator W can be written

$$W|\varphi\rangle = \sum_k |f_k\rangle\langle e_k|\varphi\rangle, \tag{5.483}$$

with $\{e_k\}$ and $\{f_k\}$ two sets of linearly independent vectors, satisfying $\langle e_k|f_j\rangle = \delta_{k,j}$. Given the finite-dimensionality of $\mathcal{H}_E$, the map Λ can be seen as an operator in the Hilbert space $\mathrm{HS}(\mathcal{H}_E)$ of Hilbert-Schmidt operators introduced in Sect. 2.2.2. The scalar product in this space is given by (2.53), so that (5.483) becomes

$$\Lambda[X] = \sum_{i=1}^{n^2} \mathrm{Tr}_E\{A_i^\dagger X\}B_i, \tag{5.484}$$

with $\{A_i\}$ and $\{B_i\}$ two sets of linearly independent operators, and the requirement of idempotency is warranted if 1 is satisfied. Hermiticity of Λ further implies that these operators have to be self-adjoint. Given that (5.484) has the same form as (5.478), they share the same Choi matrix (5.482), so that complete positivity is warranted by condition 3. Finally, trace preservation holds if

$$\mathrm{Tr}_E\left\{\sum_{i=1}^{n^2} B_i\, \mathrm{Tr}_E\{A_i X\}\right\} = \mathrm{Tr}_E\left\{\left[\sum_{i=1}^{n^2} \mathrm{Tr}_E\{B_i\}A_i\right] X\right\} \tag{5.485}$$

$$= \mathrm{Tr}_E\{X\} \tag{5.486}$$

for all operators X, thus leading to the requirement 2. $\qquad\square$

In this framework, the standard projection operator onto a factorized state (5.472) is obtained for the choice

$$A = \mathbb{1}_E, \qquad B = \sigma_E \tag{5.487}$$

with σ_E a fixed environmental state

$$\mathcal{P}\rho = \mathrm{Tr}_E[\rho] \otimes \sigma_E = \rho_S \otimes \sigma_E. \tag{5.488}$$

Note that, as stressed by the notation used, at the r.h.s. we do not have the product of the marginals of ρ, but rather the product of the first marginal with a fixed environmental state. In this case the map Λ appearing in (5.475) actually coincides with the idempotent completely positive trace-preserving map $\mathcal{A}_{\sigma_E} \circ \mathrm{Tr}_E$ introduced in (3.58). An example of correlated projection operator can be obtained considering an orthogonal decomposition of the identity $\{\Pi_i\}$ on the space $\mathcal{H}_E$ as in (2.73) and identifying

$$A_i = \Pi_i, \qquad B_i = \sigma_E^i = \frac{\Pi_i \sigma_E \Pi_i}{\mathrm{Tr}_E\{\Pi_i \sigma_E\}} \qquad i = 1, \ldots, \dim \mathcal{H}_E \qquad (5.489)$$

with σ_E^i the collection of statistical operators obtained from a fixed environmental state σ_E upon the action of a map which implements a von Neumann instrument according to Sect. 3.2.7.1. Note that while Theorem 5.2 was proven making reference to a finite-dimensional Hilbert space $\mathcal{H}_E$, the projections (5.472) and (5.474) corresponding to (5.487) and (5.489) are well-defined for an arbitrary environment.

5.5.2 Projected Equations of Motion

We suppose that system and environment altogether follow a unitary evolution determined by an Hamiltonian as in (5.5). We can then write the Liouville von Neumann equation for system and environment in the form

$$\frac{\mathrm{d}}{\mathrm{d}t}\rho(t) = \mathcal{L}(t)\rho(t), \qquad (5.490)$$

where we have defined the Liouvillian operator $\mathcal{L}(t)$ as

$$\mathcal{L}(t)[\sigma] = -\frac{i}{\hbar}\lambda[V(t), \sigma], \qquad (5.491)$$

with λ a dimensionless parameter that will help in keeping track of the perturbation order. Equation (5.490) is written in interaction picture with respect to $H_S + H_E$, though to keep the notation simple we omit a further subscript I in the expression of $\rho(t)$ and $V(t)$, which is given by

$$V(t) = e^{\frac{i}{\hbar}(H_S+H_E)t} V e^{-\frac{i}{\hbar}(H_S+H_E)t}. \qquad (5.492)$$

We now consider an operator $\mathcal{P}$ as in (5.475), together with the complementary projection operator

$$Q = \mathbb{1} - \mathcal{P}, \qquad (5.493)$$

in order to write the statistical operator as the sum of a relevant and irrelevant part

$$\rho = \mathcal{P}\rho + Q\rho. \qquad (5.494)$$

Due to (5.490) the two contributions obey the equations

$$\frac{d}{dt}\mathcal{P}\rho(t) = \mathcal{P}\mathcal{L}(t)(\mathcal{P}\rho(t) + \mathcal{Q}\rho(t)) \tag{5.495}$$

$$\frac{d}{dt}\mathcal{Q}\rho(t) = \mathcal{Q}\mathcal{L}(t)(\mathcal{P}\rho(t) + \mathcal{Q}\rho(t)). \tag{5.496}$$

It is now convenient to introduce the operator

$$\overleftarrow{\mathcal{G}}_{\mathcal{Q}}(t, t_1) = \overleftarrow{T}\,\exp\left(\int_{t_1}^{t} dt_2 \mathcal{Q}\mathcal{L}(t_2)\right), \tag{5.497}$$

that obeys the homogeneous equation

$$\frac{d}{dt}\overleftarrow{\mathcal{G}}_{\mathcal{Q}}(t, t_1) = \mathcal{Q}\mathcal{L}(t)\,\overleftarrow{\mathcal{G}}_{\mathcal{Q}}(t, t_1), \tag{5.498}$$

and such that

$$\overleftarrow{\mathcal{G}}_{\mathcal{Q}}(t, t) = \mathbb{1}. \tag{5.499}$$

Recalling that the differential equation

$$\frac{d}{dt}a(t) = b(t)(c(t) + a(t)) \tag{5.500}$$

has the general solution

$$a(t) = e^{\int_0^t d\tau\, b(\tau)} a(0) + \int_0^t d\tau e^{\int_\tau^t d\sigma\, b(\sigma)} b(\tau)c(\tau), \tag{5.501}$$

equation (5.496) for the irrelevant part can be checked to be solved by

$$\mathcal{Q}\rho(t) = \overleftarrow{\mathcal{G}}_{\mathcal{Q}}(t, 0)\mathcal{Q}\rho(0) + \int_0^t dt_1 \overleftarrow{\mathcal{G}}_{\mathcal{Q}}(t, t_1)\mathcal{Q}\mathcal{L}(t_1)\mathcal{P}\rho(t_1). \tag{5.502}$$

When we substitute this expression in (5.495) we obtain

$$\frac{d}{dt}\mathcal{P}\rho(t) = \mathcal{P}\mathcal{L}(t)\mathcal{P}\rho(t) + \mathcal{P}\mathcal{L}(t)\overleftarrow{\mathcal{G}}_{\mathcal{Q}}(t, 0)\mathcal{Q}\rho(0) \tag{5.503}$$

$$+ \int_0^t dt_1 \mathcal{P}\mathcal{L}(t)\overleftarrow{\mathcal{G}}_{\mathcal{Q}}(t, t_1)\mathcal{Q}\mathcal{L}(t_1)\mathcal{P}\rho(t_1).$$

We now consider two simplifying assumptions of general validity. First we suppose that the inhomogeneous term vanishes, so that

$$\mathcal{Q}\rho(0) = 0, \tag{5.504}$$

which means that the initial state is an eigenoperator of the projection on the relevant part. In the case of the standard projection (5.472) this corresponds to a factorized initial state, which grants as discussed in Sect. 5.2 the existence of the reduced dynamics of the system. The second assumption is the vanishing of contributions with an odd number of Liouvillians in between two projections

$$\mathcal{P}\mathcal{L}(t_1)\ldots\mathcal{L}(t_{2n+1})\mathcal{P} = 0. \tag{5.505}$$

This condition is equivalently written as

$$\mathrm{Tr}_E[V(t_1)\ldots V(t_{2n+1})\rho_E] = 0 \tag{5.506}$$

for the standard projection. We have for example

$$\mathcal{P}\mathcal{L}(t)\mathcal{P}\rho = \mathrm{Tr}_E[\mathcal{L}(t)\mathcal{P}\rho] \otimes \rho_E \tag{5.507}$$

$$= \frac{i}{\hbar}(\rho_S \, \mathrm{Tr}_E[\rho_E V(t)] - \mathrm{Tr}_E[V(t)\rho_E]\rho_S) \otimes \rho_E, \tag{5.508}$$

where $\mathrm{Tr}_E[V(t)\rho_E]$ is an operator on $\mathcal{H}_S$ and we have denoted reduced density operators as in (4.1). Equation (5.503) thus takes the form

$$\frac{\mathrm{d}}{\mathrm{d}t}\mathcal{P}\rho(t) = \int_0^t \mathrm{d}t_1 \mathcal{P}\mathcal{L}(t)\overleftarrow{\mathcal{G}}_Q(t, t_1)\mathcal{L}(t_1)\mathcal{P}\rho(t_1), \tag{5.509}$$

where we have exploited the fact that thanks to (5.505) the operator Q of (5.493) appearing in the integrand of (5.503) can be replaced with the identity.

5.5.3 Nakajima-Zwanzig Master Equation

We now introduce the integral kernel

$$\mathcal{K}_{\mathrm{NZ}}(t, t_1) = \mathcal{P}\mathcal{L}(t)\overleftarrow{\mathcal{G}}_Q(t, t_1)\mathcal{L}(t_1)\mathcal{P}, \tag{5.510}$$

which allows to rewrite (5.509) in the more compact form

$$\frac{\mathrm{d}}{\mathrm{d}t}\mathcal{P}\rho(t) = \int_0^t \mathrm{d}t_1 \mathcal{K}_{\mathrm{NZ}}(t, t_1)\mathcal{P}\rho(t_1), \tag{5.511}$$

known as Nakajima-Zwanzig master equation. It is often the case that the kernel depends on the difference in the time arguments, so that with the notation of (2.353) the master equation can be written in the form of a convolution

$$\frac{\mathrm{d}}{\mathrm{d}t}\mathcal{P}\rho(t) = (\mathcal{K}_{\mathrm{NZ}} \star \mathcal{P}\rho)(t). \tag{5.512}$$

Despite its complexity, due to the fact that its solution would be the exact expression for the reduced dynamics, this integro-differential master equation has the advantage that it naturally leads to a perturbative expansion arising from an expansion of the operator in (5.497). Different choices of $\mathcal{P}$ thanks to (5.471) all allow to obtain the exact expression for the reduced state ρ_S. They however lead to evolution equations for different relevant states, and therefore in particular to different expansions.

We consider the expansion of (5.497) relying on smallness of the parameter λ appearing in the expression of the Liouvillian (5.491)

$$\overleftarrow{\mathcal{G}}_Q(t, t_1) = \mathbb{1} + \int_{t_1}^{t} dt_2 \mathcal{Q}\mathcal{L}(t_2)$$
$$+ \int_{t_1}^{t} dt_2 \int_{t_1}^{t_2} dt_3 \mathcal{Q}\mathcal{L}(t_2)\mathcal{Q}\mathcal{L}(t_3) + \dots \qquad (5.513)$$

leading to write

$$\mathcal{K}_{\mathrm{NZ}} = \mathcal{K}_{\mathrm{NZ}}^{(2)} + \mathcal{K}_{\mathrm{NZ}}^{(4)} + \dots, \qquad (5.514)$$

where the superscript denotes the perturbation order corresponding to powers of λ. By making repeatedly use of (5.505) we obtain

$$\mathcal{K}_{\mathrm{NZ}}^{(2)}(t, t_1) = \mathcal{P}\mathcal{L}(t)\mathcal{L}(t_1)\mathcal{P} \qquad (5.515)$$

and

$$\mathcal{K}_{\mathrm{NZ}}^{(4)}(t, t_1) = \int_{t_1}^{t} dt_2 \int_{t_1}^{t_2} dt_3 [\mathcal{P}\mathcal{L}(t)\mathcal{L}(t_2)\mathcal{L}(t_3)\mathcal{L}(t_1)\mathcal{P} - \mathcal{P}\mathcal{L}(t)\mathcal{L}(t_2)\mathcal{P}\mathcal{L}(t_3)\mathcal{L}(t_1)\mathcal{P}],$$
$$(5.516)$$

where time-ordering is reflected in the integration order $t \geqslant t_2 \geqslant t_3 \geqslant t_1$. The first non-trivial contribution is obtained replacing $\overleftarrow{\mathcal{G}}_Q(t, t_1)$ with the identity, thus obtaining, upon setting λ equal to 1,

$$\frac{d}{dt}\mathcal{P}\rho(t) = -\frac{1}{\hbar^2}\int_0^t dt_1 \mathcal{P}[V(t), [V(t_1), \mathcal{P}\rho(t_1)]], \qquad (5.517)$$

and amounts to a weak-coupling second order or Born approximation, as it is usually called in scattering theory [102].

In order to obtain the master equation for the reduced state $\rho_S(t)$ we have to specify the projection $\mathcal{P}$. We first consider the standard projection (5.472) to obtain

$$\frac{d}{dt}\rho_S(t) \otimes \rho_E = -\frac{1}{\hbar^2}\int_0^t dt_1 \mathrm{Tr}_E\left[[V(t), [V(t_1), \rho_S(t_1) \otimes \rho_E]]\right] \otimes \rho_E, \quad (5.518)$$

where to comply with the constraint (5.504) the operator $\sigma_E \to \rho_E$ is exactly the marginal of the initial state $\rho_{SE}(0) = \rho_S(0) \otimes \rho_E$ assumed to be factorized in order to warrant the existence of a reduced dynamics in the sense of (5.9). Further taking the partial trace with respect to the environmental degrees of freedom, we obtain the following closed evolution equation for the reduced system statistical operator

$$\frac{\mathrm{d}}{\mathrm{d}t}\rho_S(t) = -\frac{1}{\hbar^2}\int_0^t \mathrm{d}t_1\,\mathrm{Tr}_E\left[[V(t),[V(t_1),\rho_S(t_1)\otimes\rho_E]]\right], \qquad (5.519)$$

which is often called generalized master equation. A different situation is encountered if we consider the projection (5.474) with σ_E commuting with the Π_i. Introducing the sub-collections

$$w_i(t) = \mathrm{Tr}_E[\Pi_i\rho(t)], \qquad (5.520)$$

which are elements of $\mathcal{T}(\mathcal{H}_S)$ with trace less or equal to one, and tracing out the environmental degrees of freedom we obtain

$$\frac{\mathrm{d}}{\mathrm{d}t}w_j(t) = -\frac{1}{\hbar^2}\int_0^t \mathrm{d}t_1\,\mathrm{Tr}_E\left[\Pi_j[V(t),[V(t_1),\sum_{i=1}^n w_i(t_1)\otimes\frac{\Pi_i}{\mathrm{Tr}_E\{\Pi_i\}}]]\right]. \qquad (5.521)$$

Exploiting the fact that $\{\Pi_i\}$ is a resolution of the identity we have

$$\rho_S(t) = \sum_j w_j(t), \qquad (5.522)$$

and therefore the master equation

$$\frac{\mathrm{d}}{\mathrm{d}t}\rho_S(t) = -\frac{1}{\hbar^2}\int_0^t \mathrm{d}t_1\,\mathrm{Tr}_E\left[[V(t),[V(t_1),\sum_{i=1}^n w_i(t_1)\otimes\frac{\Pi_i}{\mathrm{Tr}_E\{\Pi_i\}}]]\right]. \qquad (5.523)$$

Note however that in accordance to the fact that the initial state is not necessarily factorized, but is only an eigenoperator of the correlated projection, there generally is no closed dynamics at the level of the reduced system only. Indeed to solve the last equation one does not simply need to know the initial state of the system, but knowledge of all the sub-collections at the initial time is required. The initial state $\rho_S(0)$ only fixes $\sum_{i=1}^n w_i(0)$, and not the single contributions.

References

1. F. Bloch, Phys. Rev. **70**, 460 (1946). https://doi.org/10.1103/PhysRev.70.460
2. E.T. Jaynes, F.W. Cummings, Proc. IEEE **51**, 89 (1963). https://doi.org/10.1109/PROC.1963.1664
3. P. Ullersma, Physica **32**, 27 (1966). https://doi.org/10.1016/0031-8914(66)90102-9
4. D. Chruściński, Phys. Rep. **992**, 1 (2022). https://doi.org/10.1016/j.physrep.2022.09.003
5. M. Reed, B. Simon, *I: Functional Analysis* (Academic Press, San Diego, 1980)
6. M. Reed, B. Simon, *II: Fourier Analysis, Self-Adjointness* (Academic Press, San Diego, 1975)

7. F. Riesz, B. Sz.-Nagy, *Functional Analysis* (Dover, 1956)
8. K. Yosida, *Functional Analysis*. Classics in Mathematics (Springer, Berlin, Heidelberg, 1995). https://doi.org/10.1007/978-3-642-61859-8
9. A. Nordsieck, W. Lamb, G. Uhlenbeck, Physica **7**, 344 (1940). https://doi.org/10.1016/S0031-8914(40)90102-1
10. D. Chruściński, S. Pascazio, Open Syst. Inf. Dyn. **24**, 1740001 (2017). https://doi.org/10.1142/S1230161217400017
11. R. Alicki, in *Dynamics of Dissipation*, ed. by P. Garbaczewski, R. Olkiewicz, Lecture Notes in Physics, vol. 597 (Springer, Berlin, Heidelberg, 2002), pp. 239–264. https://doi.org/10.1007/3-540-46122-1_10
12. R.S. Ingarden, Rep. Math. Phys. **10**, 43 (1976). https://doi.org/10.1016/0034-4877(76)90005-7
13. V. Gorini, A. Kossakowski, E.C.G. Sudarshan, J. Math. Phys. **17**, 821 (1976). https://doi.org/10.1063/1.522979
14. G. Lindblad, Commun. Math. Phys. **48**, 119 (1976). https://doi.org/10.1007/BF01608499
15. A. Kossakowski, Rep. Math. Phys. **3**, 247 (1972). https://doi.org/10.1016/0034-4877(72)90010-9
16. A. Kossakowski, Bull. Acad. Pol. Sci. Serie Math. Astr. Phys. **20**, 1021 (1972)
17. D. Chruściński, Open Syst. Inf. Dyn. **28**, 2150015 (2021). https://doi.org/10.1142/S1230161221500153
18. A. Bassi, G. Ghirardi, Phys. Rep. **379**, 257 (2003). https://doi.org/10.1016/S0370-1573(03)00103-0
19. R. Alicki, K. Lendi, *Quantum Dynamical Semigroups and Applications*. Lecture Notes in Physics, vol. 286 (Springer, Berlin, 1987)
20. K.R. Parthasarathy, *An Introduction to Quantum Stochastic Calculus*. Monographs in Mathematics 85 (Birkhäuser Basel, 1992)
21. P. Hayden, J. Sorce, J. Phys. A: Math. Theor. **55**, 225302 (2022). https://doi.org/10.1088/1751-8121/ac65c2
22. A. Colla, H.-P. Breuer, Phys. Rev. A **105**, 052216 (2022). https://doi.org/10.1103/PhysRevA.105.052216
23. K. Mølmer, Y. Castin, J. Dalibard, J. Opt. Soc. Am. B **10**, 524 (1993). https://doi.org/10.1364/JOSAB.10.000524
24. H. Carmichael, *An Open Systems Approach to Quantum Optics* (Springer, Berlin, 1993)
25. H.-P. Breuer, F. Petruccione, *The Theory of Open Quantum Systems* (Oxford University Press, Oxford, 2002)
26. A. Barchielli, M. Gregoratti, *Quantum Trajectories and Measurements in Continuous Time*. Lecture Notes in Physics, vol. 782 (Springer, Berlin, 2009). https://doi.org/10.1007/978-3-642-01298-3
27. L. Diósi, W.T. Strunz, Phys. Lett. A **235**, 569 (1997). https://doi.org/10.1016/S0375-9601(97)00717-2
28. W.T. Strunz, L. Diósi, N. Gisin, Phys. Rev. Lett. **82**, 1801 (1999). https://doi.org/10.1103/PhysRevLett.82.1801
29. J. Piilo, S. Maniscalco, K. Härkönen, K.A. Suominen, Phys. Rev. Lett. **100**, 180402 (2008). https://doi.org/10.1103/PhysRevLett.100.180402
30. A. Smirne, M. Caiaffa, J. Piilo, Phys. Rev. Lett. **124**, 190402 (2020). https://doi.org/10.1103/PhysRevLett.124.190402
31. V. Link, K. Luoma, W.T. Strunz, New J. Phys. **25**, 093006 (2023). https://doi.org/10.1088/1367-2630/aceff3
32. P.P.G. Dyke, *An Introduction to Laplace Transforms and Fourier Series* (Springer, London 2001). https://doi.org/10.1007/978-1-4471-6395-4
33. K. Lendi, Phys. Rev. A **33**, 3358 (1986). https://doi.org/10.1103/PhysRevA.33.3358
34. F.J. Dyson, Phys. Rev. **75**, 486 (1949). https://doi.org/10.1103/PhysRev.75.486

35. R.P. Feynman, Phys. Rev. **84**, 108 (1951). https://doi.org/10.1103/PhysRev.84.108
36. R. Karplus, J. Schwinger, Phys. Rev. **73**, 1020 (1973). https://doi.org/10.1103/PhysRev.73.1020
37. D.J. Evans, G. Morriss, Chem. Phys. **87**, 451 (1984). https://doi.org/10.1016/0301-0104(84)85125-3
38. B. Vacchini, K. Hornberger, Phys. Rep. **478**, 71 (2009). https://doi.org/10.1016/j.physrep.2009.06.001
39. E. Joos, H.D. Zeh, C. Kiefer, D. Giulini, J. Kupsch, I.O. Stamatescu, *Decoherence and the Appearance of a Classical World in Quantum Theory*, 2nd ed. (Springer, Berlin, 2003). https://doi.org/10.1007/978-3-662-05328-7
40. M. Schlosshauer, *Decoherence and the Quantum-to-Classical Transition* (Springer, Berlin, 2007). https://doi.org/10.1007/978-3-540-35775-9
41. K. Hornberger, in *Entanglement and Decoherence*, ed. by A. Buchleitner, C. Viviescas, M. Tiersch, Lecture Notes in Physics (Springer, Berlin, 2009), pp. 221–276. https://doi.org/10.1007/978-3-540-88169-8_5
42. W.T. Strunz, in *Coherent Evolution in Noisy Environments*, ed. by A. Buchleitner, K. Hornberger (Springer, Berlin, 2002), Lecture Notes in Physics, pp. 199–233. https://doi.org/10.1007/3-540-45855-7_5
43. K. Hornberger, S. Uttenthaler, B. Brezger, L. Hackermüller, M. Arndt, A. Zeilinger, Phys. Rev. Lett. **90**, 160401 (2003). https://doi.org/10.1103/PhysRevLett.90.160401
44. D.A. Kokorowski, A.D. Cronin, T.D. Roberts, D.E. Pritchard, Phys. Rev. Lett. **86**, 2191 (2001). https://doi.org/10.1103/PhysRevLett.86.2191
45. J.J. Sakurai, *Modern Quantum Mechanics* (Addison-Wesley, Reading, MA, 1993)
46. B. Vacchini, Phys. Rev. Lett. **95**, 230402 (2005). https://doi.org/10.1103/PhysRevLett.95.230402
47. E. Joos, H.D. Zeh, Z. Phys. B **59**, 223 (1985). https://doi.org/10.1007/BF01725541
48. W.H. Zurek, in *Quantum Decoherence: Poincaré Seminar 2005*, ed. by B. Duplantier, J.M. Raimond, V. Rivasseau (Birkhäuser Basel, Basel, 2007), pp. 1–31. https://doi.org/10.1007/978-3-7643-7808-0_1
49. K. Hornberger, Phys. Rev. A **73**, 052102 (2006). https://doi.org/10.1103/PhysRevA.73.052102
50. A. Smirne, B. Vacchini, Phys. Rev. A **82**, 042111 (2010). https://doi.org/10.1103/PhysRevA.82.042111
51. A. Streltsov, H. Kampermann, S. Wölk, M. Gessner, D. Bruß, New J. Phys. **20**, 053058 (2018). https://doi.org/10.1088/1367-2630/aac484
52. A.S. Holevo, Rep. Math. Phys. **32**, 211 (1993). https://doi.org/10.1016/0034-4877(93)90014-6
53. B. Vacchini, J. Math. Phys. **42**, 4291 (2001). https://doi.org/10.1063/1.1386409
54. L. Diósi, Europhys. Lett. **30**, 63 (1995). https://doi.org/10.1209/0295-5075/30/2/001
55. F. Petruccione, B. Vacchini, Phys. Rev. E **71**, 046134 (2005). https://doi.org/10.1103/PhysRevE.71.046134
56. K. Hornberger, Phys. Rev. Lett. **97**, 060601 (2006). https://doi.org/10.1103/PhysRevLett.97.060601
57. H.-P. Breuer, B. Vacchini, Phys. Rev. E **76**, 036706 (2007). https://doi.org/10.1103/PhysRevE.76.036706
58. M. Busse, P. Pietrulewicz, H.-P. Breuer, K. Hornberger, Phys. Rev. E **82**, 026706 (2010). https://doi.org/10.1103/PhysRevE.82.026706
59. A.O. Caldeira, A.J. Leggett, Physica A **121**, 587 (1983). https://doi.org/10.1016/0378-4371(83)90013-4
60. S. Gnutzmann, F. Haake, Z. Phys. B **101**, 263 (1996). https://doi.org/10.1007/s002570050208
61. A. Barchielli, Nuovo Cimento B **74**, 113 (1983). https://doi.org/10.1007/BF02721671

62. A. Isar, A. Sandulescu, H. Scutaru, E. Stefanescu, W. Scheid, Int. J. Mod. Phys. E **03**, 635 (1994). https://doi.org/10.1142/S0218301394000164

63. L. Van Hove, Phys. Rev. **95**, 249 (1954). https://doi.org/10.1103/PhysRev.95.249

64. V.F. Sears, *Neutron Optics* (Oxford University Press, Oxford, 1989)

65. L. Pitaevskii, S. Stringari, *Bose-Einstein Condensation* (Oxford University Press, Oxford, 2003)

66. J. Rau, Phys. Rev. **129**, 1880 (1963). https://doi.org/10.1103/PhysRev.129.1880

67. V. Scarani, M. Ziman, P. Štelmachovič, N. Gisin, V. Bužek, Phys. Rev. Lett. **88**, 097905 (2002). https://doi.org/10.1103/PhysRevLett.88.097905

68. V. Giovannetti, G.M. Palma, Phys. Rev. Lett. **108**, 040401 (2012). https://doi.org/10.1103/PhysRevLett.108.040401

69. F. Ciccarello, G.M. Palma, V. Giovannetti, Phys. Rev. A **87**, 040103 (2013). https://doi.org/10.1103/PhysRevA.87.040103

70. B. Vacchini, Int. J. Quantum Inf. **12**, 1461011 (2014). https://doi.org/10.1142/S0219749914610115

71. S. Campbell, B. Vacchini, EPL **133**, 60001 (2021). https://doi.org/10.1209/0295-5075/133/60001

72. F. Ciccarello, S. Lorenzo, V. Giovannetti, G.M. Palma, Phys. Rep. **954**, 1 (2022). https://doi.org/10.1016/j.physrep.2022.01.001

73. G.M. Palma, K.A. Suominen, A. Ekert, Proc. Roy. Soc. A **452**, 567 (1996). https://doi.org/10.1098/rspa.1996.0029

74. G.D. Mahan, *Many-Particle Physics* (Springer, Berlin, 2000). https://doi.org/10.1007/978-1-4757-5714-9

75. W. Magnus, Commun. Pure Appl. Math. **7**, 649 (1954). https://doi.org/10.1002/cpa.3160070404

76. D.F. Walls, G.J. Milburn, *Quantum Optics* (Springer, Berlin, 1994). https://doi.org/10.1007/978-3-540-28574-8

77. R. Alicki, Open Syst. Inf. Dyn. **11**, 53 (2004). https://doi.org/10.1023/B:OPSY.0000024755.58888.ac

78. A.W. Chin, A. Rivas, S.F. Huelga, M.B. Plenio, J. Math. Phys. **51**, 092109 (2010). https://doi.org/10.1063/1.3490188

79. M. Nielsen, I. Chuang, *Quantum Computation and Quantum Information* (Cambridge University Press, Cambridge, 2000)

80. G. Clos, H.-P. Breuer, Phys. Rev. A **86**, 012115 (2012). https://doi.org/10.1103/PhysRevA.86.012115

81. D. Tamascelli, A. Smirne, J. Lim, S.F. Huelga, M.B. Plenio, Phys. Rev. Lett. **123**, 090402 (2019). https://doi.org/10.1103/PhysRevLett.123.090402

82. S. Mukamel, *Principles of Nonlinear Optical Spectroscopy* (Oxford University Press, 1995)

83. R. Martinazzo, B. Vacchini, K.H. Hughes, I. Burghardt, J. Chem. Phys. **134**, 011101 (2011). https://doi.org/10.1063/1.3532408

84. D. Gribben, A. Strathearn, J. Iles-Smith, D. Kilda, A. Nazir, B.W. Lovett, P. Kirton, Phys. Rev. Res. **2**, 013265 (2020). https://doi.org/10.1103/PhysRevResearch.2.013265

85. A. Garg, J.N. Onuchic, V. Ambegaokar, J. Chem. Phys. **83**, 4491 (1985). https://doi.org/10.1063/1.449017

86. M.P. Woods, R. Groux, A.W. Chin, S.F. Huelga, M.B. Plenio, J. Math. Phys. **55**, 032101 (2014). https://doi.org/10.1063/1.4866769

87. L.A. Correa, B. Xu, B. Morris, G. Adesso, J. Chem. Phys. **151**, 094107 (2019). https://doi.org/10.1063/1.5114690

88. J. Iles-Smith, N. Lambert, A. Nazir, Phys. Rev. A **90**, 032114 (2014). https://doi.org/10.1103/PhysRevA.90.032114

89. I.S. Gradshteyn, I.M. Ryzhik, *Table of Integrals, Series and Products* (Academic Press, 1965)

90. G.L. Ingold, in *Coherent Evolution in Noisy Environments*, ed. by A. Buchleitner, K. Hornberger (Springer, Berlin, 2002), pp. 1–53. https://doi.org/10.1007/3-540-45855-7_5

91. J.M. Taylor, C.M. Marcus, M.D. Lukin, Phys. Rev. Lett. **90**, 206803 (2003). https://doi.org/10.1103/PhysRevLett.90.206803

92. N. Zhao, S.W. Ho, R.B. Liu, Phys. Rev. B **85**, 115303 (2012). https://doi.org/10.1103/PhysRevB.85.115303

93. J. Jing, L.A. Wu, Sci. Rep. **8**, 1471 (2018). https://doi.org/10.1038/s41598-018-19977-9

94. W.A. Coish, D. Loss, Phys. Rev. B **70**, 195340 (2004). https://doi.org/10.1103/PhysRevB.70.195340

95. H.-P. Breuer, D. Burgarth, F. Petruccione, Phys. Rev. B **70**, 045323 (2004). https://doi.org/10.1103/PhysRevB.70.045323

96. E. Ferraro, H.-P. Breuer, A. Napoli, M.A. Jivulescu, A. Messina, Phys. Rev. B **78**, 064309 (2008). https://doi.org/10.1103/PhysRevB.78.064309

97. B.M. Garraway, Phys. Rev. A **55**, 2290 (1997). https://doi.org/10.1103/PhysRevA.55.2290

98. S. Nakajima, Prog. Theor. Phys. **20**, 948 (1958). https://doi.org/10.1143/PTP.20.948

99. R. Zwanzig, J. Chem. Phys. **33**, 1338 (1960). https://doi.org/10.1063/1.1731409

100. H.-P. Breuer, Phys. Rev. A **75**, 022103 (2007). https://doi.org/10.1103/PhysRevA.75.022103

101. H.-P. Breuer, in *Theoretical Foundations of Quantum Information Processing and Communication*, ed. by E. Bruening, F. Petruccione, Lecture Notes in Physics (Springer, Berlin, 2010), pp. 125–139. https://doi.org/10.1007/978-3-642-02871-7_5

102. J.R. Taylor, *Scattering Theory* (Wiley, New York, 1972)

Weak-Coupling Master Equation 6

Abstract

We explore the conditions under which, starting from a microscopic approach within the Nakajima–Zwanzig projection operator technique, a master equation leading to a Markov semigroup evolution for the reduced dynamics can be recovered. The obtained weak-coupling master equation has a number of interesting properties that will be investigated. We will consider in particular the case of a two-level system interacting with bosonic degrees of freedom, recovering the so-called quantum Bloch equations originated in the field of nuclear magnetic resonance and widely used in atomic physics and quantum optics. We will put into evidence typical quantum features emerging in this framework, such as the structure of the spectral peaks and photon anti-bunching in the light emitted by resonance fluorescence.

6.1 Introduction

In Sect. 5.3 we have put into evidence an important class of quantum dynamical maps, whose mathematical structure is well-characterized, namely quantum dynamical semigroups. The key constraint allowing for the characterization besides complete positivity was the requirement of a semigroup composition law forward in time. In Sect. 5.4 we have clarified by means of examples, in which an analytic treatment was feasible to a large extent, that this is indeed not the general situation, so that in Sect. 5.5 a general strategy to obtain a master equation describing the dynamics of the system starting from a microscopic Hamiltonian and considering a perturbative expansion was outlined. We will now investigate the approximations that lead within this microscopic approach to a dynamics described by a generator in Lindblad form, that is a semigroup evolution law. This allows to obtain a microscopically motivated expression for operators and coefficients appearing in the mathematical expression

B. Vacchini, *Open Quantum Systems*, Graduate Texts in Physics,
https://doi.org/10.1007/978-3-031-58218-9_6

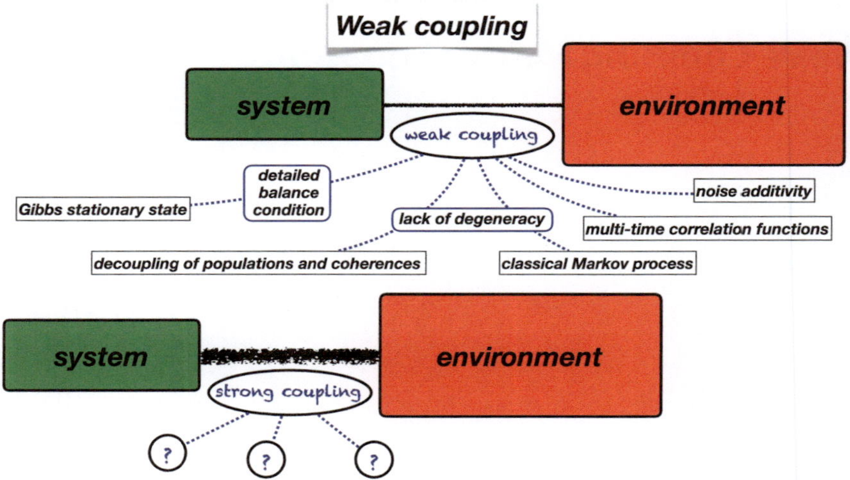

Fig. 6.1 In the presence of weak coupling between system and environment a number of physically meaningful properties can be shown to hold, whose validity is not generally warranted in the case of arbitrary strong coupling. These properties, investigated in Sect. 6.3, include the existence of a stationary state in Gibbs form, the approach to equilibrium, the connection to a classical Markovian description for the populations in the energy eigenbasis, the additivity of the effect of different noise sources, the validity of the regression formula for multi-time correlation functions

of the generator. This expression describes a dynamics with many physically relevant properties, as sketched in Fig. 6.1 and discussed in Sect. 6.3.

We will take as starting point the Nakajima–Zwanzig master equation (5.511), though other strategies can be considered, such as the time-convolutionless projection operator technique [1–3], or a standard interaction-picture perturbation expansion [4]. An important advantage of projection operator techniques is that they provide a convenient tool to deal in a systematic way with higher order contributions and allow for phenomenologically motivated Ansatz for the memory kernel. The rigorous mathematical derivation of a master equation in Lindblad form in the weak-coupling limit is due to Davies [5].

6.2 Born–Markov Approximation

We take as starting point (5.519), corresponding to the second order term in the expansion of the Nakajima–Zwanzig kernel obtained when considering a standard projection. We perform a change of variables $t_1 = t - \tau$ in the integral thus obtaining the expression

$$\frac{d\rho_S(t)}{dt} = -\frac{1}{\hbar^2} \int_0^t d\tau \, \mathrm{Tr}_E \left[[V(t), [V(t-\tau), \rho_S(t-\tau) \otimes \rho_E]] \right]. \tag{6.1}$$

A natural approximation is introduced assuming a separation of time-scales between a typical time-scale τ_E of the environment and a typical time-scale τ_R of the reduced system, namely

$$\tau_E \ll \tau_R. \tag{6.2}$$

The time-scale τ_E is determined by the decay time of the relevant correlation functions of the environment, that implicitly appear in (6.1) due to the presence of the average over the environmental state of a bilinear expression in the interaction operator, as we shall detail in Sect. 6.2.1. Relying on the time-scale separation expressed by (6.2) we approximate $\rho_S(t-\tau) \approx \rho_S(t)$, neglecting the small retardation time. We thus obtain, instead of an integro-differential, a time-local master equation, in which at the r.h.s. we have a time-dependent map acting on the statistical operator evaluated at the same time

$$\frac{d\rho_S(t)}{dt} = -\frac{1}{\hbar^2} \int_0^t d\tau \, \mathrm{Tr}_E \left[[V(t), [V(t-\tau), \rho_S(t) \otimes \rho_E]] \right]. \tag{6.3}$$

A general discussion of the features of time-local master equations can be found in Sect. 7.4.2. The master equation (6.3) is known as Redfield master equation. It was first introduced in the energy eigenbasis and used for the study of relaxation processes [6,7]. We can further simplify this expression extending the integral to the whole positive real line. The latter approximation is known as Markov approximation. Again, the basic justification behind this approximation is the separation of time-scales (6.2), so that the integrand can be considered appreciably different from zero only on a short time interval. The extension of the integration to all positive times should therefore scarcely affect it, albeit with the advantage of removing the dependence on the time integration interval. We finally obtain

$$\frac{d\rho_S(t)}{dt} = -\frac{1}{\hbar^2} \int_0^\infty d\tau \, \mathrm{Tr}_E \left[[V(t), [V(t-\tau), \rho_S(t) \otimes \rho_E]] \right], \tag{6.4}$$

known as master equation in the Born–Markov approximation. We will now further analyze this expression putting it in a more specific context allowing to introduce further approximations and obtain the so-called weak-coupling master equation. A word of warning is in place here, since there is no perfect consensus in the wide literature on the names associated to the different master equations at the different stages of approximation. Furthermore the impact and relevance of each approximation in the description of different aspects of the dynamics is still the object of intense research, see e.g. [8–14] and references therein.

6.2.1   Dressed Interaction Operators and Correlation Functions

Following [5, 15, 16] we will now derive relevant expressions of master equation that can be obtained from (6.4) for the case in which H_S has purely discrete spectrum. We consider a general interaction term in the form

$$V = \hbar \sum_\mu A_\mu \otimes X_\mu, \tag{6.5}$$

with $\{A_\mu\}$ and $\{X_\mu\}$ acting on the Hilbert space of system and environment respectively, and V self-adjoint. In order to sort out different contributions to the dynamics according to their time dependence it is convenient to decompose the interaction term into eigenoperators of the free system Hamiltonian H_S. Given the purely discrete spectrum of H_S we can write

$$H_S = \sum_n \hbar \omega_n P_n, \tag{6.6}$$

where $\hbar \omega_n$ denote the eigenvalues of the free system Hamiltonian and P_n the projections on the corresponding eigenspaces. The P_n are not necessarily one dimensional, but provide an orthogonal resolution of the identity in the sense of (2.73). We can thus introduce the following decomposition for the system operators appearing in the interaction term

$$A_\mu = \left(\sum_{\omega_n} P_n \right) A_\mu \left(\sum_{\omega_m} P_m \right) \tag{6.7}$$

$$= \sum_\omega \sum_{\omega_m - \omega_n = \omega} P_n A_\mu P_m \tag{6.8}$$

$$= \sum_\omega A_\mu(\omega), \tag{6.9}$$

where the real quantities ω do correspond to all possible frequency differences among distinct levels of the system, thus taking on both positive and negative values. Equation (6.9) defines the collection of operators

$$A_\mu(\omega) = \sum_{\omega_m - \omega_n = \omega} P_n A_\mu P_m, \tag{6.10}$$

satisfying

$$A_\mu(\omega)^\dagger = A_\mu^\dagger(-\omega). \tag{6.11}$$

These operators together with their adjoints are by construction eigenoperators of the free system Hamiltonian according to the relations

$$[H_S, A_\mu(\omega)] = -\omega A_\mu(\omega) \tag{6.12}$$

$$[H_S, A_\mu(\omega)^\dagger] = +\omega A_\mu(\omega)^\dagger. \tag{6.13}$$

The index μ here appears as a degeneracy index for the eigenoperators corresponding to the eigenvalues $\pm\omega$. Exploiting these commutation relations we have

$$[H_S, A_\mu(\omega)^\dagger A_\nu(\omega)] = 0. \tag{6.14}$$

In interaction picture these operators are only modified by a time dependent phase, namely

$$e^{iH_S t} A_\mu(\omega) e^{-iH_S t} = e^{-i\omega t} A_\mu(\omega), \tag{6.15}$$

so that in this picture the interaction term of (6.5) can be written as

$$V(t) = \hbar \sum_\mu \sum_\omega e^{-i\omega t} A_\mu(\omega) \otimes X_\mu(t), \tag{6.16}$$

where $X_\mu(t)$ are the interaction picture environment operators

$$X_\mu(t) = e^{iH_E t} X_\mu e^{-iH_E t} \tag{6.17}$$

satisfying

$$X_\mu(t)^\dagger = X_\mu^\dagger(t). \tag{6.18}$$

Exploiting self-adjointness of the interaction term we can write it also in the equivalent form

$$V(t) = \hbar \sum_\mu \sum_\omega e^{+i\omega t} A_\mu(\omega)^\dagger \otimes X_\mu^\dagger(t). \tag{6.19}$$

The condition (5.505) used to obtain the second-order Nakajima–Zwanzig master equation in the form (5.519), here used as our starting point, implies

$$\mathcal{P}\mathcal{L}(t)\mathcal{P} = 0, \tag{6.20}$$

that in this framework translates into the requirement

$$\mathrm{Tr}_E\{X_\mu(t)\rho_E\} = \langle X_\mu(t)\rangle_{\rho_E} = 0. \tag{6.21}$$

The master equation (6.4) can be written in the more compact form

$$\frac{d\rho_S(t)}{dt} = \frac{1}{\hbar^2} \int_0^\infty d\tau\, \mathrm{Tr}_E \left[V(t-\tau)\rho_S(t) \otimes \rho_E V(t) - V(t)V(t-\tau)\rho_S(t) \otimes \rho_E + h.c. \right],$$

$$(6.22)$$

so that if we introduce the explicit expressions (6.16) and (6.19) for the operators $V(t-\tau)$ and $V(t)$, using dummy labels ν, ω and μ, ω' respectively, we obtain

$$\frac{d\rho_S(t)}{dt} = \sum_{\substack{\omega,\,\omega' \\ \mu,\,\nu}} e^{i(\omega'-\omega)t} \int_0^\infty d\tau e^{i\omega\tau}\, \mathrm{Tr}_E\{X_\mu^\dagger(t)X_\nu(t-\tau)\rho_E\}$$

$$\times [A_\nu(\omega)\rho_S(t)A_\mu(\omega')^\dagger - A_\mu(\omega')^\dagger A_\nu(\omega)\rho_S(t)] + h.c., \qquad (6.23)$$

where we have used the cyclic property of the trace and commutation of system and environment operators. In view of (6.23), it is natural to introduce the following $\mathbb{C}$-number functions

$$\Gamma_{\mu\nu}(\omega) = \int_0^\infty d\tau e^{i\omega\tau}\, \mathrm{Tr}_E\{X_\mu^\dagger(t)X_\nu(t-\tau)\rho_E\}. \qquad (6.24)$$

If the environmental state is invariant under the free evolution, so that

$$[H_E, \rho_E] = 0, \qquad (6.25)$$

the correlation functions appearing in (6.24) are stationary, and we thus obtain the time independent expression

$$\Gamma_{\mu\nu}(\omega) = \int_0^\infty d\tau e^{i\omega\tau} \langle X_\mu^\dagger(\tau)X_\nu(0)\rangle_{\rho_E}, \qquad (6.26)$$

where $\langle \ldots \rangle_{\rho_E}$ denotes the expectation value with respect to the stationary state ρ_E, corresponding to the one-sided Fourier transform of the bath correlation function. Thanks to the introduction of the dressed operators (6.10), each term of the sum appearing in (6.23) puts into evidence an explicit time dependence through the oscillating functions $\exp[i(\omega'-\omega)t]$. We now perform a so-called secular approximation, consisting in keeping only terms with $\omega = \omega'$, under the guiding idea that the terms with $\omega \neq \omega'$ do not give a significant contribution to the time evolution if they oscillate very rapidly. This approximation thus requires that

$$\tau_S \ll \tau_R, \qquad (6.27)$$

where τ_S is a typical time-scale of the free system evolution, which can be estimated as $\tau_S \approx 1/|\omega - \omega'|$, while τ_R is a typical time-scale of the dynamics of the system coupled to the environment, roughly of the order of the inverse of the coupling constants appearing in the master equation. In the course of the derivation we have thus put into evidence the existence of three time-scales, that should be well separated: τ_E is associated to the bath correlation functions; τ_S is associated to the free system dynamics; τ_R is associated to the system dynamics when coupled to the environment, thus including decoherence and dissipative effects. The condition $\tau_E \ll \tau_R$ of (6.2) is related to the Markov approximation, while the condition $\tau_S \ll \tau_R$ of (6.27) is needed for the secular approximation, sometimes called rotating-wave approximation. With respect to terminology, the same warning expressed below (6.4) applies, so that one has to pay attention to the different usage of the same word in the literature. It is further convenient, in order to better put into evidence the operator structure of the master equation, to introduce the Hermitian and anti-Hermitian part of the matrix of functions (6.26) using the notation

$$\gamma_{\mu\nu}(\omega) = \Gamma_{\mu\nu}(\omega) + \Gamma^*_{\nu\mu}(\omega) \tag{6.28}$$

and

$$S_{\mu\nu}(\omega) = \frac{1}{2i}[\Gamma_{\mu\nu}(\omega) - \Gamma^*_{\nu\mu}(\omega)], \tag{6.29}$$

so that

$$\Gamma_{\mu\nu}(\omega) = \frac{1}{2}\gamma_{\mu\nu}(\omega) + i S_{\mu\nu}(\omega), \tag{6.30}$$

with $\gamma_{\mu\nu}(\omega)$ and $S_{\mu\nu}(\omega)$ Hermitian matrices. Rearranging terms we obtain

$$\frac{d\rho_S(t)}{dt} = -\frac{i}{\hbar}\left[\sum_\omega \sum_{\mu,\nu} S_{\mu\nu}(\omega) A_\mu(\omega)^\dagger A_\nu(\omega), \rho_S(t)\right]$$
$$+ \sum_\omega \sum_{\mu,\nu} \gamma_{\mu\nu}(\omega)\left[A_\nu(\omega)\rho_S(t)A_\mu(\omega)^\dagger - \frac{1}{2}\{A_\mu(\omega)^\dagger A_\nu(\omega), \rho_S(t)\}\right], \tag{6.31}$$

where according to their explicit expression

$$\gamma_{\mu\nu}(\omega) = \int_{-\infty}^{+\infty} d\tau\, e^{i\omega\tau} \langle X^\dagger_\mu(\tau) X_\nu(0)\rangle_{\rho_E} \tag{6.32}$$

the coefficients appearing in the master equation, that keep track of the effect of the environment, are given by the Fourier transform of the coupling operator auto-correlation function.

6.2.2 Lindblad form and Complete Positivity

The very structure of (6.31) complies with (4.127) of Sect. 4.4.2, so that it is an Hermiticity-preserving and trace-annihilating map. Most importantly, it turns out that the r.h.s. of (6.31) is indeed a sum of contributions in Lindblad form. This follows from the fact that the matrix $\gamma_{\mu\nu}(\omega)$, corresponding to the Kossakowski matrix introduced in Sect. 5.3.1, is positive for any value of ω. To prove this we show that the function

$$\mathfrak{h}(\omega) = \sum_{\mu,\nu} v_\mu^* \gamma_{\mu\nu}(\omega) v_\nu \tag{6.33}$$

is positive for all collections of complex numbers $\{v_\mu\}$. Indeed, $\mathfrak{h}(\omega)$ is the Fourier transform of the positive-definite function

$$\mathfrak{f}(\tau) = \sum_{\mu,\nu} v_\mu^* \langle X_\mu^\dagger(\tau) X_\nu(0) \rangle_{\rho_E} v_\nu, \tag{6.34}$$

hence positive as a consequence of Bochner's theorem [17] as explained in Remark 6.1.

> ### 6.1 Bochner and Bernstein Theorems
>
> We present in the following Bochner's theorem, connecting positive-definite functions with characteristic functions and positive measures by means of the Fourier transform. To this aim we first introduce the notion of positive-definite function [18]. A continuous bounded function
>
> $$\mathfrak{f} : \mathbb{R} \to \mathbb{C} \tag{6.35}$$
>
> is said to be positive-definite if it is Hermitian, in the sense that
>
> $$\mathfrak{f}(t)^* = \mathfrak{f}(-t), \tag{6.36}$$
>
> and if
>
> $$\sum_{i,j=1}^{n} v_i^* \mathfrak{f}(t_i - t_j) v_j \geqslant 0 \tag{6.37}$$
>
> for any integer n and any collection of real numbers $t_1, \ldots, t_n$, as well as of complex numbers $v_1, \ldots, v_n$. The condition (6.37) can be equivalently formulated asking the $n \times n$ matrices $a_{ij} = \mathfrak{f}(t_i - t_j)$ to be positive for any integer

n and any collection of real numbers $t_1, \ldots, t_n$. In particular for $n = 1, 2$ we recover the constraints $\mathfrak{f}(0) \geqslant 0$ and $|\mathfrak{f}(t)| \leqslant \mathfrak{f}(0)$. From the condition (6.37) it follows that the Fourier transform of a positive function is positive-definite, but also the converse holds, as shown in a theorem by Bochner [19].

Theorem 6.1 (Bochner, 1933) A function $\mathfrak{f} : \mathbb{R} \to \mathbb{C}$ can be written in the form

$$\mathfrak{f}(t) = \int_{-\infty}^{+\infty} e^{i\omega t} \, dF(\omega), \tag{6.38}$$

with $dF(\omega)$ a positive bounded measure iff it is positive-definite.

Each such measure identifies a positive non-decreasing function $F(\omega)$

$$F : \mathbb{R} \to \mathbb{R}_+, \tag{6.39}$$

given by

$$F(\omega) = \int_{-\infty}^{\omega} dF(\omega'), \tag{6.40}$$

and called distribution function. If the measure $dF(\omega)$ admits a density, in the sense that $dF(\omega) = f(\omega)d\omega$, then $f(\omega)$ is a positive function. If the distribution function is bounded by one, it is called cumulant distribution function, a key object in probability theory. In particular, it defines a probability measure. In this case (6.38), which can also be seen as the Riemann–Stieltjes integral of the complex exponential function with respect to the cumulant distribution function

$$\Phi(t) = \int_{-\infty}^{+\infty} e^{i\omega t} \, dF(\omega), \tag{6.41}$$

is known as characteristic function, and it uniquely identifies the probability distribution. If $dF(\omega)$ admits a density $f(\omega)$, which is then a probability density, the characteristic function is its Fourier transform. We can therefore say (up to a normalization factor) that a function is positive-definite iff it is a characteristic function.

We now show that the function $\mathfrak{f}(\tau)$ defined in (6.34) is definite positive, thus warranting positivity of the matrix (6.32). We define

$$B = \sum_{\nu} X_\nu v_\nu, \tag{6.42}$$

so that

$$\mathfrak{f}(\tau) = \langle B^\dagger(\tau)B(0)\rangle_{\rho_E}, \tag{6.43}$$

and therefore using (6.17) together with (6.25) and the cyclic property of the trace

$$\sum_{i,j=1}^{n} v_i^* \mathfrak{f}(t_i - t_j)v_j = \sum_{i,j=1}^{n} v_i^* \mathrm{Tr}_E\{e^{iH_E t_i} B^\dagger e^{-iH_E t_i} e^{iH_E t_j} B e^{-iH_E t_j} \rho_E\}v_j \tag{6.44}$$

$$= \mathrm{Tr}_E\left\{\left(\sum_{i=1}^{n} B(t_i)v_i\right)^\dagger \left(\sum_{j=1}^{n} B(t_j)v_j\right)\rho_E\right\}, \tag{6.45}$$

which is the expectation value of a positive operator, and therefore positive. Note that in a similar way, for any Hamiltonian H_S, the expectation value of the associated evolution operator $U(t)$ with an arbitrary state $\mathrm{Tr}\{\rho_S U(t)\}$ is a positive-definite function. We have in fact, considering the standard orthogonal decomposition of a state (2.72) and the group composition law (5.16)

$$\sum_{i,j=1}^{n} v_i^* \mathrm{Tr}\{\rho_S U(t_i - t_j)\}v_j = \sum_{k} p_k \|\psi_k\|^2 \tag{6.46}$$

with the identification

$$\psi_k = \sum_{j=1}^{n} v_j U(t_j)^\dagger \varphi_k. \tag{6.47}$$

If we consider in particular a pure state ψ, the considered quantity is the so-called survival amplitude

$$a_\psi(t) = \langle \psi|U(t)\psi\rangle, \tag{6.48}$$

which is therefore a positive-definite function. This remark is at the basis of a proof of Stone's theorem [20].

The representation (6.38) has an interesting analog in terms of the Laplace transform, of relevance when considering master equations with a memory kernel, as discussed in Sect. 7.5.2. We introduce the class of completely monotone functions [21], that is functions

$$\mathfrak{g} : \mathbb{R}_+ \to \mathbb{R} \tag{6.49}$$

that are infinitely differentiable and satisfy the conditions

$$(-)^n \mathfrak{g}^{(n)}(t) \geqslant 0 \quad \forall n \in \mathbb{N} \quad \forall t \in [0, +\infty), \tag{6.50}$$

where $\mathfrak{g}^{(n)}$ denotes the n-th derivative of the function. In particular, the conditions for $n = 1, 2, 3$ imply that completely monotone functions are positive, decreasing and convex. These functions admit an interesting integral representation analog to (6.38), as Laplace transform of a positive function. Indeed, we have the following result by Bernstein [22].

Theorem 6.2 (Bernstein, 1922) A function $\mathfrak{g} : \mathbb{R}_+ \to \mathbb{R}$ can be written in the form

$$\mathfrak{g}(t) = \int_0^\infty e^{-\omega t} dG(\omega), \tag{6.51}$$

with $dG(\omega)$ a positive measure iff it is completely monotone.

We can rephrase the representation (6.51) saying that a completely monotone function can be expressed as a continuous mixture of real exponential functions. If the measure $dG(\omega)$ admits a density $g(\omega)$, this probability density provides the weights of the different components in the mixture. Bernstein's theorem thus provides an integral representation of completely monotone functions in terms of exponential functions. An important class of completely monotone functions is given by the Laplace transform of waiting time distributions, that is probability densities over the positive real line, that allow to evaluate the time one has to wait before a given event takes place. These functions appear in connection with open quantum system dynamics described by memory-kernel master equations, as we shall see in Sect. 7.5.2.1.

To make the notation more compact, we introduce the Hamiltonian contribution

$$H_{\mathrm{LS}} = \sum_\omega \sum_{\mu,\nu} S_{\mu\nu}(\omega) A_\mu(\omega)^\dagger A_\nu(\omega), \tag{6.52}$$

where the subscript LS stands for Lamb shift, a notation used to stress that the contribution (6.52) induces a coherent contribution to the dynamics equivalent to a level shift, together with the incoherent contribution

$$\mathcal{D}[\sigma] = \sum_\omega \sum_{\mu,\nu} \gamma_{\mu\nu}(\omega) \left[A_\nu(\omega)\sigma A_\mu(\omega)^\dagger - \frac{1}{2}\{A_\mu(\omega)^\dagger A_\nu(\omega), \sigma\} \right], \tag{6.53}$$

corresponding to the dissipator of (5.47). In this notation (6.31) reads

$$\frac{\mathrm{d}\rho_S(t)}{\mathrm{d}t} = -\frac{i}{\hbar}[H_{\mathrm{LS}}, \rho_S(t)] + \mathcal{D}[\rho_S(t)], \tag{6.54}$$

which thanks to positivity of the matrix $\gamma_{\mu\nu}(\omega)$ is indeed of the Lindblad form (5.27), so that the associated solution is given by a semigroup of completely positive quantum dynamical maps.

Note that, thanks to the fact that the operators $A_\nu(\omega)$ are eigenoperators of the system free Hamiltonian, the dissipator $\mathcal{D}$ is covariant in the sense of (2.323) with respect to the unitary transformation generated by H_S, namely

$$\mathcal{D}[e^{-\frac{i}{\hbar}H_S t}\rho_S(t)e^{+\frac{i}{\hbar}H_S t}] = e^{-\frac{i}{\hbar}H_S t}\mathcal{D}[\rho_S(t)]e^{+\frac{i}{\hbar}H_S t}. \tag{6.55}$$

This property together with (6.14) implies that the Schrödinger picture master equation corresponding to the interaction picture master equation (6.54) is only modified by the addition of a commutator term with H_S.

6.3 General Properties of the Master Equation

We now put into evidence some important properties exhibited by the master equation under suitable further hypotheses. This analysis by far does not exhaust the appealing physical features of the weak-coupling master equation, that was the starting point for seminal work on quantum thermodynamics, such as the study of irreversible entropy production by Spohn in [23] or the analysis of heat engines by Alicki in [24]. See [25,26] for a concise presentation of recent developments in quantum thermodynamics as well as the reviews [27–29] and references therein.

6.3.1 Detailed Balance and Stationary State

We assume that not only (6.25) holds, but the environment is a thermal reservoir described by the canonical equilibrium state (5.277). In this case we can verify that the reduced system state

$$\rho^{\mathrm{th}} = \frac{e^{-\beta H_S}}{\mathrm{Tr}_S\{e^{-\beta H_S}\}} \tag{6.56}$$

is a stationary solution of (6.54), where β is the same inverse temperature $\beta = 1/(k_B T)$ characterizing the equilibrium state of the environment. To verify this fact we first observe that from (6.14) we have

$$[H_S, H_{\mathrm{LS}}] = 0 \tag{6.57}$$

and therefore

$$[\rho^{\text{th}}, H_{\text{LS}}] = 0. \tag{6.58}$$

Moreover, we observe that due to (6.12) and (6.13), in analogy to (6.15) we have

$$\rho^{\text{th}} A_\mu(\omega) = e^{+\beta\hbar\omega} A_\mu(\omega)\rho^{\text{th}} \tag{6.59}$$

$$\rho^{\text{th}} A_\mu(\omega)^\dagger = e^{-\beta\hbar\omega} A_\mu(\omega)^\dagger \rho^{\text{th}}. \tag{6.60}$$

Exploiting (6.59) and (6.60) we thus obtain from (6.53)

$$\mathcal{D}[\rho^{\text{th}}] = \sum_\omega \sum_{\mu,\nu} [\gamma_{\mu\nu}(\omega)e^{-\beta\hbar\omega} A_\nu(\omega)A_\mu(\omega)^\dagger - \gamma_{\mu\nu}(\omega)A_\mu(\omega)^\dagger A_\nu(\omega)]\rho^{\text{th}}.$$

$$\tag{6.61}$$

As shown in Remark 6.2, for a thermal reservoir validity of the β-KMS condition
implies for the matrix of dissipation coefficients the relation (6.73) and therefore

$$\sum_\omega \sum_{\mu,\nu} \gamma_{\mu\nu}(\omega)e^{-\beta\hbar\omega} A_\nu(\omega)A_\mu(\omega)^\dagger = \sum_\omega \sum_{\mu,\nu} \int_{-\infty}^{+\infty} d\sigma\, e^{-i\omega\sigma}$$

$$\times \langle X_\nu(\sigma) X_\mu^\dagger(0)\rangle_{\text{th}} A_\nu(\omega)A_\mu(\omega)^\dagger \tag{6.62}$$

so that exploiting at the r.h.s. the crucial property of self-adjointness of the interaction
term (6.16)

$$\sum_\omega \sum_{\mu,\nu} \gamma_{\mu\nu}(\omega)e^{-\beta\hbar\omega} A_\nu(\omega)A_\mu(\omega)^\dagger = \sum_\omega \sum_{\mu,\nu} \int_{-\infty}^{+\infty} d\sigma\, e^{i\omega\sigma}$$

$$\times \langle X_\nu^\dagger(\sigma) X_\mu(0)\rangle_{\text{th}} A_\nu(\omega)^\dagger A_\mu(\omega) \tag{6.63}$$

$$= \sum_\omega \sum_{\mu,\nu} \gamma_{\nu\mu}(\omega)A_\nu(\omega)^\dagger A_\mu(\omega), \tag{6.64}$$

implying

$$\sum_\omega \sum_{\mu,\nu} [\gamma_{\mu\nu}(\omega)e^{-\beta\hbar\omega} A_\nu(\omega)A_\mu(\omega)^\dagger - \gamma_{\mu\nu}(\omega)A_\mu(\omega)^\dagger A_\nu(\omega)] = 0 \tag{6.65}$$

and therefore stationarity of ρth

$$\mathcal{D}[\rho^{\text{th}}] = 0. \tag{6.66}$$

When the environment is in the canonical equilibrium state, or more precisely when
(6.69) holds, the weak-coupling master equation (6.31) predicts that a state in the form
(6.56) is a stationary solution of the dynamics. As a result, for a system coupled to a
thermal bath at temperature $\beta = 1/(k_B T)$ a Gibbs state with the same temperature

of the bath and the free system Hamiltonian H_S is indeed a stationary state for the system. A natural and interesting question is whether this state is also an equilibrium state, that is, it is reached as a consequence of the interaction between system and bath. If this happens for an arbitrary initial state and the stationary state is unique the considered semigroup dynamics is said to be relaxing. Sufficient conditions to ensure this behavior have been proven in the literature, both in the finite and infinite-dimensional case [30]. These conditions are related to strict positivity of the dissipation matrix $\gamma_{\mu\nu}(\omega)$ or low enough degeneracy of its zero eigenvalue, as well as to the fact that only trivial operators proportional to the identity commute with the operators appearing in the master equation, as we shall see in an explicit example in Sect. 6.5.4. The physical interpretation of these conditions is that all transitions of the system are coupled to the reservoir.

6.2 β-KMS Condition

In the general case the environment is not finite-dimensional, and an operator of the form (5.277) cannot be a trace class operator. The proper formulation of the notion of thermal bath in this setting is given by a requirement on the correlation functions, the so-called β-KMS condition. In the notation of (5.316) we introduce the correlation functions of the bath

$$G_{\mu\nu}(t) = \langle X_{\mu}^{\dagger}(t) X_{\nu}(0) \rangle_{\rho_E}, \tag{6.67}$$

as well as the complex functions

$$F_{\mu\nu}(z) = \langle X_{\nu}(0) X_{\mu}^{\dagger}(z) \rangle_{\rho_E}. \tag{6.68}$$

We can now formulate the β-KMS condition, named after Kubo, Martin and Schwinger [31], asking that each function $F_{\mu\nu}$ is analytic in the strip $0 < \Im z < \beta\hbar$, the functions $F_{\mu\nu}$ and $G_{\mu\nu}$ are continuous and bounded for z on the real axis, and $G_{\mu\nu}$ is the boundary value of $F_{\mu\nu}$ when $z \to t + i\beta\hbar$. In this case the state of the environment is said to be thermal and we compactly write

$$\langle X_{\mu}^{\dagger}(t) X_{\nu}(0) \rangle_{\text{th}} = \langle X_{\nu}(0) X_{\mu}^{\dagger}(t + i\beta\hbar) \rangle_{\text{th}}. \tag{6.69}$$

This condition is formally directly verified by inspection if we use the expression

$$\langle X_{\mu}^{\dagger}(t) X_{\nu}(0) \rangle_{\text{th}} = \text{Tr}_E \left\{ X_{\mu}^{\dagger}(t) X_{\nu}(0) \frac{e^{-\beta H_E}}{Z_E} \right\}. \tag{6.70}$$

Starting from the explicit expression (6.32) of the dissipation coefficients and building on (6.69) we have

$$\gamma_{\mu\nu}(-\omega) = \int_{-\infty}^{+\infty} d\tau \, e^{-i\omega\tau} \langle X_\nu(0) X_\mu^\dagger(\tau + i\beta\hbar) \rangle_{\text{th}}, \qquad (6.71)$$

so that relying on analyticity of the functions $F_{\mu\nu}(z)$ of (6.68) we are left with

$$\gamma_{\mu\nu}(-\omega) = e^{-\beta\hbar\omega} \int_{-\infty}^{+\infty} d\sigma \, e^{-i\omega\sigma} \langle X_\nu(0) X_\mu^\dagger(\sigma) \rangle_{\text{th}}, \qquad (6.72)$$

and exploiting stationarity implied by (6.25), so that the correlation functions only depend on the time difference, we finally obtain

$$\gamma_{\mu\nu}(\omega) e^{-\beta\hbar\omega} = \int_{-\infty}^{+\infty} d\sigma \, e^{-i\omega\sigma} \langle X_\nu(\sigma) X_\mu^\dagger(0) \rangle_{\text{th}}, \qquad (6.73)$$

corresponding to the so-called quantum detailed balance condition. For the case in which the coupling operators are self-adjoint, the relation (6.73) takes the simpler form

$$\gamma_{\mu\nu}(-\omega) = e^{-\beta\hbar\omega} \gamma_{\nu\mu}(\omega). \qquad (6.74)$$

Note that the coupling term (6.5) can always be rewritten as sum of self-adjoint operators [32], though this is generally not the natural physical expression. Generators of quantum dynamical semigroups of the form (6.54) and obeying the quantum detailed balance condition (6.74) are often called Davies generators [5, 33].

6.3.2 Dynamics of Populations and Coherences

We now take a closer look at the behavior of the matrix elements of the master equation in the eigenbasis of the system Hamiltonian (6.6). In particular, following standard usage, we will call populations the diagonal matrix elements in this basis, and coherences the off-diagonal matrix elements. The populations provide a probability distribution over the possible free energy eigenvalues, while the coherences describe typical quantum states and phenomena.

A particularly interesting situation arises if the discrete spectrum of H_S is non-degenerate, so that we have

$$P_i = |\varphi_i\rangle\langle\varphi_i|, \qquad (6.75)$$

and each eigenvector corresponds to a distinct eigenvalue. In order to evaluate the diagonal matrix elements of (6.54) in the energy eigenbasis we observe that due to (6.10), and thanks to orthonormality of the $\{\varphi_i\}$ we have

$$\langle \varphi_m | A_\mu(\omega)\varphi_n \rangle = \sum_{\omega_j - \omega_i = \omega} \langle \varphi_m | \varphi_i \rangle \langle \varphi_i | A_\mu \varphi_j \rangle \langle \varphi_j | \varphi_n \rangle \qquad (6.76)$$

$$= \delta_{\omega_n - \omega_m, \omega} \langle \varphi_m | A_\mu \varphi_n \rangle. \qquad (6.77)$$

Considering the expression (6.52) this implies

$$\langle \varphi_n | [H_{\mathrm{LS}}, \rho_S(t)]\varphi_n \rangle = 0. \qquad (6.78)$$

For the dissipative part we have to consider

$$\langle \varphi_n | A_\nu(\omega)\rho_S(t) A_\mu(\omega)^\dagger \varphi_n \rangle = \sum_{r,s} \langle \varphi_n | A_\nu(\omega)\varphi_r \rangle \langle \varphi_r | \rho_S(t)\varphi_s \rangle \langle \varphi_s | A_\mu(\omega)^\dagger \varphi_n \rangle$$

$$= \sum_r \delta_{\omega_r - \omega_n, \omega} \langle \varphi_n | A_\nu \varphi_r \rangle \langle \varphi_n | A_\mu \varphi_r \rangle^* \langle \varphi_r | \rho_S(t)\varphi_r \rangle, \qquad (6.79)$$

where only diagonal matrix elements appear. Dealing in the same way with the anti-commutator part we obtain

$$\langle \varphi_n | \mathcal{D}[\rho_S(t)]\varphi_n \rangle = \sum_r \left(\sum_{\mu,\nu} \gamma_{\mu\nu}(\omega_r - \omega_n) \langle \varphi_n | A_\nu \varphi_r \rangle \langle \varphi_n | A_\mu \varphi_r \rangle^* \right) \langle \varphi_r | \rho_S(t)\varphi_r \rangle$$

$$- \sum_r \left(\sum_{\mu,\nu} \gamma_{\mu\nu}(\omega_r - \omega_n) \langle \varphi_r | A_\nu \varphi_n \rangle \langle \varphi_r | A_\mu \varphi_n \rangle^* \right) \langle \varphi_n | \rho_S(t)\varphi_n \rangle, \qquad (6.80)$$

where we have used the non-degeneracy of the spectrum. We can now consider the quantities

$$W(n|r) = \sum_{\mu,\nu} \gamma_{\mu\nu}(\omega_r - \omega_n) \langle \varphi_n | A_\nu \varphi_r \rangle \langle \varphi_n | A_\mu \varphi_r \rangle^*, \qquad (6.81)$$

which are actually positive, in that for any fixed r and n they are a particular instance of (6.33), so that they can be interpreted as transition rates. Introducing the notation

$$\langle \varphi_n | \rho_S(t)\varphi_n \rangle = p_n(t) \qquad (6.82)$$

for the diagonal matrix elements of the statistical operator, that define a time dependent probability distribution, we finally have from (6.54) considering (6.78) and (6.80)

$$\frac{\mathrm{d}}{\mathrm{d}t} p_n(t) = \sum_r [W(n|r) p_r(t) - W(r|n) p_n(t)]. \tag{6.83}$$

This equation is also known as Pauli master equation. It is the master equation for a classical Markovian process, as we shall discuss in Sect. 7.2.1, with transition rates $W(n|r)$. In the absence of degeneracy, therefore, the populations in the energy eigenbasis obey closed evolution equations that admit a classical interpretation, the quantum features only appearing in the detailed expression of the transition rates in terms of suitable matrix elements of the coupling operators as given by (6.81). If the β-KMS condition is satisfied, so that the matrix of dissipation coefficients obeys the detailed balance condition (6.74), the classical rates (6.81) obey the detailed balance condition

$$W(n|r)\mathrm{e}^{-\beta \hbar \omega_r} = W(r|n)\mathrm{e}^{-\beta \hbar \omega_n}. \tag{6.84}$$

Validity of this condition ensures that the Boltzmann distribution

$$p_n^{\mathrm{eq}} = \frac{\mathrm{e}^{-\beta \hbar \omega_n}}{\sum_m \mathrm{e}^{-\beta \hbar \omega_m}} \tag{6.85}$$

is a stationary solution of (6.83). In the stationary state the occupation of the different energy levels thus follows the Boltzmann law.

6.4 Quantum Optical Master Equation

We now move from the general expression of the weak-coupling master equation to a definite model of environment and system. We take the environment to be a quantized electromagnetic field. In analogy to Sect. 5.4.1 we start considering a discrete number of modes, leaving a suitable continuous limit to a later stage.

6.4.1 Coupling to the Electromagnetic Field

We assume that the electromagnetic field is confined in a cubical region of volume V and consider periodic boundary conditions. Apart from an infinite $\mathbb{C}$-number contribution related to normal ordering, the free electromagnetic Hamiltonian can be written

$$H_E = \sum_{\boldsymbol{k}} \sum_{\lambda=1,2} \hbar \omega_{\boldsymbol{k}} b_\lambda^\dagger(\boldsymbol{k}) b_\lambda(\boldsymbol{k}), \tag{6.86}$$

where the sum is over wavevector k and polarization state λ of the photons, with frequencies fixed by the dispersion relation

$$\omega_k = c|k|. \tag{6.87}$$

The creation and annihilation operators are assumed to satisfy the canonical commutation relations

$$[b_\lambda(k), b_{\lambda'}^\dagger(k')] = \delta_{\lambda,\lambda'}\delta_{k,k'} \tag{6.88}$$

$$[b_\lambda(k), b_{\lambda'}(k')] = 0 \tag{6.89}$$

$$[b_\lambda^\dagger(k), b_{\lambda'}^\dagger(k')] = 0, \tag{6.90}$$

so that we have a bosonic environment. We denote by $\boldsymbol{\varepsilon}_\lambda(k)$ the unit polarization vectors, that must lie in the plane orthogonal to the wave vector, and are chosen so as to form a canonical basis

$$k \cdot \boldsymbol{\varepsilon}_\lambda(k) = 0 \tag{6.91}$$

$$\boldsymbol{\varepsilon}_\lambda(k) \cdot \boldsymbol{\varepsilon}_{\lambda'}(k) = \delta_{\lambda,\lambda'}. \tag{6.92}$$

Expressing an orthogonal resolution of the identity operator in this three-dimensional linear space by means of this orthogonal basis we can write

$$\mathbb{1} = |\boldsymbol{\varepsilon}_1(k)\rangle\langle\boldsymbol{\varepsilon}_1(k)| + |\boldsymbol{\varepsilon}_2(k)\rangle\langle\boldsymbol{\varepsilon}_2(k)| + |\hat{k}\rangle\langle\hat{k}|, \tag{6.93}$$

with $\hat{k} = k/|k|$. It will be useful to consider the projection of a vector on its transverse component with respect to the direction k, given by

$$P_{k^\perp} = \mathbb{1} - |\hat{k}\rangle\langle\hat{k}|, \tag{6.94}$$

so that taking matrix elements of (6.93) we come to the identity

$$\sum_{\lambda=1,2} (\boldsymbol{\varepsilon}_\lambda(k))_i \, (\boldsymbol{\varepsilon}_\lambda(k))_j = \delta_{ij} - \frac{k_i k_j}{|k|^2}. \tag{6.95}$$

An explicit expression of this Cartesian basis can be given in terms of the standard polar coordinates as follows

$$\hat{k} = (\sin\theta\cos\varphi, \sin\theta\sin\varphi, \cos\theta)$$

$$\boldsymbol{\varepsilon}_1(k) = (-\cos\theta\cos\varphi, -\cos\theta\sin\varphi, \sin\theta) \tag{6.96}$$

$$\boldsymbol{\varepsilon}_2(k) = (\sin\varphi, -\cos\varphi, 0).$$

Having in mind to identify the system with the levels of an atomic system, we consider a coupling term of the form

$$V = - \sum_{i=1,2,3} D_i \otimes E_i \tag{6.97}$$

$$= -\boldsymbol{D} \cdot \boldsymbol{E}, \tag{6.98}$$

that complies with the general expression (6.5), where the sum is now over the three Cartesian directions. The triple of operators D_i are the components of the dipole operator for the system, while E_i are the three Cartesian components of the quantized electric field. This term properly describes the interaction between atom and electromagnetic field in the electric dipole approximation, as detailed in Remark 6.3.

6.3 Dipole Approximation

We provide here a compact justification of the expression (6.98) for the coupling term [34]. We consider the Hamiltonian for a particle interacting with a classically described electromagnetic field obtained according to the correspondence rule

$$H = \frac{1}{2m} \left(\hat{\boldsymbol{p}} - \frac{q}{c} \boldsymbol{A}(\hat{\boldsymbol{x}}, t) \right)^2 + q\Phi(\hat{\boldsymbol{x}}, t), \tag{6.99}$$

where $\boldsymbol{A}$ and Φ denote the electromagnetic potentials and q the charge of the considered particle. We now consider the time-dependent unitary transformation

$$\bar{\psi} = O\psi \tag{6.100}$$

with

$$O = \exp \left(\frac{i}{\hbar} \frac{q}{c} g(\hat{\boldsymbol{x}}, t) \right), \tag{6.101}$$

and $g(\boldsymbol{x}, t)$ a real function. The vectors ψ and $\bar{\psi}$, being connected by a unitary transformation, provide the same physical description, and if ψ denotes a solution of the Schrödinger equation determined by (6.99), then $\bar{\psi}$ obeys

$$i\hbar \frac{\mathrm{d}\bar{\psi}(t)}{\mathrm{d}t} = \bar{H}\bar{\psi}(t) \tag{6.102}$$

with

$$\bar{H} = OHO^{\dagger} - i\hbar O \frac{\partial O^{\dagger}}{\partial t}. \tag{6.103}$$

Starting from the expression (6.101) and exploiting the fundamental commutation relations we obtain

$$\hat{\boldsymbol{p}} O^{\dagger} = O^{\dagger} \hat{\boldsymbol{p}} - i\hbar \nabla O^{\dagger} \tag{6.104}$$

$$= O^{\dagger} \hat{\boldsymbol{p}} - O^{\dagger} \frac{q}{c} \nabla g(\hat{\boldsymbol{x}}, t), \tag{6.105}$$

and therefore

$$\left(\hat{\boldsymbol{p}} - \frac{q}{c} \boldsymbol{A}(\hat{\boldsymbol{x}}, t) \right) O^{\dagger} = O^{\dagger} \left(\hat{\boldsymbol{p}} - \frac{q}{c} (\boldsymbol{A}(\hat{\boldsymbol{x}}, t) + \nabla g(\hat{\boldsymbol{x}}, t)) \right) \tag{6.106}$$

so that iterating

$$\left(\hat{\boldsymbol{p}} - \frac{q}{c} \boldsymbol{A}(\hat{\boldsymbol{x}}, t) \right)^2 O^{\dagger} = O^{\dagger} \left(\hat{\boldsymbol{p}} - \frac{q}{c} (\boldsymbol{A}(\hat{\boldsymbol{x}}, t) + \nabla g(\hat{\boldsymbol{x}}, t)) \right)^2. \tag{6.107}$$

Further noting

$$i\hbar O \frac{\partial O^{\dagger}}{\partial t} = \frac{q}{c} \frac{\partial g(\hat{\boldsymbol{x}}, t)}{\partial t}, \tag{6.108}$$

equation (6.103) explicitly reads

$$\bar{H} = \frac{1}{2m} \left(\hat{\boldsymbol{p}} - \frac{q}{c} (\boldsymbol{A}(\hat{\boldsymbol{x}}, t) + \nabla g(\hat{\boldsymbol{x}}, t)) \right)^2 + q \left(\Phi(\hat{\boldsymbol{x}}, t) - \frac{1}{c} \frac{\partial g(\hat{\boldsymbol{x}}, t)}{\partial t} \right). \tag{6.109}$$

For the Hamiltonian (6.99) the unitary operator (6.101) thus induces a transformation that has the same form of a gauge transformation of the electromagnetic field. We now consider the case of vanishing scalar potential $\Phi = 0$, and introduce the dipole approximation by neglecting the position dependence in the vector potential so that

$$\boldsymbol{A}(\hat{\boldsymbol{x}}, t) \simeq \boldsymbol{A}(t). \tag{6.110}$$

This approximation is justified if the radiation wavelength is much larger than the atomic dimensions. The unitary transformation (6.101) corresponding to the choice

$$g(\hat{\boldsymbol{x}}, t) = -\hat{\boldsymbol{x}} \cdot \boldsymbol{A}(t) \tag{6.111}$$

then leads to

$$\bar{H} = \frac{\hat{p}^2}{2m} - q\hat{x} \cdot E(t), \tag{6.112}$$

with

$$E(t) = -\frac{1}{c}\frac{\partial A(t)}{\partial t} \tag{6.113}$$

the electric field. An alternative treatment, in which the dipole approximation is taken only after explicitly expanding the Hamiltonian (6.109) for the choice of gauge (6.111), that provides an alternative insight in the dipole approximation, can be found in [35].

We start from the reference expression of the electromagnetic field vector potential in a volume V in the interaction picture [35,36]

$$A(t) = c\sqrt{\frac{2\pi\hbar}{V}} \sum_{k} \sum_{\lambda=1,2} \frac{1}{\sqrt{\omega_k}} \boldsymbol{\varepsilon}_\lambda(k)[b_\lambda(k)e^{-i\omega_k t} + b_\lambda^\dagger(k)e^{+i\omega_k t}] \tag{6.114}$$

to obtain, according to (6.113)

$$E(t) = i\sqrt{\frac{2\pi\hbar}{V}} \sum_{k} \sum_{\lambda=1,2} \sqrt{\omega_k}\,\boldsymbol{\varepsilon}_\lambda(k)[b_\lambda(k)e^{-i\omega_k t} - b_\lambda^\dagger(k)e^{+i\omega_k t}]. \tag{6.115}$$

Note that in this case the constraint (6.21) originating from (6.20) reads

$$\langle E(t) \rangle_{\rho_E} = 0. \tag{6.116}$$

Still assuming (6.25), so that the bath autocorrelation functions only depend on the time difference, (6.26) now becomes

$$\Gamma_{rs}(\omega) = \frac{1}{\hbar^2} \int_0^\infty d\tau\, e^{i\omega\tau} \langle E_r(\tau) E_s(0) \rangle_{\rho_E}. \tag{6.117}$$

6.4.2 Thermal Environment

Considering the electromagnetic field in a thermal state with temperature $T = 1/(k_B\beta)$, so that in particular (6.116) is verified, we have the identities

$$\langle b_\lambda(\mathbf{k})b_{\lambda'}^\dagger(\mathbf{k}')\rangle_{\rho_E} = \delta_{\lambda,\lambda'}\delta_{\mathbf{k},\mathbf{k}'}(1 + N_\beta(\omega_k)) \tag{6.118}$$

$$\langle b_\lambda^\dagger(\mathbf{k})b_{\lambda'}(\mathbf{k}')\rangle_{\rho_E} = \delta_{\lambda,\lambda'}\delta_{\mathbf{k},\mathbf{k}'}N_\beta(\omega_k), \tag{6.119}$$

where $N_\beta(\omega)$ is the Bose-Einstein distribution

$$N_\beta(\omega) = \frac{1}{e^{\beta\hbar\omega} - 1}, \tag{6.120}$$

and the vacuum state is recovered for $N_\beta(\omega) \to 0$. The other expectation values are equal to zero

$$\langle b_\lambda(\mathbf{k})b_{\lambda'}(\mathbf{k}')\rangle_{\rho_E} = \langle b_\lambda^\dagger(\mathbf{k})b_{\lambda'}^\dagger(\mathbf{k}')\rangle_{\rho_E} = 0, \tag{6.121}$$

and therefore do not contribute. Inserting (6.115) in (6.117) and using the identities (6.118), (6.119) and (6.121) we obtain the explicit expression

$$\Gamma_{ij}(\omega) = \frac{2\pi}{\hbar V}\sum_{\mathbf{k}}\omega_k\sum_{\lambda=1,2}(\boldsymbol{\varepsilon}_\lambda(\mathbf{k}))_i(\boldsymbol{\varepsilon}_\lambda(\mathbf{k}))_j$$
$$\times \{(1 + N_\beta(\omega_k))\varphi(\omega - \omega_k) + N_\beta(\omega_k)\varphi(\omega + \omega_k)\}, \tag{6.122}$$

where we have introduced the function

$$\varphi(\omega) = \int_0^\infty d\tau e^{i\omega\tau}, \tag{6.123}$$

to be interpreted in the sense of distributions. We will give meaning to integrals of this quantity with a smooth function, according to (6.136). We now exploit (6.95) and take a continuum limit, so as to consider the electromagnetic field in free space rather than in a finite volume, adopting the replacement

$$\frac{1}{V}\sum_{\mathbf{k}} \to \int\frac{d^3\mathbf{k}}{(2\pi)^3} = \frac{1}{(2\pi)^3c^3}\int_0^\infty d\omega_k\omega_k^2\int d\Omega, \tag{6.124}$$

with ω_k as in (6.87). Further using (6.96) we have

$$\int d\Omega\left(\delta_{ij} - \frac{k_ik_j}{|\mathbf{k}|^2}\right) = \frac{8}{3}\pi\delta_{ij}, \tag{6.125}$$

so that the matrix (6.26) now becomes proportional to the identity and we thus set

$$\Gamma_{ij}(\omega) = \Gamma(\omega)(\mathbb{1})_{ij}, \tag{6.126}$$

with

$$\Gamma(\omega) = \frac{2}{3\pi\hbar c^3} \int_0^\infty d\varpi\, \varpi^3 \{(1 + N_\beta(\varpi))\varphi(\omega - \varpi) + N_\beta(\varpi)\varphi(\omega + \varpi)\}. \tag{6.127}$$

To evaluate the integrals appearing in this expression we rely on the formula (6.136) considered in Remark 6.4, so that further recalling (6.28) and (6.126) we obtain

$$\gamma(\omega) = 2\Re\Gamma(\omega) \tag{6.128}$$

$$= \frac{4}{3\hbar c^3} \int_0^\infty d\varpi\, \varpi^3 \{(1 + N_\beta(\varpi))\delta(\omega - \varpi) + N_\beta(\varpi)\delta(\omega + \varpi)\}, \tag{6.129}$$

as well as

$$S(\omega) = \Im\Gamma(\omega) \tag{6.130}$$

$$= \frac{2}{3\pi\hbar c^3}\, \mathrm{PV} \int_0^\infty d\varpi\, \varpi^3 \left\{ \frac{1 + N_\beta(\varpi)}{\omega - \varpi} + \frac{N_\beta(\varpi)}{\omega + \varpi} \right\}. \tag{6.131}$$

6.4 Sokhotski–Plemelj Formula

We briefly recall the so-called Sokhotski–Plemelj formula for the real line [37], namely the expression

$$\lim_{\varepsilon \to 0} \int_a^b dx\, \frac{f(x)}{x - y \mp i\varepsilon} = \mathrm{PV} \int_a^b dx\, \frac{f(x)}{x - y} \pm i\pi f(y), \tag{6.132}$$

where PV denotes the Cauchy principal value. The formula is valid for a function $f(z)$, holomorphic and bounded in a strip $0 < \Im z < \eta$, with η a positive constant and $a < y < b$. In the physical literature the formula is often written as

$$\frac{1}{x \mp i\varepsilon} = \mathrm{PV}\,\frac{1}{x} \pm i\pi\delta(x). \tag{6.133}$$

Exploiting this formula we have for the integral of the function (6.123) on the positive real axis

$$\int_0^\infty d\varpi\, f(\varpi)\varphi(\omega \pm \varpi) = \lim_{\varepsilon \to 0^+} \int_0^\infty d\varpi\, f(\varpi) \int_0^\infty d\tau\, e^{i(\omega \pm \varpi + i\varepsilon)\tau} \tag{6.134}$$

$$= i \lim_{\varepsilon \to 0^+} \int_0^\infty d\varpi\, \frac{f(\varpi)}{\omega \pm \varpi + i\varepsilon}, \tag{6.135}$$

so that (6.132) leads to the identity

$$\int_0^\infty d\varpi\, f(\varpi)\varphi(\omega \mp \varpi) = \pi f(\pm\omega) + i\,\mathrm{PV}\int_0^\infty d\varpi\,\frac{f(\varpi)}{\omega \pm \varpi}. \tag{6.136}$$

We can simplify the expression of (6.128) noting that from (6.120) and according to (5.304) we have

$$2N_\beta(\omega) + 1 = \coth\left(\frac{\beta}{2}\hbar\omega\right), \tag{6.137}$$

so that odd parity of the hyperbolic cotangent implies

$$1 + N_\beta(-\omega) = -N_\beta(\omega). \tag{6.138}$$

Whatever the sign of ω we thus obtain from (6.129)

$$\gamma(\omega) = \frac{4\omega^3}{3\hbar c^3}(1 + N_\beta(\omega)). \tag{6.139}$$

Upon the identification

$$J(\omega) = \frac{4\omega^3}{3\hbar c^3} \tag{6.140}$$

this quantity can be written in the form (5.355). Indeed, for an odd function $J(\omega)$ we have from (5.355) and (6.137) for any real ω the identity

$$J_{\mathrm{eff}}^\beta(\omega) = J(\omega)(1 + N_\beta(\omega)). \tag{6.141}$$

The Hamiltonian contribution (6.52) appearing in the master equation (6.54) thus takes the explicit form

$$H_{\mathrm{LS}} = \sum_\omega \hbar S(\omega)\boldsymbol{D}(\omega)^\dagger \cdot \boldsymbol{D}(\omega), \tag{6.142}$$

where as in (6.98) the dot denotes summation over Cartesian indexes and according to (6.10) we have introduced the dressed dipole operators

$$D_i(\omega) = \sum_{\omega_m - \omega_n = \omega} P_n D_i P_m. \tag{6.143}$$

Note that with the identification (6.140) we have

$$S(\omega) = \text{PV}\, \frac{1}{2\pi} \int_0^\infty d\varpi\, J(\varpi) \left\{ \frac{1 + N_\beta(\varpi)}{\omega - \varpi} + \frac{N_\beta(\varpi)}{\omega + \varpi} \right\}. \tag{6.144}$$

This term according to (6.52) produces a shift in the energy levels, including both a thermal and a vacuum contribution [16,38]. In the same notation the dissipator in (6.54) reads

$$\mathcal{D}[\rho_S(t)] = \sum_\omega J_{\text{eff}}^\beta(\omega) \left[\boldsymbol{D}(\omega) \cdot \rho_S(t) \boldsymbol{D}(\omega)^\dagger - \frac{1}{2}\{\boldsymbol{D}(\omega)^\dagger \cdot \boldsymbol{D}(\omega), \rho_S(t)\} \right]. \tag{6.145}$$

Given that the components of the dipole operator are self-adjoint

$$D_i = D_i^\dagger, \tag{6.146}$$

from (6.11) we have

$$D_i(\omega)^\dagger = D_i(-\omega), \tag{6.147}$$

so that further using (6.138) and (6.141) we can conveniently split the contributions in (6.145) in two sums according to

$$\mathcal{D}[\rho_S(t)] = \sum_{\omega>0} J(\omega)N_\beta(\omega) \left[\boldsymbol{D}(\omega)^\dagger \cdot \rho_S(t) \boldsymbol{D}(\omega) - \frac{1}{2}\{\boldsymbol{D}(\omega) \cdot \boldsymbol{D}(\omega)^\dagger, \rho_S(t)\} \right] \tag{6.148}$$

$$+ \sum_{\omega>0} J(\omega)(1 + N_\beta(\omega)) \left[\boldsymbol{D}(\omega) \cdot \rho_S(t) \boldsymbol{D}(\omega)^\dagger - \frac{1}{2}\{\boldsymbol{D}(\omega)^\dagger \cdot \boldsymbol{D}(\omega), \rho_S(t)\} \right]. \tag{6.149}$$

In this representation only positive frequencies appear and the role of the Planck distribution (6.120), providing the mean number of excitations present in the bosonic environment at the considered temperature and frequency, is clearly spelled out. The two contributions in (6.148) and (6.149) can be interpreted respectively as transfer of an excitation from the environment to the system and vice versa. To clarify this statement consider an eigenvector φ_n of the system Hamiltonian H_S corresponding to the eigenvalue E_n, so that

$$H_S\varphi_n = E_n\varphi_n. \tag{6.150}$$

Due to (6.143), that warrants (6.12) and (6.13), we have that $D_i(\omega)\varphi_n$ still is an eigenvector corresponding to the eigenvalue $E_n - \omega$, smaller than the original one since we now only consider positive values of ω, namely

$$H_S D_i(\omega)\varphi_n = D_i(\omega)(H_S - \omega\mathbb{1})\varphi_n \tag{6.151}$$

$$= (E_n - \omega)D_i(\omega)\varphi_n, \tag{6.152}$$

and similarly for $D_i(\omega)^\dagger \varphi_n$, corresponding to an eigenvector with eigenvalue $E_n + \omega$. The action of the Lindblad operators appearing in (6.149) thus removes an excitation from the system, and the rate corresponding to this event is proportional to $1 + N_\beta(\omega)$, so that this contribution is non-vanishing even if the electromagnetic field is in the vacuum. Correspondingly the terms in (6.148) describe the inverse process and vanish if the field is in the vacuum state, so that no excitations can be transferred to the system. A similar situation has been encountered in Sect. 5.4.2, where due to the fact that the environment starts in the vacuum state, the exact master equation (5.424) only contains a contribution corresponding to loss of an excitation of the system to the environment.

6.5 Bloch Equations

We now further specify the structure of the weak-coupling master equation (6.54) by taking as system interacting with the electromagnetic field a two-level atom coherently stimulated by a monochromatic laser described in classical terms. The expression is meant to describe an atom or any other quantum system responsive to the electromagnetic field in which with good accuracy only two levels are involved in the considered dynamics. A system with these features is often called driven damped two-level system, and the equations describing it are known as optical Bloch equations. The name is due to the fact that equations of this form for a quantum system undergoing a dissipative evolution were first introduced phenomenologically by Bloch in a seminal paper [39]. The aim was the description of the dynamics of a nuclear spin coupled to a driving magnetic field in the presence of an environment.

For the case of a two-level system the free Hamiltonian can be taken to be of the form (5.227), so that the decomposition (6.6) reads

$$H_S = \frac{\hbar\omega_0}{2} P_+ - \frac{\hbar\omega_0}{2} P_-, \tag{6.153}$$

with

$$P_+ = |+\rangle\langle+| \tag{6.154}$$

and

$$P_- = |-\rangle\langle-| \tag{6.155}$$

projections on excited and ground state respectively. The only two dressed operators are thus given by

$$D_i(\omega_0) = P_- D_i P_+ \tag{6.156}$$

and

$$D_i(-\omega_0) = P_+ D_i P_-, \tag{6.157}$$

connected via (6.147), so that we can write

$$\boldsymbol{D}(\omega_0) = \boldsymbol{d}\sigma_- \tag{6.158}$$

and

$$\boldsymbol{D}(-\omega_0) = \boldsymbol{d}^*\sigma_+, \tag{6.159}$$

with σ_+ and σ_- as in (2.87) and (2.88) of Remark 2.1. Here $\boldsymbol{d}$ denotes the dipole moment of the atom with respect to the two considered electronic levels

$$\boldsymbol{d} = \langle -|\boldsymbol{D}|+\rangle = \langle -|e\hat{\boldsymbol{x}}|+\rangle, \tag{6.160}$$

with e electron charge and $\hat{\boldsymbol{x}}$ the relevant triple of position operators [40]. The two states $|+\rangle$ and $|-\rangle$ therefore denote the two states of the atom, labelled by suitable quantum numbers, that are relevant for the considered dynamics. According to (6.12) and (6.13) these operators satisfy

$$[H_S, \boldsymbol{D}(\pm\omega_0)] = \pm\omega_0 \boldsymbol{D}(\pm\omega_0). \tag{6.161}$$

The Hamiltonian contribution (6.52) now reads

$$H_{\mathrm{LS}} = \hbar|\boldsymbol{d}|^2(S(\omega_0)P_+ + S(-\omega_0)P_-), \tag{6.162}$$

while the dissipator of (6.145) can be written as

$$\mathcal{D}[\rho_S(t)] = \gamma_0(1 + N_\beta(\omega_0))\left[\sigma_-\rho_S(t)\sigma_+ - \frac{1}{2}\{\sigma_+\sigma_-, \rho_S(t)\}\right]$$
$$+ \gamma_0 N_\beta(\omega_0)\left[\sigma_+\rho_S(t)\sigma_- - \frac{1}{2}\{\sigma_-\sigma_+, \rho_S(t)\}\right], \tag{6.163}$$

where we have introduced the rate

$$\gamma_0 = J(\omega_0)|\boldsymbol{d}|^2, \tag{6.164}$$

called spontaneous emission rate. Note that the expression (6.163) corresponds to (5.47) for the case in which we have just two Lindblad operators, namely

$$L_1 = \sigma_- \tag{6.165}$$
$$L_2 = \sigma_+, \tag{6.166}$$

with corresponding rates

$$\gamma_1 = \gamma_0(1 + N_\beta(\omega_0)) \tag{6.167}$$

$$\gamma_2 = \gamma_0 N_\beta(\omega_0). \tag{6.168}$$

To assess the validity of the approximations leading to the weak-coupling master equation (6.54) the time-scale identified by this rate, namely $\tau_R \approx 1/\gamma_0$, should be much larger than the time-scale identified by a typical frequency of the electromagnetic field $\tau_E \approx 1/\omega_0$.

We further consider the coupling of the two-level atom to a classical monochromatic electromagnetic field in the electric dipole and rotating-wave approximation. The driving electric field can be written as

$$\boldsymbol{E} = i\boldsymbol{\varepsilon}\sqrt{\frac{2\pi\hbar}{V}}\sqrt{\omega_L}(A_L e^{-i\omega_L t} - A_L^* e^{+i\omega_L t}) \tag{6.169}$$

with A_L the amplitude of the laser and $\boldsymbol{\varepsilon}$ its polarization vector [40]. In the absence of a permanent dipole moment the system Hamiltonian is modified by the time-dependent interaction term

$$H_L(t) = -i\boldsymbol{d} \cdot \boldsymbol{\varepsilon}\sqrt{\frac{2\pi\hbar}{V}}\sqrt{\omega_L} A_L e^{-i\omega_L t}\sigma_+ + h.c. \tag{6.170}$$

of the form (6.98). It is now convenient to define the quantity

$$\Omega = -i\boldsymbol{d} \cdot \boldsymbol{\varepsilon}\sqrt{\frac{2\pi}{\hbar V}}\sqrt{\omega_L} A_L^*, \tag{6.171}$$

which has the inverse dimensions of time and can always be taken to be positive, by making a suitable choice of phase in the definition of the considered electronic states $|+\rangle$ and $|-\rangle$ of the atom, so that we can write

$$H_L(t) = -\frac{\hbar\Omega}{2}(e^{-i\omega_L t}\sigma_+ + e^{i\omega_L t}\sigma_-). \tag{6.172}$$

The quantity Ω is called Rabi frequency, and plays an important role in the characterization of the dynamical evolution.

6.5.1 Rotating Frame

The weak-coupling master equation for the considered driven two-level system interacting in the dipole approximation with the electromagnetic field thus reads

$$\frac{d\rho_S(t)}{dt} = -\frac{i}{\hbar}[H_S + H_{LS} + H_L(t), \rho_S(t)] + \mathcal{D}[\rho_S(t)], \tag{6.173}$$

with Hamiltonians and dissipator explicitly given by (6.153), (6.162), (6.172) and (6.163). The contribution H_{LS} has the only effect of modifying the system frequency, so that we will neglect it in the following. The expression (6.173) refers to the Schrödinger picture. According to the discussion at the end of Sect. 6.2.2, moving from the interaction to the Schrödinger picture has not affected the dissipator, thanks to its covariance expressed by (6.55). With abuse of notation we still denote the system state in the Schrödinger picture with $\rho_S(t)$. To cope with the time dependence in the Hamiltonian term it is convenient to move to a new reference frame, obtained by a rotation around the z direction with the frequency ω_L of the laser. We thus consider a time dependent unitary transformation described by the operators

$$\mathsf{U}(\omega_L t) = e^{i\omega_L t \frac{\sigma_z}{2}}. \tag{6.174}$$

We denote the statistical operator in this new reference frame as $\varrho_S(t)$, just changing the character font, that is related as follows to the Schrödinger picture statistical operator in the original frame

$$\varrho_S(t) = \mathsf{U}(\omega_L t)\rho_S(t)\mathsf{U}(\omega_L t)^\dagger, \tag{6.175}$$

so that

$$\frac{d\varrho_S(t)}{dt} = i\left[\frac{\omega_L}{2}\sigma_z, \varrho_S(t)\right] + \mathsf{U}(\omega_L t)\frac{d\rho_S(t)}{dt}\mathsf{U}(\omega_L t)^\dagger, \tag{6.176}$$

where note that this transformation only affects the coherences, that is the off-diagonal matrix elements of the statistical operator. In order to obtain the time evolution equation for the statistical operator in the rotating frame, we exploit the fact that H_S and H_{LS} are invariant with respect to the action of the unitary transformation (6.174), while according to the relations

$$\mathsf{U}(\theta)\sigma_\pm\mathsf{U}(\theta)^\dagger = e^{\pm i\theta}\sigma_\pm \tag{6.177}$$

we have

$$\mathsf{U}(\omega_L t)H_L(t)\mathsf{U}(\omega_L t)^\dagger = -\frac{\hbar\Omega}{2}\sigma_x. \tag{6.178}$$

We further exploit covariance of the dissipator in the sense of Remark 2.9 with respect to (6.174), so that similarly to (6.55) we have

$$\mathcal{D}[\mathsf{U}(\omega_L t)\rho_S(t)\mathsf{U}(\omega_L t)^\dagger] = \mathsf{U}(\omega_L t)\mathcal{D}[\rho_S(t)]\mathsf{U}(\omega_L t)^\dagger. \tag{6.179}$$

Starting from (6.173), (6.175) and (6.176), using (6.178) and (6.179) we obtain

$$\frac{d\varrho_S(t)}{dt} = -i\left[\frac{\Delta}{2}\sigma_z - \frac{\Omega}{2}\sigma_x, \varrho_S(t)\right] + \mathcal{D}[\varrho_S(t)], \tag{6.180}$$

$$= \mathcal{L}[\varrho_S(t)], \tag{6.181}$$

where as anticipated we have neglected the Lamb shift contribution only renormalizing the value of the original system frequency. The quantity

$$\Delta = \omega_0 - \omega_L \tag{6.182}$$

is called detuning and provides information on how close the laser frequency is to the system frequency. As we shall see this master equation can be used to describe, besides the dynamics of populations and coherences, the spectrum of emitted radiation and the statistics of photon detection.

6.5.2 Matrix Representation

We now exploit the tools introduced in Sect. 3.3 to obtain from the operator equation (6.180) a set of coupled linear differential equations. According to (3.393), (3.409) and (3.410) we can express the action of the map $\mathcal{L}$ in the form

$$\mathcal{L}[\varrho(t)] = \sum_{i,j=1}^{4} \mathcal{L}_{ij} E_{ij}(\varrho(t)), \tag{6.183}$$

with

$$\mathcal{L}_{ij} = \mathrm{Tr}\{\Sigma_i \mathcal{L}[\Sigma_j]\} \tag{6.184}$$

and

$$E_{ij}(\varrho(t)) = \Sigma_i \,\mathrm{Tr}\{\Sigma_j \varrho(t)\}. \tag{6.185}$$

Given the expression (3.380) of the Bloch basis in $\mathcal{B}(\mathbb{C}^2)$ we explicitly obtain from (6.184)

$$\mathcal{L}_{ij} = \frac{1}{2}\,\mathrm{Tr}\{\sigma_i \mathcal{L}[\sigma_j]\}, \tag{6.186}$$

with $\{\sigma_i\} = \{\mathbb{1}, \boldsymbol{\sigma}\}$, namely the identity operator in $\mathbb{C}^2$ and the three Pauli matrices. Knowledge of the matrix $\mathcal{L}_{ij}$ of (3.382) implies full knowledge of the action of the map $\mathcal{L}$. For the case at hand, starting from the expression (6.180) and using the relevant identities for Pauli matrices given in Remark 2.1 we have

$$\begin{aligned}
\mathcal{L}[\mathbb{1}] &= -\gamma_0\sigma_z \\
\mathcal{L}[\sigma_x] &= -\tfrac{\gamma}{2}\sigma_x + \Delta\sigma_y \\
\mathcal{L}[\sigma_y] &= -\Delta\sigma_x - \tfrac{\gamma}{2}\sigma_y - \Omega\sigma_z \\
\mathcal{L}[\sigma_z] &= \Omega\sigma_y - \gamma\sigma_z,
\end{aligned} \tag{6.187}$$

leading to

$$\mathcal{L}_{ij} = \begin{pmatrix}
0 & 0 & 0 & 0 \\
0 & -\tfrac{\gamma}{2} & -\Delta & 0 \\
0 & \Delta & -\tfrac{\gamma}{2} & \Omega \\
-\gamma_0 & 0 & -\Omega & -\gamma
\end{pmatrix}, \tag{6.188}$$

where according to standard usage we have denoted with

$$\gamma = \gamma_0(1 + 2N_\beta(\omega_0)) \tag{6.189}$$

the sum of the two rates appearing in the dissipator (6.163). Expressing a state in $\mathbb{C}^2$ in terms of the associated Bloch vector according to (2.79) we write

$$\varrho(t) = \frac{1}{2}(\mathbb{1} + \langle\boldsymbol{\sigma}(t)\rangle \cdot \boldsymbol{\sigma}), \tag{6.190}$$

so that (6.183) explicitly becomes

$$\mathcal{L}[\varrho(t)] = \sum_{i,j=1}^{4} \mathcal{L}_{ij}\sigma_i\langle\sigma_j(t)\rangle, \tag{6.191}$$

and therefore (6.181) leads to

$$\frac{d}{dt}\langle\sigma_i(t)\rangle = \sum_{j=1}^{4} \mathcal{L}_{ij}\langle\sigma_j(t)\rangle, \tag{6.192}$$

for $i = 1, 2, 3$. These equations explicitly read

$$\begin{aligned}
\frac{d}{dt}\langle\sigma_x(t)\rangle &= -\frac{\gamma}{2}\langle\sigma_x(t)\rangle - \Delta\langle\sigma_y(t)\rangle \\
\frac{d}{dt}\langle\sigma_y(t)\rangle &= +\Delta\langle\sigma_x(t)\rangle - \frac{\gamma}{2}\langle\sigma_y(t)\rangle + \Omega\langle\sigma_z(t)\rangle \\
\frac{d}{dt}\langle\sigma_z(t)\rangle &= -\Omega\langle\sigma_y(t)\rangle - \gamma\langle\sigma_z(t)\rangle - \gamma_0,
\end{aligned} \tag{6.193}$$

and provide a closed set of equations for the mean values. The equivalent set of equations directly focusing on the behavior of populations, that is to say diagonal matrix elements of the statistical operator in the energy basis, and coherences, namely off-diagonal matrix elements, reads

$$\frac{d}{dt}\langle\sigma_+(t)\rangle = \left(i\Delta - \frac{\gamma}{2}\right)\langle\sigma_+(t)\rangle + i\frac{\Omega}{2}(\langle P_+(t)\rangle - \langle P_-(t)\rangle) \tag{6.194}$$

$$\frac{d}{dt}\langle P_+(t)\rangle = i\frac{\Omega}{2}(\langle\sigma_+(t)\rangle - \langle\sigma_-(t)\rangle) - \frac{\gamma}{2}(\langle P_+(t)\rangle - \langle P_-(t)\rangle) - \frac{\gamma_0}{2}. \tag{6.195}$$

Note that according to Sect. 6.3.2 in the absence of driving, that is for $\Omega = 0$, the evolution of coherences and populations as given by (6.194) and (6.195) is decoupled. Equations (6.193) are known as optical Bloch equations.

6.5.3 Coherent Evolution

To grasp the role of the Rabi frequency, we first consider the situation in which the system is not coupled to the environment, so that $\gamma = \gamma_0 = 0$. The corresponding unitary evolution is given by

$$\begin{aligned}
\langle\sigma_x(t)\rangle &= \left[1 + \frac{\Delta^2}{\varpi^2}[\cos(\varpi t) - 1]\right]\sigma_x(0) \\
&\quad - \frac{\Delta}{\varpi}\sin(\varpi t)\sigma_y(0) + \frac{\Delta\Omega}{\varpi^2}[\cos(\varpi t) - 1]\sigma_z(0) \\
\langle\sigma_y(t)\rangle &= \frac{\Delta}{\varpi}\sin(\varpi t)\sigma_x(0) + \cos(\varpi t)\sigma_y(0) + \frac{\Omega}{\varpi}\sin(\varpi t)\sigma_z(0) \\
\langle\sigma_z(t)\rangle &= \frac{\Delta\Omega}{\varpi^2}[\cos(\varpi t) - 1]\sigma_x(0) \\
&\quad - \frac{\Omega}{\varpi}\sin(\varpi t)\sigma_y(0) + \left[1 + \frac{\Omega^2}{\varpi^2}[\cos(\varpi t) - 1]\right]\sigma_z(0),
\end{aligned} \tag{6.196}$$

corresponding to an oscillating behavior with frequency

$$\varpi = \sqrt{\Omega^2 + \Delta^2}. \tag{6.197}$$

If the laser is exactly in resonance with the system frequency, so that $\Delta = 0$, $\varpi = \Omega$ and we observe the so-called Rabi oscillations, that can be described in the Bloch sphere as rotations around the x axis with frequency Ω, namely

$$\langle\boldsymbol{\sigma}(t)\rangle = R_x(\Omega t)\langle\boldsymbol{\sigma}(0)\rangle, \tag{6.198}$$

where $R_x(\theta) \in SO(3)$ is the matrix describing a clockwise rotation by an angle θ around the x axis. Denoting as

$$\langle\boldsymbol{\sigma}(t)\rangle_\varrho = \mathrm{Tr}\{\boldsymbol{\sigma}\varrho(t)\} \tag{6.199}$$

the solution in the rotating frame and as

$$\langle \boldsymbol{\sigma}(t) \rangle_\rho = \mathrm{Tr}\{\boldsymbol{\sigma}\rho(t)\} \tag{6.200}$$

the solution in the Schrödinger picture in the original frame, so that according to (6.175) the difference is in the change of font, we have

$$\langle \boldsymbol{\sigma}(t) \rangle_\rho = R_z(\omega_0 t) \langle \boldsymbol{\sigma}(t) \rangle_\varrho \tag{6.201}$$

$$= R_z(\omega_0 t) R_x(\Omega t) \langle \boldsymbol{\sigma}(0) \rangle. \tag{6.202}$$

The effect of the interaction of the laser, with intensity proportional to Ω^2, is therefore a pure driving affecting the evolution in time of both coherences and populations.

6.5.4 General Solution and Stationary State

In order to consider the solutions of (6.194) and (6.195) in the presence of an environment it is convenient to exploit the results of Remark 3.18 to express (6.192) as an affine transformation of the Bloch vector given by

$$\frac{\mathrm{d}}{\mathrm{d}t} \langle \boldsymbol{\sigma}(t) \rangle = -T \langle \boldsymbol{\sigma}(t) \rangle + \boldsymbol{b} \tag{6.203}$$

according to the identification

$$\mathcal{L} = \begin{pmatrix} 0 & \boldsymbol{0} \\ \boldsymbol{b}^T & -T \end{pmatrix}, \tag{6.204}$$

with $(\mathcal{L})_{ij} = \mathcal{L}_{ij}$ the matrix of coefficients introduced in (6.188). We note that

$$\det T = \frac{\gamma}{4}(\gamma^2 + 2\Omega^2 + 4\Delta^2), \tag{6.205}$$

so that for $\gamma > 0$ the matrix is invertible. While the purely coherent evolution has already been considered in Sect. 6.5.3, we restrict here to the case $\gamma > 0$, corresponding to a non-negligible interaction with the environment. Equation (6.203) is of the form (5.500), so that its solution according to (5.501) reads

$$\langle \boldsymbol{\sigma}(t) \rangle = \mathrm{e}^{-Tt} \langle \boldsymbol{\sigma}(0) \rangle + \int_0^t \mathrm{d}s\, \mathrm{e}^{-T(t-s)} \boldsymbol{b} \tag{6.206}$$

and exploiting the fact that T is invertible can also be written

$$\langle \boldsymbol{\sigma}(t) \rangle = \mathrm{e}^{-Tt} \langle \boldsymbol{\sigma}(0) \rangle + (\mathbb{1} - \mathrm{e}^{-Tt}) T^{-1} \boldsymbol{b}. \tag{6.207}$$

Moreover, though not Hermitian, the matrix T is a normal operator in the sense that

$$[T, T^\dagger] = 0, \tag{6.208}$$

so that it can be diagonalized.

6.5.4.1 Stationary State

The exact expression of the eigenvalues on resonance is given by

$$\left\{ \frac{\gamma}{2}, \frac{3}{4}\gamma \pm i\sqrt{\Omega^2 - \left(\frac{\gamma}{4}\right)^2} \right\}, \tag{6.209}$$

while in the general case calls for the solution of an algebraic equation of the third order. Their real part is however always positive, so that independently of the initial condition the system reaches a state identified by the Bloch vector

$$\langle \boldsymbol{\sigma} \rangle^{\mathrm{eq}} = T^{-1}\boldsymbol{b} \tag{6.210}$$

$$= -\frac{\gamma_0}{4 \det T} \begin{pmatrix} -4\Omega\Delta \\ 2\Omega\gamma \\ 4\Delta^2 + \gamma^2 \end{pmatrix}. \tag{6.211}$$

The semigroup dynamics is therefore relaxing in the sense of Sect. 6.3.1, with stationary equilibrium state

$$\varrho^{\mathrm{eq}} = \frac{1}{2}(\mathbb{1} + \langle \boldsymbol{\sigma} \rangle^{\mathrm{eq}} \cdot \boldsymbol{\sigma}). \tag{6.212}$$

In the present case the semigroup is relaxing due to the fact that the only operators commuting with the Hamiltonian (6.153) and the two Lindblad operators (6.165) and (6.166) are multiples of the identity.

6.5.4.2 Undriven Case

In the absence of driving, so that $\Omega = 0$, and according to (6.182) $\Delta = \omega_0$, we are left with the canonical stationary solution (6.56) determined by the detailed balance condition according to (6.84) and (6.85). It takes the explicit form

$$\varrho^{\mathrm{eq}} = \frac{1}{2\,\mathrm{Ch}\left(\frac{\beta\hbar\omega}{2}\right)} \begin{pmatrix} e^{-\beta\frac{\hbar\omega_0}{2}} & 0 \\ 0 & e^{+\beta\frac{\hbar\omega_0}{2}} \end{pmatrix}, \tag{6.213}$$

corresponding to

$$\langle \boldsymbol{\sigma} \rangle^{\mathrm{eq}} = \left(0, 0, -\frac{\gamma_0}{\gamma} \right), \tag{6.214}$$

where according to (6.189) $\gamma_0/\gamma = 1/(1 + 2N_\beta(\omega_0))$.

6.5.4.3   Driven Resonant Case

The driving affects the populations, so that taking for simplicity the resonant case
we obtain

$$\langle \sigma_z \rangle^{\mathrm{eq}} = -\frac{\gamma_0}{\gamma} \frac{1}{1 + 2(\Omega/\gamma)^2}, \tag{6.215}$$

and it further leads to non-zero coherences in the stationary state, according to

$$\langle \sigma_+ \rangle^{\mathrm{eq}} = -i \frac{\gamma_0 \Omega}{2\Omega^2 + \gamma^2}. \tag{6.216}$$

In the presence of very strong driving we have $\langle \sigma_z \rangle^{\mathrm{eq}} \approx 0$, so that ground and excited
states are equally populated.

6.5.5   Bloch Equations for Nuclear Magnetic Resonance

The equations (6.193) were originally introduced [39,41] to describe the dynamics
of nuclear induction, experimentally first observed in [42,43], namely to study the
effect of an external magnetic field and the molecular surroundings on the nuclear
polarization. These equations have a fundamental role in the theoretical treatment
of nuclear magnetic resonance spectroscopy, that has found a wealth of applications
[44,45]. To more closely show the connection between the obtained optical Bloch
equations (6.193) and the magnetic Bloch equations used to describe the evolution
of a nuclear spin we express the nuclear magnetic moment as

$$\boldsymbol{\mu} = \zeta \hbar \boldsymbol{I}, \tag{6.217}$$

where $\boldsymbol{I}$ is the triple of nuclear spin operators and ζ denotes the proportionality factor
between angular momentum and magnetic moment, namely the gyromagnetic ratio.
Assuming the value 1/2 for the nuclear spin, we recast (6.193) in the form

$$\frac{\mathrm{d}}{\mathrm{d}t} \langle I_{x,y}(t) \rangle = \zeta(\langle \boldsymbol{I}(t) \rangle \times \boldsymbol{B})_{x,y} - \frac{\langle I_{x,y}(t) \rangle}{T_2} \tag{6.218}$$

$$\frac{\mathrm{d}}{\mathrm{d}t} \langle I_z(t) \rangle = \zeta(\langle \boldsymbol{I}(t) \rangle \times \boldsymbol{B})_z - \frac{\langle I_z(t) \rangle - \langle I_z \rangle^{\mathrm{eq}}}{T_1}, \tag{6.219}$$

where we have set

$$\frac{1}{T_1} = \gamma \tag{6.220}$$

$$\frac{1}{T_2} = \frac{\gamma}{2} \tag{6.221}$$

together with

$$\langle I_z \rangle^{\mathrm{eq}} = -\frac{\gamma_0}{\gamma} = -\,\mathrm{th}\left(\frac{\beta\hbar\omega_0}{2}\right), \tag{6.222}$$

according to (6.189), further using the identification

$$\zeta \boldsymbol{B} = (\Omega, 0, \omega_L - \omega_0). \tag{6.223}$$

In the last component we have used the very definition of detuning frequency (6.182) to put into evidence two distinct terms, namely the contribution ω_L due to the driving field and ω_0, which is now to be interpreted as the Larmor precession frequency in the laboratory frame. In the laboratory frame the spin is subject to a strong static field along the z direction, determining the energy splitting according to $\omega_0 = |\zeta B_z|$. A time varying field, whose strength is determined by the Rabi frequency, is then applied in the plane perpendicular to the static contribution, and is responsible for excitation of the two-level system. Observation of the decaying behavior of the system provides important information on its surroundings. The relaxation times T_1 and T_2 were introduced to this aim on a phenomenological basis, while in the present treatment they are determined from a microscopic model. Exploiting the proportionality between the spin and the magnetic moment these results lead to a vectorial equation for the mean magnetization vector $\boldsymbol{M}$ in the rotating frame, namely

$$\frac{\mathrm{d}}{\mathrm{d}t}\boldsymbol{M}(t) = \zeta \boldsymbol{M}(t) \times \boldsymbol{B} - R[\boldsymbol{M}(t) - \boldsymbol{M}^{\mathrm{eq}}], \tag{6.224}$$

where the relaxation matrix R is in diagonal form

$$R = \mathrm{diag}\left(\frac{1}{T_2}, \frac{1}{T_2}, \frac{1}{T_1}\right). \tag{6.225}$$

To clarify the meaning of these constants we consider the situation in which there is no driving field, so that we can exploit (6.202) with $\Omega \to 0$. The constant T_2 is known as transverse or spin-spin relaxation time, because it is the rate describing the exponential decay of the transverse components

$$M_x(t) = [M_x(0)\cos(\omega_0 t) - M_y(0)\sin(\omega_0 t)]\mathrm{e}^{-t/T_2} \tag{6.226}$$

and

$$M_y(t) = [M_x(0)\sin(\omega_0 t) + M_y(0)\cos(\omega_0 t)]e^{-t/T_2}, \tag{6.227}$$

while the constant T_1 is known as longitudinal or spin-lattice relaxation time, describing the relaxation to the equilibrium value of the magnetization along the direction of the bias field

$$M_z(t) = M_z(0)e^{-t/T_1} + M_z^{\mathrm{eq}}(1 - e^{-t/T_1}). \tag{6.228}$$

In the microscopic description these relaxation rates are determined by details of interaction and bath, typically encoded in the spectral density appearing in (6.145). The relaxation rate $1/T_1$ describes the behavior of the populations and can be understood classically in terms of relaxation to equilibrium due to energy exchanges. It is the smallest relaxation rate, usually determined by the order of magnitude of the Larmor frequency. The relaxation rate $1/T_2$ is stronger, indeed T_1 values are typically about 5–10 times larger than T_2 values. It describes the loss of coherence taking place both because of energy exchanges and other phenomena, including field inhomogeneities, so that it is an inherent quantum phenomenon. In the present model we have the exact relation

$$2T_1 = T_2, \tag{6.229}$$

while in Remark 6.5 we show that the requirement of complete positivity can generally be satisfied if

$$2T_1 \geqslant T_2 > 0, \tag{6.230}$$

thus complying with the experimental evidence [44,45].

6.5 Constraints on Relaxation Rates

We here consider the constraints on the relaxation rates induced by the requirement of a completely positive semigroup evolution law. To this aim we consider an effective dissipator replacing in (6.180) the explicit expression (6.163) with the general structure of the dissipative contribution in the generator of a quantum dynamical semigroup, that according to (5.27) for the case of $\mathbb{C}^2$ can be written as

$$\mathcal{D}^{\mathrm{eff}}[\varrho_S(t)] = \sum_{k=2}^{4} \gamma_k \left[L_k \varrho_S(t) L_k^\dagger - \frac{1}{2}\{L_k^\dagger L_k, \varrho_S(t)\} \right], \tag{6.231}$$

with $\gamma_k \geqslant 0$. As discussed in Sect. 5.3.1 we take the Lindblad operators L_k to be linearly independent, and a convenient choice is given by

$$L_k \to \{\sigma_+, \sigma_-, \sigma_z\}, \tag{6.232}$$

with associated non negative coefficients

$$\gamma_k \to \{\gamma_+, \gamma_-, \gamma_z\}. \tag{6.233}$$

We therefore consider the dynamics described by the generator

$$\mathcal{L}^{\mathrm{eff}}[\varrho_S(t)] = -i\left[\frac{\Delta}{2}\sigma_z - \frac{\Omega}{2}\sigma_x, \varrho_S(t)\right] + \mathcal{D}^{\mathrm{eff}}[\varrho_S(t)], \tag{6.234}$$

and verify under which conditions it reproduces the Bloch equations (6.218) and (6.219). We follow the strategy of Sect. 6.5.2 and identify the matrix representation of the generator $\mathcal{L}^{\mathrm{eff}}$ defined by (6.234), which according to (6.186) is given by

$$\mathcal{L}^{\mathrm{eff}} = \begin{pmatrix} 0 & 0 & 0 & 0 \\ 0 & -\frac{\gamma_+ + \gamma_- + 4\gamma_z}{2} & -\Delta & 0 \\ 0 & \Delta & -\frac{\gamma_+ + \gamma_- + 4\gamma_z}{2} & \Omega \\ \gamma_+ - \gamma_- & 0 & -\Omega & -(\gamma_+ + \gamma_-) \end{pmatrix}. \tag{6.235}$$

We now impose validity of (6.218) and (6.219), which in view of (6.193) leads to the following identifications

$$\frac{1}{T_1} = \gamma_+ + \gamma_- \tag{6.236}$$

$$\frac{1}{T_2} = \frac{1}{2}(\gamma_+ + \gamma_- + 4\gamma_z) \tag{6.237}$$

$$\frac{\langle I_z \rangle^{\mathrm{eq}}}{T_1} = \gamma_+ - \gamma_-. \tag{6.238}$$

with $\langle I_z \rangle^{\mathrm{eq}}$ given by (6.222). Inverting these relations we obtain

$$\gamma_+ = \frac{1 - \mathrm{th}\left(\frac{\beta\hbar\omega_0}{2}\right)}{2T_1} \tag{6.239}$$

$$\gamma_- = \frac{1 + \mathrm{th}\left(\frac{\beta\hbar\omega_0}{2}\right)}{2T_1} \tag{6.240}$$

$$\gamma_z = \frac{2T_1 - T_2}{4T_1 T_2}. \tag{6.241}$$

Positivity of these coefficients, which according to Theorem 5.1 is a necessary and sufficient conditions for complete positivity of the time evolution, thus leads to the constraint (6.230), first put into evidence in [46,47]. For the microscopic model considered in Sects. 6.4 and 6.5 this constraint is obeyed with the strict equality sign. We stress that the relevance of these constraints depends on the validity of the approximations leading to the considered reduced semigroup dynamical description.

6.5.6 Explicit Solution for Zero Detuning

We complete the analysis of Sect. 6.5.4, by considering the explicit time dependence of the solutions. To this aim we take the case of a resonant interaction, corresponding to $\Delta = 0$, which allows for an exact analytic treatment and therefore an explicit insight in the time dependence. Solving the coupled linear equations in (6.193) for a resonant interaction we obtain

$$\langle \sigma_x(t) \rangle = e^{-\frac{\gamma}{2}t} \langle \sigma_x(0) \rangle \tag{6.242}$$

together with

$$
\begin{aligned}
\langle \sigma_y(t) \rangle =& \frac{\Omega}{\mu} e^{-\frac{3}{4}\gamma t} \sin(\mu t) \langle \sigma_z(0) \rangle \\
&+ e^{-\frac{3}{4}\gamma t} \left[\cos(\mu t) + \frac{1}{4} \frac{\gamma}{\mu} \sin(\mu t) \right] \langle \sigma_y(0) \rangle \\
&+ \frac{2\Omega\gamma_0}{2\Omega^2 + \gamma^2} \left\{ e^{-\frac{3}{4}\gamma t} \left[\cos(\mu t) + \frac{3}{4} \frac{\gamma}{\mu} \sin(\mu t) \right] - 1 \right\}
\end{aligned}
\tag{6.243}
$$

and

$$
\begin{aligned}
\langle \sigma_z(t) \rangle =& -\frac{\Omega}{\mu} e^{-\frac{3}{4}\gamma t} \sin(\mu t) \langle \sigma_y(0) \rangle \\
&+ e^{-\frac{3}{4}\gamma t} \left[\cos(\mu t) - \frac{1}{4} \frac{\gamma}{\mu} \sin(\mu t) \right] \langle \sigma_z(0) \rangle \\
&+ \frac{\gamma\gamma_0}{2\Omega^2 + \gamma^2} \left\{ e^{-\frac{3}{4}\gamma t} \left[\cos(\mu t) - \frac{8\Omega^2 + \gamma^2}{4\gamma\mu} \sin(\mu t) \right] - 1 \right\},
\end{aligned}
\tag{6.244}
$$

where

$$\mu = \sqrt{\Omega^2 - \left(\frac{\gamma}{4}\right)^2}. \tag{6.245}$$

These quantities allow to determine the evolution of populations and coherences that are fixed respectively by

$$
\begin{aligned}
\langle P_+(t)\rangle =& i\frac{\Omega}{2\mu}\mathrm{e}^{-\frac{3}{4}\gamma t}\sin(\mu t)(\langle \sigma_+(0)\rangle - \langle \sigma_-(0)\rangle) \\
& + \mathrm{e}^{-\frac{3}{4}\gamma t}\left[\cos(\mu t) - \frac{1}{4}\frac{\gamma}{\mu}\sin(\mu t)\right]\langle P_+(0)\rangle \\
& + \frac{2\Omega^2 + \gamma(\gamma - \gamma_0)}{2(2\Omega^2 + \gamma^2)} \\
& \times \left\{1 - \mathrm{e}^{-\frac{3}{4}\gamma t}\left[\cos(\mu t) - \frac{1}{4}\frac{\gamma}{\mu}\left(1 - \frac{(\gamma_0/\gamma)8\Omega^2}{(2\Omega^2 + \gamma(\gamma - \gamma_0))}\right)\sin(\mu t)\right]\right\}
\end{aligned}
\tag{6.246}
$$

and

$$
\begin{aligned}
\langle \sigma_+(t)\rangle =& \frac{1}{2}\mathrm{e}^{-\frac{\gamma}{2}t}(\langle \sigma_+(0)\rangle + \langle \sigma_-(0)\rangle) \\
& + \frac{1}{2}\mathrm{e}^{-\frac{3}{4}\gamma t}\left[\cos(\mu t) + \frac{1}{4}\frac{\gamma}{\mu}\sin(\mu t)\right](\langle \sigma_+(0)\rangle - \langle \sigma_-(0)\rangle) \\
& + i\frac{\Omega}{\mu}\mathrm{e}^{-\frac{3}{4}\gamma t}\sin(\mu t)\langle P_+(0)\rangle \\
& - i\frac{\Omega\gamma_0}{2\Omega^2 + \gamma^2}\left\{1 - \mathrm{e}^{-\frac{3}{4}\gamma t}\left[\cos(\mu t) + \frac{1}{4\mu}\left(3\gamma - \frac{2\Omega^2 + \gamma^2}{\gamma_0/2}\right)\sin(\mu t)\right]\right\}.
\end{aligned}
\tag{6.247}
$$

According to its definition (6.245) the quantity μ can be either positive or purely imaginary depending on the ratio between the Rabi frequency Ω and the total damping rate γ as defined respectively in (6.171) and (6.189). For $\Omega > \gamma/4$, so that the driving dominates, we are in the so-called underdamped regime, in which μ is a real quantity. For $\Omega < \gamma/4$ we are in the so-called over-damped regime, in which the oscillating functions are replaced by hyperbolic functions and the damping effects due to the interaction with the environment hide the effect of driving. In the long time limit the populations reach the asymptotic value

$$
\langle P_+(\infty)\rangle = \langle P_+\rangle^{\mathrm{eq}}
\tag{6.248}
$$

$$
= \frac{2\Omega^2 + \gamma(\gamma - \gamma_0)}{2(2\Omega^2 + \gamma^2)},
\tag{6.249}
$$

while coherences tend to the value

$$
\langle \sigma_+(\infty)\rangle = \langle \sigma_+\rangle^{\mathrm{eq}}
\tag{6.250}
$$

$$
= -i\frac{\Omega\gamma_0}{2\Omega^2 + \gamma^2}.
\tag{6.251}
$$

The different regimes together with the long-time values are visualized in Fig. 6.2 for the populations, and in Fig. 6.3 for the coherences.

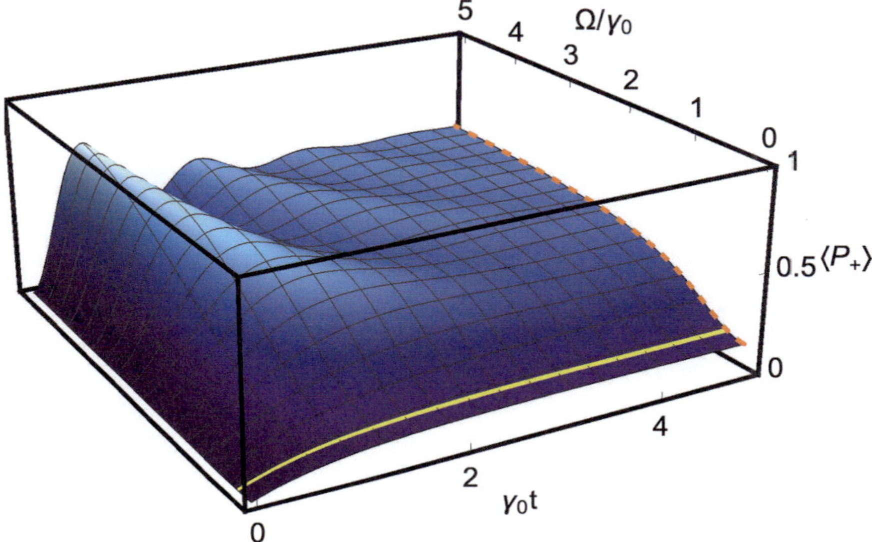

Fig. 6.2 Behavior of populations as given by (6.246) as a function of rescaled time and ratio between driving and damping strength, for a low temperature case characterized by $N_\beta(\omega_0) = 0.2$. The solid (yellow) line corresponds to the transition between under-damped and over-damped regime, the (red) dashed line provides the asymptotic value of the populations. The system starts in the ground state

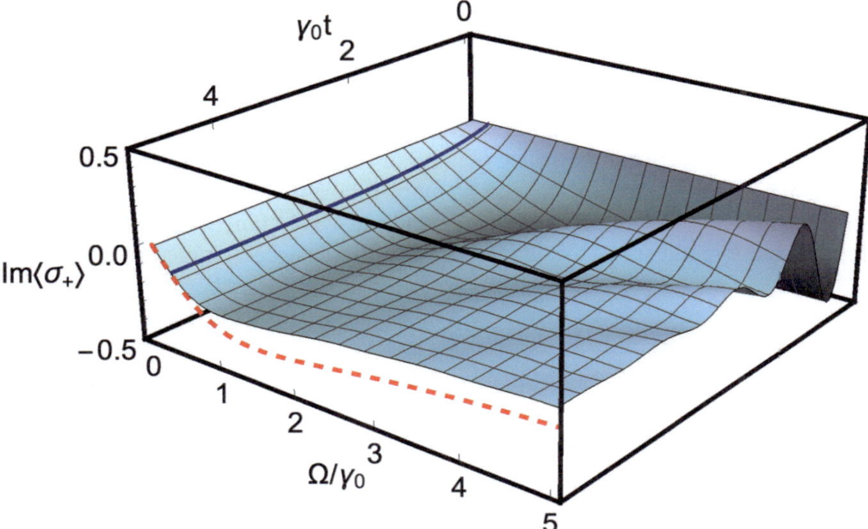

Fig. 6.3 Behavior of the imaginary part of the coherences as given by (6.247), in the same notation and the same conditions of Fig. 6.2. The solid (blue) line corresponds to the transition between under-damped and over-damped regime, the (red) dashed line provides the asymptotic value of the coherences

In the case of very strong driving $\Omega \gg \gamma/4$, (6.246) and (6.247) are well approximated by the simpler expressions

$$\langle P_+(t) \rangle \approx \frac{1}{2}\left[1 - e^{-\frac{3}{4}\gamma t}\cos(\Omega t)(1 - 2\langle P_+(0)\rangle)\right]$$
$$+ i\frac{1}{2}e^{-\frac{3}{4}\gamma t}\sin(\Omega t)(\langle \sigma_+(0)\rangle - \langle \sigma_-(0)\rangle) \tag{6.252}$$

and

$$\langle \sigma_+(t)\rangle = \frac{1}{2}e^{-\frac{\gamma}{2}t}(\langle \sigma_+(0)\rangle + \langle \sigma_-(0)\rangle) + \frac{1}{2}e^{-\frac{3}{4}\gamma t}\cos(\Omega t)(\langle \sigma_+(0)\rangle - \langle \sigma_-(0)\rangle)$$
$$- \frac{i}{2}e^{-\frac{3}{4}\gamma t}\sin(\Omega t)(1 - 2\langle P_+(0)\rangle), \tag{6.253}$$

so that for long times $\langle P_+(t)\rangle \to 1/2$ and coherences die out. We will exploit these results in Sects. 6.6.2 and 6.6.3 to study respectively the quantum description of atom fluorescence in the resonant case and the statistics of emitted photons.

6.6 System Correlation Functions

So far we have only considered the time evolution of mean values of the system observables, that have been completely characterized. It is of interest to consider the behavior in time of higher-order correlation functions, allowing to predict typical quantum features that can be observed in the behavior of the system. Having a proper handle on higher-order correlation functions for our model of a two-level atom interacting in dipole approximation with the electromagnetic field, we will consider the fluorescence spectrum in Sect. 6.6.2 and the statistics of the emitted photons in Sect. 6.6.3. These predictions depend on multi-time correlation functions of second and fourth order respectively. To evaluate these correlation functions we will consider a regression hypothesis as introduced in [48], that can be assumed to hold in the Markovian framework considered so far. For a more comprehensive treatment on the meaning of regression relations we refer to [15, 16], as well as to [49, 50] for the connection with Markovian features of the dynamics.

6.6.1 Multi-time Correlation Functions and Regression

We consider a quantum dynamical semigroup with generator $\mathcal{L}$, that well approximates the reduced dynamics of a system S interacting with an environment E according to a microscopic Hamiltonian H of the form (5.5), for a suitable choice of initial environmental state ρ_E. For any pair of system operators A_S, B_S, we can introduce their two-time correlation function as the expectation value with respect to the total

initial state of the product of the two system operators, in the Heisenberg picture
with respect to the total Hamiltonian H

$$\langle A_S(t+\tau)B_S(t)\rangle = \text{Tr}\{e^{iH(t+\tau)}A_S e^{-iH(t+\tau)}e^{iHt}B_S e^{-iHt}\rho_{SE}(0)\}. \quad (6.254)$$

Thanks to the cyclic property of the trace and the fact that the operators act in the
system Hilbert space only we have

$$\langle A_S(t+\tau)B_S(t)\rangle = \text{Tr}_S\{A_S \text{Tr}_E[e^{-iH\tau}B_S\rho_{SE}(t)e^{+iH\tau}]\} \quad (6.255)$$
$$= \text{Tr}_S\{A_S \text{Tr}_E[X_t(\tau)]\}, \quad (6.256)$$

where we have defined the time-dependent operator $X_t \in \mathcal{T}(\mathcal{H}_S \otimes \mathcal{H}_E)$

$$X_t = B_S(e^{-iHt}\rho_{SE}(0)e^{+iHt}), \quad (6.257)$$

appearing in (6.256) in the Schrödinger picture with respect to H, and we have used
the short-hand notation $A_S = A_S \otimes \mathbb{1}_E$ and $B_S = B_S \otimes \mathbb{1}_E$. Assuming an initial
factorized state (5.6) for system and environment, so as to ensure the existence of
a reduced system dynamics as discussed in Sect. 5.2, X_t obeys the Liouville–von
Neumann equation

$$\frac{d}{d\tau}X_t(\tau) = -\frac{i}{\hbar}[H, X_t(\tau)] \quad (6.258)$$

with initial condition $X_t(0)$ given by (6.257). Moreover, according to (5.10) we have

$$\text{Tr}_E[X_t] = B_S \text{Tr}_E[e^{-iHt}\rho_S(0) \otimes \rho_E e^{+iHt}] \quad (6.259)$$
$$= B_S\Phi(t, 0)[\rho_S(0)], \quad (6.260)$$

and we assume that $\Phi(t, 0)$ is well approximated by a quantum dynamical semigroup.
Invoking the same physical approximations, we assume that the dynamics of the
operator $\text{Tr}_E[X_t(\tau)]$ in its τ dependence is given by the same map [51,52], namely

$$\text{Tr}_E[X_t(\tau)] = \Phi(t + \tau, t)[\text{Tr}_E[X_t(0)]], \quad (6.261)$$

and therefore using (6.257)

$$\text{Tr}_E[X_t(\tau)] = \Phi(t + \tau, t)[B_S\Phi(t, 0)[\rho_S(0)]]. \quad (6.262)$$

Inserting in (6.256) we thus come to

$$\langle A_S(t+\tau)B_S(t)\rangle = \text{Tr}_S\{A_S\Phi(t + \tau, t)[B_S\Phi(t, 0)[\rho_S(0)]]\}, \quad (6.263)$$

where the square brackets denote the operators the map is acting on. This expression
has the very important feature of allowing to evaluate the multi-time correlation
function in terms of the dynamical map $\Phi(t, s)$ which provides the time evolution

of the mean values. On a similar footing, we can consider the reverse order in time, corresponding to the correlation function

$$\langle B_S(t) A_S(t+\tau) \rangle = \mathrm{Tr}\{e^{iHt} B_S e^{-iHt} e^{iH(t+\tau)} A_S e^{-iH(t+\tau)} \rho_{SE}(0)\}, \quad (6.264)$$

that using (6.257) takes the form

$$\langle B_S(t) A_S(t+\tau) \rangle = \mathrm{Tr}_S\{A_S \, \mathrm{Tr}_E[X_t^\dagger(\tau)]\}. \tag{6.265}$$

Under the above-mentioned approximations we exploit (6.262) to obtain

$$\langle B_S(t) A_S(t+\tau) \rangle = \mathrm{Tr}_S\{A_S \Phi(t+\tau, t)[\Phi(t, 0)[\rho_S(0)] B_S]\}. \tag{6.266}$$

This procedure can be extended to consider arbitrary correlation functions of the form

$$\langle A_1(s_1) \ldots A_j(s_j) B_i(t_i) \ldots B_1(t_1) \rangle = \mathrm{Tr}\{B_i(t_i) \ldots B_1(t_1) \rho_{SE} A_1(s_1) \ldots A_j(s_j)\}, \tag{6.267}$$

with ordered times $t_i > \cdots > t_1 \geqslant 0$, $s_j > \cdots > s_1 \geqslant 0$, where all operators are considered in Heisenberg picture with respect to the full Hamiltonian. Depending on the specific choice of operators $\{A_i\}$ and $\{B_i\}$ and times, the expression (6.267) can recover a number of interesting expressions. If $B_i = A_i^\dagger$ we recover a sequence of POVMs, if $A_i = \mathbb{1}$ or similarly $B_i = \mathbb{1}$ we recover multi-time correlation functions on the side of the system or of the environment, depending to the Hilbert space on which the $\{A_i\}$ and the $\{B_i\}$ act.

If the mean values of a set of system operators, evolving according to the reduced dynamics, obey closed homogeneous evolution equations, these relationships can be used to directly obtain the expression of multi-time correlation functions. We thus assume validity of the equations

$$\frac{\mathrm{d}}{\mathrm{d}t}\langle C_j(t) \rangle = \sum_k \Lambda_{jk}\langle C_k(t) \rangle, \tag{6.268}$$

where $\{C_j\}$ are system operators, Λ_{jk} is a matrix of coefficients, and the expectation value can be expressed as

$$\langle C_k(t) \rangle = \mathrm{Tr}_S\{C_k \Phi(t, 0)[\rho_S(0)]\}. \tag{6.269}$$

According to (6.192) this is true for the Bloch equations considered in Sect. 6.5, as a consequence of the fact that the Pauli matrices form a Lie algebra. Taking the time derivative of the identity (6.269) at the initial time we obtain the relation

$$\frac{\mathrm{d}}{\mathrm{d}t}\langle C_j(t) \rangle\big|_{t=0} = \mathrm{Tr}_S\{C_j \mathcal{L}[\rho_S(0)]\} \tag{6.270}$$

$$= \mathrm{Tr}_S\{\mathcal{L}'[C_j]\rho_S(0)\} \tag{6.271}$$

valid for an arbitrary initial condition, so that we can consider the operator identity

$$\mathcal{L}'[C_j] = \sum_k \Lambda_{jk} C_k, \tag{6.272}$$

where the map $\mathcal{L}'$, adjoint of $\mathcal{L}$ in the sense of (3.4), describes the Heisenberg evolution of the observables. Taking the time derivative of the regression formula (6.263) for the multi-time correlation function we obtain for an arbitrary system operator D

$$\frac{\mathrm{d}}{\mathrm{d}\tau}\langle C_j(t+\tau)D(t)\rangle = \mathrm{Tr}_S\{C_j\mathcal{L}[\Phi(t+\tau,t)[D\Phi(t,0)[\rho_S(0)]]]\} \tag{6.273}$$

$$= \mathrm{Tr}_S\{\mathcal{L}'[C_j]\Phi(t+\tau,t)[D\Phi(t,0)[\rho_S(0)]]\}, \tag{6.274}$$

and therefore exploiting (6.272)

$$\frac{\mathrm{d}}{\mathrm{d}\tau}\langle C_j(t+\tau)D(t)\rangle = \sum_k \Lambda_{jk}\langle C_k(t+\tau)D(t)\rangle. \tag{6.275}$$

In the same way, considering two arbitrary system operators D and E we obtain

$$\frac{\mathrm{d}}{\mathrm{d}\tau}\langle E(s)C_j(t+\tau)D(t)\rangle = \sum_k \Lambda_{jk}\langle E(s)C_k(t+\tau)D(t)\rangle. \tag{6.276}$$

These equations are known as quantum regression formulae. As a result, if the exploited assumptions can be assumed to be valid, the higher-order correlation functions have the same time dependence of the mean values and can be obtained once their time evolution is known. This powerful result is known as quantum regression theorem.

6.6.2   Resonance Fluorescence

We now focus on the spectrum of the emitted radiation in response to the driving field, in the long-time limit in which the system reaches the equilibrium state, introduced in Sect. 6.5.4. The phenomenon of light scattering from resonantly or nearly resonantly excited systems is known as resonance fluorescence. As we shall see, for a low intensity driving field only elastic scattering appears, characterized by a narrow linewidth determined by the damping constant γ. When the intensity grows, and the Rabi frequency Ω associated to the driving field becomes comparable or larger than the atomic line-width, sidebands appear in the spectrum of the emitted radiation, as first predicted by Mollow [53] and later experimentally confirmed. This effect is determined by quantum fluctuations and becomes relevant in the presence of a strong driving field, when the atom has the same probability to be in the ground and in the excited state, so that quantum fluctuations dominate and the field autocorrelation functions are far from factorizing. The very expression of the spectrum describing resonance fluorescence is determined by a two-point correlation function of the system.

To describe this behavior we consider the expression of the radiation field generated by a classical point dipole [54]

$$E(x, t) = \frac{\omega_0^2}{c^2 r} \left(n \times d \left(t - \frac{r}{c} \right) \right) \times n, \tag{6.277}$$

and recall that the dipole is now described by the operator (6.158), so that the positive frequency component of the retarded electromagnetic field radiated by the atomic dipole at a point x is given by the expression [38]

$$E^{(+)}(x, t) = \frac{\omega_0^2}{c^2 r} (n \times d) \times n \sigma_- \left(t - \frac{r}{c} \right), \tag{6.278}$$

where $x = nr$ so that $r = |x|$. The key assignment (6.278) is obtained considering the interaction of a two-level system with quantized radiation modes in the rotating-wave approximation, namely a model as in Sect. 5.4.1 though with coupling given by

$$V = \hbar \sum_k (g_k b_k^\dagger \sigma_+ + g_k^* b_k \sigma_-), \tag{6.279}$$

rather then (5.232) and (5.233). Solving the Heisenberg equations of motion for this model, we can obtain the expression of the annihilation operator in terms of the operator σ_- in Heisenberg picture, leading to (6.278) as detailed in [38]. Thanks to stationarity of the two-point correlation function in the equilibrium state, we rely on the Wiener–Khintchine Theorem 6.3 discussed in Remark 6.6, so that according to (6.311) the spectrum of the process is given by the Fourier transform of the autocorrelation function of the field. Assuming to observe the radiation along a fixed direction we treat the field operators as scalars and come to the expression

$$\frac{\mathrm{d}I}{\mathrm{d}\Omega}(\nu) = \frac{cr^2}{4\pi} \int_{-\infty}^{+\infty} \frac{\mathrm{d}\tau}{2\pi} e^{i\nu\tau} \langle E^{(-)}(t, x) E^{(+)}(t + \tau, x) \rangle \tag{6.280}$$

$$= I_0(x) \Sigma(\nu), \tag{6.281}$$

for the spectral density radiated by the atomic dipole per unit solid angle, where

$$I_0(x) = \frac{\omega_0^4}{4\pi^2 c^3} |(n \times d) \times n|^2 \tag{6.282}$$

determines the spatial dependence, while the quantity

$$\Sigma(\nu) = \frac{1}{2\pi} \int_{-\infty}^{+\infty} \mathrm{d}\tau e^{i(\omega_0 - \nu)\tau} \langle \sigma_+(\tau)\sigma_-(0) \rangle^{\mathrm{eq}}, \tag{6.283}$$

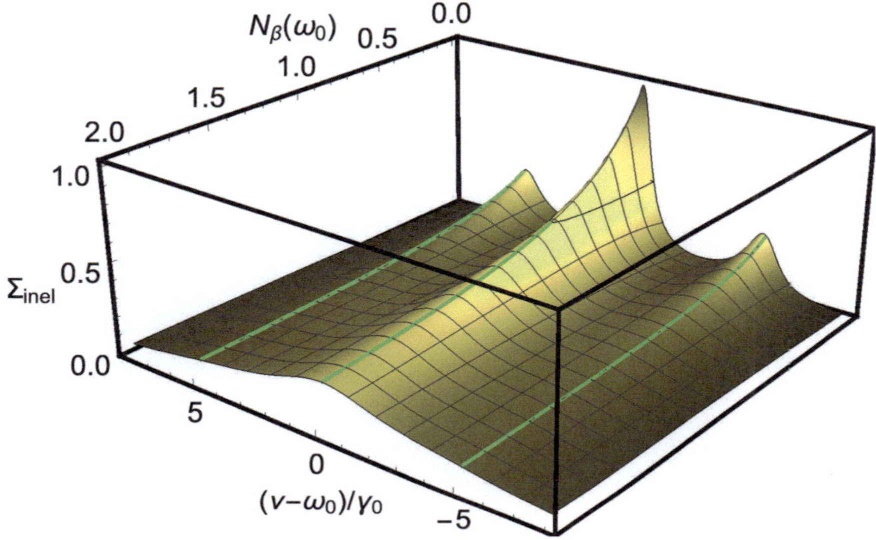

Fig. 6.4 Plot of the inelastic part of the spectrum in its dependence on the temperature of the environment, expressed by the mean photon number $N_\beta(\omega_0)$, for $\Omega/\gamma_0 = 5$. The spectrum is plotted in units of $1/\gamma_0$ as a function of the frequency difference with respect to the resonant frequency ω_0. The solid lines put into evidence the three Mollow peaks described in Sect. 6.6.2, that quickly disappear for growing temperature

characterizes the spectrum of the emitted radiation. The factor $e^{i\omega_0 t}$ comes from moving back to the system reference frame, recalling that we are considering the case of zero detuning. Note that due to stationarity retardation effects do not play any role.

The system correlation function appearing in the integral according to (6.263) is given by

$$\langle \sigma_+(t)\sigma_-(0)\rangle^{\mathrm{eq}} = \mathrm{Tr}_S\{\sigma_+ e^{t\mathcal{L}}[\sigma_- \rho^{\mathrm{eq}}]\}, \tag{6.284}$$

with $\mathcal{L}$ as in (6.181), and is defined for positive times. The expression (6.283) has to be understood as

$$\Sigma(\nu) = \frac{1}{2\pi}\int_{-\infty}^{+\infty} dt\, e^{i(\omega_0-\nu)t}[\chi_{[0,+\infty)}(t)\langle \sigma_+(t)\sigma_-(0)\rangle^{\mathrm{eq}}$$

$$+ \chi_{(-\infty,0)}(t)(\langle \sigma_+(0)\sigma_-(t)\rangle^{\mathrm{eq}})^*], \tag{6.285}$$

where $\chi_{(a,b)}$ denotes the characteristic function of the interval. The spectrum is thus given by the two-times dipole-dipole atomic quantum correlation function at

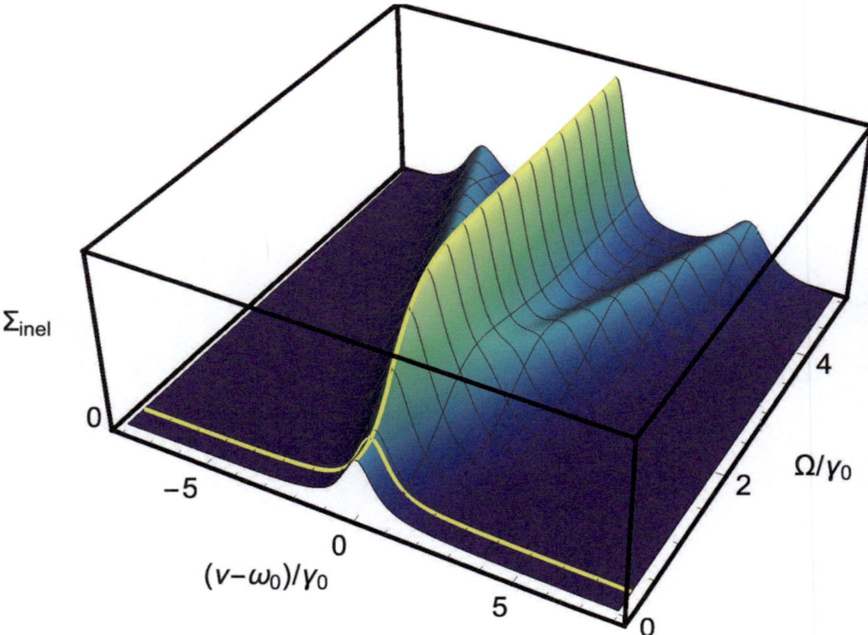

Fig. 6.5 Plot of the inelastic part of the spectrum in its dependence on the driving strength Ω, for a very low temperature environment, characterized by $N_\beta(\omega_0) = 0.1$. All frequencies are in units of the damping rate γ_0. The solid line puts into evidence the transition from the weak to the strong driving regime

equilibrium. According to the fact that $\mathcal{L}$ is Hermiticity preserving from (6.284) we have for negative times

$$(\langle\sigma_+(0)\sigma_-(t)\rangle^{\text{eq}})^* = \text{Tr}_S\{e^{|t|\mathcal{L}}[\rho^{\text{eq}}\sigma_+]\sigma_-\}, \tag{6.286}$$

equation (6.285) can thus be written in a compact explicit form as

$$\Sigma(\nu) = 2\Re \int_0^{+\infty} dt\, e^{i(\omega_0-\nu)t} \langle\sigma_+(t)\sigma_-(0)\rangle^{\text{eq}}. \tag{6.287}$$

Assuming validity of the quantum regression formula (6.275) as discussed in Sect. 6.6.1, in order to evaluate the relevant correlation function (6.284) we exploit (6.247) with the following replacement of initial conditions:

$$\langle\sigma_+(0)\rangle \to \langle\sigma_+\sigma_-\rangle^{\text{eq}} \tag{6.288}$$

$$= \frac{2\Omega^2 + \gamma(\gamma - \gamma_0)}{2(2\Omega^2 + \gamma^2)} \tag{6.289}$$

according to (6.249), together with

$$\langle \sigma_-(0) \rangle \to \langle \sigma_-^2 \rangle^{\mathrm{eq}} \tag{6.290}$$

$$= 0 \tag{6.291}$$

and

$$\langle P_+(0) \rangle \to \langle P_+ \sigma_- \rangle^{\mathrm{eq}} \tag{6.292}$$

$$= 0. \tag{6.293}$$

For the last contribution of (6.247) we have to consider the replacement

$$\mathbb{1} \to \langle \sigma_- \rangle^{\mathrm{eq}} \tag{6.294}$$

$$= i\frac{\Omega \gamma_0}{2\Omega^2 + \gamma^2}, \tag{6.295}$$

that follows from (6.251). We thus obtain after some algebra the expression

$$\langle \sigma_+(t)\sigma_-(0) \rangle^{\mathrm{eq}} = \frac{\Omega^2 \gamma_0^2}{(2\Omega^2 + \gamma^2)^2} + e^{-\frac{\gamma}{2}t}\frac{2\Omega^2 + \gamma(\gamma - \gamma_0)}{4(2\Omega^2 + \gamma^2)}$$

$$+ \frac{e^{-\frac{3}{4}\gamma t}}{8(2\Omega^2 + \gamma^2)^2}\{e^{i\mu t}\Gamma_-(\Omega, \gamma_0, N) + e^{-i\mu t}\Gamma_+(\Omega, \gamma_0, N)\}, \tag{6.296}$$

where we have moved from trigonometric to exponential functions of the frequency μ, defined in (6.245), and we have introduced the complex amplitudes

$$\Gamma_\pm(\Omega, \gamma_0, N) = \pm\frac{i}{4\mu}[\gamma(2\Omega^2 + \gamma^2)^2 + \gamma_0(6\Omega^2(\gamma^2 - 2\gamma\gamma_0) + 16\Omega^4 - \gamma^4)]$$

$$+ (2\Omega^2 + \gamma^2)(2\Omega^2 + \gamma(\gamma - \gamma_0)) - 4\Omega^2\gamma_0^2. \tag{6.297}$$

The dependence on the temperature is encoded in the relation (6.189). To obtain the spectrum we evaluate the integral (6.287) exploiting the formula

$$\Re \int_0^{+\infty} dt\, \Gamma e^{i(\omega_0 - \nu - \varpi)t - \varsigma t} = \Re \left(\frac{\Gamma}{\varsigma - i(\omega_0 - \nu - \varpi)} \right) \tag{6.298}$$

with $\Re \varsigma > 0$. In view of (6.296), where the first term is a constant, it is natural to write the spectrum as the sum of an elastic and inelastic contribution

$$\Sigma(\nu) = \Sigma_{\mathrm{el}}(\nu) + \Sigma_{\mathrm{inel}}(\nu). \tag{6.299}$$

The spectrum depends on the driving strength via the Rabi frequency Ω, and on the features of the environment determining the stationary state via the damping constant γ_0 and the temperature dependent mean photon number $N_\beta(\omega_0)$. The elastic contribution corresponds to Rayleigh scattering and is given by

$$\Sigma_{\text{el}}(\nu) = \frac{\Omega^2 \gamma_0^2}{(2\Omega^2 + \gamma^2)^2} 2\pi \delta(\nu - \omega_0), \tag{6.300}$$

the only relevant contribution in the weak driving regime $\Omega \ll \gamma/4$. The inelastic contribution providing the actual fluorescence spectrum can be written as the sum of three terms

$$\Sigma_{\text{inel}}(\nu) = \Sigma^0_{\text{inel}}(\nu) + \Sigma^+_{\text{inel}}(\nu) + \Sigma^-_{\text{inel}}(\nu), \tag{6.301}$$

with

$$\Sigma^0_{\text{inel}}(\nu) = \frac{2\Omega^2 + \gamma(\gamma - \gamma_0)}{4(2\Omega^2 + \gamma^2)} \frac{\gamma}{(\gamma/2)^2 + (\omega - \omega_0)^2} \tag{6.302}$$

and

$$\Sigma^\pm_{\text{inel}}(\nu) = \frac{\Omega^2}{2(2\Omega^2 + \gamma^2)^2}$$
$$\times \Re\left(\Gamma_\pm(\Omega, \gamma_0, N_\beta(\omega_0)) \frac{\frac{3}{4}\gamma + i(\omega_0 - \nu \mp \mu)}{\left(\frac{3}{4}\gamma\right)^2 + (\omega_0 - \nu \mp \mu)^2}\right). \tag{6.303}$$

The inelastic contributions are plotted in Fig. 6.4 in the strong driving regime as a function of the environmental temperature. For low enough temperature, together with a central peak in correspondence to the system frequency ω_0, two other peaks appear, displaced by $\pm\mu$, giving rise to the so-called Mollow triplet [53]. This behavior is examined more closely in Fig. 6.5, where the spectrum is plotted at low temperature as a function of the driving strength. For strong enough driving, the resonance peak splits into three contributions, with two sidebands of equal height. In the low temperature and strong driving regime, $N_\beta(\omega_0) \ll 1$ and $\Omega/\gamma \gg 1/4$, the spectrum is well approximated by the sum of three Lorentzian curves according to

$$\Sigma_{\text{inel}}(\nu) \approx \frac{1}{2} \frac{\frac{\gamma_0}{2}}{\left(\frac{\gamma_0}{2}\right)^2 + (\nu - \omega_0)^2}$$
$$+ \frac{1}{4} \frac{\frac{3}{4}\gamma_0}{\left(\frac{3}{4}\gamma_0\right)^2 + (\nu - \omega_0 + \Omega)^2} + \frac{1}{4} \frac{\frac{3}{4}\gamma_0}{\left(\frac{3}{4}\gamma_0\right)^2 + (\nu - \omega_0 - \Omega)^2}. \tag{6.304}$$

In this limit the coherent elastically scattered light is inversely proportional to the intensity of the driving field. The height of the lateral maxima is one-third of the central maximum, whose integrated intensity is twice the intensity of each of the other two maxima.

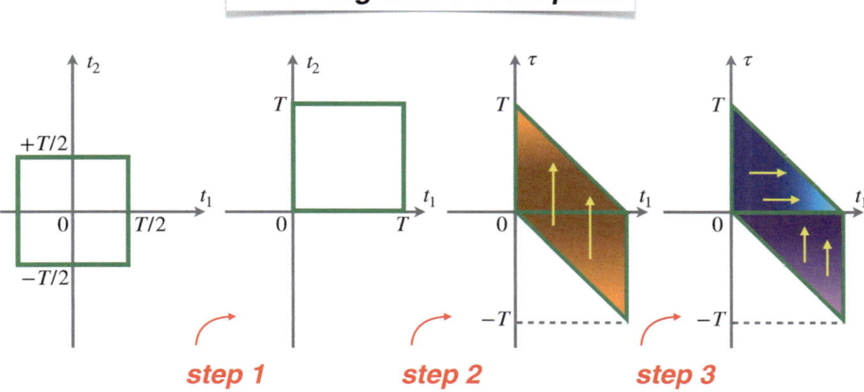

Fig. 6.6 Visualization of the intermediate steps leading to (6.315). The thick solid line denotes the integration region and the arrows the direction of integration

6.6 Wiener–Khintchine Theorem

We consider a stochastic process $X(t)$ taking values in $\mathbb{C}$ [16,55], that is a time-dependent collection of random variables defined over a common sample space, as discussed in Sect. 1.2. We further denote with $\mathbb{E}[\cdot]$ the expectation value over the sample space. We say that such a process is wide-sense or second-order stationary if the one and two dimensional distributions only depend on the time difference, so that we have for all times

$$\mathbb{E}[X(t)] = \mathbb{E}[X(0)] \tag{6.305}$$

$$\mathbb{E}[X(t_2)X^*(t_1)] = \mathbb{E}[X(t_2 - t_1)X^*(0)]. \tag{6.306}$$

More generally, a strictly stationary stochastic process is such that all finite-dimensional distributions only depend on the difference of the time arguments. We thus denote the autocorrelation function of the process as

$$G_X(\tau) = \mathbb{E}[X(\tau)X^*(0)], \tag{6.307}$$

and suppose for the sake of simplicity that both $G_X(\tau)$ and its Fourier transform are absolutely integrable. For wide-sense stationary processes the spectral decomposition of the autocorrelation function then takes a particularly meaningful and useful form [34,56].

Theorem 6.3 (Wiener & Khintchine, 1930 & 1934) Let $\{X(t)\}_t$ be a wide-sense stationary process. Then its autocorrelation function $G_X(t)$ has the spectral decomposition

$$G_X(t) = \int_{-\infty}^{+\infty} d\omega\, S_X(\omega) e^{+i\omega t}, \tag{6.308}$$

where $S_X(\omega)$ is the power spectrum of the process defined as

$$S_X(\omega) = \lim_{T\to\infty} \frac{1}{T} \mathbb{E}[|\hat{X}_T(\omega)|^2], \tag{6.309}$$

with

$$\hat{X}_T(\omega) = \int_{-\frac{T}{2}}^{+\frac{T}{2}} dt\, X(t) e^{-i\omega t} \tag{6.310}$$

truncated Fourier transform of the process. The power spectrum of the process can thus be obtained as

$$S_X(\omega) = \int_{-\infty}^{+\infty} dt\, G_X(t) e^{-i\omega t}. \tag{6.311}$$

The Wiener–Khintchine theorem states that the limit for large T of the random variable $\mathbb{E}[|\hat{X}_T(\omega)|^2]/T$ exists and is given by the Fourier transform of the autocorrelation function of the process. The functions $G_X(t)$ and $S_X(\omega)$ constitute a Fourier couple, that is the power spectrum and the autocorrelation function of a stationary process are related via Fourier transform. The power spectrum is often called power spectral density or simply spectral density. Note that due to its very expression (6.306) the autocorrelation function of a stationary process is positive-definite according to (6.37), so that Bochner's Theorem 6.1 warrants positivity of the so defined power spectrum. In particular, the value at $t = 0$ of the autocorrelation function provides the integrated power spectrum

$$G_X(0) = \int_{-\infty}^{+\infty} d\omega\, S_X(\omega). \tag{6.312}$$

Note further that ergodicity of the process is not invoked.

Proof We first consider the very expression of the r.h.s. of (6.309), that according to the definition of truncated Fourier transform (6.310) reads

$$\frac{1}{T}\mathbb{E}[|\hat{X}_T(\omega)|^2] = \frac{1}{T}\mathbb{E}\left[\int_{-T/2}^{+T/2} dt \int_{-T/2}^{+T/2} d\tau\, X(t) X^*(\tau) e^{-i\omega(t-\tau)}\right],$$

$$\tag{6.313}$$

so that, after exchanging the finite time integration with the expectation value, the stationarity property of the process leads to

$$\frac{1}{T}\mathbb{E}[|\hat{X}_T(\omega)|^2] = \frac{1}{T}\int_{-T/2}^{+T/2} \mathrm{d}t \int_{-T/2}^{+T/2} \mathrm{d}\tau\, G_X(t-\tau)\mathrm{e}^{-i\omega(t-\tau)}. \tag{6.314}$$

We now exploit the identity

$$\int_{-T/2}^{+T/2}\mathrm{d}t_2 \int_{-T/2}^{+T/2}\mathrm{d}t_1\, f(t_2-t_1) = \int_{-T}^{T}\mathrm{d}\tau\,(T-|\tau|)f(\tau), \tag{6.315}$$

to obtain the expression

$$\frac{1}{T}\mathbb{E}[|\hat{X}_T(\omega)|^2] = \int_{-T}^{T}\mathrm{d}\tau\left(1-\frac{|\tau|}{T}\right)G_X(\tau)\mathrm{e}^{-i\omega\tau}, \tag{6.316}$$

which we rewrite as the Fourier transform of the function

$$G_{X,T}(\tau) = \begin{cases} \left(1-\frac{|\tau|}{T}\right)G_X(\tau) & |\tau| < T \\ 0 & |\tau| \geqslant T \end{cases}, \tag{6.317}$$

namely

$$\frac{1}{T}\mathbb{E}[|\hat{X}_T(\omega)|^2] = \int_{-\infty}^{+\infty}\mathrm{d}\tau\, G_{X,T}(\tau)\mathrm{e}^{-i\omega\tau}. \tag{6.318}$$

We now note that the functions

$$f_n(\tau) = G_{X,n}(\tau)\mathrm{e}^{-i\omega\tau} \tag{6.319}$$

provide a sequence converging to

$$f(\tau) = G_X(\tau)\mathrm{e}^{-i\omega\tau}. \tag{6.320}$$

The sequence is dominated by the integrable function

$$g(\tau) = |G_X(\tau)|, \tag{6.321}$$

in the sense that

$$|G_{X,n}(\tau)\mathrm{e}^{-i\omega\tau}| \leqslant |G_X(\tau)| \tag{6.322}$$

for all n. We thus have from (6.316) and (6.317)

$$S_X(\omega) = \lim_{T \to \infty} \frac{1}{T} \mathbb{E}[|\hat{X}_T(\omega)|^2] \tag{6.323}$$

$$= \lim_{T \to \infty} \int_{-\infty}^{+\infty} d\tau\, G_{X,T}(\tau) e^{-i\omega\tau}, \tag{6.324}$$

but considering the sequence (6.319) and the dominating function (6.321) thanks to Lebesgue's dominated convergence theorem [57] we can exchange limit and integral, thus obtaining the result (6.311).

As a last step we prove the identity (6.315) used in the proof. We do this in three steps as depicted in Fig. 6.6. We first move the corner of the square of integration to the origin by a translation along both directions as in step 1.

$$\int_{-T/2}^{+T/2} dt_2 \int_{-T/2}^{+T/2} dt_1\, f(t_2 - t_1) = \int_0^T dt_2 \int_0^T dt_1\, f(t_2 - t_1). \tag{6.325}$$

In step 2 we introduce the new variable $\tau = t_2 - t_1$ and transform the square of integration in a rhombus on the (τ, t_1)-plane with corners $\{(0, 0), (T, 0), (0, T), (-T, T)\}$, thus coming to the identity

$$\int_0^T dt_2 \int_0^T dt_1\, f(t_2 - t_1) = \int_0^T dt_1 \int_{-t_1}^{T-t_1} d\tau\, f(\tau), \tag{6.326}$$

where for fixed t_1 one has a vertical integration along the τ axis. We now split the integration area in two triangles, corresponding to the area above and below the horizontal axis, and perform the integral in the upper part in the horizontal direction, while the one in the lower part in the vertical direction, according to step 3 as visualized in Fig. 6.6, thus obtaining

$$\int_0^T dt_1 \int_0^{T-t_1} d\tau\, f(\tau) + \int_0^T dt_1 \int_{-t_1}^0 d\tau\, f(\tau)$$
$$= \int_0^{T-\tau} dt_1 \int_0^T d\tau\, f(\tau) + \int_{-\tau}^T dt_1 \int_{-T}^0 d\tau\, f(\tau). \tag{6.327}$$

The integrations over t_1 can now be directly performed and combining the two terms we obtain (6.315). $\qquad\square$

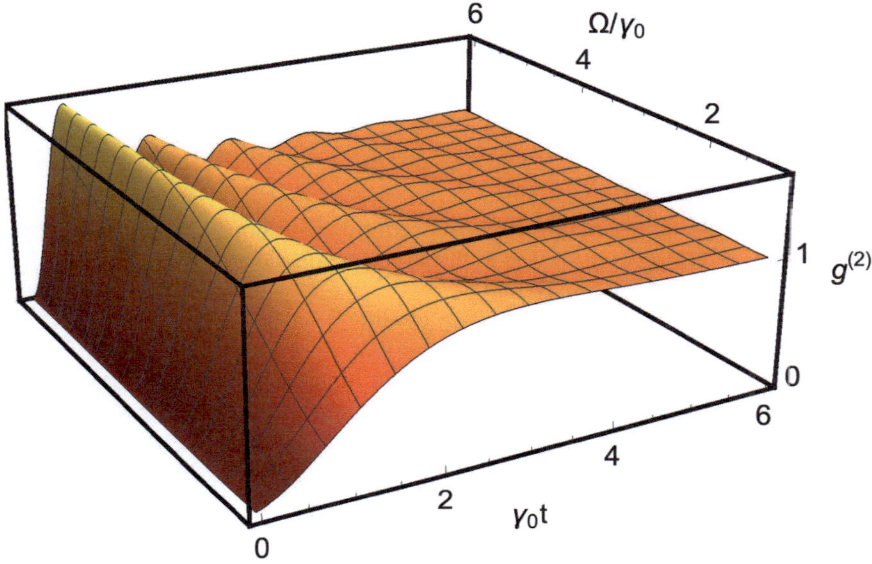

Fig. 6.7 Plot of the normalized second-order correlation function $g^{(2)}$ given by (6.336) as a function of rescaled delay time $\gamma_0 t$ and of rescaled driving strength Ω/γ_0, for a very low temperature environment characterized by $N_\beta(\omega_0) = 0.1$. It clearly appears the increase of the function at the initial time

6.6.3 Photon Anti-bunching

We consider here another important feature of the scattered field in resonance fluorescence, allowing to put into evidence a non-classical feature, also relying on the evaluation of a higher-order correlation function of the reduced system. We introduce the normalized second-order correlation function considered in optics to describe the statistical properties of the light field

$$g^{(2)}(t, x) = \frac{\langle E^{(-)}(x, s) E^{(-)}(x, s + t) E^{(+)}(x, s + t) E^{(+)}(x, s) \rangle}{\langle E^{(-)}(x, s) E^{(+)}(x, s) \rangle \langle E^{(-)}(x, t) E^{(+)}(x, t) \rangle}, \quad (6.328)$$

that relying on (6.278) in the considered equilibrium state takes the form

$$g^{(2)}(t) = \frac{\langle \sigma_+(0)\sigma_+(t)\sigma_-(t)\sigma_-(0) \rangle^{\mathrm{eq}}}{(\langle \sigma_+(0)\sigma_-(0) \rangle^{\mathrm{eq}})^2}, \quad (6.329)$$

where t is interpreted as the delay between subsequent detection events.

This intensity-intensity correlation function characterizes the field fluctuations and in the present treatment depends on the quantum fluctuations of the field, with a behavior determined by the driving strength. In particular, we will see the emergence of the phenomenon of anti-bunching, that is the increase of $g^{(2)}(t)$ for short delays with respect to the initial value $g^{(2)}(0) = 0$. This behavior cannot be exhibited by classical light, which rather shows photon bunching. Photon bunching means that detection events will take place in bunches, close together in time rather than far apart, so that

$$g^{(2)}(t) \leqslant g^{(2)}(0). \tag{6.330}$$

This constraint also holds in the quantum description of a coherent or thermal source of light. Anti-bunching corresponds to the opposite situation

$$g^{(2)}(t) \geqslant g^{(2)}(0), \tag{6.331}$$

which cannot be met by classical light [58]. In both cases a form of bunching is preferred to random arrivals. As in Sect. 6.6.2, to obtain (6.329) we assume validity of the quantum regression formula (6.276) introduced in Sect. 6.6.1. Recalling that $\sigma_+(t)\sigma_-(t) = P(t)$, we exploit (6.246) with a suitable replacement of initial conditions. The only non-zero contribution is given by

$$\mathbb{1} \rightarrow \langle \sigma_+\sigma_- \rangle^{\mathrm{eq}} \tag{6.332}$$

$$= \frac{2\Omega^2 + \gamma(\gamma - \gamma_0)}{2(2\Omega^2 + \gamma^2)}, \tag{6.333}$$

where we have used (6.249). The initial conditions replacing $\langle \sigma_\pm(0) \rangle$ and $\langle P_+(0) \rangle$ are instead equal to zero, due to

$$\langle \sigma_+\sigma_\pm\sigma_- \rangle^{\mathrm{eq}} = \langle \sigma_+ P_+ \sigma_- \rangle^{\mathrm{eq}} = 0. \tag{6.334}$$

We are thus left with

$$\langle \sigma_+(0)\sigma_+(t)\sigma_-(t)\sigma_-(0) \rangle^{\mathrm{eq}} =$$
$$\langle \sigma_+\sigma_- \rangle^{\mathrm{eq}} \left\{ 1 - \mathrm{e}^{-\frac{3}{4}\gamma t} \left[\cos(\mu t) - \frac{1}{4}\frac{\gamma}{\mu} \left(1 - \frac{(\gamma_0/\gamma)8\Omega^2}{(2\Omega^2 + \gamma(\gamma - \gamma_0))} \right) \sin(\mu t) \right] \right\}, \tag{6.335}$$

so that according to (6.329) we have

$$g^{(2)}(t) = 1 - \mathrm{e}^{-\frac{3}{4}\gamma t} \left[\cos(\mu t) - \frac{1}{4}\frac{\gamma}{\mu} \left(1 - \frac{(\gamma_0/\gamma)8\Omega^2}{(2\Omega^2 + \gamma(\gamma - \gamma_0))} \right) \sin(\mu t) \right], \tag{6.336}$$

which for zero temperature, so that $\gamma = \gamma_0$, simplifies to

$$g^{(2)}_{T=0}(t) = 1 - \mathrm{e}^{-\frac{3}{4}\gamma t} \left[\cos(\mu t) + \frac{3}{4}\frac{\gamma}{\mu} \sin(\mu t) \right]. \tag{6.337}$$

A confirmation of these results derived relying on the regression hypothesis can be obtained looking for the probability to be in the excited state starting from the ground state, that is $\langle P_+(t)\rangle$ of (6.246) evaluated with initial condition $\langle P_+(0)\rangle = \langle \sigma_+(0)\rangle = \langle \sigma_-(0)\rangle = 0$, divided by the equilibrium probability, which indeed recovers (6.336). This probability indeed captures the physical mechanism underlying the effect of anti-bunching, namely after the emission of a photon the atom must be excited again before a new emission can take place. The behavior of the intensity-intensity correlation function (6.329) is plotted in Fig. 6.7 in its dependence on delay time and driving strength. Note the increase of $g^{(2)}$ for short delay times, thus featuring anti-bunching as given by (6.331), signature of a quantum source of light.

References

1. N. Hashitsumae, F. Shibata, M. Shingu, J. Stat. Phys. **17**, 155 (1977). https://doi.org/10.1007/BF01040099
2. H.-P. Breuer, B. Kappler, F. Petruccione, Ann. Phys. (N.Y.) **291**, 36 (2001). https://doi.org/10.1006/aphy.2001.6152
3. M. Ban, S. Kitajima, F. Shibata, Phys. Lett. A **374**, 2324 (2010). https://doi.org/10.1016/j.physleta.2010.03.066
4. D.F. Walls, G.J. Milburn, *Quantum Optics* (Springer, Berlin, 1994). https://doi.org/10.1007/978-3-540-28574-8
5. E.B. Davies, Comm. Math. Phys. **39**, 91 (1974). https://doi.org/10.1007/BF01608389
6. A.G. Redfield, IBM, J. Res. Dev. **1**, 19 (1957). https://doi.org/10.1147/rd.11.0019
7. A. Redfield, in *Advances in Magnetic Resonance*, vol. 1, ed. by J.S. Waugh (Academic, 1965), pp. 1–32. https://doi.org/10.1016/B978-1-4832-3114-3.50007-6
8. Á. Rivas, A.D.K. Plato, S.F. Huelga, M.B. Plenio, New J. Phys. **12**, 113032 (2010). https://doi.org/10.1088/1367-2630/12/11/113032
9. P.P. Hofer, M. Perarnau-Llobet, L.D.M. Miranda, G. Haack, R. Silva, J.B. Brask, N. Brunner, New J. Phys. **19**, 123037 (2017). https://doi.org/10.1088/1367-2630/aa964f
10. D. Farina, V. Giovannetti, Phys. Rev. A **100**, 012107 (2019). https://doi.org/10.1103/PhysRevA.100.012107
11. R. Hartmann, W.T. Strunz, Phys. Rev. A **101**, 012103 (2020). https://doi.org/10.1103/PhysRevA.101.012103
12. M. Cattaneo, G.L. Giorgi, S. Maniscalco, R. Zambrini, Phys. Rev. A **101**, 042108 (2020). https://doi.org/10.1103/PhysRevA.101.042108
13. A. Trushechkin, Phys. Rev. A **103**, 062226 (2021). https://doi.org/10.1103/PhysRevA.103.062226
14. A. D'Abbruzzo, V. Cavina, V. Giovannetti, SciPost Phys. **15**, 117 (2023). https://doi.org/10.21468/SciPostPhys.15.3.117
15. C.W. Gardiner, P. Zoller, *Quantum Noise* (Springer, New York, 2000)
16. H.-P. Breuer, F. Petruccione, *The Theory of Open Quantum Systems* (Oxford University Press, Oxford, 2002)
17. M. Reed, B. Simon, *I: Functional Analysis* (Academic, San Diego, 1980)
18. W. Rudin, *Functional Analysis* (McGraw-Hill, New York, 1973)
19. S. Bochner, Math. Ann. **108**, 378 (1933). https://doi.org/10.1007/BF01452844
20. F. Riesz, B.Sz.-Nagy, *Functional Analysis* (Dover, 1956)
21. W. Feller, *An Introduction to Probability Theory and Its Applications*, vol. II (Wiley, New York, 1971)
22. S. Bernstein, Acta Math. **52**, 1 (1929). https://doi.org/10.1007/BF02592679

23. H. Spohn, J. Math. Phys. **19**, 1227 (1978). https://doi.org/10.1063/1.523789
24. R. Alicki, J. Phys. A: Math. Gen. **12**, L103 (1979). https://doi.org/10.1088/0305-4470/12/5/007
25. J. Gemmer, M. Michel, G. Mahler, in *Lecture Notes in Physics*, vol. 784, 2nd edn. (Springer, Heidelberg, 1999)
26. S. Deffner, S. Campbell, *Quantum Thermodynamics* (Morgan & Claypool Publishers, 2019), pp. 2053–2571. https://doi.org/10.1088/2053-2571/ab21c6
27. J. Goold, M. Huber, A. Riera, L. del Rio, P. Skrzypczyk, J. Phys. A: Math. Theor. **49**, 143001 (2016). https://doi.org/10.1088/1751-8113/49/14/143001
28. S. Vinjanampathy, J. Anders, Contemp. Phys. **57**, 545 (2016). https://doi.org/10.1080/00107514.2016.1201896
29. G.T. Landi, M. Paternostro, Rev. Mod. Phys. **93**, 035008 (2021). https://doi.org/10.1103/RevModPhys.93.035008
30. H. Spohn, J.L. Lebowitz, Adv. Chem. Phys. **38**, 109 (1978). https://doi.org/10.1002/9780470142578.ch2
31. R. Haag, *Local Quantum Physics* (Springer, Berlin, Heidelberg, 1996). https://doi.org/10.1007/978-3-642-61458-3
32. A. Rivas, S.F. Huelga, *Open Quantum Systems: An Introduction* (Springer, Berlin, 2012). https://doi.org/10.1007/978-3-642-23354-8
33. W. Roga, M. Fannes, K. Zyczkowski, Rep. Math. Phys. **66**, 311 (2010). https://doi.org/10.1016/S0034-4877(11)00003-6
34. M. Weissbluth, *Photon-atom Interactions* (Academic, Burlington, MA, London, 1989)
35. W.H. Louisell, *Quantum Statistical Properties of Radiation* (Wiley-VCH, 1990)
36. C. Cohen-Tannoudji, B. Diu, F. Laloe, *Quantum Mechanics*, vol. I (Sons, New York, 1992)
37. N.I. Muskhelishvili, *Singular Integral Equation* (P. Noordhoff, Groningen, 1953)
38. M.O. Scully, M.S. Zubairy, *Quantum Optics* (Cambridge University Press, Cambridge, 1997)
39. F. Bloch, Phys. Rev. **70**, 460 (1946). https://doi.org/10.1103/PhysRev.70.460
40. C. Cohen-Tannoudji, B. Diu, F. Laloe, *Quantum Mechanics*, vol. II (Wiley, New York, 1992)
41. R.K. Wangsness, F. Bloch, Phys. Rev. **89**, 728 (1953). https://doi.org/10.1103/PhysRev.89.728
42. E.M. Purcell, H.C. Torrey, R.V. Pound, Phys. Rev. **69**, 37 (1946). https://doi.org/10.1103/PhysRev.69.37
43. F. Bloch, W.W. Hansen, M. Packard, Phys. Rev. **70**, 474 (1946). https://doi.org/10.1103/PhysRev.70.474
44. A. Abragam, *The Principles of Nuclear Magnetism* (Oxford University Press, 1961)
45. R.R. Ernst, G. Bodenhausen, A. Wokaun, *Principles of Nuclear Magnetic Resonance in One and Two Dimensions* (Oxford University Press, 1987)
46. V. Gorini, A. Frigerio, M. Verri, A. Kossakowski, E. Sudarshan, Rep. Math. Phys. **13**, 149 (1978). https://doi.org/10.1016/0034-4877(78)90050-2
47. A. Amann, U. Müller-Herold, *Offene Quantensysteme: Die Primas Lectures* (Springer, Berlin, 2011). https://doi.org/10.1007/978-3-642-05187-6
48. M. Lax, Phys. Rev. **129**, 2342 (1963). https://doi.org/10.1103/PhysRev.129.2342
49. G. Guarnieri, A. Smirne, B. Vacchini, Phys. Rev. A **90**, 022110 (2014). https://doi.org/10.1103/PhysRevA.90.022110
50. D. Lonigro, D. Chruściński, J. Phys. A: Math. Theor. **55**, 225308 (2022). https://doi.org/10.1088/1751-8121/ac6a2d
51. S. Swain, J. Phys. A: Math. Gen. **14**, 2577 (1981). https://doi.org/10.1088/0305-4470/14/10/013
52. R. Dümcke, J. Math. Phys. **24**, 311 (1983). https://doi.org/10.1063/1.525681
53. B.R. Mollow, Phys. Rev. **188**, 1969 (1969). https://doi.org/10.1103/PhysRev.188.1969
54. J.D. Jackson, *Classical Electrodynamics* (Wiley, 1962)
55. C.W. Gardiner, *Handbook of Stochastic Methods for Physics, Chemistry and the Natural Sciences*, *Springer Series in Synergetics*, vol. 13, 3rd edn. (Springer, Berlin, 2004)

56. A. Barchielli, M. Gregoratti, *Quantum Trajectories and Measurements in Continuous Time*, *Lecture Notes in Physics*, vol. 782 (Springer, Berlin, 2009). https://doi.org/10.1007/978-3-642-01298-3
57. W. Rudin, *Real and Complex Analysis* (McGraw-Hill, New York, 1921)
58. H.J. Kimble, M. Dagenais, L. Mandel, Phys. Rev. Lett. **39**, 691 (1977). https://doi.org/10.1103/PhysRevLett.39.691

Memory Effects

7

Abstract

We consider the characterization of quantum dynamical maps that describe the reduced dynamics of open quantum systems and embody the quantum counterpart of a classical stochastic evolution. The distinction of classical stochastic processes in Markovian and non-Markovian provides a natural reference, so that we discuss the enforcement of this perspective in the quantum setting. We especially highlight an approach to compare the different dynamics based on the analysis of the information exchange between system and environment. We address insofar these characterizations of the dynamics can be introduced with reference to the quantum dynamical map itself or to the associated evolution equations. We are thus lead to consider the notion of divisible map, as well as of time-local and memory-kernel master equations. We illustrate these concepts by means of several examples.

7.1 Introduction

In Chaps. 5 and 6 we have discussed the time evolution of several quantum systems, starting from both microscopic and phenomenological approaches. We first considered the very important class of quantum dynamical semigroup evolutions, that as shown in Sect. 5.3 thanks to the requirement of complete positivity admits a thorough and phenomenologically very useful characterization of the associated master equation. The semigroup composition law forward in time brings with itself two interesting features. On the one hand, it accounts for irreversibility introducing a preferred direction in time. On the other hand, it allows to determine the future state of the system just relying on knowledge of the state at the given time. However, this class of dynamics does not cover the most general reduced dynamical evolution, as we have put into evidence in Sect. 5.4, considering simple examples that allow for detailed treatments. Formally exact evolution equations for the general case have

© The Author(s), under exclusive license to Springer Nature Switzerland AG 2024   355
B. Vacchini, *Open Quantum Systems*, Graduate Texts in Physics,
https://doi.org/10.1007/978-3-031-58218-9_7

been further addressed in Sect. 5.5 relying on a projection operator technique. We discussed in particular the Nakajima–Zwanzig approach, leading to master equations with a memory kernel. Conditions under which the obtained exact evolution equation can be well approximated by a semigroup evolution have been discussed in Chap. 6. The validity of these approximations is typically warranted in the presence of a separation of time-scales between system and environment, so that the time evolution of the environment can be essentially neglected on the time-scale over which the system varies in an appreciable way. So far, however, no specific features have been introduced to distinguish and characterize the possible different non-unitary dynamics arising due to the interaction of an open system with an external environment. In the classical framework, the dynamics of open systems is formulated in terms of stochastic processes, which allow to properly formalize the randomicity added to the deterministic dynamics of the system by the interaction with degrees of freedom which are only effectively taken into account. In this framework, a class of stochastic processes was put into evidence by Andrej Markov [1] on the basis of a mathematical definition, the so-called Markov condition, that involves information on sampling of the process at an arbitrary number of points in time and encodes lack of memory of the process. Roughly speaking, knowledge of the previous history is not necessary to predict future outcomes. A number of properties and useful results have been obtained for Markov processes. Classical stochastic processes not enjoying this property are termed non-Markovian, and very little is known about them. In the quantum framework, it appears natural to look for a similar characterization for the reduced quantum dynamics, that provide the quantum counterpart of classical stochastic dynamics. This topic involves the very definition of a Markovian or memoryless quantum dynamics, and it is often addressed as the study of quantum non-Markovianity. This is a highly non-trivial problem, given that the very association of a value to an observable calls for a measurement, which as discussed in Sect. 3.2.3 in quantum mechanics by necessity influence the state of the system and therefore subsequent observations [2]. We will first introduce the definition of classical Markov processes, pointing to different features of these processes that can be used to introduce in the quantum setting convenient definitions of quantum non-Markovian dynamics. We will then put into evidence an approach to characterize a quantum dynamics based on the analysis of the information exchange between system and environment. Following seminal work [3] non-Markovian dynamics is associated to a backflow of information from the environment to the system. This information was previously stored in the establishment of correlations between system and environment or in changes in the environmental state conditioned on the system state. The guiding idea is sketched in Fig. 7.1, while the different facets of the formalization of this approach are visually summarized in Fig. 7.2. We will further discuss how this viewpoint relates to different approaches more directly connected to the classical definition. We finally discuss the possible general structure of master equations leading to non-Markovian dynamics and provide the analysis of several physical models in view of the considered characterizations of the deviation from a memoryless quantum dynamics.

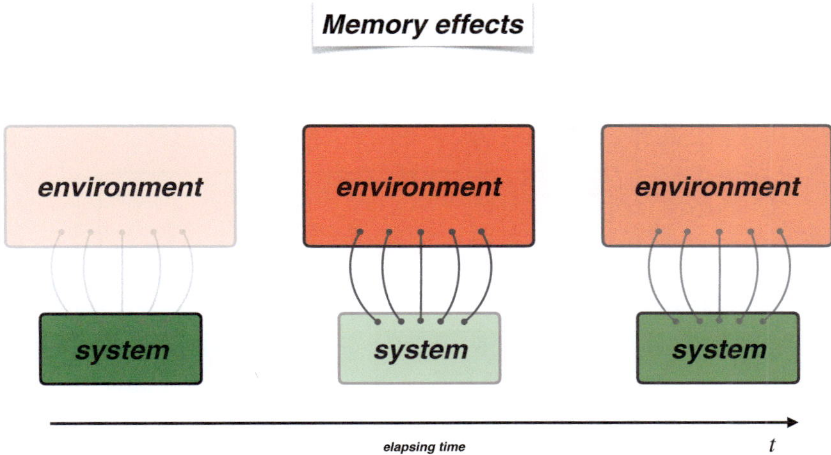

Fig. 7.1 Schematic visualization of the notion of memory in the open quantum system context. Information is visualized as brightness, encoded in either the system, the environment or the correlations between them, depicted as joining cords. With elapsing time the information initially stored in the system is transferred to correlations and environmental degrees of freedom. If this information comes back we speak of memory effects

Different possible notions of quantum Markovian dynamics distinct but related to the present treatment has been suggested in [4–9], reviewed and compared in [10–16]. It is further important to stress that utterly different approaches can be considered in the spirit of defining quantum processes of non-Markovian type, as put forward originally in [17, 18], and more recently advocated in [19, 20], that involve statements about multi-time observations. Experiments for the demonstration of the different approaches has also been performed [21–23]. It is further worth mentioning that also the possible connection between a non-Markovian dynamics and irreversible entropy production has proven an interesting research path, see [24–31] for a review and further references.

7.2 Classical Memoryless Stochastic Dynamics

With the aim to provide a convenient and possibly useful way of classifying quantum reduced evolutions, especially in view of the characterization of memory effects, we first consider in the classical framework the formalization of a dynamics that exhibits memory. This characterization is based on the notion of stochastic process. We now brief recall in a sketchy way some basic aspects in the definition and characterization of classical stochastic processes, referring the reader to [32–37] for more extensive and informative treatments along different perspectives.

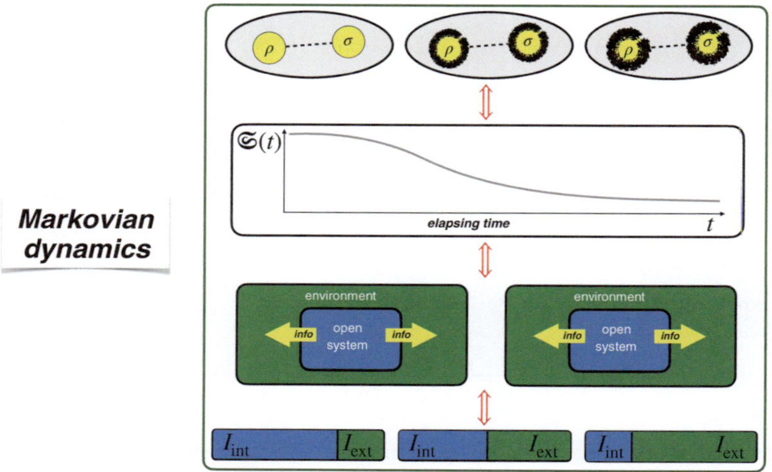

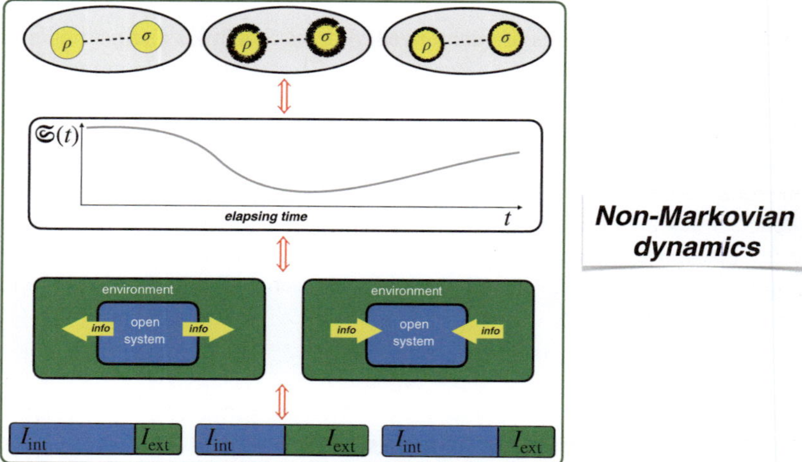

Fig. 7.2 Schematic representation of the introduced notion of Markovian versus non-Markovian dynamics. The dynamics is termed Markovian if the distinguishability of two initial states ρ and σ, as quantified by $\mathfrak{S}(\rho, \sigma)$ obeying properties **P1** to **P3**, decreases monotonically with time. This is due to the flow of information from the system to the environment, so that I_{int}, identified with $\mathfrak{S}(\rho, \sigma)$ according to (7.80), decreases at the expenses of I_{ext}, given by (7.81). To evoke this behavior, in the scheme the identification of the two states within the convex set containing them gets blurred, so that the length of the joining line decreases. Correspondingly, the fraction of external information I_{ext} increases, the overall amount remaining fixed according to (7.84). If this behavior is inverted at some point in time, then the dynamics is termed non-Markovian. The distinguishability shows a revival and there is a backflow of information from the environment to the system, leading to a regrowth of I_{int}, and therefore $\mathfrak{S}(\rho, \sigma)$, at the expenses of I_{ext}

7.2.1 Classical Markov Process

A classical stochastic process is a collection of random variables $\{X(t)\}_{t\in\mathbb{R}_+}$ defined on a common probability space [34,37], that for the case of interest can describe the values at different times of a physical quantity not undergoing an arbitrary evolution. The common underlying probability space allows to associate a probability distribution to each random variable $X(t)$, and the correlations between the values assumed by the quantity at different times are expressed by means of the joint probability densities determining the process. We are interested in stochastic processes taking real values, and we assume for the sake of notational simplicity a discrete outcome space. According to Kolmogorov's theorem [38], a stochastic process can be defined introducing a collection of joint probabilities

$$p_n(x_n, t_n; x_{n-1}, t_{n-1}; \ldots; x_1, t_1; x_0, t_0) \qquad t_n \geqslant t_{n-1} \geqslant \cdots \geqslant t_1 \geqslant t_0, \qquad (7.1)$$

that express the probability that the random variable assumes the values $\{x_i\}_{i=0,\ldots,n}$ at times $\{t_i\}_{i=0,\ldots,n}$, for any $n \in \mathbb{N}$. The subscript n underlines the fact that this probability depends on n entries. These joint probabilities have further to be invariant under permutation of their arguments. Most importantly, in view of the comparison with the quantum framework, these joint probabilities must obey the Kolmogorov condition given by the consistency requirement

$$\sum_{x_j} p_n(x_n, t_n; \ldots; x_{j+1}, t_{j+1}; x_j, t_j; x_{j-1}, t_{j-1}; \ldots; x_0, t_0) =$$

$$p_{n-1}(x_n, t_n; \ldots; x_{j+1}, t_{j+1}; x_{j-1}, t_{j-1}; \ldots; x_0, t_0), \qquad (7.2)$$

for arbitrary $n > j > 0$ and arbitrary time $t_n > t_j > t_0$. The sum in (7.2) runs over all possible outcomes of the random variable, and the identity conveys the following message. In a classical description the n-point joint probability for outcomes at a specific sequence of times can be obtained from any higher-order joint probability by ignoring, that is summing over, the outcomes at other intermediate times. Crucially, averaging over the information on values assumed by the random variable at intermediate times does not influence the following assumed values. Fulfillment of this condition allows for the definition of a classical stochastic process, that is, it allows to express the joint probabilities of (7.1) as arising from a time-dependent collection of random variables on a common probability space.

The most important and well-known family of classical stochastic processes is given by Markov processes, that embody the notion of a random evolution without memory. To characterize them we need to introduce the conditional probabilities of the process, that we denote as

$$p_{m+1|n}(x_{n+m}, t_{n+m}; \ldots; x_n, t_n | x_{n-1}, t_{n-1}; \ldots; x_0, t_0), \qquad (7.3)$$

and are defined from the relation

$$p_{n+m+1}(x_{n+m}, t_{n+m}; \ldots; x_0, t_0) = p_{m+1|n}(x_{n+m}, t_{n+m}; \ldots; x_n, t_n | x_{n-1}, t_{n-1}; \ldots; x_0, t_0)$$
$$\times \, p_n(x_{n-1}, t_{n-1}; \ldots; x_0, t_0), \tag{7.4}$$

with $n, m \in \mathbb{N}$, $n, m \geqslant 1$. A stochastic process is defined to be Markovian if the collection of its conditional probabilities satisfies

$$p_{1|n}(x_n, t_n | x_{n-1}, t_{n-1}; \ldots; x_0, t_0) = p_{1|1}(x_n, t_n | x_{n-1}, t_{n-1}) \quad \forall n \in \mathbb{N} \tag{7.5}$$

for any arbitrary collection $\{x_i\}_{i=0,\ldots,n}$ in the outcome space and any arbitrary sequence of ordered times $\{t_i\}_{i=0,\ldots,n}$. These conditions can be interpreted as lack of memory of the process. The probability for the considered random variable to assume the value x_n at time t_n, given it has taken on the values $\{x_i\}_{i=0,\ldots,n-1}$ at previous times $\{t_i\}_{i=0,\ldots,n-1}$, only depends on the last assumed value, and not on previous ones. As stressed in (7.5), the Markov condition actually involves an infinite number of constraints, since it has to be ascertained for conditional probabilities of arbitrary high order [39, 40].

7.2.2 Consequences of the Classical Markov Condition

The condition (7.5) has many important implications. We first stress that repeated application of (7.4) thanks to (7.5) leads to express any joint probability in the form

$$p_{n+1}(x_n, t_n; \ldots; x_0, t_0) = \prod_{i=1}^{n} p_{1|1}(x_i, t_i | x_{i-1}, t_{i-1}) p_1(x_0, t_0), \tag{7.6}$$

where $p_{1|1}(x, t | y, s)$ is the so-called conditional transition probability, that for a Markov process only depends on the pairs x, t and y, s. Indeed, iterating (7.4), the very definition of conditional probability implies the identity

$$p_{n+1}(x_n, t_n; \ldots; x_0, t_0) = p_{1|n}(x_n, t_n | x_{n-1}, t_{n-1}; \ldots; x_0, t_0)$$
$$\times \, p_{1|n-1}(x_{n-1}, t_{n-1} | x_{n-2}, t_{n-2}; \ldots; x_0, t_0)$$
$$\cdots$$
$$\times \, p_{1|1}(x_1, t_1 | x_0, t_0) p_1(x_0, t_0), \tag{7.7}$$

that upon validity of (7.5) leads to (7.6). A Markov process is therefore uniquely determined by its conditional transition probability and the initial distribution of the random variable. The Markov condition (7.5) has two further important consequences on one- and two-point probabilities. The most well-known implication is the fact that the conditional transition probability $p_{1|1}$ obeys the Chapman–Kolmogorov equation [41,42]

$$p_{1|1}(x,t|y,s) = \sum_z p_{1|1}(x,t|z,\tau)p_{1|1}(z,\tau|y,s), \tag{7.8}$$

with $t \geqslant \tau \geqslant s$. This equation states that a transition from y at epoch s to x at epoch t occurs via a point z at an intermediate epoch τ. Note the striking difference with respect to the notion of inserting a completeness in a transition amplitude in quantum mechanics. Here we are summing over probabilities determined by the realization of intermediate events, rather than upon amplitudes. To prove it we exploit (7.7) for $n = 2$

$$p_3(x,t;z,\tau;y,s) = p_{1|2}(x,t|z,\tau;y,s)p_{1|1}(z,\tau|y,s)p_1(y,s), \tag{7.9}$$

and sum over z. Exploiting the Kolmogorov condition (7.2) together with the Markov condition (7.5) we are left with

$$p_2(x,t;y,s) = \sum_z p_{1|1}(x,t|z,\tau)p_{1|1}(z,\tau|y,s)p_1(y,s), \tag{7.10}$$

and therefore (7.8) upon dividing by $p_1(y,s)$. We stress that the Markov condition also implies another interesting property for the one-point probability $p_1(x,t)$. Starting from (7.10), which is a consequence of the Markov property (7.5), and summing over y we are left with

$$p_1(x,t) = \sum_z p_{1|1}(x,t|z,\tau)p_1(z,\tau), \tag{7.11}$$

thanks to the Kolmogorov condition (7.2) and the very definition of conditional probability (7.4), that further implies

$$\sum_x p_{1|1}(x,t|y,s) = 1. \tag{7.12}$$

We can now consider two Markov processes determined by the same transition probability $p_{1|1}$ and only differing by the initial probability distributions, that we denote as $p_1^I(x,t_0)$ and $p_1^{II}(x,t_0)$. We then have relying on (7.11) and (7.12)

$$\sum_x |p_1^I(x,t_0) - p_1^{II}(x,t_0)| = \sum_x \left| \sum_y p_{1|1}(x,t|y,s)(p_1^I(y,s) - p_1^{II}(y,s)) \right| \tag{7.13}$$

$$\leqslant \sum_{x,y} p_{1|1}(x,t|y,s)|p_1^I(y,s) - p_1^{II}(y,s)| \tag{7.14}$$

$$= \sum_y |p_1^I(y,s) - p_1^{II}(y,s)|. \tag{7.15}$$

In analogy to (3.211) of Sect. 3.2.5.1 we can introduce the quantity

$$D_K(p, q) = \frac{1}{2} \sum_x |p(x) - q(x)| \qquad (7.16)$$

known as Kolmogorov or variational distance [43,44] and given apart from a normalization factor by the l_1-norm between the two probability distributions

$$D_K(p, q) = \frac{1}{2} \|p - q\|_1. \qquad (7.17)$$

This distance has the same properties discussed in Sect. 3.2.5.1, and provides a way to compare two probability distributions, its value being 0 iff they coincide and 1 iff they assign non-zero weights only to distinct outcomes. For the one-point probability of a Markov process we thus have the inequality

$$D_K(p_1^I(t), p_1^{II}(t)) \leqslant D_K(p_1^I(s), p_1^{II}(s)) \qquad \forall t \geqslant s, \qquad (7.18)$$

telling that the Kolmogorov distance between two distinct initial probability distributions that follow the same Markovian stochastic process is monotonically decreasing with time. The dependence on the initial condition is less and less detectable.

For the sake of comparison with the quantum treatment, we express the previous results in a more compact way, considering a process with a finite outcome set of cardinality N. We denote with $p(t)$ the probability vector fixing the one-point probability of the process at time t

$$p(t) = (p_1(x_1, t), \ldots, p_1(x_N, t)), \qquad (7.19)$$

and with $\Lambda(t, s)$ the matrix corresponding to the finite collection of conditional transition probabilities between times s and t, with $t \geqslant s$, namely

$$(\Lambda(t, s))_{ij} = p_{1|1}(x_i, t|x_j, s). \qquad (7.20)$$

The relation (7.11) between the probability vectors of a Markov process at different times then reads

$$p(t) = \Lambda(t, s)p(s), \qquad (7.21)$$

while the normalization condition for the transition probability (7.12) becomes

$$\sum_i (\Lambda(t, s))_{ij} = 1, \qquad (7.22)$$

so that each $\Lambda(t, s)$ is a stochastic matrix. These matrices obey the Chapman–Kolmogorov equation (7.8) that takes the form

$$\Lambda(t, s) = \Lambda(t, \tau)\Lambda(\tau, s), \tag{7.23}$$

with $t \geqslant \tau \geqslant s$. The other relation implied by the Markov condition, namely the monotonic decrease in time of the Kolmogorov distance between probability vectors corresponding to distinct initial conditions, now becomes

$$D_K(\boldsymbol{p}^I(t), \boldsymbol{p}^{II}(t)) \leqslant D_K(\boldsymbol{p}^I(s), \boldsymbol{p}^{II}(s)) \qquad \forall t \geqslant s \geqslant 0, \tag{7.24}$$

following from (7.21) and the fact that each $\Lambda(t, s)$ is a stochastic matrix.

Given this framework, there is a natural connection between the classical description of a stochastic dynamics and the quantum treatment of an open system dynamics, where an additional level of stochasticity, in addition to the intrinsic quantum probabilistic description, comes from the interaction with other environmental quantum degrees of freedom. In the standard presentation of quantum mechanics for a massive particle, the step from a deterministic classical description to quantum mechanics is performed associating to the relevant quantities, such as position and momentum, operators in a suitable space. In the same spirit, the present framework for the description of a classical stochastic dynamics can be connected to the formalism developed for the description of open quantum systems in Chap. 5. In this case probability vectors are replaced by statistical operators, and stochastic matrices by completely positive trace-preserving maps [45]. This probabilistic correspondence rule, replacing the mechanic correspondence rule, provides a very convenient starting point to describe and understand a general quantum dynamics, and well combines with the formulation of quantum mechanics as a probability theory advocated in Chap. 1.

7.2.3   Classical Semigroup Evolution

We now describe in more detail the connection between Markov processes and classical semigroups. To this aim we consider a Markovian process homogeneous in time, namely such that its conditional transition probability $p_{1|1}$ only depends on the difference between the two epochs

$$p_{1|1}(x, t|y, s) = \mathfrak{T}_{t-s}(x, y), \tag{7.25}$$

and therefore thanks to (7.6) all joint probabilities only depend on time differences. We have here denoted with $\mathfrak{T}_t$ the conditional transition probability for a time-homogenous Markov process, and for the sake of simplicity we will further assume a discrete outcome set. These conditional transition probabilities allow to introduce a semigroup of transformations on functions taking values on the set of the possible outcomes according to

$$(\mathfrak{F}_t f)(x) = \sum_y \mathfrak{T}_t(x, y) f(y), \qquad (7.26)$$

that associate to f the function

$$f_t = \mathfrak{F}_t f \qquad (7.27)$$

translated forward in time. These transformations form a convolution semigroup in that, thanks to the Chapman–Kolmogorov equation (7.8), they compose according to

$$\mathfrak{F}_t \circ \mathfrak{F}_s = \mathfrak{F}_{t+s} \qquad (7.28)$$

for any $t, s \geqslant 0$, with $\mathfrak{F}_0$ the identity operator. The collection $\{\mathfrak{F}_t\}$ provides in particular a contraction semigroup, since each transformation thanks to (7.12) is a contraction with respect to the l_1-norm, namely

$$\|\mathfrak{F}_t f\|_1 \leqslant \|f\|_1 \qquad \forall t \geqslant 0. \qquad (7.29)$$

A semigroup of contraction operators admits a generator characterized by the Hille–Yosida and Lumer–Phillips theorems, as discussed in Sect. 5.3, so that we can write

$$\mathfrak{F}_t = e^{t\mathfrak{L}}. \qquad (7.30)$$

In the considered case we have in particular

$$\|\mathfrak{F}_t p\|_1 = \|p\|_1, \qquad (7.31)$$

corresponding to preservation of probability if $\{p(x)\}$ is a probability distribution. The change in time of the conditional transition probability associated to the time-homogeneous Markov process then takes the form

$$\frac{\mathrm{d}p_t}{\mathrm{d}t} = \mathfrak{L}p_t, \qquad (7.32)$$

where

$$p_t = \mathfrak{F}_t p_0, \qquad (7.33)$$

with p_0 the probability distribution at the initial time t_0

$$p_0(x) = p_1(x, t_0). \qquad (7.34)$$

Equation (7.32) is known as master equation and expresses in differential form the Chapman–Kolmogorov equation. For the case of a finite set of possible outcomes we set

$$p_1(x_n, t) = \wp_n(t),\tag{7.35}$$

and (7.32) can be written componentwise

$$\frac{\mathrm{d}}{\mathrm{d}t}\wp_n(t) = \sum_r [W(n|r)\wp_r(t) - W(r|n)\wp_n(t)].\tag{7.36}$$

We have thus seen how a time-homogeneous Markov process determines a semigroup of transformations as in (7.28), whose evolution in time can be expressed by means of the master equation (7.32). Vice versa, under suitable continuity assumptions, a master equation in the form (7.32) with $\mathcal{L}$ as in (7.36) determines through its solution a classical Markov semigroup. An equation exactly of this form has already been encountered in Sect. 6.3.2, namely (6.83) obtained considering the behavior in time of the occupation probability of different energy levels of a non-degenerate quantum system weakly coupled to an environment in equilibrium. Equation (7.36) provides the evolution of the conditional transition probability of a classical Markov process and is known as master equation, or Pauli master equation. The positive quantities $W(r|n)$ have the role of transition rates. This result thus justifies the recovery of a classical Markov evolution of suitable degrees of freedom for an open quantum system mentioned in Sect. 6.3.2.

For the case of so-called compound Poisson processes we can write the generator in the form [33,46]

$$\mathfrak{F}_t = \mathrm{e}^{-\Gamma(\mathbb{1}-\mathfrak{U})t},\tag{7.37}$$

so that it also admits the representation

$$\mathfrak{F}_t = \mathrm{e}^{-\Gamma t}\sum_{n=0}^{\infty}\frac{(\Gamma t)^n}{n!}\mathfrak{U}^n,\tag{7.38}$$

where Γ is a positive rate and $\mathfrak{U}$ a stochastic matrix, that is

$$\sum_x \mathfrak{U}(x, y) = \mathbb{1}\tag{7.39}$$

and in this case in (7.36) the identification

$$W(r|n) = \Gamma\mathfrak{U}(x_r, x_n)\tag{7.40}$$

holds. Note the analogy between (7.37) and (5.142), that describes a semigroup evolution in the quantum realm, given by a quantum dynamical semigroup as introduced

in Sect. 5.3. The equivalent expression (7.38) further suggests that the quantum deco-
herence dynamics described Sect. 5.3.2 can be understood as a quantum compound
Poisson process [47–50], that is to say it arises from a collection of independent
Poisson distributed quantum events. The events take place in time according to the
Poisson distribution

$$p_n(t) = \frac{(\Gamma t)^n}{n!} e^{-\Gamma t},$$
(7.41)

and correspond to the repeated action of the transformation $\mathfrak{U}$. The latter transformation describes the effect of a single collision, according to the correspondence

$$\mathfrak{U} p_t \to \int d^3 q \mathcal{P}(q) U(q) \rho_S(t) U(q)^\dagger.$$
(7.42)

The transformation associated to n random collisions is then obtained iterating the
completely positive trace-preserving map of (7.42), leading to the transformation

$$\mathfrak{U}^n p_t \to \int d^3 q (\mathcal{P} * \cdots * \mathcal{P})(q) U(q) \rho_S(t) U(q)^\dagger,$$
(7.43)

that describes momentum kicks distributed according to the nth convolution of the
original probability density $\mathcal{P}(q)$, as first envisaged to describe the distribution of
atomic momentum in resonance fluorescence [51].

7.3 Behavior in Time of Quantum Distinguishability

So far we have recalled how the notion of memory in a classical stochastic dynamics is based on the independence of the conditional probabilities on the previously
assumed values of the random variable, but for the last. We now face the definition
of a quantum process featuring memory effects. As already discussed at the end of
Sect. 7.2.2, we can take an open quantum system dynamics as a quantum process.
Indeed, the influence of the environment induces an additional stochasticity element
on the dynamics of the quantum system, on top of the intrinsic probabilistic nature
of the theory itself. It is thus natural to look for a characterization of quantum processes that can be considered to have memory. With respect to the classical situation,
we have the crucial difference that the determination of the value of an observable,
playing the role of a random variable, requires a measurement and as discussed
in Sect. 3.2.3 each measurement unavoidably influences subsequent ones. The very
definition of the joint probability densities (7.1), and a fortiori of the conditional probabilities (7.3), is therefore non-trivial, and depends on information on the adopted
measurement procedure [2]. With this insight in mind, we now present an alternative general framework for the characterization of a quantum reduced dynamics, in
terms of features of the information exchange that takes place between system and

reservoir. This approach was established in the seminal papers [3,52], and exposed in more detail in [53]. We will also make reference to the presentation and the further developments in [54–56].

7.3.1 Quantum Markovian Behavior

We now provide a definition of Markovian dynamics in the quantum setting. To this aim we consider quantifiers of the distinguishability between pairs of quantum states ρ and σ, to be denoted as $\mathfrak{S}(\rho, \sigma)$, that should be symmetric and satisfy the following three properties for arbitrary choice of statistical operators ρ, σ and τ:

P1

$$0 \leqslant \mathfrak{S}(\rho, \sigma) \leqslant 1 \tag{7.44}$$

together with

$$\mathfrak{S}(\rho, \sigma) = 1 \quad \Leftrightarrow \quad \rho \perp \sigma \tag{7.45}$$

and

$$\mathfrak{S}(\rho, \sigma) = 0 \quad \Leftrightarrow \quad \rho = \sigma \tag{7.46}$$

P2

$$\mathfrak{S}(\Phi[\rho], \Phi[\sigma]) \leqslant \mathfrak{S}(\rho, \sigma) \tag{7.47}$$

with Φ an arbitrary completely positive trace-preserving map

P3

$$\mathfrak{S}(\sigma, \rho) - \mathfrak{S}(\tau, \rho) \leqslant \phi(\mathfrak{S}(\sigma, \tau)) \tag{7.48}$$

with ϕ a concave function, $\phi(x) \geqslant 0$ and $\phi(0) = 0$ iff $x = 0$.

According to **P1**, the distinguishability $\mathfrak{S}(\rho, \sigma)$ between two states is a non-negative bounded quantity, equal to 0 iff the two states coincide, and to its maximum value 1 iff the supports of the two states, that is the subspaces spanned by their eigenvectors, are orthogonal. Two identical states cannot be distinguished, while all others can, at least up to a certain extent. Distinguishing orthogonal states is a task that can be perfectly accomplished. The normalization to 1 guarantees a fair comparison among different quantifiers. **P2** expresses the fact that a quantum dynamical map, or more generally a transformation with a measurement character as discussed in Sects. 3.2.5 and 3.2.6, can only decrease the distinguishability between two states. Crucial consequences of this property are considered in Remark 7.1. The inequality (7.48) appearing in **P3** will be called triangle-like inequality, in that it provides a generalizations of the usual triangle inequality and recovers it if ϕ is the identity function. For the sake of simplicity we have assumed $\mathfrak{S}$ to be symmetric in its arguments, a feature that can always be enforced thanks to **P1**. The above-mentioned properties therefore respectively warrant boundedness and normalization, contractivity and validity of triangle-like inequality.

7.1 Invariances Induced by Contractivity

We now put into evidence two important properties satisfied by any distinguishability quantifier $\mathfrak{S}$ between quantum states that is a contraction under the action of completely positive trace-preserving maps according to (7.47). We first recall that as discussed in Sect. 3.2.1 a unitary transformation is a completely positive trace-preserving map, together with its inverse, so that in analogy to (3.238) we have

$$\begin{aligned}
\mathfrak{S}(\rho_S^1, \rho_S^2) &= \mathfrak{S}(\mathcal{U}^{-1} \circ \mathcal{U}[\rho_S^1], \mathcal{U}^{-1} \circ \mathcal{U}[\rho_S^2]) \\
&\leqslant \mathfrak{S}(\mathcal{U}[\rho_S^1], \mathcal{U}[\rho_S^2]) \\
&\leqslant \mathfrak{S}(\rho_S^1, \rho_S^2),
\end{aligned} \tag{7.49}$$

that implies

$$\mathfrak{S}(\mathcal{U}[\rho_S^1], \mathcal{U}[\rho_S^2]) = \mathfrak{S}(\rho_S^1, \rho_S^2) \tag{7.50}$$

for any unitary transformation $\mathcal{U}$. We further consider the assignment map $\mathcal{A}_{\rho_E}$ of (3.50), together with the partial trace operation Tr_E of (3.29) that acts as its left inverse. Both maps are completely positive and trace preserving, so that exploiting (7.47) twice we have

$$\begin{aligned}
\mathfrak{S}(\rho_S^1, \rho_S^2) &= \mathfrak{S}(\mathrm{Tr}_E \circ \mathcal{A}_{\rho_E}[\rho_S^1], \mathrm{Tr}_E \circ \mathcal{A}_{\rho_E}[\rho_S^2]) \\
&\leqslant \mathfrak{S}(\mathcal{A}_{\rho_E}[\rho_S^1], \mathcal{A}_{\rho_E}[\rho_S^2]) \\
&\leqslant \mathfrak{S}(\rho_S^1, \rho_S^2),
\end{aligned} \tag{7.51}$$

so that we have

$$\mathfrak{S}(\mathcal{A}_{\rho_E}[\rho_S^1], \mathcal{A}_{\rho_E}[\rho_S^2]) = \mathfrak{S}(\rho_S^1, \rho_S^2) \tag{7.52}$$

or equivalently

$$\mathfrak{S}(\rho_S^1 \otimes \rho_E, \rho_S^2 \otimes \rho_E) = \mathfrak{S}(\rho_S^1, \rho_S^2). \tag{7.53}$$

We thus see that contractivity of a distinguishability quantifier $\mathfrak{S}$ under completely positive trace-preserving maps implies invariance under both unitary transformations and composition with another common system.

In Sect. 3.2.5 we already introduced different quantities which obey **P2**, namely contractivity under the action of completely positive trace-preserving maps, and that quantify the capability to distinguish states according to a well-defined task. This task defines the operational meaning of the quantity, that is the kind of experimental setting in which it provides relevant estimates. More specifically, we discussed examples of divergences, namely the trace distance in Sect. 3.2.5.1, the quantum relative entropy in Sect. 3.2.5.2, and the quantum Jensen–Shannon divergence in Remark 3.14. While contractivity under the action of a stochastic map is one of the constitutive properties of classical divergences, corresponding to contractivity under the action of completely positive trace-preserving maps for their quantum counterpart, as suggested by the very name, boundedness is not always ensured. The requirement **P1**, in particular (7.44), only holds for the trace distance and the Jensen–Shannon divergence. The trace distance further obeys the triangle inequality, thus fulfilling **P3** with ϕ the identity. As we will detail in Remark 7.2, actually for a larger class of entropic distinguishability quantifiers, the Jensen–Shannon divergence fulfills (7.48) with the function $\phi(x) = \sqrt[4]{2/(\log 2)^3}\,\sqrt[4]{x}$.

We now define a quantum dynamics described by the collection of completely positive trace-preserving maps $\Phi(t)$ according to (5.9) to be Markovian if

$$\mathfrak{S}(\rho_S^1(t), \rho_S^2(t)) \leqslant \mathfrak{S}(\rho_S^1(s), \rho_S^2(s)) \qquad \forall t \geqslant s \geqslant 0 \tag{7.54}$$

for any pair of initial states $\rho_S^1(0)$ and $\rho_S^2(0)$. A Markovian dynamics is therefore such that the distinguishability between two states is a monotonically nonincreasing function of time. We have in particular that the quantum Markov condition (7.54) is obeyed by any unitary dynamics, as well as by the quantum dynamical semigroup evolution described in Sect. 5.3. Both statements follow from (7.47). The first thanks to (7.50), the second relying on (5.19), according to which

$$\begin{aligned}
\mathfrak{S}(\rho_S^1(t), \rho_S^2(t)) &= \mathfrak{S}(\Phi(t)\rho_S^1(0), \Phi(t)\rho_S^2(0)) \\
&= \mathfrak{S}(\Phi(t-s) \circ \Phi(s)\rho_S^1(0), \Phi(t-s) \circ \Phi(s)\rho_S^2(0)) \\
&\leqslant \mathfrak{S}(\Phi(s)\rho_S^1(0), \Phi(s)\rho_S^2(0)) \\
&= \mathfrak{S}(\rho_S^1(s), \rho_S^2(s))
\end{aligned} \tag{7.55}$$

for any $t \geqslant s$. Note that for the case at hand according to (5.20) we have the explicit representation $\Phi(t) = e^{\mathcal{L}t}$.

7.2 Triangle-Like Inequality for Relative Entropies

We now show validity of the triangle-like inequality (7.48) for a large class of entropic distinguishability quantifiers, obtained from the quantum relative entropy introduced in Sect. 3.2.5.2, including as special and important case the Jensen–Shannon divergence introduced in Remark 3.14. We already observed that the quantum relative entropy is generally unbounded, according to its very definition (3.246). This feature depends on the relationship between the supports of the statistical operators to be compared and hinders the existence of triangle inequalities. To cure this difficulty we start from the inequality

$$S(\rho, \mu\rho + (1 - \mu)\sigma) \leqslant \log(1/\mu), \tag{7.56}$$

where S denotes the quantum relative entropy as defined in (3.246) and μ is a mixing parameter strictly different from 0 and 1, that is

$$0 < \mu < 1. \tag{7.57}$$

The inequality relies on operator monotonicity of the logarithm, according to which

$$\log A \leqslant \log B \tag{7.58}$$

for A and B positive operators such that $B \geqslant A$. Note that the monotonicity is strict, so that for strictly positive operators $\log A < \log B$ iff $B > A$ [57,58]. According to operator monotonicity we have

$$\begin{aligned}
S(\rho, \mu\rho + (1 - \mu)\sigma) &= \mathrm{Tr}\{\rho \log \rho\} - \mathrm{Tr}\{\rho \log(\mu\rho + (1 - \mu)\sigma)\} \tag{7.59}\\
&\leqslant \mathrm{Tr}\{\rho \log \rho\} - \mathrm{Tr}\{\rho \log(\mu\rho)\} \tag{7.60}\\
&= -\log \mu, \tag{7.61}
\end{aligned}$$

thus proving (7.56). The relative entropy between a given state ρ and its mixture with another one, namely $\mu\rho + (1 - \mu)\sigma$, is therefore always finite. In particular, thanks to Klein's inequality (2.122) we have

$$S(\rho, \mu\rho + (1 - \mu)\sigma) = 0 \quad \Leftrightarrow \quad \rho = \sigma. \tag{7.62}$$

Furthermore

$$S(\rho, \mu\rho + (1 - \mu)\sigma) = \log(1/\mu) \quad \Leftrightarrow \quad \rho \perp \sigma. \tag{7.63}$$

Indeed, if $\rho \perp \sigma$ they can be diagonalized together, so that calling ρ_i and σ_i the eigenvalues in the common eigenbasis we have

$$\mathrm{Tr}\{\rho \log(\mu\rho + (1 - \mu)\sigma)\} = \sum_i \rho_i \log(\mu\rho_i + (1 - \mu)\sigma_i) \tag{7.64}$$

$$= \sum_i \rho_i \log(\mu\rho_i), \tag{7.65}$$

so that $S(\rho, \mu\rho + (1 - \mu)\sigma) = -\log\mu$. If $\rho \not\perp \sigma$, then from

$$\mu\rho + (1 - \mu)\sigma > \mu\rho \qquad (7.66)$$

and strict monotonicity of the logarithm the inequality sign in (7.60) would be strict. We further rely on two crucial upper bounds to the variation of the quantum relative entropy upon modification of either of its arguments [56, 59,60], expressed in terms of a function of the trace distance between two statistical operators introduced in (3.211).

Theorem 7.1 (Audenaert, 2011) *For $\mu \in (0, 1)$ and $\sigma, \rho_1, \rho_2 \in \mathcal{S}(\mathcal{H})$ we have the following bounds on the variation of the quantum relative entropy upon modification of its arguments*

$$S(\sigma, \mu\sigma + (1 - \mu)\rho_1) - S(\sigma, \mu\sigma + (1 - \mu)\rho_2) \leqslant$$
$$\log\left(1 + \frac{1 - \mu}{\mu} D(\rho_1, \rho_2)\right) \qquad (7.67)$$

and

$$S(\rho_1, \mu\rho_1 + (1 - \mu)\sigma) - S(\rho_2, \mu\rho_2 + (1 - \mu)\sigma) \leqslant$$
$$D(\rho_1, \rho_2)\log\left(1 + \frac{1 - \mu}{\mu}\frac{1}{D(\rho_1, \rho_2)}\right), \qquad (7.68)$$

Both inequalities are saturated for $\sigma \perp \rho_1$ and $\rho_2 = D(\rho_1, \rho_2)\sigma + (1 - D(\rho_1, \rho_2))\rho_1$.

We are now in the position to introduce entropic distinguishability quantifiers satisfying properties **P1** to **P3** of Sect. 7.3.1. We consider the quantities

$$S_\mu(\rho_1, \rho_2) = \frac{\mu}{\log(1/\mu)} S(\rho_1, \mu\rho_1 + (1 - \mu)\rho_2)$$
$$+ \frac{1 - \mu}{\log(1/(1 - \mu))} S(\rho_2, (1 - \mu)\rho_2 + \mu\rho_1), \qquad (7.69)$$

and

$$K_\mu(\rho_1, \rho_2) = \frac{\mu}{H(\{\mu, 1 - \mu\})} S(\rho_1, \mu\rho_1 + (1 - \mu)\rho_2)$$
$$+ \frac{1 - \mu}{H(\{1 - \mu, \mu\})} S(\rho_2, (1 - \mu)\rho_2 + \mu\rho_1) \qquad (7.70)$$

where H according to (2.101) denotes the Shannon entropy of the probability distribution $\{\mu, 1 - \mu\}$. These quantities are called quantum skew divergence and Holevo skew divergence [56,61]. They both arise as quantum versions of classical distinguishability quantifiers known as divergences, as discussed in Sect. 3.2.5, and they are characterized by the skewing parameter $\mu \in (0, 1)$. The Holevo skew divergence apart from a normalization factor corresponds to the so-called Holevo quantity, well-known in quantum information [62], considered for the special case of an ensemble of two states only. Thanks to (7.56) both quantities are bounded and in the range [0, 1] as required by (7.44). Equations (7.45) and (7.46) hold thanks to (7.62) and (7.63). **P1** is therefore warranted, together with **P2** that follows from the corresponding contractivity property (3.251) for the quantum relative entropy. We further note that the Pinsker inequality [45]

$$D^2(\rho_1, \rho_2) \leqslant \frac{1}{2} S(\rho_1, \rho_2) \tag{7.71}$$

in turn implies after some algebra [56]

$$D^2(\rho_1, \rho_2) \leqslant \frac{H(\{\mu, 1 - \mu\})}{2\mu(1 - \mu)} K_\mu(\rho_1, \rho_2), \tag{7.72}$$

together with

$$D^2(\rho_1, \rho_2) \leqslant \frac{\log(\mu) \log(1 - \mu)}{2\mu(1 - \mu) H(\{\mu, 1 - \mu\})} S_\mu(\rho_1, \rho_2). \tag{7.73}$$

Combining the operator inequalities (7.67) and (7.68) with the Pinsker-like inequalities (7.72) and (7.73), and using the inequality

$$\log(1 + x) \leqslant \sqrt{x}, \tag{7.74}$$

we have that also **P3** of (7.48) is obeyed for both distinguishability quantifiers S_μ and K_μ. For S_μ we have

$$\phi(x) \to \varsigma_\mu \sqrt[4]{x} \tag{7.75}$$

with $\varsigma_\mu = -\log(\mu(1 - \mu))\sqrt[4]{\mu(1 - \mu)/2 \, H(\{\mu, 1 - \mu\}) \log^3(\mu) \log^3(1 - \mu)}$, while for K_μ we have

$$\phi(x) \to \kappa_\mu \sqrt[4]{x} \tag{7.76}$$

with $\kappa_\mu = \sqrt[4]{8\mu(1 - \mu)H(\{\mu, 1 - \mu\})^3}$. For the special case $\mu = 1/2$ (7.69) and (7.70) coalesce to the quantum Jensen–Shannon divergence of (3.260)

$$S_{1/2}(\rho_1, \rho_2) = K_{1/2}(\rho_1, \rho_2) = J(\rho_1, \rho_2), \tag{7.77}$$

whose properties have already been discussed in Remark 3.14, and we have

$$\phi(x) \to \sqrt[4]{2/(\log 2)^3}\,\sqrt[4]{x}. \tag{7.78}$$

The square root of this quantity $\sqrt{J}$ has been further shown to obey the standard triangle inequality [63,64]

$$\sqrt{J(\sigma, \rho)} \leqslant \sqrt{J(\sigma, \tau)} + \sqrt{J(\tau, \rho)}, \tag{7.79}$$

corresponding to $\phi(x) = x$ in (7.48). The square root of the Jensen–Shannon divergence therefore provides a metric between probability distributions, called Jensen–Shannon distance [45].

The condition (7.54) is naturally to be seen as the quantum analogue of (7.24). The latter condition is implied by the Markov condition, so that its validity is necessary but not sufficient to identify a classical Markov process. Following [3], we take it here as the very definition of a quantum Markov process, given the special role played by one-point probabilities in the quantum setting. They are the only measurement outcomes not asking for knowledge of how the measurement has been performed, while its specification is necessary for any higher-order probability density, as discussed in Sect. 3.2.3. Note that even though the monotonicity condition (7.24) for the classical case was written with reference to a specific divergence, namely the Kolmogorov distance, given that it follows from (7.21) with $\Lambda(t, s)$ stochastic matrices, it would hold for any classical divergence [45]. As a final important remark we stress that the quantum Markov condition (7.54) only requires information on the states evolved according to the considered dynamics, not on the map or possible evolution equations describing it. The experimental assessment of the state of the system as a function of time provides all the necessary information to ascertain validity or failure of the quantum Markov condition, see [65,66] for an analysis also discussing different experimental investigations [67–71].

7.3.2   Interpretation in Terms of Information Backflow

We now want to provide a deeper motivation for the definition of quantum Markovian dynamics given in Sect. 7.3.1. At variance with the classical stochastic processes considered in Sect. 7.2, in the open quantum system framework we assume that the overall set of system and environment degrees of freedom evolves according to a unitary evolution, determined by some microscopic interaction Hamiltonian. We are thus faced with a bipartite setting in which the environmental degrees of freedom affect the system dynamics, thus adding a layer of randomicity in its evolution. We can therefore consider both the reduced system state $\rho_S(t)$ and the total state $\rho_{SE}(t)$, and the proposed definition of Markovian dynamics will now be connected to the information exchange between these two states.

We introduce a notion of internal information by identifying it with the distinguishability between system states

$$I_{\mathrm{int}}(t) = \mathfrak{S}(\rho_S^1(t), \rho_S^2(t)), \tag{7.80}$$

that can be evaluated by measurements performed on the system only. We further consider the complementary expression

$$I_{\mathrm{ext}}(t) = \mathfrak{S}(\rho_{SE}^1(t), \rho_{SE}^2(t)) - \mathfrak{S}(\rho_S^1(t), \rho_S^2(t)), \tag{7.81}$$

that quantifies the information about the distinguishability of the total states that cannot be obtained by performing measurements on the system only, namely the external information. As said, we assume that the overall set of degrees of freedom undergoes a unitary evolution, so that the quantity

$$I_{\mathrm{int}}(t) + I_{\mathrm{ext}}(t) = \mathfrak{S}(\rho_{SE}^1(t), \rho_{SE}^2(t)) \tag{7.82}$$

is constant according to (7.50). In order to warrant the existence of a reduced dynamics we assume as discussed in Sect. 5.2 the initial states to be factorized, and since the reduced evolution acting on $\rho_S^1(0)$ and on $\rho_S^2(0)$ is the same we have according to (5.6)

$$\rho_{SE}^{1,2}(0) = \rho_S^{1,2}(0) \otimes \rho_E(0), \tag{7.83}$$

so that (7.53) implies that the constant value is given by

$$I_{\mathrm{int}}(t) + I_{\mathrm{ext}}(t) = \mathfrak{S}(\rho_S^1(0), \rho_S^2(0)). \tag{7.84}$$

The quantum Markov condition (7.54) can now be reformulated as

$$I_{\mathrm{int}}(t) \leqslant I_{\mathrm{int}}(s), \tag{7.85}$$

for any $t \geqslant s$, expressing the fact that the internal information, namely the capability to distinguish the system states after interaction with the environment, is steadily decreasing for any pair of initial states. Note that according to (7.84)

$$I_{\mathrm{int}}(t) - I_{\mathrm{int}}(s) = -(I_{\mathrm{ext}}(t) - I_{\mathrm{ext}}(s)), \tag{7.86}$$

implying a corresponding increase of the external information.

According to the equivalent conditions (7.54) and (7.85) a quantum dynamics is Markovian if the distinguishability of two initial states evolving according to this dynamics never increases, or equivalently the internal information is steadily decreasing in favor of the external one. In this context information is flowing away from the system. A quantum dynamics that violates this condition is said to be non-

Markovian. This happens when at least for a pair of initial states there is a point in time in which information flows back to the system, so that the internal information increases at the expense of the external one, or equivalently there is a revival in distinguishability. This viewpoint is sketched in Fig. 7.2. We will consider explicit examples of this behavior in Sect. 7.5.

7.3.3 Bound on Distinguishability Revivals

We now justify the association of a notion of memory to the violation of the quantum Markov condition formulated in (7.54) or equivalently in (7.85), that express revivals in distinguishability or the equivalent information backflow respectively. The very definition of Markovian dynamics, and the introduction of the notion of information backflow to justify it, relied on **P2** of $\mathfrak{S}$, namely contractivity. **P1** and **P3**, namely boundedness of $\mathfrak{S}$ and the existence of triangle-like inequalities, allow to clarify the mechanism behind the revivals in distinguishability. These revivals can take place when information has been stored outside the system, in well-identified physical degrees of freedom, so that it can be later locally retrieved. To this aim we observe that thanks to **P1** and **P3** of $\mathfrak{S}$ we can derive, as shown in Remark 7.3, the following important upper bound on the possible revivals in time of the distinguishability

$$\mathfrak{S}(\rho_S^1(t), \rho_S^2(t)) - \mathfrak{S}(\rho_S^1(s), \rho_S^2(s)) \leqslant \phi \circ \phi(\mathfrak{S}(\rho_E^1(s), \rho_E^2(s)))$$
$$+ \sum_{i=1,2} \phi(\mathfrak{S}(\rho_{SE}^i(s), \rho_S^i(s) \otimes \rho_E^i(s)) \, ,$$

$$(7.87)$$

with ϕ the positive concave function identified by the functions in (7.75) and (7.76) for the case of entropic distinguishability quantifiers. If $\mathfrak{S}$ is a distance obeying the standard triangle inequality, the function ϕ is simply the identity. Thanks to **P1** and **P3** the r.h.s. of (7.87) is the sum of non-negative quantities, so that the l.h.s. can be strictly positive, i.e., the quantum Markov condition (7.54) can fail, only if at least one of the quantities at the r.h.s. is strictly positive. According to **P1** of $\mathfrak{S}$ this can happen iff either

$$\rho_E^1(s) \neq \rho_E^2(s) \tag{7.88}$$

or

$$\rho_{SE}^i(s) \neq \rho_S^i(s) \otimes \rho_E^i(s), \tag{7.89}$$

for at least one of the indexes identifying the different initial conditions. A non-Markovian behavior according to the definition (7.54), that is a revival in distinguishability at time t, can therefore only take place if at a previous time s information was stored in the different ways in which different system initial conditions affect the environment, leading to (7.88), and/or in correlations between system and environment leading to (7.89). The backflow of this information to the system degrees of freedom then leads to an increase of distinguishability in time. Memory is therefore associated to a dynamics violating the Markov condition in that information that has been stored outside the system degrees of freedom, so that it is no more accessible by performing local measurements, comes back and can be again accessed by performing measurements on the system degrees of freedom only.

7.3 Upper Bound on Revivals

We now prove (7.87) relying on properties **P1** to **P3**. Thanks to **P2**, that implies in particular (7.50) and (7.53), we have

$$\mathfrak{S}(\rho_S(t), \sigma_S(t)) - \mathfrak{S}(\rho_S(s), \sigma_S(s)) \leqslant$$
$$\mathfrak{S}(\rho_{SE}(s), \sigma_{SE}(s)) - \mathfrak{S}(\rho_S(s) \otimes \rho_E(s), \sigma_S(s) \otimes \rho_E(s)), \quad (7.90)$$

since the partial trace discussed in Remark 3.1 is a completely positive trace-preserving map, and the total dynamics is assumed to be unitary. We now add and subtract the term $\mathfrak{S}(\rho_S(s) \otimes \rho_E(s), \sigma_{SE}(s))$ at the r.h.s. and use twice the triangle-like inequality (7.48) to obtain

$$\mathfrak{S}(\rho_S(t), \sigma_S(t)) - \mathfrak{S}(\rho_S(s), \sigma_S(s)) \leqslant$$
$$\phi(\mathfrak{S}(\rho_{SE}(s), \rho_S(s) \otimes \rho_E(s))) + \phi(\mathfrak{S}(\sigma_{SE}(s), \sigma_S(s) \otimes \rho_E(s))), \quad (7.91)$$

equation (7.48) of **P3** together with (7.50) further implies

$$\mathfrak{S}(\sigma_{SE}(s), \sigma_S(s) \otimes \rho_E(s)) \leqslant$$
$$\mathfrak{S}(\sigma_{SE}(s), \sigma_S(s) \otimes \sigma_E(s)) + \phi(\mathfrak{S}(\rho_E(s), \sigma_E(s))) \quad (7.92)$$

that inserted in (7.91), thanks to the properties of ϕ required in **P3** that imply its monotonicity and subadditivity leads to (7.87).

7.3.4 Measure of Distinguishability Revivals

For the case of a dynamics violating the Markov condition (7.54), it is of interest to provide a quantifier of this degree of violation, often called non-Markovianity measure. It is useful in estimating the dependence of the degree of non-Markovian behavior on the parameters of the considered physical system. According to Sect. 7.3.1 we have a non-Markovian behavior when there are revivals in the internal information $I_{\text{int}}(t)$ identified with the local distinguishability $\mathfrak{S}(\rho_S(t), \sigma_S(t))$ in (7.80). These revivals do take place when the time derivative of the distinguishability has a positive slop, and for a fixed pair of initial states their total amount is given by

$$M(\Phi, \mathfrak{S}, \rho_S^{1,2}(0)) = \int_{\dot{\mathfrak{S}}>0} dt \dot{\mathfrak{S}}(\rho_S(t), \sigma_S(t)), \tag{7.93}$$

where the integration is restricted to the regions in which $\dot{\mathfrak{S}}(\rho_S(t), \sigma_S(t)) > 0$, so that it can be equivalently written as

$$M(\Phi, \mathfrak{S}, \rho_S^{1,2}(0)) = \sum_n [\mathfrak{S}(\rho_S^1(t_f^n), \rho_S^2(t_f^n)) - \mathfrak{S}(\rho_S^1(t_i^n), \rho_S^2(t_i^n))], \tag{7.94}$$

with t_i^n and t_f^n initial and final time of the nth revival. Following [3,52] we therefore define the quantity

$$M(\Phi, \mathfrak{S}) = \sup_{\rho_S^{1,2}(0) \in \mathcal{S}(\mathcal{H})} M(\Phi, \mathfrak{S}, \rho_S^{1,2}(0)) \tag{7.95}$$

as measure of non-Markovianity associated to the quantum dynamical map Φ. The measure keeps into account the dependence on the considered pair of initial states, since different states are differently affected by the environmental degrees of freedom and not all of them exploit the potential capability of the dynamics to store information outside the system degrees of freedom. Note that thanks to (7.44) of **P1** the measure remains finite unless there are an infinite number of revivals. For the case in which the distinguishability quantifier is identified with the trace distance D of (3.211), it has been shown that the supremum in (7.95) is attained on an orthogonal pair of states, called optimal pair [72], that as shown in Sect. 2.2.3.1 must lie on the border of the set of states, e.g., on the surface of the Bloch sphere for a two-dimensional system. We thus have for the trace distance

$$M(\Phi, D) = \sup_{\rho_S^1(0) \perp \rho_S^2(0)} M(\Phi, D, \rho_S^{1,2}(0)). \tag{7.96}$$

7.4 Characterization of Quantum Dynamical Maps

In Sect. 7.3 we have introduced a definition of quantum Markovian dynamics, given by (7.54) with $\mathfrak{S}$ a distinguishability quantifier obeying the properties **P1** to **P3** of Sect. 7.3.1. This definition calls for knowledge of the evolved statistical operator for different initial conditions. It is therefore a much less demanding constraint than the classical Markov condition (7.5), that relies on knowledge about all the conditional probability distributions. It is at the same time naturally related to the classical condition. Indeed, as discussed in Sect. 7.2.2, the classical Markov condition implies in particular the Chapman–Kolmogorov equation (7.23), as well as the monotonicity constraint (7.24). The quantum Markov condition (7.54) is the natural quantum counterpart of (7.24). In both cases any two states, be they classical or quantum, that evolve out of two different initial conditions lose distinguishability in a monotonic way. Moreover, we have seen in (7.55) that a quantum process characterized by a semigroup composition law is Markovian. The two classical necessary conditions (7.24) and (7.23) have thus respectively become the defining quantum condition (7.54), and the sufficient condition for having a quantum Markovian dynamics (7.55). We now better investigate the interplay between these two related constraints in the quantum framework. As argued in Sects. 7.3.2 and 7.3.3, the quantum Markov condition (7.54) has a direct physical interpretation. Moreover, it relies on the one-point probability only, that is to say the expectation values of the statistical operator. This feature is very important, since as discussed in Sect. 3.2.3 in the quantum framework the higher-order probability densities considered in the classical Markov condition (7.5) would unavoidably depend on the intermediate measurement transformation. A quantum counterpart of the Chapman–Kolmogorov equation (7.23), on the other hand, requires information on the quantum dynamical map describing the quantum evolution and calls for a proper formulation in view of the different mathematical frameworks of classical and quantum probability. These quantum dynamical maps, in turn, arise as solution of evolution equations, that provide the natural starting point in describing a physical phenomenon. It is therefore an important task to characterize the possible structure of master equations admitting as solutions legitimate time evolutions, possibly directly providing information on their behavior with respect to the quantum Markov condition. This step would amount to a generalization of the GKSL result of Sect. 5.3.1 that specifies a structure of master equation whose solutions are known to be quantum dynamical maps obeying the quantum Markov condition (7.54).

7.4.1 Composition Law and Divisibility Properties

We now address the characterization of a quantum counterpart of the classical Chapman–Kolmogorov equation making reference to (7.23), that expresses (7.8) for the case of a finite number of outcomes. The condition states that any conditional transition probability can be expressed as the composition of conditional transition probabilities over intermediate shorter time intervals. It therefore introduces a notion of divisibility according to the terminology used in probability theory [33]. In order to consider an analog property in the quantum framework, we consider a reduced dynamics described, as discussed in Sect. 5.2, by a collection of completely positive trace-preserving maps. We do not necessarily assume a dynamics that is homogenous in time, and we denote these maps as $\{\Phi(t, t_0)\}_t$, with t_0 the initial time. We then define a quantum reduced dynamics to be divisible if the identity

$$\Phi(t, t_0) = \Phi(t, s) \circ \Phi(s, t_0) \tag{7.97}$$

holds for $t \geqslant s \geqslant t_0$, with $\Phi(t, s)$ completely positive trace-preserving maps. In view of this specific constraint on $\Phi(t, s)$, the dynamics is often said to be CP-divisible [15]. Note that if $\Phi(t, t_0)$ is invertible these maps can be identified as [73–75]

$$\Phi(t, s) = \Phi(t, t_0) \circ \Phi(s, t_0)^{-1}. \tag{7.98}$$

The constraint (7.97) then corresponds to the existence of a composition law for the time evolution map and implies in particular

$$\Phi(t, s) = \Phi(t, \tau) \circ \Phi(\tau, s), \tag{7.99}$$

for $t \geqslant \tau \geqslant s \geqslant t_0$, which provides the direct quantum counterpart of the Chapman–Kolmogorov equation (7.23), and was already encountered in (5.96) of Sect. 5.3.1. Indeed, starting from (7.97) with $s \to \tau$ and applying $\Phi(s, t_0)^{-1}$ from the right we recover (7.99). The important fact is that each of the terms in (7.99) transforms states into states, in that it is given by a completely positive trace-preserving map. On the same footing, in (7.23) each term was a stochastic matrix sending probability vectors to probability vectors. Note that divisibility of a map in the sense of (7.97) warrants the quantum Markov condition (7.54), thanks to requirement **P2** of (7.47). Divisibility is therefore generally a stronger requirement on the dynamics with respect to monotone contractivity.

Violation of monotone contractivity was shown to be connected to a notion of memory in the sense of retrieval of information stored in degrees of freedom external with respect to the system in Sects. 7.3.2 and 7.3.3. In turn, the divisibility condition (7.99) can be phrased saying that knowledge of the state at an arbitrary intermediate time, not necessarily the initial one, suffices to determine it at a later time

$$\rho_S(t) = \Phi(t, s)\rho_S(s). \tag{7.100}$$

Its violation is therefore also connected to a notion of memory. In the classical framework both the Chapman–Kolmogorov equation (7.8) and the requirement of monotonic loss of distinguishability (7.18) are consequences of the classical Markov condition (7.5), and are not equivalent to it in full generality. Indeed, classical non-Markov processes can be devised whose conditional transition probabilities satisfy the Chapman–Kolmogorov equation [76], and given that validity of (7.8) is a stronger condition than (7.18), all the more so for fulfillment of the monotonicity condition. This lack of equivalence is to be traced back to the fact that the classical Markov condition is a constraint on the whole collection of conditional probabilities. In the quantum framework we have assumed as definition of Markovian dynamics the monotonicity condition (7.54), that only involves the statistical operator and therefore the statistics of measurements at a single point in time, without the need to include information on the measurement procedure, as discussed in Sect. 3.2.3. As already mentioned, this condition is generally weaker than the quantum Chapman–Kolmogorov equation (7.99). It it therefore of special interest to better clarify the relationship between these two notions. We will see that their connection depends on the specific distinguishability quantifier $\mathfrak{S}$, as well as on the kind of divisibility considered. Indeed, we can consider a weaker divisibility notion, known as P-divisibility, asking that (7.97) holds with $\Phi(t, s)$ positive trace-preserving maps, and not necessarily completely positive ones. This condition still warrants (7.100), but thanks to the weakening generally reduces the gap between the constraints (7.54) and (7.99).

As stressed in Sect. 7.3.1, assessment of the quantum Markov condition (7.54) based on monotone contractivity relies on knowledge of the evolving state as a function of time, and does not require explicit informations on the evolution map. This fact marks a crucial difference, fixing the experimental resources needed to ascertain the Markov condition to quantum state tomography [71]. Satisfaction or failure of this condition can however also be inferred from knowledge of the quantum evolution map $\Phi(t, t_0)$, or even only on its divisibility properties. In this respect knowledge of the evolution equation for the reduced system statistical operator can be very useful. As clarified in Sect. 5.2, for an initially factorized system-environment state the existence of a reduced dynamics is warranted and given by a collection of quantum dynamical maps. Their structure and properties are however generally unknown. An explicit characterization was possible in Sect. 5.3.1 just starting from the requirement of the semigroup composition law (5.19), thus obtaining a class of dynamics of paramount importance, namely quantum dynamical semigroups, that provide the quantum counterpart of the classical semigroup evolution considered in Sect. 7.2.3. Also in the quantum case these dynamics are considered memoryless, as shown in (7.55). In this case the semigroup composition law for the maps was equivalent to the expression (5.27) for the master equation, so that CP-divisibility, enforcing the quantum Markov condition (7.54), could be inferred directly from the expression of the master equation. It therefore appears that knowledge of the structure of master equations whose solutions are warranted to be completely positive trace-preserving evolution maps, beyond the already encountered (5.27), can be of great interest. On the one hand it would provide a larger inventory for a phenomenological approach, beyond the semigroup constraint. On the other hand, their expression might directly provide information on divisibility properties of the map, without knowledge of the explicit solution. This situation was already encountered in Sect. 5.3.1, where a master equation of the GKSL form but with positive time-dependent coefficients as in (5.80) was shown to lead to a time evolution fulfilling the composition law (5.96), and therefore CP-divisible according to (7.97).

We are thus faced with two important questions: *(i)* how to determine master equations whose solutions are warranted to be well-defined dynamical evolutions; *(ii)* how to assess validity or violation of the quantum Markov condition from the expression of the master equation itself, without the need to obtain explicit solutions. Both questions are still open, and only partial results are available. In this perspective, we will now provide some general hints about the structure of legitimate master equations, making reference to the two basic forms that they can assume, namely time-local or with a memory kernel, later providing examples in Sect. 7.5.

7.4.2 Time-Local Master Equations

We first consider the case of time-local master equations [77], that is to say evolution equations for the statistical operator that can be written in the form

$$\frac{d\rho_S(t)}{dt} = \mathcal{L}(t)[\rho_S(t)], \tag{7.101}$$

where $\mathcal{L}(t)$ denotes a linear map usually called generator. Preservation in time of Hermiticity and trace of the solutions of the equation, as discussed in Sect. 4.4.2, implies that $\mathcal{L}(t)$ has to be Hermiticity preserving and trace-annihilating, thus taking the form (4.127)

$$\mathcal{L}(t)[\rho_S(t)] = \sum_{j=1}^{n^2} \gamma_j(t) \left[V_j(t)\rho_S(t)V_j^\dagger(t) - \frac{1}{2}\{V_j^\dagger(t)V_j(t), \rho_S(t)\} \right], \tag{7.102}$$

where we have assumed the Hilbert space of the system to have finite dimension n, and the $\gamma_j(t)$ are real coefficients. As we have shown at the end of Sect. 5.3.1 asking for positivity of the real coefficients, that is

$$\gamma_j(t) \geqslant 0 \qquad \forall t, \tag{7.103}$$

warrants that (7.102) admits as solution the collection of completely positive trace-preserving maps

$$\Phi(t, 0) = \overleftarrow{T} \exp\left(\int_0^t d\tau \mathcal{L}(\tau) \right), \tag{7.104}$$

where $\overleftarrow{T}$ denotes chronological time-ordering. This time evolution is in particular CP-divisible according to (5.96), and hence describes a Markovian behavior. These dynamics have their classical counterpart in time non-homogenous Markovian stochastic processes, and were already suggested in [78]. This condition, or equivalently positivity of the coefficients $\gamma_j(t)$, is also necessary [73]. Indeed, assuming the maps

$$\Phi(t, s) = \overleftarrow{T} \exp\left(\int_s^t d\tau \mathcal{L}(\tau) \right) \tag{7.105}$$

to be completely positive trace-preserving for any $t \geqslant s$, we can take them as time evolutions starting at s with generator that according to (5.21) reads

$$\lim_{t \to 0^+} \frac{\Phi(t + s, s) - \mathbb{1}}{t} = \mathcal{L}(s) \tag{7.106}$$

and according to Theorem 5.1 must be in the GKSL form, which calls for positivity of the coefficients $\gamma_j(s)$. We have thus seen that, when the evolution equations are written in time-local form according to (7.101), CP-divisibility can be read directly from a condition on the coefficients appearing in (7.102).

Also the weaker property of P-divisibility, mentioned in Sect. 7.4.1 and corresponding to divisibility according to (7.97) with $\Phi(t, s)$ positive for $t \geqslant s$, can be translated into a necessary and sufficient condition if the master equation is written in the time-local form. As proven in the study of generators of positive semigroups [46,79] the dynamics is P-divisible iff the condition

$$\sum_{j=1}^{n^2} \gamma_j(t)|\langle u_m|V_j(t)u_n\rangle|^2 \geqslant 0 \tag{7.107}$$

holds with $n \neq m$ for any choice of basis $\{u_n\}$ in the system Hilbert space.

We have thus provided necessary and sufficient conditions on the structure of the generator in (7.101) that warrant the existence of a well-defined dynamics, beyond the fundamental result Theorem 5.1 about generators of quantum dynamical semigroups. These dynamics furthermore obey the composition rule (7.97), implying the quantum counterpart of the Chapman–Kolmogorov equation (7.8). In this respect they still lead to quantum Markovian dynamics in the sense of (7.54).

A general characterization of weaker constraints on the time-dependent coefficients in (7.102), warranting the solutions of the master equations to be completely positive trace-preserving maps without obeying a particular composition law, is not known. In general the time-dependent coefficients might also become negative, as we learn from the exact results obtained in Sect. 5.4. More specifically, the master equations (5.351) of Sect. 5.4.1, (5.424) of Sect. 5.4.2 and (5.465) of Sect. 5.4.3, all exhibiting temporarily negative coefficients. See [15] for a more comprehensive treatment. The characterization of time-local master equations describing a non-Markovian dynamics is therefore presently an open problem, except for special cases in low dimensionality [15] or with constraints such as Gaussianity preservation [80]. Time-local evolution equations of the form (7.101) can be obtained by means of the Shibata-Hashitsumae time-convolutionless projection operator technique [81–83]. Similarly to the developments in Sect. 5.5, this strategy allows for a perturbation expansion of the equations of motions in time-local form starting from the expression of the underlying microscopic interaction Hamiltonian. The expansion in the coupling strength, however, does not provide any hint about complete positivity or even only positivity of the solution of the obtained master equation.

We stress that generators of dynamics with a well-defined divisibility property do form a positive cone, in the sense that if $\mathcal{L}_1(t)$ and $\mathcal{L}_2(t)$ obey (7.107) or (7.103), so that their solutions are P-divisible or CP-divisible maps respectively, then the master equation

$$\frac{\mathrm{d}\rho_S(t)}{\mathrm{d}t} = (\mu\mathcal{L}_1(t) + \lambda\mathcal{L}_2(t))[\rho_S(t)] \qquad \mu, \lambda \geqslant 0, \tag{7.108}$$

also describes a well-defined dynamics with the same divisibility properties. The addition of environmental disturbances leading to divisible evolutions therefore still corresponds to a physical environment. This fact is generally no more true for generic quantum non-Markovian dynamics. Note however that at the level of quantum dynamical maps the set of Markovian dynamics is not convex, so that mixing Markovian dynamics might well result in a non-Markovian dynamics [54,84]. As a paradigmatic example we mention the random mixture of unitary evolutions considered in Sect. 5.4.3, whose non-Markovian behavior is investigated in Sect. 7.5.1.3.

7.4.3 Memory-Kernel Master Equations

We now dwell with memory-kernel master equations, namely equations that can be written in the form

$$\frac{\mathrm{d}\rho_S(t)}{\mathrm{d}t} = \int_0^t \mathrm{d}\tau \mathcal{K}(t - \tau)[\rho_S(\tau)], \tag{7.109}$$

where due to Hermiticity and trace preservation of the statistical operator the kernel $\mathcal{K}(t)$ can still be taken of the form (7.102). Evolution equations of this form have been obtained in Sect. 5.5.3 for an arbitrary open system, with the only requirement of an initially factorized system-environment state, relying on the Nakajima–Zwanzig memory-kernel projection operator technique. As mentioned above, an analog strategy, namely the Shibata–Hashitsumae time-convolutionless projection operator technique [81–83], allows to obtain time-local evolution equations of the form (7.101). In both cases the obtained result is formally exact and liable to a perturbation expansion. However, while the solutions of the exact dynamics are completely positive trace-preserving maps, complete positivity is not stable under approximations, so that it is not warranted when one considers intermediate approximate expressions, as we have also explicitly seen in Sect. 6.2. Indeed, the weak-coupling master equation (6.54), whose solutions are completely positive trace-preserving time evolutions, was obtained starting from a second-order approximation of the Nakajima–Zwanzig memory kernel, but introducing further approximations to compensate for the loss of complete positivity originating from the finite perturbative order. The two techniques introduced by Nakajima & Zwanzig and Shibata & Hashitsumae lead to different expansions of the evolutions equations, that might perform differently in many respects, namely complete positivity of the solutions, radius of convergence of the perturbation expansion, accuracy with respect to the exact dynamics. We will consider these aspects by means of example in Sect. 7.5.2.2.

At variance with the case of time-local master equations discussed in Sect. 7.4.2, no general explicit conditions that allow to infer about positivity properties of the solutions directly from the expression of the operator $\mathcal{K}(t)$ are known for the case of memory-kernel master equations of the form (7.109). The known sufficient conditions rely on the splitting of the kernel in gain and loss term as in (4.127) and cannot be simply formulated as a constraint on time-dependent coefficients, but rather puts into evidence a subtle relationship between the different contributions in the operator structure, including the Hamiltonian contribution [85–88]. Also in this case, interesting results have been obtained building on quantum counterpart of structures appearing in classical probability, rather than classical mechanics. This is the case of quantum semi-Markov processes [87,89–93], that is dynamics obtained as solutions of memory-kernel master equations constructed with reference to classical semi-Markov processes [33,94], that is a family of non-Markovian classical stochastic processes obtained combining Markov chains and renewal processes [47].

Given the lack of general results, we provide more detail on the structure of memory-kernel master equations by means of examples in Sect. 7.5.2. The different structure of master equations and their relationship are schematized in Fig. 7.11. As stressed in [95] the relationship between the two descriptions is highly non-trivial, typically a simple time-local master equation corresponds to an involved memory kernel and vice versa. General strategies to relate memory-kernel and time-local master equations have been devised in [95–99], while paradigmatic examples of their relationship will be worked out in Sect. 7.5.2.

7.5   Analysis of Memory Effects in Exactly Solvable Models

We will now exploit the theoretical developments of Sects. 7.3 and 7.4 to analyze and provide further insight on the dynamics of open quantum systems. In Sect. 7.5.1 we will make reference to the exactly solvable models introduced in Sect. 5.4, for which we have information on both the behavior in time of the solutions and the exact equations they obey, while paradigmatic physical models in which an analytic approach is not amenable have been considered in [100,101]. In the different models, we will highlight various features connected to the violation of the quantum Markov condition (7.54) discussed in Sect. 7.3. In Sect. 7.5.2 we will put into evidence the different possible descriptions of the same dynamics in terms of time-local and memory-kernel master equations, pointing to the possible different insights gained from these two distinct descriptions. A pictorial description of the different formulations of the equations of motion when going beyond the GKSL master equation is given in Fig. 7.11, putting into evidence the conditions recovering a semigroup evolution.

7.5.1   Markovian and Non-Markovian Regimes

We revisit here the models considered in Sect. 5.4 in view of the characterization of quantum Markovian evolutions given by (7.54). We will see how the variation of some model parameters, associated to specific physical features, might induce the transition from a Markovian to a non-Markovian behavior, thus clarifying the conditions under which memory can emerge. Also in this case we make reference to situations in which an analytic treatment is feasible to a large extent, so as to better put into evidence the basic features of the approach.

7.5.1.1 Spin-Boson Dephasing Model and Non-Markovianity Measure

We first analyze the Markov condition for the dephasing dynamics of a two-level system interacting with a bosonic bath considered in Sect. 5.4.1. The only modification in time of the statistical operator is the reduction of the coherences by a factor $e^{-\Gamma(t)}$, with $\Gamma(t)$ a positive function determined by the spectral density of the model and given by (5.311) for an environment starting in a thermal state. The quantum dynamical map is then given by the explicit expression (5.314). In view of Sects. 3.2.5 and 7.3 we consider as distinguishability quantifiers the trace distance and the Jensen–Shannon distance [102]. Exploiting the results of Remark 3.11 we have according to (3.245) that the trace distance as a function of time reads

$$D(\rho_S^1(t), \rho_S^2(t)) = \sqrt{\delta p^2(0) + e^{-2\Gamma(t)}|\delta c(0)|^2}, \tag{7.110}$$

where $\delta p(0)$ and $\delta c(0)$ denote the difference in populations and coherences respectively between the two initial states, with time derivative

$$\frac{d}{dt}D(\rho_S^1(t), \rho_S^2(t)) = -\frac{e^{-2\Gamma(t)}|\delta c(0)|^2}{\sqrt{\delta p^2(0) + e^{-2\Gamma(t)}|\delta c(0)|^2}}\frac{d}{dt}\Gamma(t). \tag{7.111}$$

The greatest slope is obtained for states such that populations coincide and the difference in initial coherences assumes the highest values 1, thus identifying optimal pairs in the sense of (7.96) with pure orthogonal states on the equator of the Bloch sphere. We are thus left with the simple expression

$$D(\rho_S^1(t), \rho_S^2(t)) = e^{-\Gamma(t)}, \tag{7.112}$$

and therefore

$$\frac{d}{dt}D(\rho_S^1(t), \rho_S^2(t)) = \frac{d}{dt}e^{-\Gamma(t)}, \tag{7.113}$$

corresponding to the fact that a non-Markovian behavior is observed in the presence of revivals in the coherence of the system state. The same conclusion can be drawn considering as distinguishability quantifier the quantum Jensen–Shannon distance. For qubits this quantity is conveniently expressed in terms of the Bloch vectors of the states to be compared according to (3.263). For the considered pure orthogonal pair on the equator of the Bloch sphere we get the expression

$$\sqrt{J(\rho_S^1(t), \rho_S^2(t))} = \sqrt{1 - h(e^{-\Gamma(t)})}, \tag{7.114}$$

where $h(x)$ given by (3.258) is the binary Shannon entropy of the distribution $\{(1 + x)/2, (1 - x)/2\}$, that is, comparing with (2.101)

$$h(x) = H(\{(1 + x)/2, (1 - x)/2\}). \tag{7.115}$$

The function h is monotonically decreasing, so that, denoting its time derivative with a dot, according to

$$\frac{d}{dt}\sqrt{J(\rho_S^1(t), \rho_S^2(t))} = -\frac{1}{2}\frac{\dot{h}(e^{-\Gamma(t)})}{\sqrt{1 - h(e^{-\Gamma(t)})}}\frac{d}{dt}e^{-\Gamma(t)} \tag{7.116}$$

the non-Markovian behavior is again observed when the time derivative of $\Gamma(t)$ becomes positive. The non-Markovianity measures associated to the two quantifiers are thus given by

$$M(\Phi, D) = \sum_n [e^{-\Gamma(t_f^n)} - e^{-\Gamma(t_i^n)}] \tag{7.117}$$

and

$$M(\Phi, \sqrt{J}) = \sum_n \left[\sqrt{1 - h(e^{-\Gamma(t_f^n)})} - \sqrt{1 - h(e^{-\Gamma(t_i^n)})}\right] \tag{7.118}$$

respectively, where the pairs $\{t_i^n, t_f^n\}_n$ denote the time windows in which the function $\Gamma(t)$ decreases. The behavior of these measures, as a function of the inverse bath temperature β and of the coupling strength κ, for the spectral density J_{Garg} of (5.361) introduced in Remark 3.14 of Sect. 5.4.1, leading to the decoherence function Γ_{Garg} of (5.374) is shown in Figs. 7.3 and 7.4 respectively. For the case at hand these two quantifiers, as appears from (7.112) and (7.114), are monotone functions of each other and have a similar behavior. This is however generally not the case [103]. The value of the non-Markovianity measure depends monotonically on the temperature, and grows when the temperature decreases. On the contrary, it exhibits a non-monotonic behaviour as a function of the coupling strength κ, decreasing for larger coupling strength, that here only acts as a multiplying factor in front of the function $\Gamma(t)$. The exact master equation for the model takes the form reported in (5.352)

$$\frac{d}{dt}\rho_S(t) = \frac{\dot{\Gamma}(t)}{2}[\sigma_z\rho_S(t)\sigma_z - \rho_S(t)]. \tag{7.119}$$

The only Lindblad operator appearing in it, namely σ_z, describes dephasing, while the relevant coefficient in the master equation is given by the time derivative of $\Gamma(t)$. As discussed in Sect. 7.4.2 the structure of the equation complies with (4.127) in order to warrant preservation of Hermiticity and trace. Furthermore, the presence of a single Lindblad operator implies that the conditions for CP-divisibility and P-divisibility, respectively (7.103) and (7.107), do coincide and correspond to positivity of the only time-dependent coefficient $\dot{\Gamma}(t)$. This condition also warrants a Markovian behavior

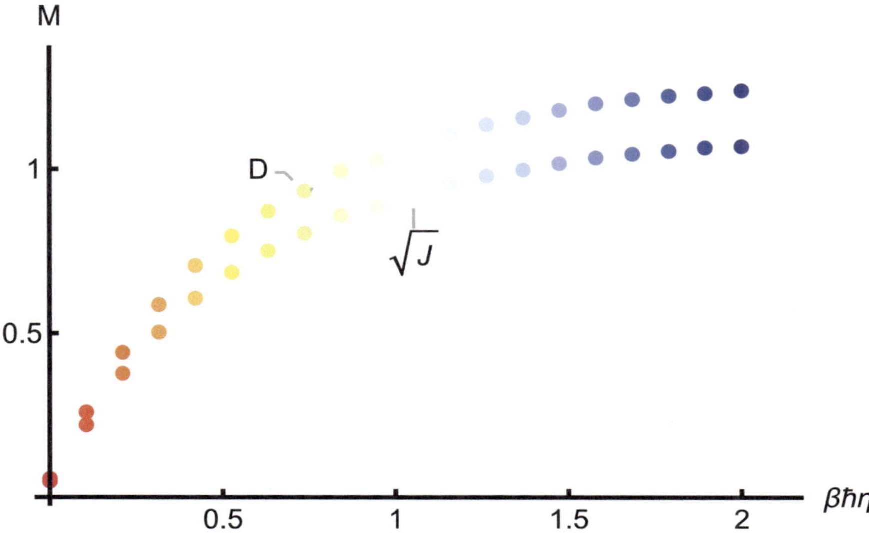

Fig. 7.3 Non-Markovianity measure for the spin-boson dephasing model evaluated for the trace distance D and the Jensen–Shannon distance $\sqrt{J}$ according to (7.117) and (7.118) respectively, considering the decoherence function Γ_{Garg} of (5.374). The quantity is plotted as a function of the inverse temperature. The non-Markovian behavior is enhanced in the low temperature regime

according to (7.113) or equivalently (7.116). Making reference to Figs. 5.7 and 5.8 we notice that, independently of the temperature, in the long time limit the function $\Gamma(t)$ becomes constant and therefore the dynamics Markovian.

7.5.1.2 Hyperfine Coupling and Markovian to Non-Markovian Transition

In the dephasing spin-boson model of Sect. 7.5.1.1 the features of the considered environment and coupling encoded in the spectral density were such that the dynamics was never Markovian, independently of the coupling strength. Indeed, the relevant function for the discrimination between a Markovian and a non-Markovian behavior, namely $\Gamma(t)$ as defined by (5.311), was simply proportional to the spectral density $J(\omega)$, in which the coupling strength appeared as multiplicative constant. The situation is utterly different for the model of Sect. 5.4.2 describing the energy exchange between a two-level system and a spin-bath in the ground state. The reduced dynamics is given by the evolution map of (5.412), just depending on the function $S(t)$ solution of (5.408), where the two-point correlation function of the environment appears. In the notation of Sect. 7.5.1.1 from the expression of the map we obtain

$$D(\rho_S^1(t), \rho_S^2(t)) = |S(t)|\sqrt{|S(t)|^2 \delta p^2(0) + |\delta c(0)|^2}, \qquad (7.120)$$

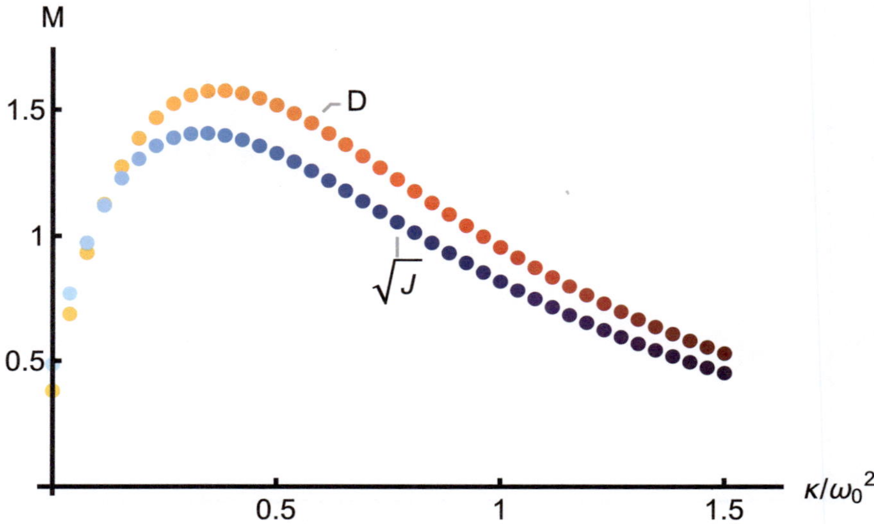

Fig. 7.4 The same non-Markovianity measure as in Fig. 7.3 but plotted as a function of the coupling strength rescaled by the system frequency, for the lowest temperature value considered in Fig. 7.3. The dependence of the measure on the coupling strength is non-monotonic, and the non-Markovian behavior is suppressed for sufficiently high coupling

and therefore for the time derivative of the trace distance

$$\frac{\mathrm{d}}{\mathrm{d}t} D(\rho_S^1(t), \rho_S^2(t)) = \frac{2|S(t)|^2 \delta p^2(0) + |\delta c(0)|^2}{\sqrt{|S(t)|^2 \delta p^2(0) + |\delta c(0)|^2}} \frac{\mathrm{d}}{\mathrm{d}t} |S(t)|. \qquad (7.121)$$

For a pair of states in the equatorial plane of the Bloch sphere, so that $\delta p^2(0) = 0$ and $|\delta c(0)|^2 = 1$, we are left with

$$\frac{\mathrm{d}}{\mathrm{d}t} D(\rho_S^1(t), \rho_S^2(t)) = \frac{\mathrm{d}}{\mathrm{d}t} |S(t)|, \qquad (7.122)$$

which since $|S(t)|^2 \leqslant 1$ maximizes the revivals in the trace distance [104]. The expression of the Jensen–Shannon distance for the same pair of states similarly to (3.263) is given by

$$\sqrt{J(\rho_S^1(t), \rho_S^2(t))} = \sqrt{1 - h(|S(t)|)}, \qquad (7.123)$$

with derivative

$$\frac{\mathrm{d}}{\mathrm{d}t} \sqrt{J(\rho_S^1(t), \rho_S^2(t))} = -\frac{1}{2} \frac{\dot{h}(|S(t)|)}{\sqrt{1 - h(|S(t)|)}} \frac{\mathrm{d}}{\mathrm{d}t} |S(t)|, \qquad (7.124)$$

so that thanks to monotonicity of the function $h(x)$ both distinguishability quantifiers have the same monotonicity properties of the function $S(t)$.

Explicit knowledge of the function $S(t)$ is obtained solving (5.408) with initial condition $S(0) = 1$. The kernel of this equation is determined by a suitable two-point correlation function of operators of the environment that can be expressed in terms of a spectral density for the model according to (5.405). As for the dephasing model worked out in Sect. 7.5.1.1, this spectral density determines the Markovian or non-Markovian behavior of the reduced dynamics. In the present case, however, just due to the different way in which this spectral density determines the reduced dynamics, we have a crucial dependence on the coupling strength. To investigate this feature we consider a spectral density exactly of Lorentzian form as in (5.366)

$$J_{\text{Lorentz}}(\omega) = \frac{\kappa \omega_0}{2} \frac{\frac{\eta}{2}}{(\omega - \omega_0)^2 + \left(\frac{\eta}{2}\right)^2}. \tag{7.125}$$

Evaluating the integral of (5.405) we have

$$\mathsf{f}(t) = \frac{\pi}{2} \kappa \omega_0 e^{-\frac{\eta}{2}|t|}, \tag{7.126}$$

and therefore for the integral equation (5.408) the kernel

$$f(t) = \frac{\pi}{2} \kappa \omega_0 e^{-\frac{\eta}{2}|t|} e^{-i\varpi t}, \tag{7.127}$$

where according to (5.400) the quantity ϖ is the sum of the values of the longitudinal coupling constants. For this kernel the analytical solution can be obtained and with the relevant initial condition reads

$$S(t) = \cosh\left(\frac{t}{2}\sqrt{\left(\frac{\eta}{2} + i\varpi\right)^2 - 2\pi\kappa\omega_0}\right) e^{-\left(\frac{\eta}{2} + i\varpi\right)\frac{t}{2}}$$

$$+ \left(\frac{\eta}{2} + i\varpi\right) \frac{\sinh\left(\frac{t}{2}\sqrt{\left(\frac{\eta}{2} + i\varpi\right)^2 - 2\pi\kappa\omega_0}\right)}{\sqrt{\left(\frac{\eta}{2} + i\varpi\right)^2 - 2\pi\kappa\omega_0}} e^{-\left(\frac{\eta}{2} + i\varpi\right)\frac{t}{2}}. \tag{7.128}$$

According to (7.122) and (7.124) the characterization of the dynamics depends on the behavior of the modulus of the function $S(t)$ given in (7.128). Non-monotonicity of $|S(t)|$ sets in for strong enough coupling, and the magnitude of the revivals is reduced with growing ϖ, as can be seen from Fig. 7.5. We therefore investigate the dependence of the non-Markovian behavior on the coupling strength considering a vanishing ϖ. In Fig. 7.6 we plot $|S(t)|$ as a function of time and of the coupling strength, while in Fig. 7.7 we evaluate the non-Markovianity measure in its dependence on the coupling strength.

It clearly appears that the non-Markovian behavior sets in when the coupling strength grows above a threshold determined by the features of the model. For the

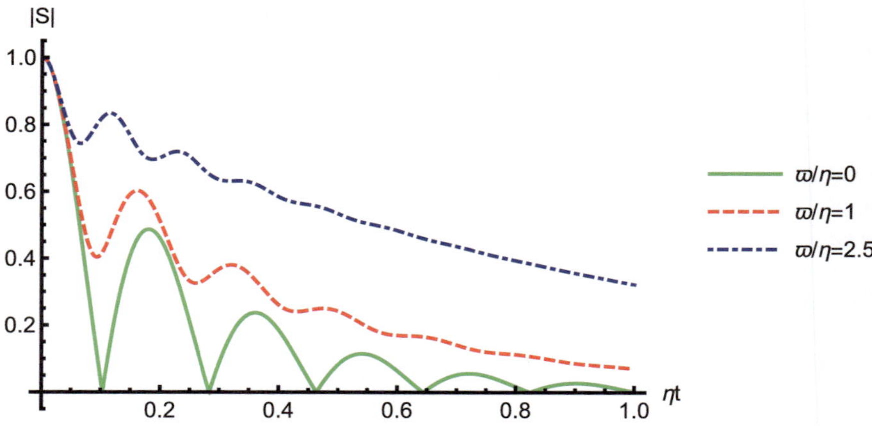

Fig. 7.5 Behavior of $|S(t)|$ for fixed coupling strength $\eta^2/(8\pi\omega_0) = 20$ and different values of the sum of the rescaled longitudinal coupling constants ϖ/η. It appears how the non-Markovian behavior is enhanced for small values of this rescaled quantity. The value $\varpi/\eta = 0$ corresponds to the one used in Fig. 7.6 to explore the dependence on the coupling strength. Note that $|S(t)|$ only depends on the modulus of ϖ

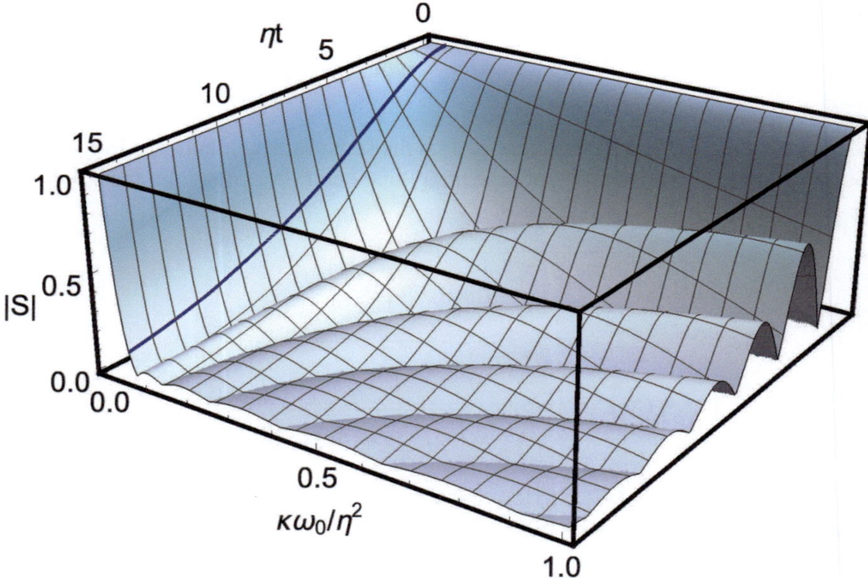

Fig. 7.6 Behavior of $|S(t)|$ as a function of rescaled time and interaction strength, with $S(t)$ as in (7.128). The average value of the longitudinal coupling constants is taken to be zero. The solid line marks the transition from a monotonically decreasing to an oscillating behavior

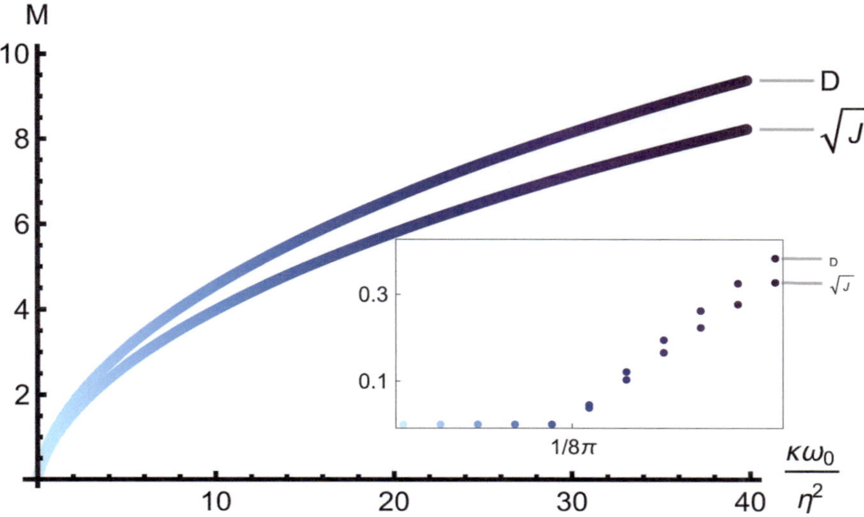

Fig. 7.7 Non-Markovianity measure for the spin-bath model of Sect. 5.4.2 evaluated for the trace distance D and the Jensen–Shannon distance $\sqrt{J}$. The measure is plotted as function of the rescaled coupling strength $\kappa\omega_0/\eta^2$, for a vanishing average value of the longitudinal coupling constants ϖ. The measure is a monotonic growing function of the coupling strength, non-zero only above the threshold value $\eta^2/(8\pi\omega_0)$, as shown in the inset

spectral density of (7.125) and for $\varpi = 0$ the behavior is non-Markovian for

$$\kappa > \frac{\eta^2}{8\pi\omega_0} \qquad (7.129)$$

and the measure increases for increasing coupling strength.

7.5.1.3   Spin-Star Decoherence Model and Distinguishability Revivals

We now consider the model of Sect. 5.4.3 in which the dynamics still only affects coherences, but the bath is composed by a finite number of spins, whose dynamics can be followed for a sufficiently small dimension of the bath. This fact will allow us to exemplify the bound (7.87), that we have introduced to justify the connection between the violation of the quantum Markov condition (7.54) and a notion of memory in the quantum framework. An example in the framework of continuous variables can be found in [105]. We will further put into evidence that different environments and couplings can lead to the very same reduced dynamics, and therefore to the same amount of information backflow, despite the different storage of information outside the open quantum system degrees of freedom, and therefore different values for the bounds. We will further put into evidence the different tightness of the bounds for different kind of distinguishability quantifiers, namely distances and entropic divergences.

We consider initial system-environment states of the product form (5.448) with the environmental spins all initialized in the same state

$$\rho_E = \frac{1}{2}\begin{pmatrix} 1 + \langle\sigma_z\rangle & c \\ c^* & 1 - \langle\sigma_z\rangle \end{pmatrix},\tag{7.130}$$

with

$$|c|^2 \leqslant 1 - \langle\sigma_z\rangle^2.\tag{7.131}$$

The evolved system-environment state, necessary to evaluate the upper bound in (7.87), can be obtained from (5.449) and takes the form

$$\rho_{SE}(t) = \begin{pmatrix} \rho_{++}(0)\bigotimes_k W_k(t) & \rho_{+-}(0)\bigotimes_k Z_k(t) \\ \rho_{-+}(0)\bigotimes_k Z_k^\dagger & \rho_{--}(0)\bigotimes_k W_k^\dagger \end{pmatrix}\tag{7.132}$$

where

$$W_k(t) = \frac{1}{2}\begin{pmatrix} 1 + \langle\sigma_z\rangle & ce^{-i\omega_\parallel^k t} \\ c^*e^{+i\omega_\parallel^k t} & 1 - \langle\sigma_z\rangle \end{pmatrix},\tag{7.133}$$

and

$$Z_k(t) = \frac{1}{2}\begin{pmatrix} (1 + \langle\sigma_z\rangle)e^{-i\omega_\parallel^k t} & c \\ c^* & (1 - \langle\sigma_z\rangle)e^{+i\omega_\parallel^k t} \end{pmatrix}.\tag{7.134}$$

The reduced system state is obtained from (7.132) by taking the partial trace with respect to the environmental degrees of freedom, so that the diagonal matrix elements of the state of the system are left unchanged, since $\mathrm{Tr}_E\{W_k(t)\} = 1$, and the off-diagonals are multiplied by the factor

$$\mathfrak{D}(t) = \prod_{k=1}^{N}\mathrm{Tr}_E\{Z_k(t)\},\tag{7.135}$$

that corresponds to (5.453) and where the only k-dependent quantity is the coupling strength $\omega_\parallel^k$ to the single spins. Note that the reduced system dynamics does not depend on the amount of coherences in the single environmental units, quantified by the value of c in (7.130), while this happens for the total state $\rho_{SE}(t)$ in (7.132) and therefore for the correlations contained in it. From (7.132) we can also obtain the marginal environmental state

$$\rho_E(t) = \rho_{++}(0)\bigotimes_k W_k(t) + \rho_{--}(0)\bigotimes_k W_k^\dagger(t),\tag{7.136}$$

that also depends on the coherences in the initial environmental state.

We are now in the position to evaluate all contributions at the r.h.s of (7.87), that quantify the changes in the environmental state and the amount of correlations established, in their dependence on the initial state of the system. We consider two distinct initial conditions

$$\rho_{SE}^i(0) = \rho_S^i(0) \otimes \bigotimes_{k=1}^{N} \rho_E \tag{7.137}$$

with system states given by the pair of orthogonal pure states on the equator of the Bloch sphere

$$\rho_S^1(0) = \frac{1}{2} \begin{pmatrix} 1 & 1 \\ 1 & 1 \end{pmatrix} \tag{7.138}$$

and

$$\rho_S^2(0) = \frac{1}{2} \begin{pmatrix} 1 & -1 \\ -1 & 1 \end{pmatrix}. \tag{7.139}$$

This pair maximizes the measure of non-Markovianity (7.95) for a dephasing dynamics, as discussed in Sect. 7.5.1.1. The behavior of the distinguishability in time of these two states, together with the bound on their revivals, is plotted in Fig. 7.8 for the case of the trace distance D of (3.211) and of the Jensen–Shannon distance $\sqrt{J}$ given by the square root of the Jensen–Shannon divergence (3.259). These bounds refer to the standard triangle inequality (3.212) and (7.79) respectively, recovered in (7.87) for $\phi(x) = x$. The upper bound is plotted for two distinct choices of environmental state ρ_E, characterized by a different amount of coherences. These states do lead to the very same dynamics, and therefore the same amount of information backflow, but differ in the established amount of correlations [106, 107]. The same analysis is performed in Fig. 7.9 for the entropic distinguishability quantifiers given by the quantum skew divergence of (7.69) and the Holevo skew divergence of (7.70). In this case the bound is given by (7.87) with $\phi(x)$ proportional to $\sqrt[4]{x}$ according to (7.75) and (7.76). The comparison between the two figures clearly shows that although for both distances and divergences the revivals follow the behavior of the bound, in the case of distances the bound is much tighter. Moreover in both cases we can see that the same revivals, and therefore the same amount of non-Markovian behavior, can take place in the presence of quite different amounts of information stored outside the system.

This fact is further put into evidence in Fig. 7.10, where the initial pair of system states is identified with $\rho_S^1(0)$ of (7.138) and

$$\rho_S^2(0) = \begin{pmatrix} \cos^2\left(\frac{\theta}{2}\right) & -\sin\left(\frac{\theta}{2}\right)\cos\left(\frac{\theta}{2}\right) \\ -\sin\left(\frac{\theta}{2}\right)\cos\left(\frac{\theta}{2}\right) & \sin^2\left(\frac{\theta}{2}\right) \end{pmatrix}, \tag{7.140}$$

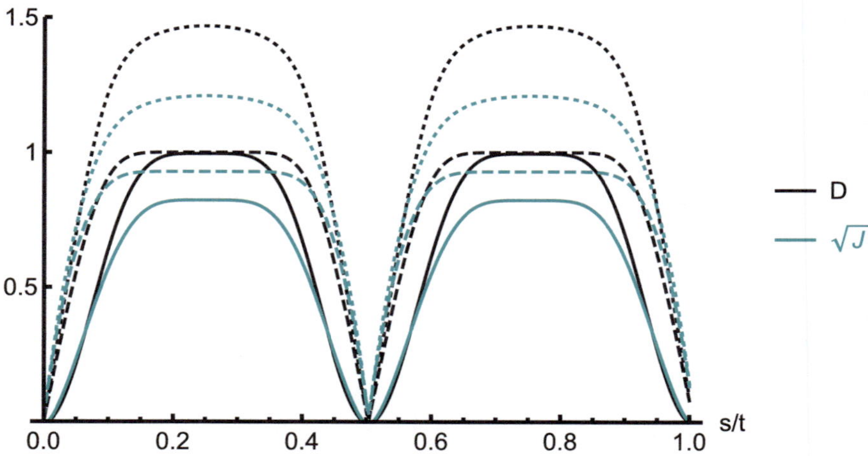

Fig. 7.8 Visualization of the bound on distinguishability revivals (7.87) for the spin-star decoherence model. The considered distinguishability quantifiers are the trace distance D (black) and the Jensen–Shannon distance $\sqrt{J}$ (green). The solid line corresponds to the l.h.s. of (7.87) as a function of s, running from 0 to a fixed reference time t equal to 4 in inverse units of the average coupling strength. For each distinguishability quantifier we consider the corresponding bound, given by the sum of the three contributions at the r.h.s. of the inequality, also plotted as a function of s. For each of the quantity we plot the bounds, corresponding to two different initial environmental states (given by (7.130) for $\langle \sigma_z \rangle = 0$, while c equals 0 and .75 respectively) leading to the very same reduced dynamics, but to a different information flow. These bounds on the revivals are plotted in the same color of the quantifier and given by dashed and dotted lines respectively. The considered environment is composed of 5 spins with uniform coupling

so that they do not have orthogonal supports any more, that is $\rho_S^1(0) \not\perp \rho_S^2(0)$. For this pair all three contributions at the r.h.s. of (7.87) are different from zero and from each other. We notice that for the optimal pair of (7.138) and (7.139) we have instead $\rho_E^1(t) = \rho_E^2(t)$, and the amount of created correlations does not depend on the initial system state, so that in this special case out of the three contributions at the r.h.s. of (7.87) one is zero and the other two are identical.

7.5.2 Memory-Kernel Versus Time-Local Master Equations

We have analyzed in Sect. 7.5.1 the time evolution of the exactly solvable models discussed in Sect. 5.4 in view of the Markovian characterization of a quantum dynamics introduced in Sect. 7.3. We now want to better investigate the interplay between the different possible evolution equations describing the reduced dynamics discussed in Sect. 7.4 and their relationship to the characterization of the dynamics itself.

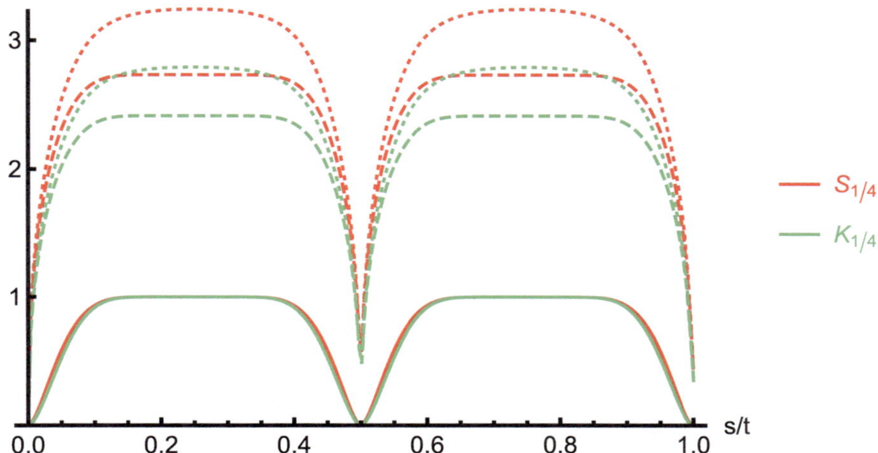

Fig. 7.9 Visualization of the bound on distinguishability revivals for the same situation as in Fig. 7.8, but for the entropic quantifiers given by the quantum skew divergence of (7.69) (red) and the Holevo skew divergence of (7.70) (green). In both cases the skewing parameter μ is set to $1/4$. The functions ϕ appearing in the bound (7.87) are given by (7.75) and (7.76) respectively. Comparison with Fig. 7.8 clearly shows that for entropic divergences the bound is looser, though still reflecting the behavior of the revivals. We recall that dashed and dotted lines correspond to upper bounds evaluated for different initial environmental states leading to exactly the same reduced dynamics

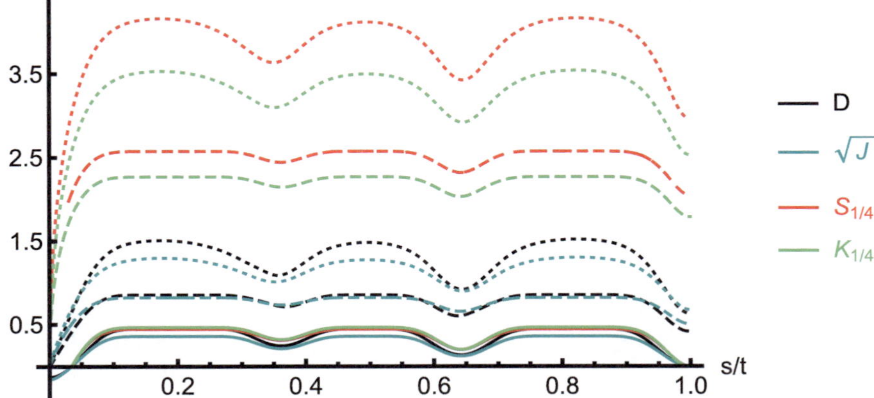

Fig. 7.10 The same quantities as in Figs. 7.8 and 7.9 but for the case of a different pair of initial states, with $\rho_S^2(0)$ now given by (7.140) for $\theta = \pi/4$. We have considered an environment still composed of 5 spins, but with random couplings. The fixed reference time t is here fixed equal to 5 in inverse units of the average coupling strength. The dependence of the bounds on the different system-environment models (corresponding to the dashed and dotted lines), all leading to the same system reduced dynamics, is more marked than in Figs. 7.8 and 7.9 since in this case also the environmental reduced states undergo different evolutions. Note the different tightness of the bounds for the case of distances and divergences

7.5.2.1   Relation Between Master Equations

In order to exemplify and put into evidence some aspects of the possible relationship between the different evolution equations discussed in Sects. 7.4.2 and 7.4.3, we consider a dynamics determined by repeated random interactions acting on top of an arbitrary continuous evolution. In the spirit of (5.66) we consider an evolution of the form

$$
\rho_S(t) = p_0(t)\mathcal{F}(t)\rho_S(0)
$$
$$
+ \sum_{n=1}^{\infty} \int_0^t dt_n \ldots \int_0^{t_2} dt_1\, p_n(t; t_n, \ldots, t_1)\mathcal{F}(t - t_n)\mathcal{E}\ldots\mathcal{E}\mathcal{F}(t_1)\rho_S(0),
$$

$$(7.141)$$

where $\mathcal{E}$ is a jump operator given by a completely positive trace-preserving map, and $\{\mathcal{F}(t)\}_{t \geq 0}$ is a family of completely positive trace-preserving maps. The functions $p_n(t; t_n, \ldots, t_1)$ are probability densities of a classical counting process and provide the probability to have n counts at fixed times $t_1, \ldots, t_n$, without further counts up to time t, so that $p_0(t)$ is the probability to have no jump up to time t, also known as survival probability. The map $\Phi(t)$ corresponding to (7.141) is itself completely positive and trace-preserving, since complete positivity is stable under composition and we have

$$
\mathrm{Tr}\{\rho_S(t)\} = \left(p_0(t) + \sum_{n=1}^{\infty} \int_0^t dt_n \ldots \int_0^{t_2} dt_1\, p_n(t; t_n, \ldots, t_1) \right) \mathrm{Tr}\{\rho_S(0)\},
$$

$$(7.142)$$

warranting trace preservation thanks to the identity

$$
p_0(t) + \sum_{n=1}^{\infty} \int_0^t dt_n \ldots \int_0^{t_2} dt_1\, p_n(t; t_n, \ldots, t_1) = \sum_{n=0}^{\infty} p_n(t), \qquad (7.143)
$$

where $p_n(t)$ is the probability for exactly n jumps up to time t. According to the discussion of Sect. 5.3.1 clarifying the emergence of (5.66), the map $\Phi(t)$ describes a situation in which the system evolves with the map $\mathcal{F}(t)$ in between transformations described by the completely positive trace-preserving map $\mathcal{E}$, that we mention as jumps and describe microscopic interaction events. At variance with (5.66) the information about the number and distribution in time of the jumps is here encoded in the probability density $p_n(t; t_n, \ldots, t_1)$. A convenient expression is obtained for the case of a so-called renewal process, in which the random times in between subsequent jumps all follow the same distribution $f(t)$, obeying

$$
f(t) \geq 0 \qquad \int_0^{+\infty} d\tau\, f(\tau) = 1, \qquad (7.144)
$$

called waiting time distribution [47]. In this case we have

$$p_n(t; t_n, \ldots, t_1) = f(t - t_n) \ldots f(t_2 - t_1)g(t_1), \tag{7.145}$$

with

$$g(t) = 1 - \int_0^t d\tau f(\tau) \tag{7.146}$$

the survival probability associated to $f(t)$, so that

$$g(t) = p_0(t), \tag{7.147}$$

leading according to (7.143) to the expression

$$p_n(t) = \int_0^t (f * \ldots * g)(\tau)d\tau, \tag{7.148}$$

where the convolution is defined as

$$(u * v)(t) = \int_0^t u(t - \tau)v(\tau)d\tau. \tag{7.149}$$

A dynamics purely driven by random interactions is recovered if we assume $\mathcal{F}(t) \rightarrow \mathbb{1}$, so that from (7.141) we immediately obtain

$$\rho_S(t) = \sum_{n=0}^\infty p_n(t)\mathcal{E}^n \rho_S(0). \tag{7.150}$$

This simple scenario allows us to compactly compare the time-local and memory-kernel evolution equations associated to dynamics described by (7.150), and connect it with the description of collisional decoherence in Sect. 5.3.2. The different evolution equations are derived in Remark 7.4 and lead to the following results. Independently from the specific expression of $\mathcal{E}$ the associated memory-kernel master equation takes the form (7.163), namely

$$\frac{d}{dt}\rho_S(t) = \int_0^t d\tau k(t - \tau)(\mathcal{E} - \mathbb{1})\rho_S(\tau), \tag{7.151}$$

where the memory kernel $k(t)$ is determined by the waiting time distribution fixing the distribution of the jumps in time according to (7.165) or equivalently (7.164). There are however many time-local master equations associated to (7.150), depending on the properties of the completely positive trace-preserving map $\mathcal{E}$. We can write them in the form,

$$\frac{\mathrm{d}}{\mathrm{d}t}\rho_S(t) = \gamma_{\mathcal{E}}(t)(\mathcal{E}-\mathbb{1})\rho_S(t), \tag{7.152}$$

where the generally time-dependent function $\gamma_{\mathcal{E}}(t)$ is determined by both the waiting time distribution $f(t)$ and the completely positive trace-preserving map $\mathcal{E}$. If $\mathcal{E}$ is idempotent, namely (7.167) holds, then

$$\gamma_{\mathcal{E}}(t) \to -\frac{\dot{g}(t)}{g(t)}, \tag{7.153}$$

with $g(t)$ the survival probability introduced in (7.146), while for $\mathcal{E}$ squaring to the identity, namely when (7.173) holds, we have

$$\gamma_{\mathcal{E}}(t) \to -\frac{1}{2}\frac{\dot{q}(t)}{q(t)}, \tag{7.154}$$

with $q(t)$ the difference between the probabilities for having an even and an odd number of jumps respectively up to time t introduced in (7.181). There is a crucial difference in the obtained expressions, indeed (7.153) is always positive, while (7.154) does not have a definite sign. As a result, while both dynamics are completely positive trace-preserving, when $\mathcal{E}$ is idempotent the dynamics is Markovian and in particular CP-divisible, while when $\mathcal{E}$ squares to the identity the dynamics is generally non-Markovian. A simple interpretation of this fact can be given as follows. When $\mathcal{E}$ is idempotent, only the first interaction matters, and the state is subsequently kept frozen. When $\mathcal{E}$ squares to the identity, repeating the action of the map acts as stochastic reset to the initial state. Importantly, while in the time-local form of the master equation these differences can be appreciated and Markovian or non-Markovian features of the dynamics can be read directly from the master equation, this is not so for the memory-kernel master equation. We realize in this very simple example that conclusions on the Markovian or non-Markovian behavior of the dynamics cannot be naively drawn from the form of the obeyed evolution equations. A memory-kernel master equation does not necessarily brings with itself a deviation from the Markovian behavior. Note that for the case of an exponential waiting time distribution

$$f(t) = \Gamma e^{-\Gamma t}, \tag{7.155}$$

with $\Gamma > 0$, the random events are Poisson distributed according to (7.41), so that (7.150) simply becomes

$$\rho_S(t) = e^{-\Gamma(\mathbb{1}-\mathcal{E})t}\rho_S(0), \tag{7.156}$$

that is independently of the completely positive trace-preserving map $\mathcal{E}$ a semigroup evolution. In this case, the memory kernel becomes a Dirac delta, $k(t-\tau) =$

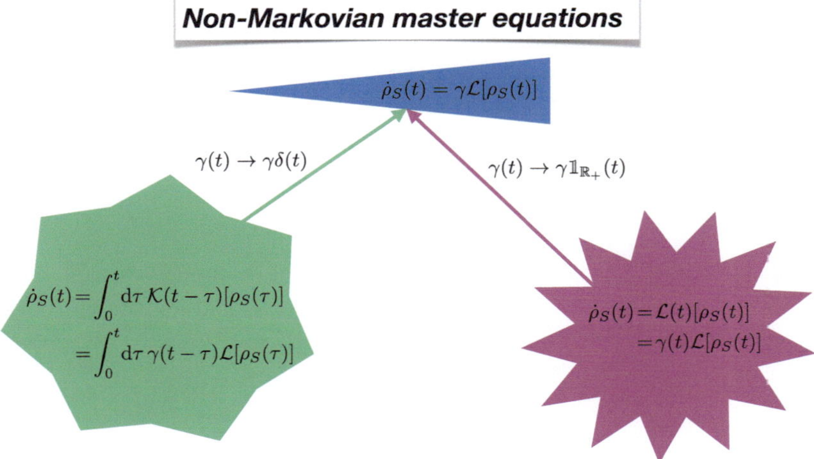

Fig. 7.11 When departing from the semigroup composition law, identified by a GKSL generator with constant and positive rates, two possible formulations of the evolution can be considered, as discussed in Sects. 7.3 and 7.4 and compared in Sect. 7.5.2.2. While the set of GKSL generators is a convex cone, as recalled by the cone-shaped coloured background, the set of time-local generators $\mathcal{L}(t)$ and memory kernels $\mathcal{K}(t)$ do not have this property, as suggested again by the star-shaped coloured background. In the scheme for simplicity it is assumed that the generators include a single Kraus operator, so that an overall coefficient can be put into evidence. The semigroup dynamics is recovered for a constant positive coefficient from a time-local formulation, or for a delta correlated coefficient from a memory-kernel formulation. The two different descriptions then coalesce in a unique one

$\Gamma\delta(t-\tau)$, while the survival probability is given by $g(t) = e^{-\Gamma t}$ and the unbalance between even and odd number of jumps reads $q(t) = e^{-2\Gamma t}$. In this case all master equations therefore converge to the Lindblad form master equation (5.27) with constant positive rate Γ, as also visualized in Fig. 7.11. Note that the exponential waiting time distribution (7.155) is the only memoryless distribution [33, 47]. In the considered example we can therefore fruitfully build on results in classical probability and the theory of classical stochastic processes to obtain results in the framework of quantum probability.

7.4 Master Equation for Random Jumps

We derive the master equation leading to the dynamics of (7.150), describing an evolution driven by interaction events randomly distributed in time and described by the action of the completely positive trace-preserving map $\mathcal{E}$. Note that the obtained results, despite their simplicity, do hold for a system described in a Hilbert space of arbitrary dimension.

We first derive the master equation in memory-kernel form. In view of the presence of time convolutions, it is quite convenient to work in terms of the Laplace transform, so that (7.150) becomes

$$\hat{\rho}_S(u) = \sum_{n=0}^{\infty} \hat{p}_n(u)\mathcal{E}^n \rho_S(0), \tag{7.157}$$

and therefore thanks to (7.146) and (7.148), exploiting the properties (5.71) and (5.72) of the Laplace transform recalled in Remark 5.2, we have from (7.148)

$$\hat{p}_n(u) = \hat{g}(u)\hat{f}^n(u), \tag{7.158}$$

and from (7.146)

$$\hat{g}(u) = \frac{1 - \hat{f}(u)}{u} \tag{7.159}$$

implying, together with a formal resummation of the geometric series, the identity

$$\hat{\rho}_S(u) = \frac{1 - \hat{f}(u)}{u} \frac{\mathbb{1}}{\mathbb{1} - \hat{f}(u)\mathcal{E}} \rho_S(0). \tag{7.160}$$

This identity can be rewritten in the form

$$u\left(\mathbb{1} - \frac{\hat{f}(u)}{1 - \hat{f}(u)}(\mathcal{E} - \mathbb{1})\right)\hat{\rho}_S(u) = \rho_S(0), \tag{7.161}$$

finally leading to

$$u\hat{\rho}_S(u) - \rho_S(0) = \frac{\hat{f}(u)}{\hat{g}(u)}(\mathcal{E} - \mathbb{1})\hat{\rho}_S(u). \tag{7.162}$$

Thanks to (5.70) and (5.71) in the time domain (7.162) corresponds to the memory-kernel master equation

$$\frac{\mathrm{d}}{\mathrm{d}t}\rho_S(t) = \int_0^t \mathrm{d}\tau\, k(t - \tau)(\mathcal{E} - \mathbb{1})\rho_S(\tau), \tag{7.163}$$

where $k(t)$ is a memory kernel uniquely determined by the waiting time distribution $f(t)$ according to

$$\hat{k}(u) = \frac{\hat{f}(u)}{\hat{g}(u)}, \tag{7.164}$$

or equivalently in the time domain

$$f(t) = \int_0^t d\tau k(t - \tau)g(\tau). \tag{7.165}$$

This expression for the quantum memory kernel was first put into evidence in [108, 109], and is the same that appears in classical generalized master equations describing non-Markovian classical processes [33, 110, 111]. The determination of memory kernels leading to well-defined time evolutions is a difficult task both in the classical and quantum case, but sufficient conditions can be expressed starting from the identity

$$\hat{f}(u) = \frac{\hat{k}(u)}{u + \hat{k}(u)} \tag{7.166}$$

and relying on the fact that a function is a waiting time distribution iff its Laplace transform is a completely monotone function, that is according to Theorem 6.2 discussed in Remark 6.1 it can be expressed as a mixture of exponential distributions. The evolution given by (7.150) is therefore obtained as a solution of (7.163), that takes the form of a memory-kernel master equation as discussed in Sect. 7.4.3.

We now consider the time-local master equation leading to the evolution (7.150). We will realize that in this case its expression depends on the properties of the completely positive trace-preserving map $\mathcal{E}$. We consider first the case in which the completely positive trace-preserving map $\mathcal{E}$ is idempotent

$$\mathcal{E}^2 = \mathcal{E}, \tag{7.167}$$

so that (7.150) thanks to normalization and the identification (7.147) becomes

$$\rho_S(t) = [\mathcal{E} - g(t)(\mathcal{E} - \mathbb{1})]\rho_S(0). \tag{7.168}$$

We note that due to (7.167) also $\mathbb{1} - \mathcal{E}$ is idempotent, and furthermore

$$(\mathbb{1} - \mathcal{E})\mathcal{E} = 0, \tag{7.169}$$

so that taking the time derivative of (7.168), exploiting idempotency and (7.169) we have

$$\frac{d}{dt}\rho_S(t) = -\dot{g}(t)(\mathcal{E} - \mathbb{1})\rho_S(0) \tag{7.170}$$

$$= -\frac{\dot{g}(t)}{g(t)}(\mathcal{E} - \mathbb{1})[\mathcal{E} - g(t)(\mathcal{E} - \mathbb{1})]\rho_S(0), \tag{7.171}$$

and in view of (7.168)

$$\frac{\mathrm{d}}{\mathrm{d}t}\rho_S(t) = -\frac{\dot{g}(t)}{g(t)}(\mathcal{E} - \mathbb{1})\rho_S(t), \tag{7.172}$$

that is to say a master equation in time-local form as in Sect. 7.4.2. From the very definition of survival probability (7.146) we can conclude that the quantity $-\dot{g}(t)/g(t)$ is always positive, which as discussed in Sect. 7.4.2 implies that the corresponding dynamics is CP-divisible and hence Markovian according to the definition (7.54). Indeed, this quantity is known in classical probability as hazard function [47]. The case of the exponential waiting time distribution (7.155) corresponds in particular to the constant value Γ, thus recovering a semigroup dynamics. We further consider a completely positive trace-preserving map $\mathcal{E}$ whose square is the identity

$$\mathcal{E}^2 = \mathbb{1}, \tag{7.173}$$

so that it coincides with its inverse. This property holds for example for the dephasing channel

$$\mathcal{E}[\rho] = \sigma_z \rho \sigma_z, \tag{7.174}$$

acting in $\mathbb{C}^2$ and determining the master equation (5.352) of the spin-boson dephasing model. Starting from (7.150) and exploiting (7.173) we have

$$\rho_S(t) = [\mathcal{E} - p_{\text{even}}(t)(\mathcal{E} - \mathbb{1})]\rho_S(0), \tag{7.175}$$

where $p_{\text{even}}(t)$ denotes the probability for having an even number of jumps up to time t. Noting that (7.173) implies

$$(\mathcal{E} - \mathbb{1})\mathcal{E} = -(\mathcal{E} - \mathbb{1}), \tag{7.176}$$

together with

$$(\mathcal{E} - \mathbb{1})^2 = -2(\mathcal{E} - \mathbb{1}), \tag{7.177}$$

we have taking the time derivative of (7.175)

$$\frac{\mathrm{d}}{\mathrm{d}t}\rho_S(t) = -\dot{p}_{\text{even}}(t)(\mathcal{E} - \mathbb{1})\rho_S(0) \tag{7.178}$$

$$= -\frac{\dot{p}_{\text{even}}(t)}{2p_{\text{even}}(t) - 1}(\mathcal{E} - \mathbb{1})[\mathcal{E} - p_{\text{even}}(t)(\mathcal{E} - \mathbb{1})]\rho_S(0), \tag{7.179}$$

and therefore from (7.175)

$$\frac{\mathrm{d}}{\mathrm{d}t}\rho_S(t) = -\frac{1}{2}\frac{\dot{q}(t)}{q(t)}(\mathcal{E} - \mathbb{1})\rho_S(t). \tag{7.180}$$

We have thus obtained a master equation in time-local form as in Sect. 7.4.2, where the quantity

$$q(t) = 2p_{\mathrm{even}}(t) - 1 = p_{\mathrm{even}}(t) - p_{\mathrm{odd}}(t) \tag{7.181}$$

is given by the difference between the probabilities for having an even and an odd number of jumps respectively up to time t, so that the time-dependent function appearing as coefficient in (7.180) does not have a definite sign. In particular, while complete positivity of the solutions of the master equation is warranted by construction, the Markovian behavior depends on the detailed features of the waiting time distribution. Also in this case, an exponential waiting time of the form (7.155), corresponding to the only classical memoryless waiting time distribution, leads to a Markov semigroup behavior. This can be observed noting that the Laplace transform of (7.181) reads thanks to (7.158)

$$\hat{q}(u) = \sum_{n=0}^{\infty}(-)^n \hat{f}^n(u)\hat{g}(u), \tag{7.182}$$

and therefore using (7.159) and resumming the geometric series we obtain

$$\hat{q}(u) = \frac{1}{u}\frac{1 - \hat{f}(u)}{1 + \hat{f}(u)}, \tag{7.183}$$

reducing to the constant Γ for the memoryless exponential waiting time distribution (7.155). Both time-local master equations (7.172) and (7.180) are characterized by time-dependent coefficients written as the product of the time derivative and the inverse of a function, and the same structure was already observed in (5.351), as well as (5.424) and (5.465). The reason lies in the fact that comparing (5.9) and (7.101) it appears that the generator in a time-local master equation is formally obtained as [15,73,97]

$$\mathcal{L}(t) = \dot{\Phi}(t)\Phi(t)^{-1}. \tag{7.184}$$

These examples, that allow for a thorough analytic treatment and a direct connection between time-local and memory-kernel form of the master equation, are naturally related to the realistic description of collisional decoherence considered in Sect. 5.3.2. In that case the dynamics, when neglecting the intermediate free evolution, was described as a sequence of collisions, Poisson distributed in time, whose effect was described by the completely positive trace-preserving map (5.134). The semigroup composition law can now be traced back to an exponential waiting time distribution for the jumps, with parameter Γ. At variance with the special cases (7.167) and (7.173), however, the effect of the repeated action of this map gives rise to transformations with different distributions of transferred momenta, obtained by taking repeated convolutions of the distribution for momentum kicks in a single collision, as described in (7.43), so that $\mathcal{E}^n \neq \mathcal{E}$, which has important and observable physical implications [51].

7.5.2.2   Comparison Between Expansions

We now consider the physical model of a central spin in a spin bath, introduced in Sect. 5.4.2, and show that an exact evaluation of both the time-local and the memory-kernel form of the master equation can be obtained. We will thus put into evidence how the operator structure may change when moving from the one to the other description. Indeed, while comparing (7.163) with either (7.172) or (7.180) in Sect. 7.5.2.1, we see that the operator acting on the statistical operator is always the same, this is generally not true [97,98]. Moreover, this model allows to compare the different validity of perturbative expansions based on the Nakajima–Zwanzig and Shibata–Hashitsumae approaches leading to memory-kernel and time-local master equations respectively. For a further comparison between the approaches see also [99].

In Sect. 5.4.2 we already obtained, starting from the expression of the time evolution map (5.412), the exact time-local master equation (5.424) for the dynamics of a central spin coupled to a spin bath, assumed to start in the ground state. The dynamics is fixed by the function $S(t)$ solution of (5.408) and by the sum of the values of the longitudinal coupling constants ϖ introduced in (5.400). To determine the corresponding memory-kernel master equation we follow the strategy considered in [112] for the study of two-level system damped by a bosonic bath starting in the vacuum. We thus make the following Ansatz for the memory kernel $\mathcal{K}(t)$ appearing in (7.109)

$$\mathcal{K}(t)[\rho_S(t)] = -i\varepsilon(t)[\sigma_+\sigma_-, \rho_S(t)]$$
$$+ w_1(t)\left[\sigma_-\rho_S(t)\sigma_+ - \frac{1}{2}\{\sigma_+\sigma_-, \rho_S(t)\}\right]$$
$$+ w_2(t)[\sigma_z\rho_S(t)\sigma_z - \rho_S(t)]. \tag{7.185}$$

The functions $w_1(t)$, $w_2(t)$ and $\varepsilon(t)$ all have to be real, in order to preserve Hermiticity. The kernel (7.185) leads to the following evolution equations for the matrix elements of the statistical operator

$$\frac{\mathrm{d}}{\mathrm{d}t}\rho_{++}(t) = -(w_1 * \rho_{++})(t) \tag{7.186}$$

and

$$\frac{\mathrm{d}}{\mathrm{d}t}\rho_{+-}(t) = -\left(\left[\frac{1}{2}(w_1 + 4w_2) + i\varepsilon\right] * \rho_{+-}\right)(t), \tag{7.187}$$

with the convolution defined as in (7.149). The unknown functions can be determined recalling that according to (5.412) we have

$$\rho_{++}(t) = \rho_{++}(0)|S(t)|^2 \tag{7.188}$$

and

$$\rho_{+-}(t) = \rho_{+-}(0)S(t)e^{i\varpi t}. \tag{7.189}$$

In order to properly match the solutions $w_1(t)$ has to obey the integral equation

$$\frac{\mathrm{d}}{\mathrm{d}t}|S(t)|^2 = -(w_1 * |S|^2)(t) \tag{7.190}$$

with initial condition $|S(0)|^2 = 1$, while (7.187) exploiting (5.406) and (5.408) determines $w_2(t)$ and $\varepsilon(t)$ to obey

$$w_2(t) = \frac{1}{4}(2\Re \mathsf{f}(t) - w_1(t)) \tag{7.191}$$

and

$$\varepsilon(t) = \Im \mathsf{f}(t) - \varpi\delta(t), \tag{7.192}$$

where the function $\mathsf{f}(t)$ given by (5.401) is the two-point correlation function of the environment determining the reduced dynamics. We finally obtain for the exact memory-kernel master equation

$$\begin{aligned}
\frac{\mathrm{d}}{\mathrm{d}t}\rho_S(t) =\ & i\varpi[\sigma_+\sigma_-, \rho_S(t)] - i\int_0^t \mathrm{d}\tau\,\Im \mathsf{f}(t-\tau)[\sigma_+\sigma_-, \rho_S(\tau)] \\
& + \int_0^t \mathrm{d}\tau\, w_1(t-\tau)\left[\sigma_-\rho_S(\tau)\sigma_+ - \frac{1}{2}\{\sigma_+\sigma_-, \rho_S(\tau)\}\right] \\
& + \int_0^t \mathrm{d}\tau\,\frac{1}{4}(2\Re \mathsf{f}(t-\tau) - w_1(t-\tau))[\sigma_z\rho_S(t)\sigma_z - \rho_S(t)], \tag{7.193}
\end{aligned}$$

where $w_1(t)$ is implicitly determined by the two-point correlation function $f(t)$ via (7.190). The comparison with (5.424) points to the appearance of a new term, determined by the self-adjoint Lindblad operator σ_z, which as seen in (5.354) of Sect. 5.4.1 in a semigroup description describes a dephasing dynamics. However, the memory-kernel master equation (7.193) and the time-local master equation (5.424) describe exactly the same dynamics. It therefore appears that at variance with the semigroup case, a direct physical interpretation of the single contributions appearing in the master equation is not applicable. The new term appearing in the second line of (7.193) cannot be taken to describe a decoherence effect on top of the damping associated to the term in the third line of the same identity. Considering the real correlation function (7.126) used in Sect. 7.5.1.2 to explore the non-Markovian behavior of the dynamics of this system, corresponding to the spectral density (7.125) with a Lorentzian shape, and assuming $\varpi = 0$ for the sake of simplicity, the function $w_1(t)$ can be determined solving (7.190) with

$$|S(t)|^2 = e^{-\frac{\eta t}{2}}\left[\cosh\left(\frac{\eta t}{4}\sqrt{1 - \frac{8\pi\kappa\omega_0}{\eta^2}}\right) \right.$$
$$\left. + \frac{1}{\sqrt{1 - \frac{8\pi\kappa\omega_0}{\eta^2}}}\sinh\left(\frac{\eta t}{4}\sqrt{1 - \frac{8\pi\kappa\omega_0}{\eta^2}}\right)\right]^2 \tag{7.194}$$

as obtained from (7.128). Considering the Laplace transform of (7.190) the solution reads

$$\hat{w}_1(u) = \frac{1 - u\widehat{|S|^2}(u)}{\widehat{|S|^2}(u)}, \tag{7.195}$$

and can be evaluated from (5.75) together with its inverse Laplace transform thanks to the fact that (7.194) is a linear combination of exponential functions. We are thus left with

$$\frac{d}{dt}\rho_S(t) = \int_0^t d\tau\, k_1(t - \tau)\left[\sigma_-\rho_S(\tau)\sigma_+ - \frac{1}{2}\{\sigma_+\sigma_-, \rho_S(\tau)\}\right]$$
$$+ \int_0^t d\tau\, k_2(t - \tau)[\sigma_z\rho_S(t)\sigma_z - \rho_S(t)], \tag{7.196}$$

with

$$k_1(t) = \pi\kappa\omega_0 e^{-\frac{3}{4}\eta t}\left[\cosh\left(\frac{\eta t}{4}\sqrt{1 - \frac{16\pi\kappa\omega_0}{\eta^2}}\right) + \frac{\sinh\left(\frac{\eta t}{4}\sqrt{1 - \frac{16\pi\kappa\omega_0}{\eta^2}}\right)}{\sqrt{1 - \frac{16\pi\kappa\omega_0}{\eta^2}}}\right] \tag{7.197}$$

and

$$k_2(t) = \frac{1}{4}\left(\pi\kappa\omega_0 e^{-\frac{\eta t}{2}} - k_1(t)\right). \tag{7.198}$$

This expression is to be compared to the one obtained from (5.424) for the very same expression (7.126) of correlation function. The function $S(t)$ of (7.190) is real for $\varpi = 0$, so that $\Im(\dot{S}(t)/S(t))$ equals zero, and we are left with

$$\frac{\mathrm{d}}{\mathrm{d}t}\rho_S(t) = \gamma(t)\left[\sigma_-\rho_S(t)\sigma_+ - \frac{1}{2}\{\sigma_+\sigma_-, \rho_S(t)\}\right] \tag{7.199}$$

where

$$\gamma(t) = \frac{4\pi\kappa\omega_0}{\eta}\frac{1}{1 + \sqrt{1 - \frac{8\pi\kappa\omega_0}{\eta^2}}\coth\left(\frac{\eta t}{4}\sqrt{1 - \frac{8\pi\kappa\omega_0}{\eta^2}}\right)}. \tag{7.200}$$

In order to compare (7.196) and (7.199) we consider an expansion of the different functions in powers of the coupling strength κ. We have

$$k_1(t) = \pi\kappa\omega_0 e^{-\frac{\eta t}{2}} + (\pi\kappa\omega_0)^2\frac{4}{\eta^2}e^{-\frac{\eta t}{2}}\left(e^{-\frac{\eta t}{2}} + \frac{\eta t}{2} - 1\right) + O(\kappa^3) \tag{7.201}$$

so that according to (7.198) the function $k_2(t)$ has the same expansion, but with opposite signs and vanishing contribution at order κ. This implies in particular that in the lowest approximation in the coupling strength (7.196) and (7.199) have the same operator structure, and the additional term in the memory kernel described by the Lindblad operator σ_z only appears at order κ^2. Importantly, the function $k_1(t)$, and therefore also $k_2(t)$, is analytic in κ, so that the Nakajima–Zwanzig memory kernel has an infinite radius of convergence. The function appearing in the time-local master equation admits the expansion

$$\gamma(t) = \pi\kappa\omega_0\frac{1}{\frac{\eta}{4}\coth\left(\frac{\eta t}{4}\right)}$$
$$+ (\pi\kappa\omega_0)^2\frac{4}{\eta^3}\left(1 - \eta t e^{-\frac{\eta t}{2}} - e^{-\eta t}\right) + O(\kappa^3), \tag{7.202}$$

but is not analytic in κ, having poles when the denominator of (7.200) goes through zero for high enough values of κ. As a result the radius of convergence of the function determining the time-local master equation is finite and determined by the first zero of the denominator in (7.200). According to this observation, the perturbation expansion obtained from the Nakajima–Zwanzig projection operator technique detailed in Sect. 5.5 and leading to the memory-kernel master equation is well-defined for arbitrary coupling strength. On the contrary, the perturbation expansion obtained from the Shibata–Hashitsumae time-convolutionless projection operator technique

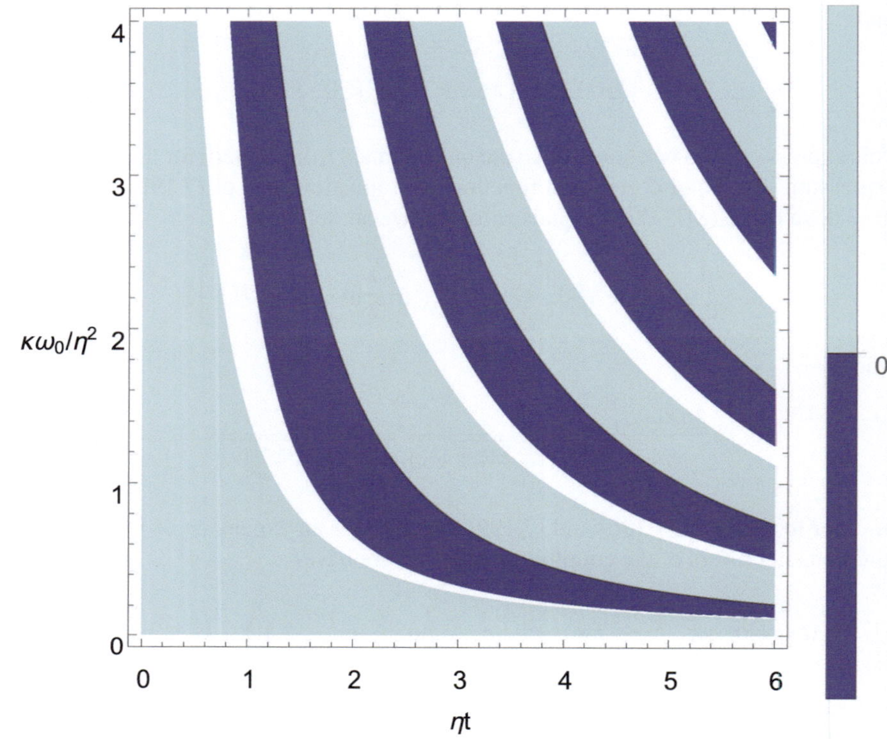

Fig. 7.12 Plot of the sign of the coefficient $\gamma(t)$ of (7.200) as a function of rescaled time and coupling strength. The white regions identify the points of singularity in the expansion, in which the coefficient has a pole and therefore diverges. It appears that in the weak-coupling limit the expression is well-defined at all times, and for short times higher coupling strengths are allowed

[81, 113], that provides a perturbative path to the exact time-local master equation, breaks down at finite coupling strength. We visualize this feature in Fig. 7.12.

References

1. G.P. Basharin, A.N. Langville, V.A. Naumov, Linear Algebra Appl. **386**, 3 (2004). https://doi.org/10.1016/j.laa.2003.12.041
2. B. Vacchini, A. Smirne, E.M. Laine, J. Piilo, H.-P. Breuer, New J. Phys. **13**, 093004 (2011). https://doi.org/10.1088/1367-2630/13/9/093004
3. H.-P. Breuer, E.M. Laine, J. Piilo, Phys. Rev. Lett. **103**, 210401 (2009). https://doi.org/10.1103/PhysRevLett.103.210401
4. A. Rivas, S.F. Huelga, M.B. Plenio, Phys. Rev. Lett. **105**, 050403 (2010). https://doi.org/10.1103/PhysRevLett.105.050403

5. S. Luo, S. Fu, H. Song, Phys. Rev. A **86**, 044101 (2012). https://doi.org/10.1103/PhysRevA.86.044101

6. S. Lorenzo, F. Plastina, M. Paternostro, Phys. Rev. A **88**, 020102 (2013). https://doi.org/10.1103/PhysRevA.88.020102

7. D. Chruściński, S. Maniscalco, Phys. Rev. Lett. **112**, 120404 (2014). https://doi.org/10.1103/PhysRevLett.112.120404

8. C. Benedetti, M.G.A. Paris, S. Maniscalco, Phys. Rev. A **89**, 012114 (2014). https://doi.org/10.1103/PhysRevA.89.012114

9. A.A. Budini, Phys. Rev. Lett. **121**, 240401 (2018). https://doi.org/10.1103/PhysRevLett.121.240401

10. C. Addis, B. Bylicka, D. Chruściński, S. Maniscalco, Phys. Rev. A **90**, 052103 (2014). https://doi.org/10.1103/PhysRevA.90.052103

11. Á. Rivas, S.F. Huelga, M.B. Plenio, Rep. Prog. Phys. **77**, 094001 (2014). https://doi.org/10.1088/0034-4885/77/9/094001

12. B. Bylicka, D. Chruściński, S. Maniscalco, Sci. Rep. **4**, 5720 (2014). https://doi.org/10.1038/srep05720

13. I. de Vega, D. Alonso, Rev. Mod. Phys. **89**, 015001 (2017). https://doi.org/10.1103/RevModPhys.89.015001

14. L. Li, M.J. Hall, H.M. Wiseman, Phys. Rep. **759**, 1 (2018). https://doi.org/10.1016/j.physrep.2018.07.001

15. D. Chruściński, Phys. Rep. **992**, 1 (2022). https://doi.org/10.1016/j.physrep.2022.09.003

16. A.A. Budini, Entropy **24** (2022). https://doi.org/10.3390/e24050649

17. L. Accardi, A. Frigerio, J.T. Lewis, Publ. RIMS, Kyoto Univ. **18**, 97 (1982). https://doi.org/10.2977/prims/1195184017

18. G. Lindblad, Comm. Math. Phys. **65**, 281 (1979). https://doi.org/10.1007/BF01197883

19. C. Giarmatzi, F. Costa, Quantum **5**, 440 (2021). https://doi.org/10.22331/q-2021-04-26-440

20. S. Milz, K. Modi, PRX Quantum **2**, 030201 (2021). https://doi.org/10.1103/PRXQuantum.2.030201

21. S. Yu, A.A. Budini, Y.T. Wang, Z.J. Ke, Y. Meng, W. Liu, Z.P. Li, Q. Li, Z.H. Liu, J.S. Xu, J.S. Tang, C.F. Li, G.C. Guo, Phys. Rev. A **100**, 050301 (2019). https://doi.org/10.1103/PhysRevA.100.050301

22. G.A.L. White, C.D. Hill, F.A. Pollock, L.C.L. Hollenberg, K. Modi, Nat. Commun. **11**, 6301 (2020). https://doi.org/10.1038/s41467-020-20113-3

23. K. Goswami, C. Giarmatzi, C. Monterola, S. Shrapnel, J. Romero, F. Costa, Phys. Rev. A **104**, 022432 (2021). https://doi.org/10.1103/PhysRevA.104.022432

24. S. Deffner, E. Lutz, Phys. Rev. Lett. **107**, 140404 (2011). https://doi.org/10.1103/PhysRevLett.107.140404

25. S. Marcantoni, S. Alipour, F. Benatti, R. Floreanini, A.T. Rezakhani, Sci. Rep. **7**, 12447 (2017). https://doi.org/10.1038/s41598-017-12595-x

26. S. Bhattacharya, A. Misra, C. Mukhopadhyay, A.K. Pati, Phys. Rev. A **95**, 012122 (2017). https://doi.org/10.1103/PhysRevA.95.012122

27. M. Popovic, B. Vacchini, S. Campbell, Phys. Rev. A **98**, 012130 (2018). https://doi.org/10.1103/PhysRevA.98.012130

28. P. Strasberg, M. Esposito, Phys. Rev. E **99**, 012120 (2019). https://doi.org/10.1103/PhysRevE.99.012120

29. A. Colla, H.-P. Breuer, Phys. Rev. A **104**, 052408 (2021). https://doi.org/10.1103/PhysRevA.104.052408

30. H.T. Şenyaşa, Ş. Kesgin, G. Karpat, B. Çakmak, Entropy **24** (2022). https://doi.org/10.3390/e24060824

31. G.T. Landi, M. Paternostro, Rev. Mod. Phys. **93**, 035008 (2021). https://doi.org/10.1103/RevModPhys.93.035008

32. W. Feller, *An Introduction to Probability Theory and Its Applications*, vol. I (Wiley, New York, 1968)
33. W. Feller, *An Introduction to Probability Theory and Its Applications*, vol. II (Wiley, New York, 1971)
34. C.W. Gardiner, *Handbook of Stochastic Methods for Physics, Chemistry and the Natural Sciences, Springer Series in Synergetics*, vol. 13, 3rd edn. (Springer, Berlin, 2004)
35. N. van Kampen, *Stochastic Processes in Physics and Chemistry* (North-Holland, Amsterdam, 1992)
36. D.T. Gillespie, *Markov Processes* (Academic, Boston, 1992)
37. H.-P. Breuer, F. Petruccione, *The Theory of Open Quantum Systems* (Oxford University Press, Oxford, 2002)
38. A.N. Kolmogorov, *Fundamental Concepts of Probability Theory* (Nauka, Moscow, 1974)
39. N. van Kampen, Braz. J. Phys. **28**, 90 (1998). https://doi.org/10.1590/S0103-97331998000200003
40. D.T. Gillespie, Am. J. Phys. **66**, 533 (1998). https://doi.org/10.1119/1.18895
41. S. Chapman, Proc. Roy. Soc. A **119**, 34 (1928). https://doi.org/10.1098/rspa.1928.0082
42. A. Kolmogoroff, Math. Ann. **104**, 415 (1931). https://doi.org/10.1007/BF01457949
43. C.A. Fuchs, J. van de Graaf, IEEE Trans. Inf. Th. **45**, 1216 (1999). https://doi.org/10.1109/18.761271
44. A. N. Kolmogorov, Sankhyā: The Indian J. Stat. Series A **25**, 159 (1963). http://www.jstor.org/stable/25049261
45. I. Bengtsson, K. Życzkowski, *Geometry of Quantum States: An Introduction to Quantum Entanglement*, 2nd edn. (Cambridge University Press, 2017)
46. A. Kossakowski, Rep. Math. Phys. **3**, 247 (1972). https://doi.org/10.1016/0034-4877(72)90010-9
47. S.M. Ross, *Introduction to Probability Models* (Academic, Burlington, MA, 2007)
48. F. Bardou, J.P. Bouchaud, A. Aspect, C. Cohen-Tannoudji, *Lévy Statistics and Laser Cooling* (Cambridge University Press, 2001). https://doi.org/10.1017/CBO9780511755668
49. M.S. Chapman, T.D. Hammond, A. Lenef, J. Schmiedmayer, R.A. Rubenstein, E. Smith, D.E. Pritchard, Phys. Rev. Lett. **75**, 3783 (1995). https://doi.org/10.1103/PhysRevLett.75.3783
50. B. Vacchini, Phys. Rev. Lett. **95**, 230402 (2005). https://doi.org/10.1103/PhysRevLett.95.230402
51. L. Mandel, J. Opt. **10**, 51 (1979). https://doi.org/10.1088/0150-536X/10/2/001
52. E.M. Laine, J. Piilo, H.-P. Breuer, Phys. Rev. A **81**, 062115 (2010). https://doi.org/10.1103/PhysRevA.81.062115
53. H.-P. Breuer, E.M. Laine, J. Piilo, B. Vacchini, Rev. Mod. Phys. **88**, 021002 (2016). https://doi.org/10.1103/RevModPhys.88.021002
54. H.-P. Breuer, G. Amato, B. Vacchini, New J. Phys. **20**, 043007 (2018). https://doi.org/10.1088/1367-2630/aab2f9
55. N. Megier, A. Smirne, B. Vacchini, Phys. Rev. Lett. **127**, 030401 (2021). https://doi.org/10.1103/PhysRevLett.127.030401
56. A. Smirne, N. Megier, B. Vacchini, Phys. Rev. A **106**, 012205 (2022). https://doi.org/10.1103/PhysRevA.106.012205
57. R. Bhatia, *Matrix Analysis* (Springer, 1997). https://doi.org/10.1007/978-1-4612-0653-8
58. T. Furuta, J. Math. Inequal. **7**, 93 (2013). https://doi.org/10.7153/jmi-07-08
59. W. Roga, M. Fannes, K. Zyczkowski, Phys. Rev. Lett. **105**, 040505 (2010). https://doi.org/10.1103/PhysRevLett.105.040505
60. K.M.R. Audenaert, arXiv:1102.3041 (2011). https://doi.org/10.48550/arXiv.1102.3041
61. K.M.R. Audenaert, J. Math. Phys. **55**, 112202 (2014). https://doi.org/10.1063/1.4901039
62. M. Nielsen, I. Chuang, *Quantum Computation and Quantum Information* (Cambridge University Press, Cambridge, 2000)
63. D. Virosztek, Adv. Math. **380**, 107595 (2021). https://doi.org/10.1016/j.aim.2021.107595

64. S. Sra, Linear Algebra Appl. **616**, 125 (2021). https://doi.org/10.1016/j.laa.2020.12.023

65. C.F. Li, G.C. Guo, J. Piilo, EPL **127**, 50001 (2019). https://doi.org/10.1209/0295-5075/127/50001

66. C.F. Li, G.C. Guo, J. Piilo, EPL **128**, 30001 (2020). https://doi.org/10.1209/0295-5075/128/30001

67. F. Fanchini, F.G. Karpat, B. Cakmak, K. Castelano, L.H. Aguilar, G.O.J. Farias, P. Walborn, S.H.S. Ribeiro, P.C. de Oliveira, M. Phys, Rev. Lett. **112**, 210402 (2014). https://doi.org/10.1103/PhysRevLett.112.210402

68. B.H. Liu, D.Y. Cao, Y.F. Huang, C.F. Li, G.C. Guo, E.M. Laine, H.-P. Breuer, J. Piilo, Sci. Rep. **3**, 1781 (2013). https://doi.org/10.1038/srep01781

69. A. Chiuri, C. Greganti, L. Mazzola, M. Paternostro, P. Mataloni, Sci. Rep. **2**, 968 (2012). https://doi.org/10.1038/srep00968

70. J.S. Tang, C.F. Li, Y.L. Li, X.B. Zou, G.C. Guo, H.-P. Breuer, E.M. Laine, J. Piilo, EPL **97**, 10002 (2012). https://doi.org/10.1209/0295-5075/97/10002

71. B.H. Liu, L. Li, Y.F. Huang, C.F. Li, G.C. Guo, E.M. Laine, H.-P. Breuer, J. Piilo, Nat. Phys. **7**, 931 (2011). https://doi.org/10.1038/nphys2085

72. S. Wißmann, A. Karlsson, E.M. Laine, J. Piilo, H.-P. Breuer, Phys. Rev. A **86**, 062108 (2012). https://doi.org/10.1103/PhysRevA.86.062108

73. H.-P. Breuer, J. Phys. B **45**, 154001 (2012). https://doi.org/10.1088/0953-4075/45/15/154001

74. D. Chruściński, A. Kossakowski, A. Rivas, Phys. Rev. A **83**, 052128 (2011). https://doi.org/10.1103/PhysRevA.83.052128

75. D. Chruściński, A. Rivas, E. Størmer, Phys. Rev. Lett. **121**, 080407 (2018). https://doi.org/10.1103/PhysRevLett.121.080407

76. J. Hachigian, Ann. Math. Statist. **34**, 233 (1963). https://doi.org/10.1214/aoms/1177704261

77. D. Chruściński, Open Syst. Inf. Dyn. **21**, 1440004 (2014). https://doi.org/10.1142/S1230161214400046

78. K. Lendi, Phys. Rev. A **33**, 3358 (1986). https://doi.org/10.1103/PhysRevA.33.3358

79. A. Kossakowski, Bull. Acad. Pol. Sci., Serie Math. Astr. Phys. **20**, 1021 (1972)

80. L. Ferialdi, Phys. Rev. Lett. **116**, 120402 (2016). https://doi.org/10.1103/PhysRevLett.116.120402

81. N. Hashitsumae, F. Shibata, M. Shingu, J. Stat. Phys. **17**, 155 (1977). https://doi.org/10.1007/BF01040099

82. C. Uchiyama, F. Shibata, Phys. Rev. E **60**, 2636 (1999). https://doi.org/10.1103/PhysRevE.60.2636

83. M. Ban, S. Kitajima, F. Shibata, Phys. Lett. A **374**, 2324 (2010). https://doi.org/10.1016/j.physleta.2010.03.066

84. A.A. Budini, Phys. Rev. A **74**, 053815 (2006). https://doi.org/10.1103/PhysRevA.74.053815

85. A. Kossakowski, R. Rebolledo, Open Syst. Inf. Dyn. **16**, 259 (2009). https://doi.org/10.1142/S1230161209000190

86. S. Campbell, A. Smirne, L. Mazzola, N. Lo Gullo, B. Vacchini, T. Busch, M. Paternostro, Phys. Rev. A **85**, 032120 (2012). https://doi.org/10.1103/PhysRevA.85.032120

87. D. Chruściński, A. Kossakowski, Phys. Rev. A **94**, 020103(R) (2016). https://doi.org/10.1103/PhysRevA.94.020103

88. B. Vacchini, Phys. Rev. Lett. **117**, 230401 (2016). https://doi.org/10.1103/PhysRevLett.117.230401

89. H.-P. Breuer, B. Vacchini, Phys. Rev. Lett. **101**, 140402 (2008). https://doi.org/10.1103/PhysRevLett.101.140402

90. H.-P. Breuer, B. Vacchini, Phys. Rev. E **79**, 041147 (2009). https://doi.org/10.1103/PhysRevE.79.041147

91. D. Chruściński, A. Kossakowski, EPL **97**, 20005 (2012). https://doi.org/10.1209/0295-5075/97/20005

92. D. Chruściński, A. Kossakowski, Phys. Rev. A **95**, 042131 (2017). https://doi.org/10.1103/PhysRevA.95.042131
93. B. Vacchini, Sci. Rep. **10**, 5592 (2020). https://doi.org/10.1038/s41598-020-62260-z
94. W. Feller, Proc. Natl. Acad. Sci. U.S.A. **51**, 653 (1964). https://doi.org/10.1073/pnas.51.4.653
95. D. Chruściński, A. Kossakowski, Phys. Rev. Lett. **104**, 070406 (2010). https://doi.org/10.1103/PhysRevLett.104.070406
96. A. Smirne, B. Vacchini, Phys. Rev. A **82**, 022110 (2010). https://doi.org/10.1103/PhysRevA.82.022110
97. N. Megier, A. Smirne, B. Vacchini, Entropy **22**, 796 (2020). https://doi.org/10.3390/e22070796
98. N. Megier, A. Smirne, B. Vacchini, New J. Phys. **22**, 083011 (2020). https://doi.org/10.1088/1367-2630/ab9f6b
99. K. Nestmann, V. Bruch, M.R. Wegewijs, Phys. Rev. X **11**, 021041 (2021). https://doi.org/10.1103/PhysRevX.11.021041
100. G. Clos, H.-P. Breuer, Phys. Rev. A **89**, 012115 (2012). https://doi.org/10.1103/PhysRevA.86.012115
101. S. Einsiedler, A. Ketterer, H.-P. Breuer, Phys. Rev. A **102**, 022228 (2020). https://doi.org/10.1103/PhysRevA.102.022228
102. B. Vacchini, Int. J. Quantum Inform., **2450007** (2024). https://doi.org/10.1142/S0219749924500072
103. F. Settimo, H.-P. Breuer, B. Vacchini, Phys. Rev. A **106**, 042212 (2022). https://doi.org/10.1103/PhysRevA.106.042212
104. Z.Y. Xu, W.L. Yang, M. Feng, Phys. Rev. A **81**, 044105 (2010). https://doi.org/10.1103/PhysRevA.81.044105
105. S. Campbell, M. Popovic, D. Tamascelli, B. Vacchini, New J. Phys. **21**, 053036 (2019). https://doi.org/10.1088/1367-2630/ab1ed6
106. A. Smirne, N. Megier, B. Vacchini, Quantum **5**, 439 (2021). https://doi.org/10.22331/q-2021-04-26-439
107. N. Megier, A. Smirne, S. Campbell, B. Vacchini, Entropy **24** (2022). https://doi.org/10.3390/e24020304
108. A.A. Budini, Phys. Rev. A **69**, 042107 (2004). https://doi.org/10.1103/PhysRevA.69.042107
109. A.A. Budini, Phys. Rev. E **72**, 056106 (2005). https://doi.org/10.1103/PhysRevE.72.056106
110. D. Gillespie, Phys. Lett. A **64**, 22 (1977). https://doi.org/10.1016/0375-9601(77)90513-8
111. R. Metzler, J. Klafter, I.M. Sokolov, Phys. Rev. E **58**, 1621 (1998). https://doi.org/10.1103/PhysRevE.58.1621
112. B. Vacchini, H.-P. Breuer, Phys. Rev. A **81**, 042103 (2010). https://doi.org/10.1103/PhysRevA.81.042103
113. H.-P. Breuer, B. Kappler, F. Petruccione, Ann. Phys. (N.Y.) **291**, 36 (2001). https://doi.org/10.1006/aphy.2001.6152

Index